Protocols for Oligonucleotides and Analogs

Methods in Molecular Biology

John M. Walker, Series Editor

20. **Protocols for Oligonucleotides and Analogs**, edited by *Sudhir Agrawal*, 1993
19. **Biomembrane Protocols, I:** *Isolation and Analysis*, edited by *John M. Graham and Joan A. Higgins*, 1993
18. **Transgenesis Techniques**, edited by *David Murphy and David A. Carter*, 1993
17. **Spectroscopic Methods and Analyses**, edited by *Christopher Jones, Barbara Mulloy, and Adrian H. Thomas*, 1993
16. **Enzymes of Molecular Biology**, edited by *Michael M. Burrell*, 1993
15. **PCR Protocols**, edited by *Bruce A. White*, 1993
14. **Glycoprotein Analysis in Biomedicine**, edited by *Elizabeth F. Hounsell*, 1993
13. **Protocols in Molecular Neurobiology**, edited by *Alan Longstaff and Patricia Revest*, 1992
12. **Pulsed-Field Gel Electrophoresis**, edited by *Margit Burmeister and Levy Ulanovsky*, 1992
11. **Practical Protein Chromatography**, edited by *Andrew Kenney and Susan Fowell*, 1992
10. **Immunochemical Protocols**, edited by *Margaret M. Manson*, 1992
9. **Protocols in Human Molecular Genetics**, edited by *Christopher G. Mathew*, 1991
8. **Practical Molecular Virology**, edited by *Mary K. L. Collins*, 1991
7. **Gene Transfer and Expression Protocols**, edited by *Edward J. Murray*, 1991
6. **Plant Cell and Tissue Culture**, edited by *Jeffrey W. Pollard and John M. Walker*, 1990
5. **Animal Cell Culture**, edited by *Jeffrey W. Pollard and John M. Walker*, 1990
4. **New Nucleic Acid Techniques**, edited by *John M. Walker*, 1988
3. **New Protein Techniques**, edited by *John M. Walker*, 1988
2. **Nucleic Acids**, edited by *John M. Walker*, 1984
1. **Proteins**, edited by *John M. Walker*, 1984

Methods in Molecular Biology • 20

Protocols for Oligonucleotides and Analogs

Synthesis and Properties

Edited by

Sudhir Agrawal

Worcester Foundation for Experimental Biology,
Shrewsbury, Massachusetts

Humana Press ※ Totowa, New Jersey

© 1993 Humana Press Inc.
999 Riverview Drive, Suite 208
Totowa, New Jersey 07512

All rights reserved.

No part of this book may be reproduced, stored in a retrieval system, or transmitted in any form or by any means, electronic, mechanical, photocopying, microfilming, recording, or otherwise without written permission from the Publisher.

Photocopy Authorization Policy:
Authorization to photocopy items for internal or personal use, or the internal or personal use of specific clients, is granted by The Humana Press Inc., **provided** that the base fee of US $3.00 per copy, plus US $00.20 per page is paid directly to the Copyright Clearance Center at 27 Congress Street, Salem, MA 01970. For those organizations that have been granted a photocopy license from the CCC, a separate system of payment has been arranged and is acceptable to The Humana Press Inc. The fee code for users of the Transactional Reporting Service is: [0-89603-247-7/93 $3.00 + $00.20].

Printed in the United States of America. 9 8 7 6 5 4 3 2 1

Library of Congress Cataloging in Publication Data
Main entry under title:

Methods in molecular biology.

Protocols for oligonucleotides and analogs ; synthesis and properties
 / edited by Sudhir Agrawal.
     p. cm. — (Methods in molecular biology ; 20)
   Includes index.
   ISBN 0-89603-247-7
   1. Oligonucleotides—Synthesis. 2. Oligonucleotides—Derivatives—
Synthesis. I. Agrawal, Sudhir. II. Series: Methods in molecular
biology (Totowa, NJ) ; 20.
QP625.047P76   1993
547.7'9—dc20   93-7121
                                                           CIP

Preface

When first conceived, not only was the aim of *Protocols for Oligonucleotides and Analogs* to provide wide coverage of the oligonucleotide chemistry field for readers who are well versed within the field, but also to give investigators just entering into the field a new perspective. The very first book on this topic was edited and published by Michael Gait in 1984, in whose laboratory I encountered the newer aspects of oligonucleotide chemistry. Since then, oligonucleotide research has developed to such an extent that its uses extend far beyond basic studies, and now find wide application throughout clinical science as well.

Until recently, the major application of oligonucleotides has been in the area of DNA-based diagnostic and "antisense oligonucleotide"-based therapeutic approaches. However, oligonucleotides are now also being used as therapeutic agents and are thus frequently found in clinical trials in humans.

Synthesis of unmodified oligonucleotides using automated synthesizers has become a common practice in numerous laboratories. However, improvements on the existing techniques and the introduction of ever newer methods for oligonucleotide synthesis is constantly driving ahead in the leading research laboratories. And several new oligonucleotide analogs have been synthesized and studied for their individual properties in recent years. The present volume strives to bring the readers the most up-to-date information on the newest aspects of synthesis of oligonucleotides and their analogs. A separate volume covers synthesis of oligonucleotide conjugates, along with most of the analytical techniques presently used for analysis of oligonucleotides.

The first chapter of *Protocols for Oligonucleotides and Analogs* provides a sense of the history of the development of oligonucleotide synthesis. This is followed by three chapters (Chapters 2–4) describing three main approaches to the synthesis of unmodified oligonucleotides. The phosphoramidite and H-phosphonate approaches have become the two main avenues of choice for synthesizing oligonucleotides. Chapters 5 and 6 provide protocols for synthesizing oligonucleotides and their 2'-analogs, respectively. There are chapters describing the synthesis of phosphate analogs of oligonucleotides including methylphosphonates (Chapter 7), phosphorothioates (Chapter 8), phosphorodithioates (Chapter 9), and phosphotriesters (Chapter 10). There is a description of several newer phosphate modifications, e.g., boronates (Chapter 11) and phosphofluoridates (Chapter 12). Chapter 13 describes the protocols for synthesizing a-oligonucleotides. The stereospecific synthesis of chiral oligonucleotides is discussed and described (Chapter 14). Oligonucleotide analogs containing nonphosphate backbones are presently being studied and their synthetic chemistry is being worked out; Chapters 15 and 16 provide the synthetic chemistry of these oligonucleotide analogs. Included are two chapters describing various aspects of large-scale oligonucleotide synthesis using the solution phase (Chapter 17) as well as the solid phase approach (Chapter 18). Also included are descriptions of some of the supports that have been used in oligonucleotide synthesis, including protocols for their derivatization and handling (Chapter 19).

I am, of course, deeply indebted to all authors of the various chapters of this book for their dedication, hard work, and patience. To them goes credit for the book's teachings and uses, since the field itself either originated from their laboratories or was contributed to significantly by it. The high quality of the various manuscripts has made my role as editor an easy one. Let me thank Drs. Paul Zamecnik, Thoru Pederson, and Dan Brown for encouragement, guidance, and discussions. I would like to thank Dr. John Walker, Methods in Molecular Biology Series Editor for his help, guidance, and encouragement at all stages of editing. Finally, many thanks go to Mr. Thomas Lanigan, Ms. Lucia Read, and Ms. Bonnie Gustafsson of Humana Press, whose hard work and diligence have made this book a reality.

Sudhir Agrawal
Worcester Foundation
for Experimental Biology
Shrewsbury, MA
Present address:
Hybridon Inc.
One Innovation Drive,
Worcester, MA

Foreword

Since the discovery of the nucleic acids in the latter part of the last century, several decades elapsed before the structures of the constituent purines and pyrimidines, the corresponding nucleosides, and the nature of the internucleotide bonds in the polynucleotides were clarified. By 1952, the chemical structures of both classes of nucleic acids, RNA and DNA, had been established. Soon followed the Watson-Crick proposal for the DNA structure and this ushered in a new era in chemical, biochemical, and molecular biological studies.

As with other classes of biological macromolecules, interest in the synthesis of oligo- and polynucleotides increased very rapidly and systematic studies on the synthesis of polynucleotides were undertaken in the mid-1950s. Indeed, strategies were successfully developed for the synthesis of oligonucleotides of defined sequences. These, in conjunction with enzymatic approaches enabled the synthesis of high molecular weight polymers that were successfully used in the elucidation of the genetic code and later in the total laboratory synthesis of genes fully functional in vivo.

In more recent years new developments in synthesis have advanced the synthetic technology to a highly efficient level. In particular, synthesis on polymer supports, a concept first developed by Merrifield in polypeptide synthesis, has resulted in enormous improvements and speed in synthesis of large size polynucleotides.

Dr. Agrawal is to be congratulated on having produced two authoritative books with contributions from experts in different aspects of polynucleotide synthesis. These monographs should be extremely useful to those who are already practitioners in the field and also those who want to enter this field without any particular background.

H. Gobind Khorana
Massachusetts Institute of Technology
Cambridge, MA

Contents

Preface v
Foreword vii
Companion Volume Contents xi
Contributors xiii

CH. 1. A Brief History of Oligonucleotide Synthesis,
 Daniel M. Brown 1
CH. 2. Oligonucleotide Synthesis: *Phosphotriester Approach*,
 Chris Christodoulou 19
CH. 3. Oligodeoxyribonucleotide Synthesis:
 Phosphoramidite Approach,
 Serge L. Beaucage 33
CH. 4. Oligodeoxynucleotide Synthesis: *H-Phosphonate Approach*,
 Brian C. Froehler 63
CH. 5. Oligoribonucleotide Synthesis: *The Silyl-Phosphoramidite Method*,
 Masad J. Damha and Kelvin K. Ogilvie 81
CH. 6. Synthesis of 2'-O-Alkyloligoribonucleotides,
 Brian S. Sproat 115
CH. 7. An Improved Method for the Synthesis and Deprotection
 of Methylphosphonate Oligonucleotides,
 Richard I. Hogrefe, Mark A. Reynolds, Morteza M. Vaghefi,
 Kevin M. Young, Timothy A. Riley, Robert E. Klem,
 and Lyle J. Arnold, Jr. 143
CH. 8. Oligonucleoside Phosphorothioates,
 Gerald Zon 165
CH. 9. Synthesis and Purification of Phosphorodithioate DNA,
 W. T. Wiesler, W. S. Marshall, and M. H. Caruthers 191
CH. 10. Oligodeoxyribonucleotide Phosphotriesters,
 Maria Koziolkiewicz and Andrzej Wilk 207
CH. 11. Oligonucleoside Boranophosphate (Borane Phosphonate),
 Barbara Ramsay Shaw, Jon Madison, Anup Sood,
 and Bernard F. Spielvogel 225

CH. 12.	Oligonucleotide Phosphorofluoridates and Fluoridites ***Wojciech Dabkowski, Jan Michalski, and Friedrich Cramer*** 245	
CH. 13.	α-Oligodeoxynucleotides, ***François Morvan, Bernard Rayner, and Jean-Louis Imbach*** 261	
CH. 14.	Stereospecific Synthesis of P-Chiral Analogs of Oligonucleotides, ***Wojciech J. Stec and Zbigniew J. Lesnikowski*** 285	
CH. 15.	Oligonucleotide Analogs with Dimethylenesulfide, -sulfoxide, and -sulfone Groups Replacing Phosphodiester Linkages, ***Zhen Huang, K. Christian Schneider, and Steven A. Benner*** 315	
CH. 16.	Oligonucleotide Analogs Containing Dephospho-Internucleoside Linkages, ***Eugen Uhlmann and Anusch Peyman*** 355	
CH. 17.	Scale-Up of Oligonucleotide Synthesis: *Solution Phase*, ***H. Seliger*** 391	
CH. 18.	Large-Scale Oligonucleotide Synthesis Using the Solid-Phase Approach, ***Nanda D. Sinha*** 437	
CH. 19.	Solid-Phase Supports for Oligonucleotide Synthesis, ***Richard T. Pon*** 465	

Index 497

Contents
Protocols for Oligonucleotide Conjugates
Synthesis and Analytical Techniques

Ch. 1. Protecting Groups in Oligonucleotide Synthesis,
 Etienne Sonveaux
Ch. 2. Incorporation of Modified Bases into Oligonucleotides,
 Rich B. Meyer, Jr.
Ch. 3. Functionalization of Oligonucleotides with Amino Groups
 and Attachment of Amino Specific Reporter Groups ,
 Sudhir Agrawal
Ch. 4. Functionalization of Oligonucleotides
 by the Incorporation of Thio-Specific Reporter Groups,
 Jacqueline A. Fidanza, Hiroaki Ozaki, and Larry W. McLaughlin
Ch. 5. Post-Synthesis Functionalization of Oligonucleotides,
 Barbara C. F. Chu and Leslie E. Orgel
Ch. 6. Oligonucleotide–Enzyme Conjugates,
 Jerry L. Ruth
Ch. 7. Oligonucleotides Containing Degenerate Bases:
 Synthesis and Uses,
 Paul Kong Thoo Lin and Daniel M. Brown
Ch. 8. Synthesis of [^{15}N]-Labeled DNA Fragments,
 Roger A. Jones
Ch. 9. Analysis and Purification of Synthetic Oligonucleotides
 by High Performance Liquid Chromatography,
 William Warren and George Vella
Ch. 10. Sequence Analysis of Oligodeoxyribonucleotides,
 Diane M. Black and Peter T. Gilham
Ch. 11. Gel-Capillary Electrophoresis Analysis of Oligonucleotides,
 Alex Andrus
Ch. 12. Nuclear Magnetic Resonance Studies of Oligonucleotides,
 Jerzy W. Jaroszewski, Siddhartha Roy, and Jack S. Cohen
Ch. 13. Mass Spectrometry of Nucleotides and Oligonucleotides,
 Thomas D. McClure and Karl H. Schram
Ch. 14. Extracting Thermodynamic Data from Equilibrium Melting Curves for Oligo-
 nucleotide Order-Disclosure Transitions,
 Kenneth J. Breslauer

Contributors

LYLE J. ARNOLD, JR. • *Genta, Inc., San Diego, CA*
SERGE L. BEAUCAGE • *Food and Drug Administration, Bethesda, MD*
STEVEN A. BENNER • *Laboratory for Organic Chemistry, Zurich, Switzerland*
DANIEL M. BROWN • *MRC Laboratory of Molecular Biology, Cambridge, England*
M. H. CARUTHERS • *Department of Chemistry, University of Colorado, Boulder, CO*
CHRIS CHRISTODOULOU • *Glaxo Group Research Ltd., Middlesex, England*
FRIEDRICH CRAMER • *Max-Planck-Institut für Experimentelle Medizin, Gottingen, Germany*
WOJCIECH DABKOWSKI • *Max-Planck-Institut für Experimentelle Medizin, Gottingen, Germany*
MASAD J. DAMHA • *Department of Chemistry, McGill University, Montreal, Quebec, Canada*
BRIAN C. FROEHLER • *Gilead Sciences, Inc., Foster City, CA*
RICHARD I. HOGREFE • *Genta, Inc., San Diego, CA*
ZHEN HUANG • *Laboratory for Organic Chemistry, Zurich, Switzerland*
JEAN-LOUIS IMBACH • *Laboratoire de Chimie Bio-Organique, Montpellier, France*
ROBERT E. KLEM • *Genta, Inc., San Diego, CA*
MARIA KOZIOLKIEWICZ • *Department of Bioorganic Chemistry, Polish Academy of Sciences, Lodz, Poland*
ZBIGNIEW J. LESNIKOWSKI • *Department of Bioorganic Chemistry, Polish Academy of Sciences, Lodz, Poland*
JON MADISON • *Department of Chemistry, Duke University, Durham, NC*
W. S. MARSHALL • *Department of Chemistry, University of Colorado, Boulder, CO*
JAN MICHALSKI • *Max-Planck-Institut für Experimentelle Medizin, Gottingen, Germany*

FRANÇOIS MORVAN • *Laboratoire de Chimie Bio-Organique, Montpellier, France*
KELVIN K. OGILVIE • *Department of Chemistry, University of Toronto, Toronto, Ontario, Canada*
ANUSCH PEYMAN • *Hoechst AG, Frankfurt, Germany*
RICHARD T. PON • *Department of Medical Biochemistry, University of Calgary, Calgary, Alberta, Canada*
BERNARD RAYNER • *Laboratoire de Chimie Bio-Organique, Montpellier, France*
MARK A. REYNOLDS • *Genta, Inc., San Diego, CA*
TIMOTHY A. RILEY • *Genta, Inc., San Diego, CA*
K. CHRISTIAN SCHNEIDER • *Laboratory for Organic Chemistry, Zurich, Switzerland*
H. SELIGER • *Polymer Section, University of Ulm, Ulm, Germany*
BARBARA RAMSAY SHAW • *Department of Chemistry, Duke University, Durham, NC*
NANDA D. SINHA • *Millipore Corp., Bedford, MA*
ANUP SOOD • *Boron Biologicals, Inc., Raleigh, NC*
BERNARD F. SPIELVOGEL • *Boron Biologicals, Inc., Raleigh, NC*
BRIAN S. SPROAT • *The European Molecular Biology Laboratory, Heidelberg, Germany*
WOJCIECH J. STEC • *Department of Bioorganic Chemistry, Polish Academy of Sciences, Lodz, Poland*
EUGEN UHLMANN • *Hoechst AG, Frankfurt, Germany*
MORTEZA M. VAGHEFI • *Genta, Inc., San Diego, CA*
W. T. WIESLER • *Department of Chemistry, University of Colorado, Boulder, CO*
ANDRZEJ WILK • *Department of Bioorganic Chemistry, Polish Academy of Sciences, Lodz, Poland*
KEVIN M. YOUNG • *Genta, Inc., San Diego, CA*
GERALD ZON • *Lynx Therapeutics, Inc., Foster City, CA*

Chapter 1

A Brief History of Oligonucleotide Synthesis

Daniel M. Brown

1. Introduction

Julian Barnes wrote *A History of the World in 10½ Chapters* (1). Oligonucleotides on this score warrant at most a comma, a literary *augenblick*, but lacking the skill of the aforementioned author and still using a broad brush, perhaps ten pages or so may suffice to paint the picture. This is an essay or perhaps commentary, not a review, so that the references regrettably do not do justice to the work of the chemists involved in oligonucleotide synthesis who have provided the single most important development in chemistry as applied to biology in the past 40 years. Where, without it, would molecular biology be today? The matter hardly bears thinking about. In a period during which synthetic organic chemistry has flourished, when more and more subtle reactions have been uncovered and applied to more and more complex structures none, or at most very few, of the great exponents have touched oligonucleotide synthesis. Why? That linearity breeds contempt, and having only four types of building blocks to deal with might have something to do with it. After all, the two major hurdles have been the development of protecting groups, their addition and removal, and phosphorylation, and, one should add, separation methods. Perhaps it was believed that this was intellectually limiting. Nevertheless, one thing seems certain: that the drive that brought the subject to its present advanced state was not that of solving crossword puzzles, but the pressure to solve precise problems, be they in carbohydrate, heterocyclic, or phosphorus chemistry, backed by demand, the requirement of the products to solve other problems. The fact is, of course, that much novel chemistry has inevitably been generated as a result of this large enterprise.

The subject began halfway through the century. Before that there was little reason to consider oligonucleotides as useful targets. DNA had just been recognized as a high polymer and RNA as isolated was anything but, although its isolation in its various manifestations in the native state followed rapidly. During the 1950s the understanding of the general chemistry and structures of the natural polynucleotides and their enzymatic degradation products developed incredibly rapidly, so the targets and the chemical understanding on which synthetic strategies might be based were both in place. In the absence of chemical synthesis, enzymatic synthesis using polynucleotide phosphorylase, developed by Grunberg-Manago and others, provided polyribonucleotide substrates for biochemical work *(2)*.

The fact is, however, that synthesis began, not as a pursuit of oligonucleotides *per se*, but as a development or offshoot of phosphorylation studies aimed more at polyphosphates, the nucleotide coenzymes, and other cofactors—a formidable project *(3,4)*. It is this beginning that may provide some historical understanding of the way that oligonucleotide or polyphosphodiester synthesis developed.

2. Phosphorylation: Early Studies

Leaving aside phosphoryl chloride, much used earlier in sugar phosphate and related syntheses, we should note parenthetically that other diester syntheses were being studied, particularly those of phospholipids, by Baer and coworkers from 1948 *(5)*. The reagents used, $(PhO)_2 PO\cdot Cl$ and $PhOPOCl_2$ (**1**), were easily available and the aryl protecting group readily removed.

If the first and second steps in Scheme 1 had been adequately separable, oligomer synthesis might have been off to a rapid start. True, the reactivity of the intermediate phosphorochloridate (**2**) is reduced by the increased electron accession from oxygen to the phosphoryl center, but not adequately; thus symmetrical diesters (**4**) are formed concomitantly with the required unsymmetrical diester (**3**). The alternative, to isolate the diester intermediate (or generate it in an alternative way)

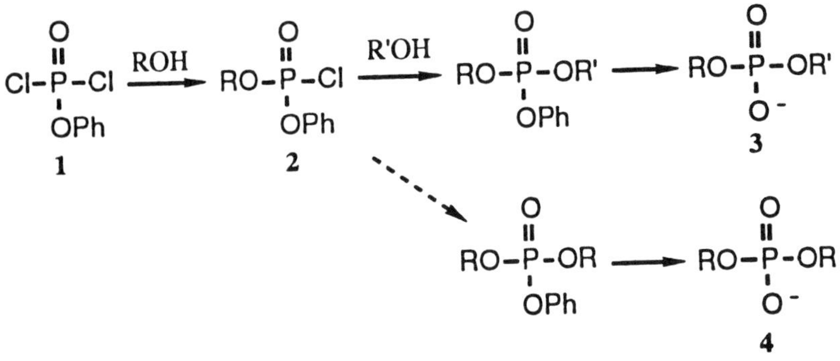

Scheme 1.

$$\underset{\mathbf{1}}{\text{Cl-P(=O)(OPh)-Cl}} \xrightarrow{\text{ROH}} \underset{\mathbf{2}}{\text{RO-P(=O)(OPh)-Cl}} \xrightarrow{\text{R'OH}} \text{RO-P(=O)(OPh)-OR'} \longrightarrow \underset{\mathbf{3}}{\text{RO-P(=O)(O}^-\text{)-OR'}}$$

$$\searrow \text{RO-P(=O)(OPh)-OR} \longrightarrow \underset{\mathbf{4}}{\text{RO-P(=O)(O}^-\text{)-OR}}$$

and then activate it for a second coupling reaction, has produced a myriad of partial or complete solutions. The important fact to grasp is, however, that in these early experiments, the phosphoryl group was fully protected.

This was the basic tenet of the Cambridge School, in which Todd and coworkers began a wide-ranging study of phosphorus chemistry ultimately directed at coenzyme synthesis (6). The benzyl group succeeded the phenyl group at this point—it was more versatile— removable by hydrogenolysis and by methods involving nucleophilic displacement on carbon (Scheme 2) (7,8). This opened the way to many other alkyl-based protecting groups and their use in synthesizing mono- and unsymmetrical disubstituted phosphates (5,6). Indeed it was at this time that much already known phosphorus chemistry (9) both at the P^V and P^{III} oxidation levels was introduced, the latter being developed much later for the synthesis of oligonucleotides and their analogs, amidates, thioates, and others (*vide infra*).

An essential point that emerged during the study of nucleoside polyphosphate (e.g., ADP, ATP) synthesis was the not surprising qualitative observation that the greater the degree of esterification, the greater the ease of hydrolysis and, more importantly, that phosphoryl exchange reactions by a series of nucleophilic displacements were increasingly facilitated, as in Scheme 3, to give the (unwanted) symmetrical product (7)(10).

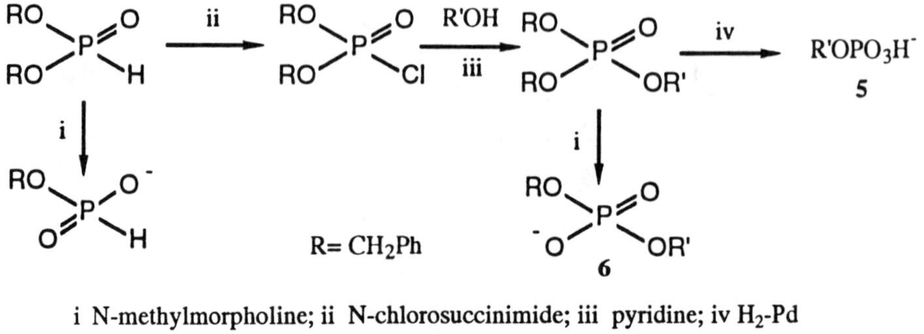

i N-methylmorpholine; ii N-chlorosuccinimide; iii pyridine; iv H$_2$-Pd

Scheme 2.

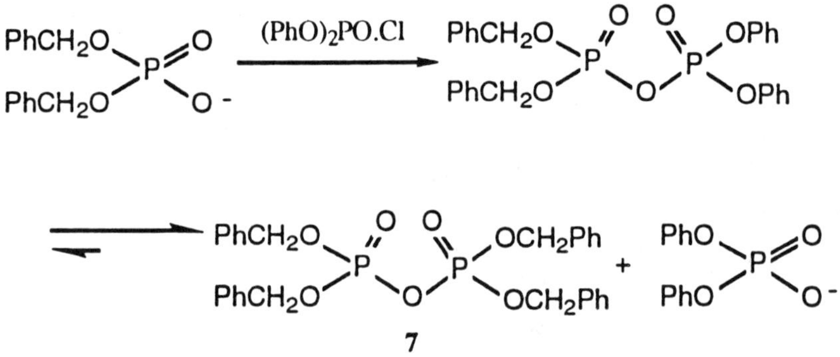

Scheme 3.

This diminished the effectiveness of fully protected intermediates unless rapid deprotection could be carried out. It was much better if a lower degree of protection was utilized. Thus, a set of methods for activation of monoesters, for example, adenosine 5'-phosphate (AMP), were developed (Scheme 4). Dicyclohexylcarbodiimide (**8**) and arenesulfonyl chlorides led rapidly to symmetrical pyrophosphates (**8**) *(11)*. Michelson found that (PhO)$_2$ PO • Cl gave activated intermediates (**9**) that were on the verge of stability, but could be put into reaction with a second component to obtain the required unsymmetrical polyphosphate (**10**) *(12)*. The complete solution to the problem depended on forming phosphoramidic acids (**11**) (and related structures), stable in themselves but capable of undergoing reaction by cleavage of the P—N bond (Scheme 4) *(6,13)*. Phosphates are better nucleophiles than

Oligonucleotide Synthesis

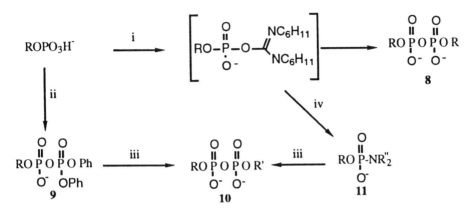

i Dicyclohexylcarbodi-imide; ii (PhO)$_2$POCl; iii R'OPO$_3$H$^-$; iv R"$_2$NH

Scheme 4.

alcohols, so the pyro- and tri-phosphate-forming reactions are much faster. Nevertheless, given the required conditions, propinquity in this case, nucleoside 5'-phosphoroimidazolides can form oligonucleotides in template-directed reactions *(14)*. Moreover, Gilham and Khorana found that carbodiimides and arenesulfonyl halides *could* achieve the required esterification of a phosphomonoester to give a diester, although slowly *(15)*.

3. Phosphodiester Chemistry

It was from this beginning in 1958 that Khorana and coworkers opened the way to synthetic sequence-defined oligonucleotides, later taken up by many other workers around the world. This vast exercise using phosphodiester chemistry, also provided the beginnings of base protection using acyl groups (on A, G, and C), dimethoxytrityl groups for nucleoside-5'-*O* protection, and more generally widened the range of protecting groups so that various strategies for chain elongation could be achieved. The early (1961) synthesis of a dinucleotide TpTp **12** is shown in Scheme 5 using DCC as coupling agent, although hindered arenesulfonyl chlorides were later seen to be more advan-tageous *(4)*. The biological problem of enormous interest in the 1960s was the complete elucidation of the genetic code. Here the painstaking synthesis, of di-, tri-, and tetranucleotides, the enzymatic conversion of these

i HO(CH₂)₂CN, DCC; ii H⁺; iii DCC; iv OH

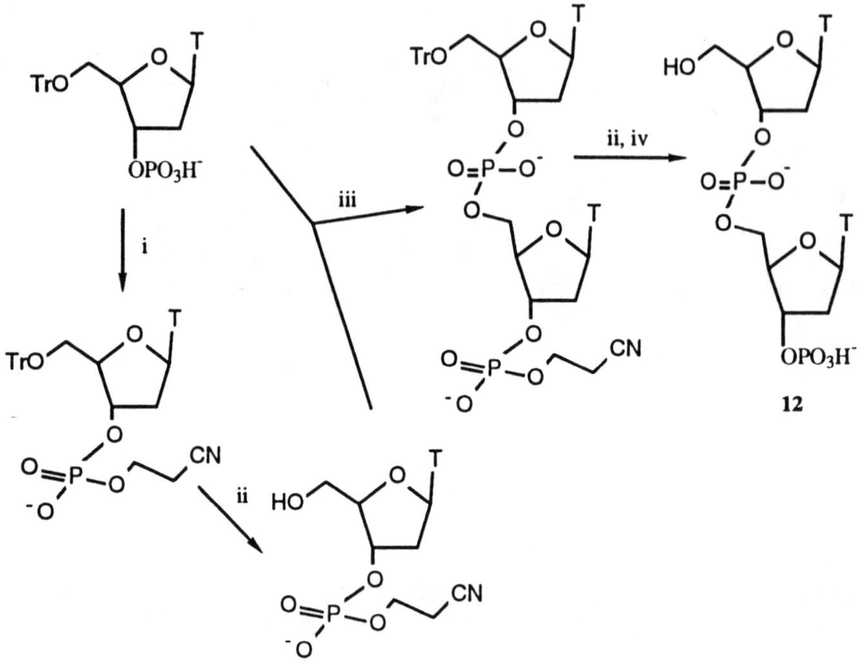

Scheme 5.

to polydeoxyribo- and thence polyribonucleotides with repeating sequences, and the use of the latter in in vitro protein synthesizing systems completed all the codon assignments. The Harvey Lecture in 1966 by Khorana reveals the excitement of the time and gives reference to the detailed chemistry involved (16).

Later, synthesis of two tRNA genes and of a transcribable fmet tRNA gene sequence was carried out (17). These are mountains, and mountains have to be climbed. Among the more recent mountaineers was the large Chinese expedition led by Yu Wang and others (18). They succeeded in synthesizing yeast alanine tRNA, including all the minor ribonucleoside residues, a remarkable achievement. Here, too, diester coupling chemistry was mainly used and then enzymatic ligation of oligomer blocks.

The nature of the chemistry involved in the phosphodiester coupling reactions with DCC and sulfonyl halides has never been wholly elucidated, but it was clear that pyrophosphate, trimetaphosphate, and higher polyphosphate esters were quickly formed and more slowly alcoholized. To complicate the matter, oligomers containing pyrophosphate linkages were formed. Anhydride exchange with acetic anhydride (to give labile acetylphosphates) satisfactorily cleaved these. It is interesting that, although protecting groups on the phosphoryl residue led to problems of stability in polyphosphate syntheses, they were not used in polydiester synthesis, where the latter problems did not in fact arise. The fact that yields were good and also presumably the greater ease, at that time, of purifying products on a charge basis—tedious but effective—gave an impetus to diester chemistry that lasted well into the 1970s.

4. Phosphotriester Chemistry

However, the first syntheses of unsymmetrical dinucleoside phosphates had been effected by triester chemistry in 1955, and after a hiatus, work continued in a number of laboratories using full protection, particularly in that of Letsinger *(19)* and later of Reese *(20)*. It was directed toward two major questions. What kind of phosphoryl group protection could be used so that neutral, more easily handed, triester products would be formed and easily deprotected? The more difficult question was how activation for coupling could be achieved without symmetrical products being formed. At the same time, strategic solutions were proposed, for example, by Catlin and Cramer, as to how to arrange nucleoside hydroxyl protection to allow chain extension in either or, both directions and, related to this, how block condensations could be carried out most efficiently *(21)*. The latter required the synthesis of terminal and nonterminal protected units; although the elaboration of these added enormously to the detailed chemistry of protecting groups, particularly in the ribonucleotide series *(20)*, in the event (although perhaps not finally), addition of monomer units sequentially became the method of choice.

The question of the phosphate-protecting group was addressed in various ways. The aryl groups were developed extensively by Reese *(20)*, and others, although they had initially unexpected disadvantages. Simply put, in a phosphotriester (such as **16** in Scheme 6), the expec-

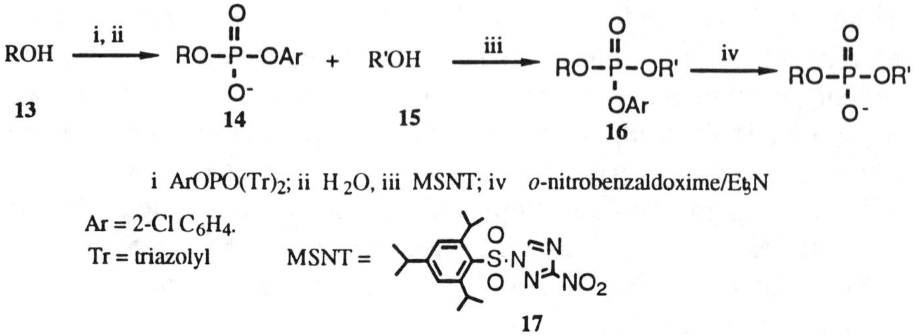

$$\text{ROH} \xrightarrow{\text{i, ii}} \text{RO-P(=O)(O}^-\text{)-OAr} + \text{R'OH} \xrightarrow{\text{iii}} \text{RO-P(=O)(OAr)-OR'} \xrightarrow{\text{iv}} \text{RO-P(=O)(O}^-\text{)-OR'}$$

13 → **14** + **15** → **16** → (product)

i ArOPO(Tr)₂; ii H₂O; iii MSNT; iv o-nitrobenzaldoxime/Et₃N

Ar = 2-Cl C$_6$H$_4$.
Tr = triazolyl   MSNT = [structure **17**: mesitylene-sulfonyl-nitrotriazolide]

Scheme 6.

tation, derived from earlier work on triester hydrolysis, was that the rate of displacement of the aryloxy residue by, say, hydroxide would be orders of magnitude greater than that of the nucleoside residues (**16**; ROH, R'OH). In fact, chain cleavage up to 3% occurred. This can be understood as resulting from pseudorotational equilibria in the pentacoordinate reaction intermediate *(22)*. By adjusting the aryl function and using displacement by aldoximate anions (antidotes for phosphorofluoridate nerve gases!), the chain cleavage was largely overcome. An example of a chain extension reaction can be visualized in Scheme 6, where a 5'-dimethoxytrityldeoxynucleoside 3'-arylphosphate (**14**) is added at the 5'-position of the growing chain (**15**).

It is as well to note here that, although displacement from a triester by P—O cleavage has the intrinsic problem outlined above, deprotection by C—O cleavage (displacement or elimination) is not affected in this way. It is for this reason that the cyanoethyl group, which was being developed concurrently, has assumed a major importance. It is curious in this connection that the unassuming methyl group, rapidly displaced by a thiolate ion, was introduced in 1977 by Daub and van Tamelin *(23)*, but not taken up until much later when it was given a major role in phosphite triester chemistry.

Many coupling reagents have been investigated, deriving from the original TPSCl; arenesulfonylazoles were shown to be especially effective, and the nitrotriazolide (**17**) became widely used in conjunction with catalytic *N*-methylimidazole. Precisely what species are formed in the phosphorylation step is not clear, but high yields and short reaction times were achieved, and rapid chain assembly became possible. It is important to recognize that the solution phase was the medium in which all of this new chemistry was opened up, contributions coming from very many laboratories.

5. Solid-Phase Chemistry

As early as 1965, Letsinger and his coworkers took two highly significant steps. The first was to reinitiate phosphotriester chemistry as mentioned earlier. The second, concurrently with Merrifields' work on peptides, was to begin the investigation of nucleotide chain extension on solid supports *(19)*. This set in motion an enormous amount of work in many laboratories, devising solid supports and methods of linking the first residue to the support *(see,* for example, Kössel and Seliger, ref. *24)*. For a while, success in synthesizing extended oligomer chains by successive coupling steps was limited; yields were not impressive. In retrospect, this was probably less related to the nature of the support than to the fact that diester chemistry, unsuited to the solid-phase, was used. In the mid-1970s, however, when triester chemistry began to be applied, remarkably, some 10 years after its first introduction, an immediate increase in yields and decrease in cycle times were obtained. Automation could begin to be applied using, for example, in the laboratories of Gait, Itakura, and Köster, the chemistry depicted in Scheme 6 and HPLC began to be used for purification of the products *(25–27)*. Oligodeoxynucleotides now became much more readily available for biochemical application. It was just at this time that there was an outburst of work demanding oligomers for probes, primers, linkers, and controlled mutagenesis as well as for gene synthesis and for the study of protein–nucleic acid interactions *(28,29)*. The modern era had begun.

Solid-phase synthesis of linear structures has one great advantage over the liquid phase. In the latter, more attention had to be given to building blocks with more complex protection, since it seemed likely that unidirectional linear addition of monomer units would not be the most efficient way of building large structures; block coupling might be necessary. With solid-phase chemistry all that was required in principle was the support with its appended nucleoside (e.g., **13**) and a set of four monomers. With triester chemistry such monomers might be, e.g. (**14**). In the simplest terms, with suitable activation, coupling occurred and removal of the DMT residue by acid from (**16,** Scheme 6) then released the 5'-hydroxyl function for a further coupling. All other protecting groups were base labile and could be removed after completion of the synthesis. Such were the cycle yields achieved (>95%) that sequential monomer addition was at the time perfectly adequate, and dimer blocks (16 of them) although used, were not pursued.

6. Phosphite and Phosphoramidite Chemistry

It has long been known that the reactivity of P^{III} derivatives is much greater than that of the corresponding P^V compounds. For example, the hydrolysis or the equivalent of PCl_3 and $P(OCH_3)_3$ compared with $POCl_3$ and $PO(OCH_3)_3$ and the corresponding comparison with sulfur and nitrogen oxyacids and esters exemplifies this. One way to decrease coupling times in oligomer synthesis would be to operate at the P^{III} oxidation level. Letsinger and coworkers in 1976, realizing this, showed that 2-chlorophenyl phosphorodichloridite (**18**) as well as the trichloroethyl ester would phosphitylate a 5'-O-protected thymidine at low temperatures, and addition of a second nucleoside component followed by oxidation gave dithymidine phosphate in good yield (Scheme 7) *(30)*

Thus began another major step in oligomer synthesis. Such phosphorochloridites were exceptionally reactive—too much so for common use. However, very shortly afterward, Beaucage and Caruthers found that nucleoside-3'-O-phosphoramidites, such as (**19**), could be synthesized, stored, and activated for coupling to give (**20**), this time by acid catalysis (compare the chemistry of the much less reactive phosphoramidates). Oxidation then led to the phosphate diester (**21;** e.g., the chain extended product, ready for the next nucleotide addition cycle) *(31)*. This has been developed into the most widely used method for automated chain assembly *(32)*. The richness of P^{III} chemistry has given rise to a great variety of modifications of the basic oligomer structure.

7. H-Phosphonate Chemistry

A further interesting development occurred soon afterward. Two groups of workers observed that a nucleoside 3'-H-phosphonate (**22**) could be activated and then coupled rapidly to give a diester H-phosphonate in high yield, and the process automated for oligomer assembly *(33,34)*. Such diester H-phosphonates (**23**) can, at will, be oxidized to the phosphate (**24**), thiated to the phosphorothioate, and aminated

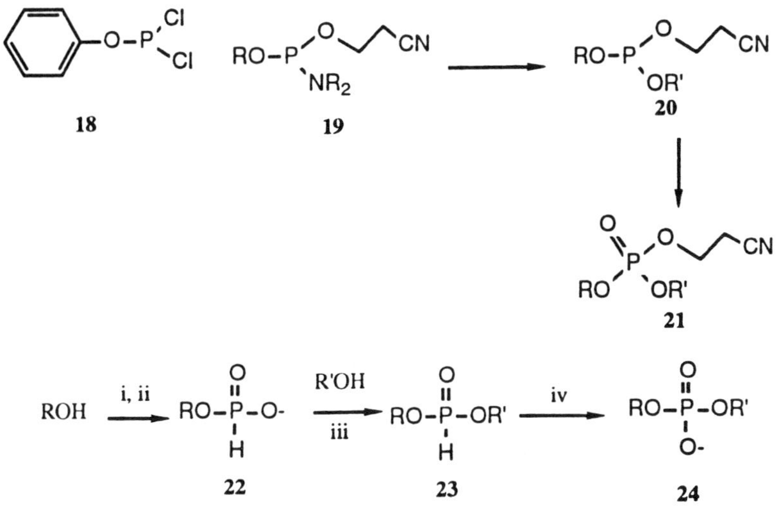

Scheme 7.

i P(triazole)$_3$; ii H$_2$O; iii R$_3$CCO.Cl; iv I$_2$/H$_2$O

to give phosphoramidates, as can (**20**), in each case very cleanly. The ramifications of oligonucleotide analog synthesis that followed from these observations were very wide. We should note, as Matteucci and Froehler did, that some of the earliest dinucleoside phosphate and other diester syntheses utilized H-phosphonates *(33)*.

It may be useful to enlarge on this. Corby and coworkers in 1952 took the view that the mixed anhydride **25** (Scheme 8) should act as a *phosphonylating* agent and showed that indeed it was so *(35)*. The synthesis of the dinucleotide pTpT (**28,** Scheme 8) set out in Scheme 8 exemplified this and included the activation of an H-phosphonate monoester (**22**—>**25**, Scheme 8) and of an unsymmetrical H-phosphonate diester (**26**—>**27,** Scheme 8), and the intermediacy of a compound (**26,** Scheme 8) with phosphorus at two oxidation levels *(36)*. In retrospect the modest yields then obtained may have had to do with exchange reactions (e.g., of **25**) as discussed before leading, in the early work, to relatively unreactive pyro- or polyphosphonates, a question that has been studied in some detail by Stawinski and colleagues *(37)*. The more recent success in utilizing H-phosphonates stems from the choice of hindered acyl halides as activating agents.

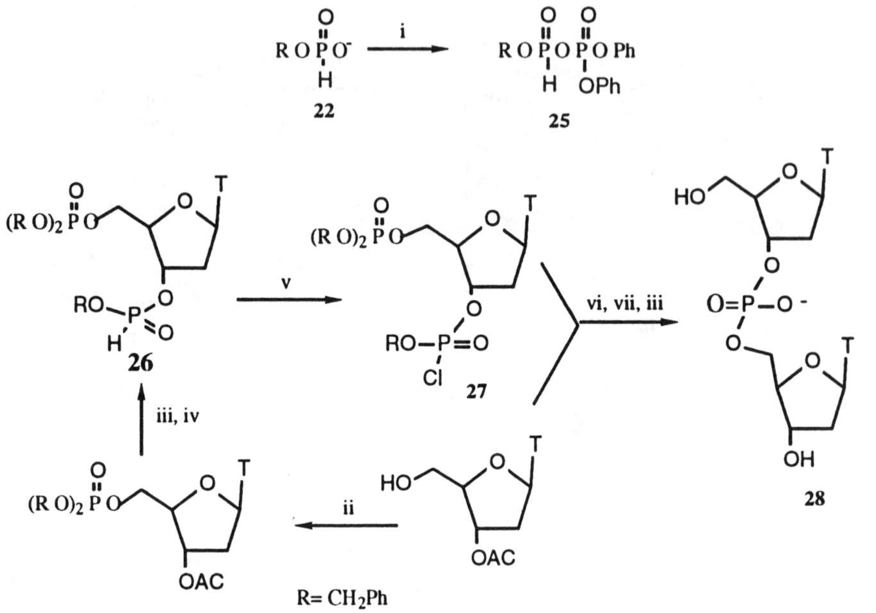

Scheme 8.

i (PhO)$_2$PO.Cl; ii (RO)$_2$PO.Cl; iii NH$_3$-CH$_3$OH; iv 25; v N-chlorosuccinimide; vi pyridine; vii H$_2$-Pd

8. Oligoribonucleotide Synthesis

In this last part, we consider oligoribonucleotide synthesis. The considerations applying to oligodeoxynucleotides that have been advanced already apply, too, in the ribo-series, with however, additional ones that have their origins in the 2'-hydroxyl group. In the first place, the 2'-hydroxyl function has to be protected, absolutely, until all other protecting groups have been removed. This applies in particular to the phosphate-protecting group since triester phosphates (or phosphites) with a neighboring hydroxyl group are inherently highly unstable. Transesterification reactions, in base, lead to chain cleavage, in acid, to cleavage and isomerization to a 2',5'-internucleotide linkage. Protecting group differentiation, as between the 5'- and 2'-positions and the phosphoryl group, has been closely investigated by very many workers (20,24). A number of acyl-protecting groups for the 5'-position have been considered, but the great advantage of the dimethoxytrityl group has made it difficult to supersede.

The differentiation of the 2'- and 3'-positions in ribonucleosides has always been difficult. Acylation on the 2'-position is kinetically favored, but facile base-catalyzed migration occurs, and generally the 3'-O-acyl isomer is thermodynamically favored; slow crystallization can therefore give high yields of the latter, convertible to a 2'-O-acetal or ketal. Silyl groups can also migrate in base, although less easily. Monosilylation of the 2',3'-diol and fractionation is used—inelegant but effective. The most valuable contribution to the problem was that of Markiewicz and Wiewiorówski, who showed that tetraisopropyl dichlorodisiloxane would form a 3',5'-bridge (**29,** Scheme 9), leaving the 2'-position free for further substitution *(38)*. Various groups for the 2'-position have been used; the prime condition was that the residue should be strongly differentiated in its rate or conditions of removal from that of the 5'-protecting group. The methoxytetrahydropyranyl and tetrahydrofuranyl groups have had some success but are clearly not adequate for more than short oligomers. In tRNA synthesis, the photolabile 2-nitrobenzyl group has been used by Ohtsuka and co-workers *(39),* as has the benzoyl group *(18).* The more complex ketal group (**30**), the most satisfactory among those removable by acid catalysis, may stay the course *(40).* At present the greater emphasis is on the *t*-butyldimethylsilyl (TBDMS) group, which is removable by nonhydrolytic means. The development of oligoribonucleotide synthesis using this 2'-protecting group (**31**) with phosphoramidite chemistry, developed by Ogilvie and coworkers, has made machine assembly possible, if not entirely routine *(41).* The question of the optimal 2'-protecting group, it must be said, is still in flux.

The second consideration that has its origins in the presence of a 2'-protected hydroxyl group is that of steric factors influencing the phosphorylation process; coupling reactions are slower than in the 2'-deoxy series. Phosphotriesters have given way to phosphoramidites and, to a much lesser extent, H-phosphonates. If steric interactions are important in the synthesis of monomers and in the phosphorylation step, it would seem reasonable that H-phosphonate chemistry would be preferred; the hydrogen atom is the smallest possible protecting group. Such a consideration could become very telling when RNAs branched at the 2'- and 3'-positions are required, with intermediates having phosphoryl groups on contiguous positions at two different oxidation levels.

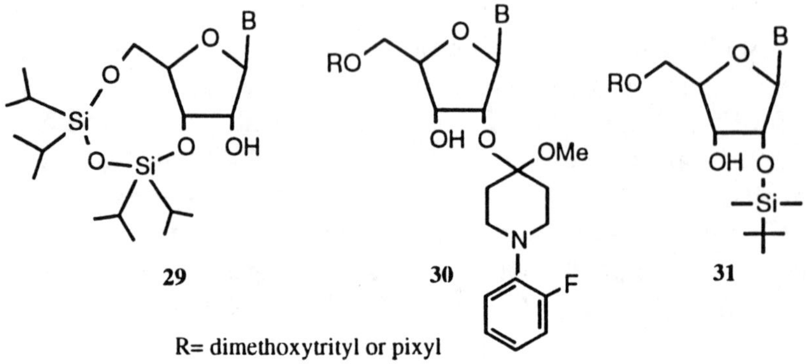

29   **30**   **31**

R= dimethoxytrityl or pixyl

Scheme 9.

9. Future Developments

Evidently, machine synthesis of DNA is adequate for most foreseeable requirements. The development in so few years is quite staggering with cycle yields of up to 99% and oligomer chain lengths of over 100. RNA synthesis and purification (with or without the 2'-protecting group in place) is intrinsically a more difficult problem, but no doubt will become routine as has DNA synthesis. Chain lengths may always lag.

At present, one of the new problems is that of scale. On the one hand, large- to very large-scale synthesis of deoxyoligomer derivatives for therapeutic use may become commonplace. It may be that, in this realm, solution-phase or phosphotriester chemistry, or oligomer block coupling will be given reconsideration. At the other end of the spectrum, a much smaller scale synthesis is of great interest and is ripe for development. Genome sequencing requires multitudes of primers. Moreover, the simultaneous synthesis of large arrays of oligomers ("chips") for diagnostic and other hybridization-based uses is being actively studied *(42)*. There is still life in oligonucleotide synthesis; the remaining chapters in this book attest to this.

References

1. Barnes, J. (1989) *A History of the World in 10½ chapters*. Jonathan Cape, London.
2. Grunberg-Manago, M., Ortiz, P. J., and Ochoa, S. (1957) Enzymic synthesis of polynucleotides I. Polynucleotide phosphorylase of *azotobacter vinelandii*. Biochim. Biophys. Acta **20**, 269–285.

3. Brown, D. M. (1963) Phosphorylation, in *Adv. Org. Chem. Methods and Results* (Raphael, R. A., Taylor, E. C., and Wynberg, H., eds.) **3**, 75–157 and references therein cited.
4. Khorana, H. G. (1961) *Some Recent Developments in the Chemistry of Phosphate Esters of Biological Interest.* John Wiley, London, and references therein cited.
5. Baer, E. (1957) The synthesis of glycerol phosphatides in phosphoric esters and related compounds. *Chem. Soc. (Lond.)* **Spec. Publ. No. 8,** 103–113.
6. Todd, A. R. (1957) Newer methods of polyphosphate synthesis, in phosphoric esters and related compounds. *Chem. Soc. (Lond.)* **Spec. Publ. No. 8,** 91–98 and references therein cited.
7. Atherton, F. R., Openshaw, H. T., and Todd, A. R. (1945) Dibenzylchlorophosphonate as a phosphorylating agent. *J. Chem. Soc. (Lond.)* 382–385.
8. Clark, V. M. and Todd, A. R. (1950) The reaction between organic bases and esters of oxy-acids of phosphorus. An interpretation based on a comparison of certain aspects of the chemistry of sulfur and phosphorus. *J. Chem. Soc. (Lond.)* 2023–2034.
9. Kosolapoff, G. M. (1950) *Organophosphorus Compounds.* John Wiley, New York.
10. Corby, N. S., Kenner, G. W., and Todd, A. R. (1952) The preparation of tetraesters and pyrophosphoric acid from diesters of phosphoric acid by means of exchange reactions. *J. Chem. Soc. (Lond.)* 1234–1243.
11. Khorana, H. G. and Todd, A. R. (1953) The reaction between carbodi-imides and acid esters of phosphoric acid. A new method for the preparation of pyrophosphates. 2257–2260.
12. Michelson, A. M. (1963) *The Chemistry of Nucleosides and Nucleotides.* Academic, London, and references therein cited.
13. Kirby, A. J. and Warren, S. G. (1967) *The Organic Chemistry of Phosphorus.* Elsevier, Amsterdam.
14. Joyce, G. F. and Orgel, L. E. (1988) Nonenzymatic template-directed synthesis on RNA random copolymers. *J. Mol. Biol.* **202,** 677–681 and references therein cited.
15. Gilham, P. T. and Khorana, H. G. (1958) An new and general method for the synthesis of the $C_{5'}$-$C_{3'}$ internucleotide linkage. Synthesis of deoxyribodinucleotides. *J. Amer. Chem. Soc.* **80,** 6212–6222.
16. Khorana, H. G. (1966–1967) Polynucleotide synthesis and the genetic code. *The Harvey Lectures,* **ser. 62,** 79–105.
17. Khorana, H. G. (1979) Total synthesis of a gene. *Science* **203,** 614–625.
18. Wang, Yu (1984) A total synthesis of yeast alanine transfer RNA. *Acc. Chem. Res.* **17,** 393–397.
19. Letsinger, R. L. and Mahadevan, V. (1965) Oligonucleotide synthesis on a polymer support. *J. Amer. Chem. Soc.* **87,** 3526–3527.
20. Reese, C. B. (1978) The chemical synthesis of oligo- and poly-nucleotides by the phosphotriester approach. *Tetrahedron* **34,** 3143–3179 and references therein cited.
21. Catlin, J. C. and Cramer, F. (1973) Deoxy oligonucleotide synthesis via the triester method. *J. Org. Chem.* **38,** 245–250.

22. Westheimer, F. H. (1968) Pseudorotation in the hydrolysis of phosphate esters. *Acc. Chem. Res.* **1,** 70–81.
23. Daub, G. W. and van Tamelen, E. E. (1977) Synthesis of oligoribonucleotides based on the facile cleavage of methyl phosphotriester intermediates. *J. Amer. Chem. Soc.* **99,** 3526–3528.
24. Kössel, H. and Seliger, H. (1975) Recent advances in polynucleotide synthesis. *Fortschritte d. Chem. Org. Naturst.* **32,** 297–508; a comprehensive review.
25. Gait, M. J., Popov, S. G., Singh, M., and Titmas, R. C. (1980) Rapid synthesis of oligonucleotides. Further studies in solid-phase synthesis of oligodeoxynucleotides through phosphotriester intermediates. *Nucl. Acids Res.* **Symp. Ser. No. 7,** 243–257.
26. Miyoshi, K. and Itakura, K. (1980) Solid-phase synthesis of polynucleotides. Synthesis of oligodeoxynucleotides by the phosphomonotriazolide method. *Nucl. Acids Res.* **Symp. Ser. No. 7,** 281–291.
27. Köster, H., Hoppe, N., Kröpelin, Kaut, H., and Kulikowski, K. (1980) Some improvements in the synthesis of DNA of biological interest. *Nucl. Acids Res.* **Symp. Ser. No. 7,** 39–60, and other contributions to this symposium.
28. Wu, R., Bahl, C. P., and Narang, S. A. (1978) Synthetic oligodeoxynucleotides for analyses of DNA structure and function. *Prog. Nucl. Acids Mol. Biol.* **21,** 101–141.
29. Engels, J. W. and Uhlmann, E. (1989) Gene synthesis. *Angew. Chem. Int. Ed. Engl.* **28,** 716–733.
30. Finnan, J. L., Varshney, A., and Letsinger, R. L. (1980) Developments in the phosphite-triester method of synthesis of oligonucleotides. *Nucl. Acids Res.* **Symp. No. 7,** 133–145 and references therein cited.
31. Beaucage, S. L. and Caruthers, M. H. (1981) Deoxynucleoside phosphoramidites—a new class of key intermediates for deoxypolynucleotide synthesis. *Tetrahedron* Lett. **22,** 1859–1862.
32. Beaucage, S. L. and Iyer, P. R. (1992) Advances in the synthesis of oligonucleotides by the phosphoramidite approach. *Tetrahedron* **48,** 2223–2311, a comprehensive review.
33. Froehler, B. C., Ng, P. G., and Matteucci, M. D. (1986) Synthesis of DNA via deoxynucleoside H-phosphonate intermediates. *Nucl. Acids Res.* **14,** 5399–5407.
34. Garegg, P. J., Lindh, I., Regberg, T., Stawinski, J., Stromberg, R., and Henrichson, C. (1986) Nucleoside H-phosphonates. Chemical synthesis of oligodeoxyribo- nucleotides by the hydrogen phosphonate approach. *Tetrahedron Lett.* **27,** 4051– 4054.
35. Corby, N. S., Kenner, G. W., and Todd, A. R. (1952) Ribonucleoside-5' phosphites. A new method for the preparation of mixed secondary phosphites. *J. Chem. Soc. (Lond.)* 3669–3675.
36. Michelson, A. M. and Todd, A. R. (1955) Synthesis of dithymidine dinucleotide containing a 3':5'-internucleotidic linkage. *J. Chem. Soc. (Lond.)* 2632–2638.
37. Garegg, P. J., Stawinski, J. and Strömberg, R. (1987) Nucleoside, H-phosphonates. Activitation of hydrogen phosphonate monoesters by chlorophosphates and arenesulphonyl derivatives. *J. Org. Chem.* **52,** 284–287.

38. Markiewicz, W. T. and Wiewiorówski, M. (1978) A new type of silyl protecting group in nucleoside chemistry. *Nucl. Acids Res.* **Spec. Publ. No. 4**, s185–188.
39. Ohtsuka, E., Markham, A. F., Tanaka, S., Miyake, T., Wakabayashi, T., Taniyama, K., Nishikawa, S., Fukamoto, R., Uemura, H., Doi, T., Tokunaga, T., and Ikehara, M. (1980) Total synthesis of tRNA fMet. *Nucl. Acids Res.* **Symp. Ser. No. 7,** 335–343.
40. Reese, C. B., Vaman Rao, M., Seraphinowska, H. T., Thompson, E. A., and Yu, P. S. (1991) Studies in the solid-phase synthesis of oligo- and poly-ribonucleotides. *Nucleosides and Nucleotides* **10**, 81–97.
41. Ogilvie, K. K., Usman, N., Nicoghosian, K., and Cedergren, R. J. (1988) Total chemical synthesis of a 77–nucleotide-long RNA sequence having methionine- acceptance activity. *Proc. Natl. Acad. Sci. USA* **85**, 5764–5768.
42. Barinaga, M. (1991) Will "DNA chip" speed genome initiative? *Science* **253**, 1489.

CHAPTER 2

Oligonucleotide Synthesis

Phosphotriester Approach

Chris Christodoulou

1. Introduction

Over the last 30 years or so, the chemical synthesis of deoxyribonucleic acid (DNA) and ribonucleic acid (RNA) has constituted one of the most challenging problems in the field of the synthetic organic chemistry of natural products. In more recent times, attention has been focused not on "natural" nucleic acid fragments, but on the preparation of nucleic acids bearing modified internucleotidic linkages or modified bases in an effort to enhance stability to nucleases and increase cellular uptake.

The advent of the "antisense" principle has inspired large numbers of research groups around the world to extend their investigations not only to the preparation of modified oligodeoxyribonucleotides, but also to the potential of small fragments of nucleic acids as therapeutic agents. Many facets of these works are discussed elsewhere in this book. The aim of this chapter is to review the phosphotriester approach to oligodeoxyribonucleotide synthesis and how it may serve the developing field of antisense oligonucleotide therapeutics.

The phosphotriester method *(1)* of oligonucleotide synthesis, that is, the realization of the need to protect the internucleotidic phosphodiester groups during synthesis, revolutionized the preparation of oligonucleotides. Adaptation of the methodology to solid-phase *(2–4)* technology led to the rapid and efficient synthesis of oligonucleotides,

From: *Methods in Molecular Biology, Vol. 20: Protocols for Oligonucleotides and Analogs*
Edited by: S. Agrawal Copyright ©1993 Humana Press Inc., Totowa, NJ

so that preparation of short strands of nucleic acids is no longer a long and laborious process and the rate-limiting step in experimental molecular biology. Moreover, recent developments in the phosphoramidite *(5)* and H-phosphonate approaches to oligonucleotide synthesis, together with automation, means that the preparation of oligonucleotides of defined sequence is now a routine process and no longer the realm of the synthetic organic chemist.

2. Materials Needed for Oligodeoxyribonucleotide Synthesis

2.1. Supports

1. DMTr dA-LCAA/Controlled pore glass (CPG)
2. DMTr dC-LCAA/CPG
3. DMTr dG-LCAA/CPG
4. DMTr T-LCAA/CPG
5. DMTr dA-succinyl-polyamide resin
6. DMTr dC-succinyl-polyamide resin
7. DMTr dG-succinyl-polyamide resin
8. DMTr T-succinyl-polyamide resin

2.2. 5'-O-Dimethoxytrityl-Protected Monomers

1. DMTr dA-3'-*O*-(2-chlorophenylphosphate)
2. DMTr dC-3'-*O*-(2-chlorophenylphosphate)
3. DMTr dG-3'-*O*-(2-chlorophenylphosphate)
4. DMTr T-3'-*O*-(2-chlorophenylphosphate)

2.3. Solvents

1. 1,2-Dichloroethane (DCE)
2. *N,N*-Dimethylformamide (DMF)
3. Dioxane
4. Pyridine

2.4. Other Reagents

1. Dichloroacetic acid (DCA)
2. 1-Mesitylene-2-sulfonyl-3-nitro-1,2,4-triazole (MSNT)
3. 1-Methylimidazole
4. Syn-2-nitro-benzaldoxime
5. 1,1,3,3-Tetramethylguanidine
6. Trichloroacetic acid (TCA)

2.5. Preparation of Materials

With the exception of purified solvents, the chemicals and reagents required for the synthesis of oligodeoxyribonucleotides by the phosphotriester approach are no longer generally available from commercial suppliers. Anyone wishing to embark on a synthesis program will need to prepare most of the reagents.

The excellent review by Gait and Sproat *(4)* provides detailed experimental notes on the preparation of the reagents for phosphorylation and coupling. These authors also included methods for the functionalization of the polydimethylacrylamide-Kieselguhr support for large-scale synthesis and the use of controlled pore glass for small-scale synthesis. The assembly of oligodeoxyribonucleotides is also described using a simple and efficient semimanual apparatus made up from the parts in a kit supplied by Ommifit Ltd (UK).

3. Methods
3.1. The Supports

The solid-phase phosphotriester method of oligodeoxyribonucleotide synthesis has largely been developed on two principal supports, controlled pore glass (CPG) *(4,6–8)* and a polydimethylacrylamide-Kieselguhr composite material *(2–4)* first used in peptides synthesis *(9)*. Controlled pore glass meets the requirements for a solid-phase synthesis support by being rigid, nonswellable, and chemically inert. The polydimethylacrylamide-Kieselguhr support, although having the disadvantage of swelling to varying degrees in different solvents and during chain assembly, has the important advantage of being functionalized to a much higher level (e.g., 90 or 180 μmol deoxyribonucleoside/g of support) to yield milligram amounts of product. In each case, the protected nucleoside is linked to the support through a succinate linkage.

3.2. The Monomers

The monomers required for the chain assembly are the appropriately protected deoxynucleoside phosphodiesters, such as the triethylammonium salts, illustrated in Fig. 1. The 5'-hydroxyl group of the 2'-deoxyribonucleosides is protected by the use of the very acid-labile protecting groups, 9-phenylxanthen-9-yl (Pixyl, Px) *(10)* or 4,4'-

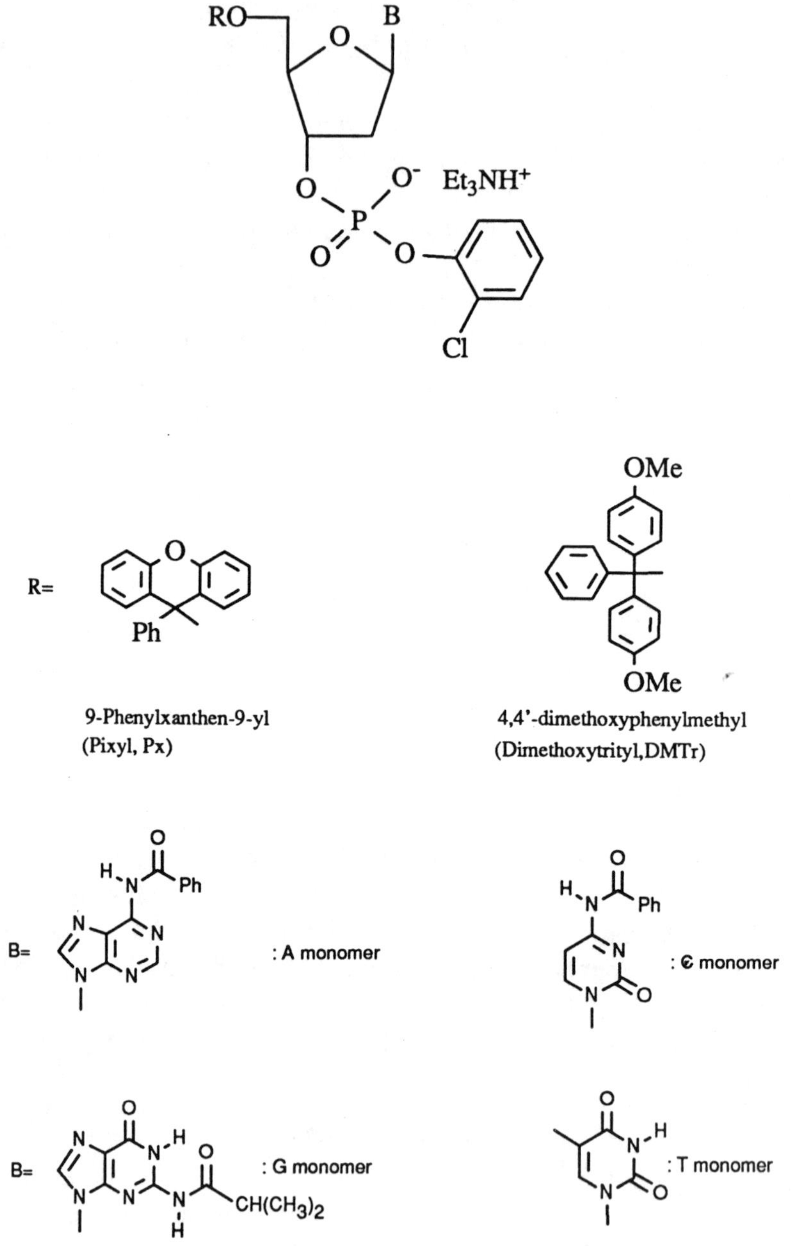

Fig. 1. Structures of the 2'-deoxyribonucleotides monomer building blocks.

dimethoxyphenylmethyl (dimethoxytrityl, DMTr) *(11)*. The 2-chlorophenyl group *(12,13)* is the protecting group of choice for the internucleotidic phosphates being readily removed by oximate ions *(13,14)* with minimal concomitant internucleotide bond cleavage. The protection of the exocyclic amino groups of adenine, cytosine, and guanine in 2'-deoxyribonucleosides is effected by *N*-acylation, as originally initiated by Khorana *(15,16)* and coworkers. Benzoyl groups are used for the protection of the 4-*N*- and 6-*N*-exocyclic amino groups of 2'-deoxycytidine and 2'-deoxyadenosine, whereas 2-*N*-isobutyryl is the protection for the 2-*N*-exocyclic amino group of 2'-deoxy-guanosine. The use of protic acids to effect the removal of the 5'-protecting groups has been shown to cause acid-promoted depurination of 6-*N*-benzoyl-2'-deoxyadenosine leading to strand cission and lowering of yields. To minimize depurination, the use of 6-*N*-phthaloyl *(17,18)* or the 5-*N*-di-*n*-butylaminomethylidene group *(19)* is to be recommended.

3.3. The Chain Assembly

The chain assembly of oligodeoxyribonucleotides by the solid-phase method is performed from the 3'- to the 5'-end in a repetitive cycle involving first removal of the 5'-protecting group (DMTr or Px) from the support bound nucleoside by a protic acid and then the condensation of an appropriately protected deoxynucleoside phosphodiester with the newly liberated 5'-hydroxyl group, in the presence of the coupling agent 1-mesitylenesulfonyl-3-nitro-1,2,4-triazole (MSNT) *(13)* and the catalyst, 1-methylimidazole. The general cycle is shown in Fig. 2. It is worth noting that depixylation (or detritylation) is faster for the purine than for the pyrimidine deoxyribonucleosides, so the acid wash time may be varied accordingly during synthesis.

The cycle is repeated until the required sequence is prepared. In general, a "capping" step (for example, treatment with acetic anhydride as in the phosphoramidite method) is not required. Truncated sequences are not normally the result of the presence of unreacted 5'-hydroxyl groups from the previous coupling reaction, but rather arising from the side reactions during coupling, such as sulfonation.

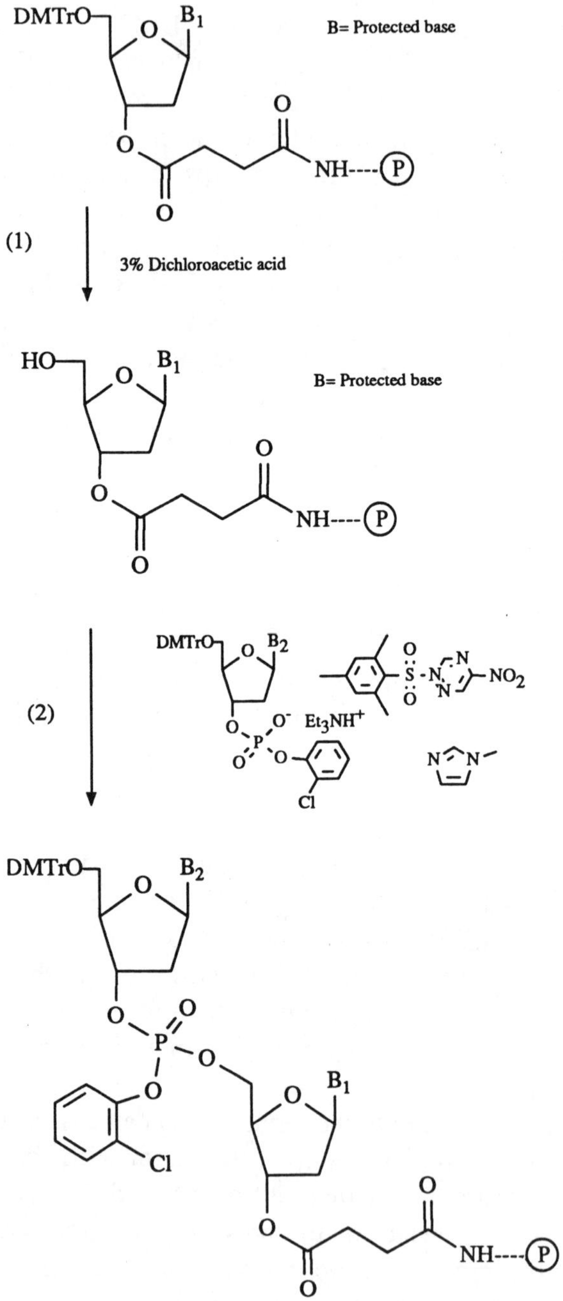

Fig. 2. The cycle of (1) deprotection and (2) condensation to form a dimer.

Table 1
The Steps for One Cycle
for a Small-Scale Synthesis on Controlled Pore Glass

Step	Reagents and solvents	Function	Time, min
1	Pyridine	Wash	2.0
2	1,2-Dichloroethane (DCE)	Wash	1.5
3	3% Dichloroacetic acid in DCE	Detritylation	0.5–1.5
4	DCE	Wash	1.5
5	Pyridine	Wash	2.5
6	Coupling (stop flow)	Addition of next nucleoside	15.0

In general, the synthesis is best performed in a semimanual apparatus (e.g., as provided by Ommifit Ltd., UK) described by Gaite *(4)*. In such a system, the solid support is contained in a small glass column linked to a manually operated valve block through which solvents are directed by a small pressure of dry argon or nitrogen. Typically, a cycle for a small-scale synthesis on controlled pore glass (ca. 25 mg support, 1.0 µmol scale) is about 25 min. The steps are outlined in Table 1.

The times shown in the cycle are for a flow rate of about 1.0 mL/min in pyridine. The coupling reaction is performed by stopping the flow in pyridine and injecting into the column a mixture of the protected nucleoside phosphodiester monomer, the coupling agent (MSNT), and the catalyst 1-methylimidazole.

The larger scale synthesis on the polydimethylacrylamide-Kieselguhr support requires correspondingly longer wash times. The rather basic support also requires a much stronger acid (10% tri-chloroacetic acid [TCA] in 1,2-dichloroethane) for the detritylation step *(3,4)*. In particular, there is an additional wash in DMF after the acid deprotection step to ensure that all traces of acid and the products of the 5'-deprotection reaction are removed. The presence of traces of acid will act as a capping agent during the coupling step and lower the yield substantially. The flow rate is again adjusted to about 1.0 mL/min in pyridine for the synthesis on the polyamide support. The steps of the cycle are outlined in Table 2.

Table 2
The Steps for One Synthesis Cycle for Large-Scale Synthesis
on the Polydimethylacrylamide-Kieselguhr Support

Step	Reagents and solvents	Function	Time, min
1	Pyridine	Wash	5.0
2	DCE	Wash	4.0
3	10% TCA in DCE	Detritylation	3.0
4	DMF	Wash	2.0
5	Pyridine	Wash	6.0
6	Coupling (stop flow)	Addition of next nucleoside	15.0

At the end of the synthesis, cleavage from the support and complete deprotection of the oligonucleotide are effected by a three-step procedure, and are the same for a synthesis on controlled pore glass or the polyamide support.

3.4. The Deprotection Procedure

1. Treat with syn-2-nitrobenzaldoxime and 1,1,3,3-tetramethylguanidine in aqueous dioxan to remove the 2-chlorophenyl-protecting groups, and free the partially protected oligonucleotide from the support.
2. Evaporate the solution from above, and redissolve in concentrated ammonia (d0.88) in a sealed flask at 55°C for 6 h to remove the N-acyl protecting groups from the bases.
3. Evaporate the ammonia solution from above, and redissolve the products in acetic acid/water (8:2 v/v) for 30 min to remove the terminal 5'-hydroxyl-protecting group.

The crude oligodeoxyribonucleotide is then purified by ion-exchange HPLC or by polyacrylamide gel electrophoresis (PAGE) followed by desalting on a column of Bio-Gel P2 or Sephadex G10 or G25. Further purification may then involve reverse phase HPLC *(4)*. Capillary gel electrophoresis *(20,21)* is becoming an increasingly powerful analytical method for oligodeoxyribonucleotides (for details, *see Protocols for Oligonucleotide Conjugates*).

4. Discussion

The phosphotriester method of synthesis may be regarded as the method of choice for the large-scale preparation of oligodeoxyribonucleotides. Although this method is perhaps not so readily adaptable to fully automated machine-aided synthesis, it is superbly amenable to scale-up processes. A major advantage of the phosphotriester method is that excesses of the phosphodiester monomer building blocks used in chain assembly can be recovered. If the consideration is to make kilogram amounts of short oligonucleotides, then the losses and cost of monomer building blocks and other reagents may be a primary factor in developing large-scale synthesis methods. Although the phosphoramidite method (and H-phosphonate method) has been adapted for use in automatic synthesis machines, and is the method of choice for general and routine synthesis of oligodeoxyribo-nucleotides, the large excesses of monomer building blocks cannot be recovered, since they undergo modifications during the synthesis.

4.1. The Coupling Reaction

Most recent advances in the phosphotriester methodology have concentrated on increasing the rate of the coupling reaction and, thus, decrease overall cycle times so that a complete synthesis is on a more convenient time scale. This has been achieved by the use of nucleophilic catalysts to enhance the rate of the MSNT-promoted coupling of the phosphodiester component with the 5'-hydroxyl group of the polymer-supported oligodeoxyribonucleotide. Whereas the condensation reaction time using MSNT in pyridine solution is around 40 min *(3)*, the addition of 1-methylimidazole more than halves the coupling time to around 14 min *(22)*. However, the most effective nucleophilic catalysts are the pyridine-*N*-oxides; 4-morpholino-pyridine-1-oxide reduces the coupling time to <4 min *(23)*.

4.2. Catalytic Phosphate-Protecting Groups

The use of nucleophilic catalysis to accelerate condensation reactions has been extended further by the introduction of catalytic phosphate-protecting groups. Here further rate enhancements are achieved as a result of nucleophilic catalysis because of the neighboring group effect.

The first such phosphate-protecting group *(24)* is the 2-(1-methylimidazol-2-yl)phenyl group (as in Fig. 3). This is an aryl phosphate-protecting group bearing an imidazole moiety in the *ortho* position confirming significant rate enhancement to the condensation reaction. Using monomers with this catalytic phosphate group, coupling times are reduced to 5–7 min *(25)*. A second series of catalytic phosphate-

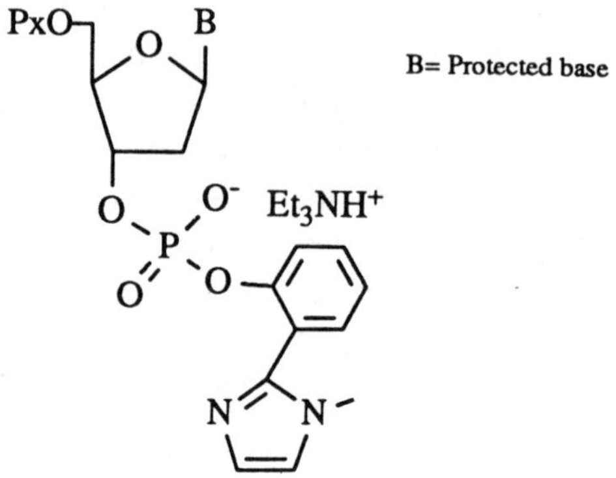

Fig. 3. Monomer building block bearing the catalytic phosphate-protecting group, 2-(1-methylimidazol-2-yl)phenyl.

protecting groups making use of the neighboring group effect has been proposed. The application of 1-oxido-4-alkoxy-2-picolyl derivatives of nucleotides *(26)* (as in Fig. 4) has led to dramatic acceleration of the rate of phosphotriester bond formation. The result is that coupling reactions of phosphodiesters bearing the 1-oxido-4-alkoxy-2-picolyl group in solid-phase synthesis proceed with 97–98% yields within 1–1.5 mins. With the reduced coupling times, overall cycle times become as rapid as cycle times in the phosphoramidite approach. The significant developments in the phosphotriester approach to oligo-deoxyribunucleotide synthesis are likely to lead to further investigation of this methodology for the large-scale synthesis of oligodeoxyribonucleotides.

5. Notes and Precautions

The most significant criteria for the successful synthesis of oligodeoxyribunucleotides by the solid-phase phosphotriester approach is the purity of solvents and reagents, and it is essential to ensure the highest purity for all reagents. The advantage of the phosphotriester method is that medium length oligodeoxyribonucleotides can be prepared reasonably quickly (about 25 min/residue) using simple and inexpensive equipment with a minimum number of steps per cycle. The technique can also be easily adapted for the preparation of larger amounts of oligodeoxyribonucleotides (e.g., >20 mg product) by the use of the high-loading polydimethylacrylamide-Kieselguhr support.

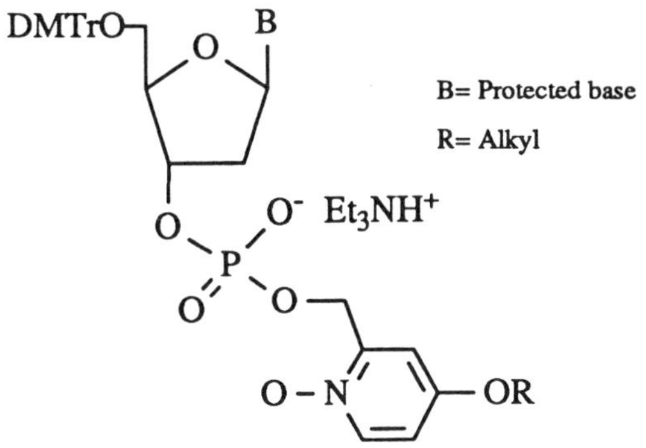

Fig. 4. Monomer building block bearing the catalytic phosphate-protecting group, 1-oxido-4-alkoxy-2-picolyl.

5.1. The Additional Protection of the Guanine and Thymine Residues

Improvements in yield can be achieved by the use of 4-O-protected thymidine and additional 6-O-protected 2-N-acyl-guanosine. The characterization of side reactions associated with the 4-O-position of the thymine *(27)* residue and the 6-O-position of the guanine *(27)* residue, particularly in synthesis of longer oligodeoxyribonucleotides with prolonged exposure to coupling agents, has prompted the development of additional base protecting groups. The protecting groups for the 6-O-position of deoxyguanosine include the diphenylcarbamoyl group *(28)*, the *p*-nitrophenylethyl group *(29)*, and aryl groups *(30,31)* (such as 2-nitrophenyl and 3-chlorophenyl). The 4-O-position of thymidine is best protected by the phenyl-protecting group. Aryl groups are favored for the additional protection of deoxyguanosine. These protecting groups are readily removed by treatment with oximate ions, and deprotection occurs in the first step of the deprotection procedure during treatment with syn-2-nitrobenzaldoxime and 1,1,3,3-tetramethylguanidine (TMG) for the removal of the 2-chlorophenyl internucleotide phosphate-protecting groups.

5.2. The Deprotection Procedure

It should be borne in mind that the three-step deprotection procedure outlined should be carried out in the order given. It is important that the terminal 5'-hydroxyl-protecting group (Px or DMTr) is not removed before treatment with syn-2-nitrobenzaldoxime and TMG for the removal of the aryl-protecting groups. The presence of a free terminal 5'-hydroxyl group under the basic conditions during this step was found to cause significant migration through a cyclic phosphate intermediate to give a product with a 5' → 5' linkage *(32)*. This is easily avoided by removing the DMTr or Px group by acid treatment as the last step of the deprotection procedure.

References

1a. Reese, C. B. (1978) *Tetrahedron* **34**, 3143.
1b. For other general reviews on the chemical synthesis of DNA, *see* refs. *4* and *5;* Sonveaux, E. (1986) *Bioorg. Chem.* **14**, 274–325; English, U. and Gauss, D. H. (1991) *Augew. Chem. Int. Ed. Engl.* **30**, 613–722; Uhlmann, E. and Peyman, A. (1990) *Chem. Rev.* **90**, 544–584; Davies, J. E. and Gassen, H. G. (1983) *Angew. Chem. Int. Ed. Engl.* **22**, 13–31.
2. Gait, M. J., Singh, M., Sheppard, R. C., Edge, M. D., Greene, A. R., Heathcliffe, G. R., Atkinson, T. C., Newton, C. R., and Markham, A. F. (1980) *Nucl. Acids Res.* **8**, 1081–1096.
3. Gait, M. J., Matthes, H. W. D., Singh, M., Sproat, B. S., and Titmas, R. C. (1982) *Nucleic Acids Res.* **10**, 6243–6254.
4. Gait, M. J. (ed.) (1984) *Oligonucleotide Synthesis. A Practical Approach.* IRL Press, Oxford Chapter 4.
5. For a review, *see* Caruthers, M. H. (1991) *Acc. Chem. Res.* **24**, 278–284.
6. Gough, G. R., Brunden, M. J., and Gilham, P. T. (1981) *Tetrahedron Lett.* **22**, 4177–4180.
7. Koster, H., Stumpe, A., and Wolter, A. (1983) *Tetrahedron Lett.* **24**, 747–750.
8. Sproat, B. S. and Bannwarth, W. (1983) *Tetrahedron Lett.* **24**, 5771–5774.
9. Atherton, E., Brown, E., Sheppard, R. C., and Rosevear, A. (1981) *J. Chem. Soc. Commun.*, 1151–1152.
10. Chattopadhyaya, J. B. and Reese, C. B. (1978) *J. Chem. Soc. Commun.* 639–640.
11. Schaller, H., Weimann, B., Lerch, B., and Khorana, H. G. (1963) *J. Am. Chem. Soc.* **85**, 3821–3827.
12. Reese, C. B. (1970) *Colloques Internationaux du C.N.R.S.* **182**, 319–328.
13. Reese, C. B., Titmas, R. C., and Yau, L. (1978) *Tetrahedron Lett.* **30**, 2727–2730.
14. Reese, C. B. and Zard, L. (1981) *Nucl. Acids Res.* **9**, 4611–4626.
15. Khorana, H. G. (1968) *Pure Appl. Chem.* **17**, 349–381.
16. Agarwal, K. L., Yamazaki, A., Cashion, P. J., and Khorana, H. G. (1972) *Angew. Chem. Int. Ed. Engl.* **11**, 451–459.
17. Kuwe, A., Sekine, M., and Hata, T. (1982) *Tetrahedron Lett.* **23**, 4365–4368.
18. Kuwe, A., Sekine, M., and Hata, T. (1983) *Chem. Lett.* 1597–1600.
19. Froehler, B. C. and Matteucci, M. D. (1983) *Nucl. Acids Res.* **11**, 8031–8036.
20. Cohen, A. S., Najaman, D. R., Paulus, A., Guttman, A., Smith, J. A., and Karger, B. L. (1988) *Proc. Natl. Acad. Sci. USA* **85**, 9660–9663.

21. Guttman, A., Cohen, A. S., Heiger, D. N., and Karger, B. L. (1990) *Anal. Chem.* **62,** 137–141.
22. Efimov, U. A., Reverdatto, S. V., and Chakhmakhcheva, O. G. (1982) *Nucl. Acids Res.* **10,** 6675–6694.
23. Ejimov, V. A., Chakhmakhcheva, O. G., and Ovchinnikov, Y. A. (1985) *Nucl. Acids Res.* **13,** 3651–3666.
24. Froehler, B. C. and Matteucci, M. D. (1985) *J. Am. Chem. Soc.* **107,** 278–279.
25. Sproat, B. S., Rider, P., and Beijer, B. (1986) *Nucl. Acids Res.* **14,** 1811–1824.
26. Efimov, V. A., Burgakova, A. A., Dubey, I. Y., Polushin, N. N., Chakhmakhcheva, O. G., and Orchinnikov, Y. A. (1986) *Nucl. Acids Res.* **14,** 6525–6540.
27. Reese, C. B. and Ubasawa, A. (1980) *Tetrahedron Lett.* **21,** 2265–2268.
28. Kamimura, T., Tsuchiya, M., Koura, K., Sekine, M., and Hata, T. (1983) *Tetrahedron Lett.* **24,** 2775–2778.
29. Trichtinger, T., Charubala, R., and Pfleiderer, W. (1983) *Tetrahedron Lett.* **24,** 711–714.
30. Jones, S. S., Reese, C. B., Sibanda, S., and Ubasawa, A. (1981) *Tetrahedron Lett.* **22,** 4755–4758.
31. Reese, C. B. and Skone, P. A. (1984) *J. Chem. Soc. Perkin Trans.* **I,** 1263–1271.
32. Boom, J. H. van, Burgers, P. M. J., Deursen, P. H. van, Rooy, J. F. M., and Reese, C. B. (1976) *J. Chem. Soc. Chem. Commun.* 167,168.

CHAPTER 3

Oligodeoxyribonucleotides Synthesis

Phosphoramidite Approach

Serge L. Beaucage

1. Introduction

The development of deoxyribonucleoside phosphoramidite derivatives for the synthesis of oligodeoxyribonucleotides was first described by Beaucage and Caruthers (1) in 1981. The conceptual basis of this methodology emerged as a modification of the "phosphorodichloridite" coupling procedure reported earlier by Letsinger and coworkers (2). Essentially, the approach involved the reaction of the protected deoxyribonucleosides **1a–d** (Fig. 1) with chloro-(N,N-dimethylamino) methoxyphosphine (**2**) in the presence of N,N-diisopropylethylamine. The rapid reaction yielded the deoxyribonucleoside phosphoramidites **3a–d** (Fig. 1), which were isolated by precipitation and stored as dry powders.

The landmark of the phosphoramidite approach entailed the conversion of the relatively stable deoxyribonucleoside phosphoramidite derivatives to reactive intermediates suitable for oligonucleotide synthesis. Specifically, the interaction of N,N-dimethylaniline hydrochloride with **3a–d** (Fig. 2) generated the corresponding deoxyribonucleoside chlorophosphites **4a–d** (Fig. 2), which upon reaction with 3'-O-levulinylthymidine afforded the (3' → 5')-dinucleoside phosphite triesters **5a–d** (Fig. 2) in near quantitative yields (1). The hygroscopic nature of tertiary amine hydrochlorides became an impediment to reliable oligonucleotide synthesis, since anhydrous conditions

From: *Methods in Molecular Biology, Vol. 20: Protocols for Oligonucleotides and Analogs*
Edited by: S. Agrawal Copyright ©1993 Humana Press Inc., Totowa, NJ

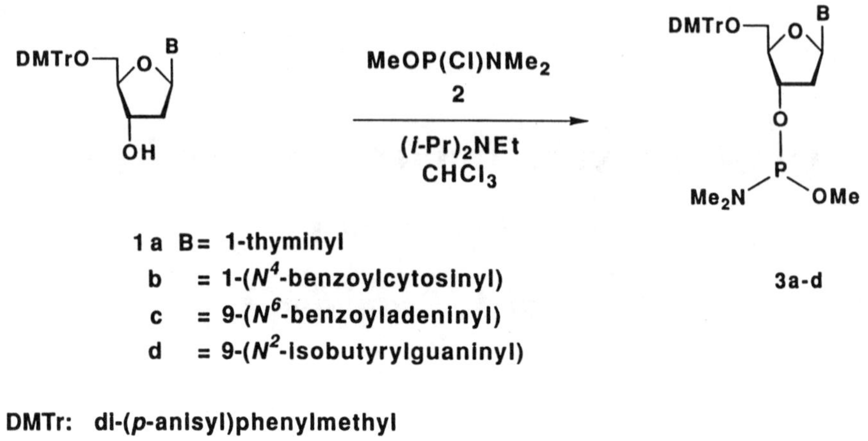
DMTrO
B
OH
MeOP(Cl)NMe2
2
(i-Pr)2NEt
CHCl3
DMTrO
B
Me2N
OMe
1 a B= 1-thyminyl
b = 1-(N4-benzoylcytosinyl)
c = 9-(N6-benzoyladeninyl)
d = 9-(N2-isobutyrylguaninyl)
3a-d
DMTr: di-(p-anisyl)phenylmethyl

Fig. 1.

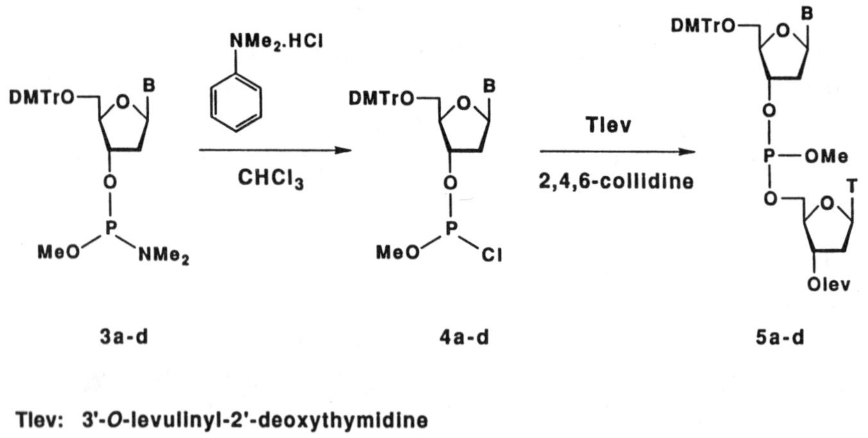
NMe2.HCl
DMTrO
B
CHCl3
MeO
NMe2
Cl
Tlev
2,4,6-collidine
P—OMe
T
Olev
3a-d
4a-d
5a-d
Tlev: 3'-O-levulinyl-2'-deoxythymidine

Fig. 2.

ment to reliable oligonucleotide synthesis, since anhydrous conditions were required for optimum coupling reactions. The search for nonhygroscopic weak acids capable of activating deoxyribonucleoside phosphoramidites led to the use of the commercially available 1*H*-tetrazole, which could be purified and dried by sublimation. The addition of 1*H*-tetrazole to **3a–d** (Fig. 2) and 3'-*O*-levulinylthymidine in dry acetonitrile generated **5a–d** (Fig. 2) in quantitative yields within a few minutes as judged by ^{31}P-NMR spectroscopy *(1)*. This strategy

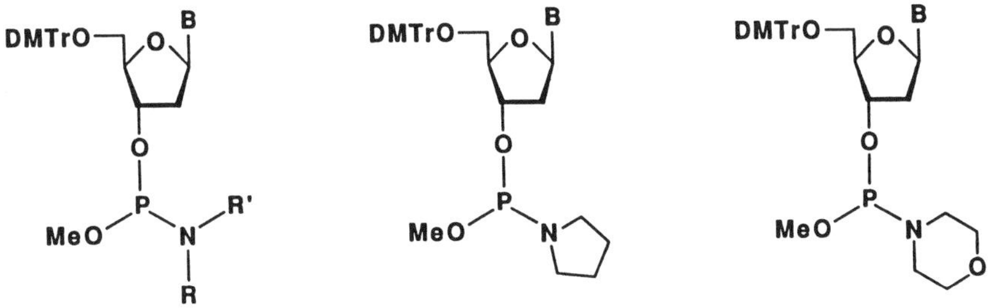

6a–d R= R'= Et	9a–d	10a–d
7a–d R= Me; R'= *i*-Pr		
8a–d R= R'= *i*-Pr		

a, B= T; b, B= C^{Bz}; c, B= A^{Bz}; d, B= G^{Ib}

Fig. 3.

has been successfully applied to the solid-phase synthesis of oligodeoxyribonucleotides of varying chain lengths *(3,4)*. However, the utilization of the deoxyribonucleoside phosphoramidites **3a–d** (Fig. 2) in automated systems has been unreliable, because the stability of these monomers in acetonitrile varied from hours to weeks depending on their purity. Deoxyribonucleoside phosphoramidites having different *N,N*-dialkylamino substituents were then investigated as potential derivatives for the automated solid-phase synthesis of oligonucleotides *(5,6)*. It was shown that **10a** (Fig. 3) was stable in acetonitrile solution for at least 42 d without significant decomposition *(5)*. Moreover, **10a–d** (Fig. 3) could be purified by silica gel chromatography, and were applied to the synthesis of deoxyribonucleotides on silica *(7)*, controlled-pore glass (CPG) *(8)*, or cellulose filter disks *(9)*. Similarly, **8a–d** (Fig. 3) did not show significant decomposition in acetonitrile after 4 d *(6)*. Under these conditions, **8b** (Fig. 3) was more stable than **6b** (Fig. 3), which in turn was more stable than **3b** *(6)*. The efficacy of the deoxyribonucleoside phosphoramidites **8a–d** (Fig. 3) was demonstrated by the synthesis of 51 mers on CPG, which, at the time, were the largest DNA segments ever chemically synthesized *(6)*.

The popularity of the deoxyribonucleoside phosphoramidites **8a–d** (Fig. 3) in the automated synthesis of oligonucleotides has nonetheless been hampered by the thiolate treatment required for the removal

```
            B
DMTrO   O
                                    OCH₂CH₂CN
       O                      Cl—P
       |                            NR₂
       P
    R₂N   OCH₂CH₂CN

    11a-d  R= I-Pr             13  R= I-Pr
    12a-d  NR₂= morpholino     14  NR₂= morpholino
```

a, B= T; b, B= N^4-(2-methylbenzoyl)cytosinyl; c, B= A^{Bz}; d, B= G^{Ib}

Fig. 4.

of the methyl phosphate-protecting groups *(10,11)*. To simplify this postsynthesis deprotection protocol, Sinha et al. *(12)* described the preparation of the phosphoramidites **11a–d** (Fig. 4) and **12a–d** (Fig. 4) from the reaction of suitably protected nucleosides with the monofunctional phosphitylating reagents **13** and **14** (Fig. 4), respectively, under conditions similar to those reported by Beaucage and Caruthers *(1)*. The phosphoramidites **11a–d** (Fig. 4) were more stable than the phosphoramidites **8a–d** (Fig. 3) in wet acetonitrile according to ^{31}P-NMR spectroscopy *(13)*, and the removal of the β-cyanoethyl phosphate-protecting groups from oligonucleotides was effected under the basic conditions required for the cleavage of the nucleobase-protecting groups. These attributes contributed to the widespread utilization of deoxyribonucleoside phosphoramidites structurally related to **11a–d** (Fig. 4) in the automated solid-phase synthesis of oligonucleotides.

An alternate strategy to the synthesis of deoxyribonucleoside phosphoramidites was reported by Beaucage *(14)* and coworkers *(15)*. The approach involved the reaction of properly protected deoxyribonucleosides (**1a–d**; Fig. 5) with *bis*-(pyrrolidino)methoxyphosphine chemoselectively activated by 4,5-dichloroimidazole. ^{31}P-NMR spectroscopy indicated that the corresponding deoxyribonucleoside phosphoramidites (**9a–d**; Fig. 3) were generated in yields exceeding 86% within 10 min. Without further purification, these monomers were immediately activated with 1*H*-tetrazole and applied to the synthesis of a deoxyribonucleotide (22 mer) *(14,15)*. This approach eliminated the problems

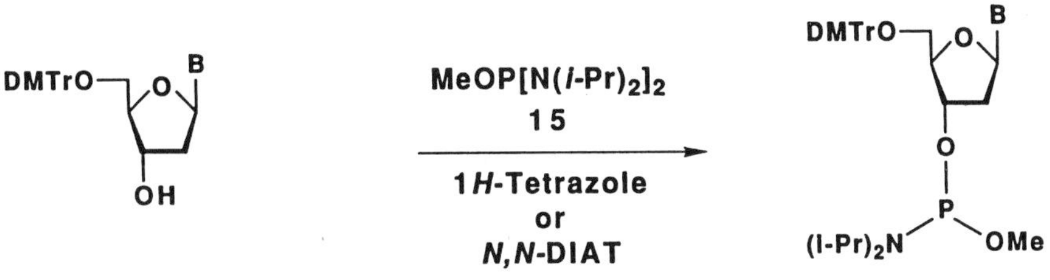

Phosphoramidite Approach

N,N-DIAT= N,N-Diisopropylammonium tetrazolide

Fig. 5.

associated with the isolation and purification of deoxyribonucleoside phosphoramidites along with those pertaining to the stability of the phosphoramidites in acetonitrile. The methodology did not, however, generate deoxyribonucleoside phosphoramidites *in situ* with complete chemoselectivity; (3' → 3')-dinucleoside methyl phosphite triester contaminants were also detected (<10%) by ^{31}P-NMR spectroscopy. The selective activation of *bis*-(N,N-dialkylamino) alkoxyphosphines was subsequently investigated by Barone et al. *(16)* and by Lee and Moon *(17)*. It was shown that the reaction of the deoxyribonucleosides **1a–d** (Fig. 5) with the phosphorodiamidite **15** (Fig. 5) and limited amounts of 1H-tetrazole or its N,N-diisopropylammonium salt afforded, within ca. 1 h, the corresponding deoxyribonucleoside phosphoramidites **8a–d** (Fig. 5) in isolated yields varying between 82–92%. Less than 1% of (3'→3')-dimers were observed. This approach became the method of choice for the preparation of deoxyribonucleoside phosphoramidite monomers given the relative stability of phosphorodiamidites in addition to the highly selective activation of these derivatives and the mildness of the phosphitylation conditions.

In this chapter, the preparation of monochlorophosphoramidite and phosphorodiamidite derivatives and their application toward a general synthesis of deoxyribonucleoside phosphoramidites will be delineated from representative literature procedures. The synthetic protocols provided herein should allow the preparation of a myriad of deoxyribonucleoside phosphoramidite monomers having any modification of the following entities: nucleobases, nucleobase-protecting groups, 5'-O-protecting groups, carbohydrate moieties, P-N and/or P-O substituents. Consequently, the incorporation of such monomers into oligonucleotides will not be discussed in detail, since specific conditions may be required for a large number of phosphoramidite analogs. Should the manual synthesis of specific oligonucleotides be undertaken, the synthetic protocols utilized by commercial DNA synthesizers and those that have been already published *(18,19)* should provide the basic guidelines necessary for the successful solid-phase synthesis of such oligonucleotides.

To facilitate the design of deoxyribonucleoside phosphoramidite analogs, information regarding the activation of deoxyribonucleoside phosphoramidites will be provided to emphasize the importance imparted by the steric bulk of the substituents attached to the phosphorus center and to specific nucleobases on the coupling rates. The capping and oxidation steps of the synthetic cycle will also be discussed in relation to the potential formation of side products during oligonucleotide synthesis.

2. Materials

1. Acetonitrile, dichloromethane, toluene, *N,N*-diisopropylamine and *N,N*-diisopropylethyl amine are dried by refluxing over calcium hydride (–40 mesh), distilled, and stored over 4-Å molecular sieves in dry amber glass containers.
2. Dry dichloromethane and deuterochloroform are passed through a column of aluminum oxide (activated, basic, Brockmann I) just prior to use.
3. Pyridine is distilled from *p*-toluenesulfonyl chloride and then calcium hydride. Dry pyridine is stored over 4-Å molecular sieves in dry amber glass bottles.
4. 3-Hydroxypropionitrile is distilled at a pressure of <1 mm Hg and is kept over 4-Å molecular sieves in dry amber glass containers.
5. Tetrahydrofuran is dried by refluxing over sodium-benzophenone under an inert atmosphere and distilled just prior to use.
6. Phosphorus trichloride is purified by distillation under the exclusion of moisture.
7. The commercially available chloro-(2-cyanoethoxy)-*N,N*-diisopropylaminophosphine (**13**; Fig. 4) and *bis*-(*N,N*-diisopropylamino)-2-cyanoethoxyphosphine are kept at 4°C over Drierite®.
8. 1*H*-Tetrazole is purified and dried by sublimation (110°C @ 0.05 mm Hg).
9. The following reagents can be purchased from reputable manufacturers and used as received:

Phosphoramidite Approach

fate, sodium carbonate, sodium chloride, Drierite® (8 mesh);
B. Protected nucleosides, such as 5'-O-di-p-methoxytrityl-2'-deoxythymidine (Fig. 1a), 5'-O-di-p-methoxytrityl-N^4-benzoyl-2'-deoxycytidine (Fig. 1b), 5'-O-di-p-methoxytrityl-N^6-benzoyl-2'-deoxyadenosine (Fig. 1c), and 5'-O-di-p-methoxytrityl-N^2-isobutyryl-2'-deoxyguanosine (Fig. 1d);
C. Reagents for gel electrophoresis (Acrylamide/BIS [19:1], urea, ammonium persulfate, N,N,N',N'-tetramethylethylene diamine [TEMED], 10X Tris-Borate-EDTA [TBE] buffer [pH 8.3], bromphenol blue) and the cation-exchange resin AG®50W-X12 hydrogen form (100–200 mesh).
10. Disposable PD-10 Sephadex® G-25M columns can be purchased from Pharmacia, Piscataway, NJ.
11. Cameo® 3N, nylon-membrane filters (3 mm, 0.22 µm) are obtained from Micron Separations Inc., Westboro, MA, and are autoclaved prior to use.

3. Methods

3.1. Preparation of Phosphorodichloridite Derivatives

3.1.1. Synthesis of 2-Cyanoethoxydichlorophosphine

Freshly distilled phosphorus trichloride (87 mL, 1 mol), anhydrous ethyl ether *(200 mL)*, and dry pyridine (81 mL, 1 mol) are mechanically stirred in a three-necked flask (1 L) under a dry argon atmosphere at –78°C. Distilled 3-hydroxypropionitrile (68 mL, 1 mol) in anhydrous ethyl ether is added dropwise over ca. 2 h to the cold mixture maintained at –78°C. The suspension is then allowed to warm to ambient temperature and stirred overnight under argon. The precipitate is filtered and washed with ethyl ether (100 mL). The filtrate is concentrated under reduced pressure, and the crude reaction product is fractionally distilled under vacuum affording 2-cyano-ethoxy-dichlorophosphine as a colorless liquid (bp 78°C at 0.6 mm Hg) in 55% yield (95 g, 0.55 mol) *(12,18)*. ^{31}P-NMR (CDCl$_3$): 178.8 ppm downfield from the external 80% H$_3$PO$_4$ standard *(20)*. ^{1}H NMR (CDCl$_3$): δ in ppm; 2.7 (*t*, 2H); 4.4 (2*t*, 2H) relative to an internal tetramethylsilane standard *(20)*.

A variety of phosphorodichloridites, including 2-cyanoethoxy-dichlorophosphine, can be prepared in the absence of pyridine *(20,21)*. Typically, a solution of the alcohol (50 mmol) in dry acetonitrile (20 mL) is added dropwise over ca. 5 min, under an inert atmosphere, to a

is added dropwise over ca. 5 min, under an inert atmosphere, to a solution of phosphorus trichloride (31 mL, 350 mmol) in dry acetonitrile (16 mL). The solution is stirred at ambient temperature for 30 min and then concentrated *in vacuo*. A large number of phosphorodichloridites obtained by this procedure could be used without further purification (*see* Note 1) in the synthesis of monochlorophosphoramidite or phosphorodiamidite derivatives.

3.2. Preparation of Monochlorophosphoramidite Derivatives

3.2.1. Synthesis of Chloro-(2-Cyanoethoxy)-N,N-Diisopropylaminophosphine (13; Fig. 4)

N,N-Diisopropylamine (28 mL, 200 mmol) or its corresponding *N*-trimethylsilyl derivative (100 mmol) in anhydrous ethyl ether (30 mL) is added over ca. 1.5 h, under an inert atmosphere, to a stirring and cool (–20°C) solution of 2-cyanoethoxydichlorophosphine (17.2 g, 100 mmol) in anhydrous ethyl ether (60 mL). The reaction mixture is then stirred at ambient temperature overnight under an inert gas. The insoluble amine hydrochloride is removed by filtration, washed with ethyl ether, and the filtrate is concentrated under low pressure. The residue is fractionally distilled under vacuum yielding 16.5 g (70 mmol) of chloro-(2-cyanoethoxy)-*N,N*-diisopropylaminophosphine (bp 103–104°C @ 0.08 mm Hg; $d = 1.061$ g/mL). ^{31}P-NMR (CH$_3$CN): 179.8 ppm downfield from the external H$_3$PO$_4$ standard. ^{1}H NMR (CDCl$_3$): δ in ppm; 4.02, 4.20 (2*t*, 2H); 3.80 (*m*, 2H); 2.77 (*t*, 2H); 1.29 (*d*, 12H) *(12)*.

This methodology has been applied to the synthesis of various monofunctional phosphitylating reagents useful in the synthesis of deoxyribonucleoside phosphoramidites. A selected number of them are listed in Table 1.

3.3. Preparation of Phosphorodiamidite Derivatives

3.3.1. Synthesis of Bis-(N,N-*Diisopropylamino)-2-Cyanoethoxyphosphine*

To a cool (–10°C) solution of 2-cyanoethoxydichlorophosphine (15 g, 87 mmol) in anhydrous ethyl ether *(200 mL)* is added *N,N*-diisopropylamine (86 g, 0.85 mol), dropwise, over 1 h under an inert

Table 1
Chlorophosphoramidite Derivatives Useful in the Synthesis of Deoxyribonucleoside Phosphoramidites

$$R'O-P(Cl)(R'')$$

R'	R''	References
Me—	—NMe$_2$	1
Me—	—N(*i*-Pr)$_2$	5,6
Me—	—N(morpholino)	5
NCCH$_2$CH$_2$—	—N(morpholino)	12
NCCH$_2$C(Me)$_2$—	—N(morpholino)	22
Cl$_3$CC(Me)$_2$—	—NMe$_2$	23
Cl$_3$CC(Me)$_2$—	—N(*i*-Pr)$_2$	23
Cl$_3$CC(Me)$_2$—	—N(morpholino)	23
Me—S(O)$_2$—CH$_2$CH$_2$—	—N(morpholino)	24
O$_2$N—C$_6$H$_4$—CH$_2$CH$_2$—	—N(morpholino)	25
O$_2$N—C$_6$H$_4$—CH$_2$CH$_2$—	—N(heptamethyleneimino)	26

atmosphere. The reaction mixture is then warmed to room temperature and stirred overnight. The amine salt is filtered, washed with ethyl ether (100 mL), and the filtrate is concentrated under reduced pressure. The crude product is purified by vacuum distillation affording 21 g (70 mmol) of *bis*-(*N,N*-diisopropylamino)-2-cyanoethoxyphosphine as a colorless liquid (bp 105–107°C @ 0.4 mm Hg *[27]*; d = 1.04 g/mL *[18]*). ^{31}P-NMR (CDCl$_3$): 123.3 ppm *(27)*.

3.3.2. Preparation of N,N-*Diisopropylammonium Tetrazolide*

N,N-Diisopropylamine (11 mL, 80 mmol) is added to a stirring solution of sublimed 1*H*-tetrazole (2.8 g, 40 mmol) in dry acetonitrile (100 mL). The precipitated product is collected by filtration, thoroughly washed with acetonitrile, and dried under vacuum. *N,N*-diisopropyl ammonium tetrazolide is obtained as a white crystalline material in quantitative yields.

The facile preparation of deoxyribonucleoside phosphoramidites from the selective activation of phosphorodiamidites by *N,N*-diisopropylammonium tetrazolide *(16)* or 1*H*-tetrazole *(17)* provides an excellent tool for the incorporation of new phosphate-protecting groups during oligonucleotide synthesis. Several phosphorodiamidites have recently been prepared, and a few of them are shown in Table 2.

3.4. General Procedure for the Preparation of Deoxyribonucleoside Phosphoramidites from Chloro-(N,N-Dialkylamino)alkoxyphosphines

5'-*O*-DMTr-*N*-protected deoxyribonucleosides are consecutively dried by coevaporation with dry pyridine, toluene, and tetrahydrofuran. To a stirring solution of a dry nucleoside (3 mmol) and anhydrous *N,N*-diisopropylethylamine (12 mmol) in dry tetrahydrofuran (15 mL) is added chloro-(2-cyanoethoxy)-*N,N*-diisopropylaminophosphine (6 mmol) by syringe over 2 min under an inert atmosphere. The reaction mixture is allowed to stir for 35 min and is then filtered to remove the precipitated amine hydrochloride. The filtrate is concentrated to a volume of ca. 5 mL under reduced pressure. Ethyl acetate (150 mL) is added to the residue, and the solution extracted twice with ice-cold aqueous 10% sodium carbonate (50 mL). The organic phase is dried over anhydrous sodium sulfate, filtered, and evaporated to dryness under low pressure. The foamy material is dissolved in 20 mL of toluene (pyrimidines) or ethyl acetate (purines), and added to stirring hexane (250 mL) cooled

Table 2
Phosphorodiamidite Derivatives
Useful in the Synthesis of Deoxyribonucleoside Phosphoramidites

$$RO-P\begin{subarray}{c}R'\\R'\end{subarray}$$

R	R'	References
NCCH$_2$CH(Me)—	—NEt$_2$	27
NCCH$_2$C(Me)$_2$—	—N(i-Pr)$_2$	27
Cl$_3$CCH$_2$—	—N(i-Pr)$_2$	28
Cl$_3$CC(Me)$_2$—	—N(i-Pr)$_2$	28
CH$_2$=CH-CH$_2$—	—N(i-Pr)$_2$	29
(CF$_3$)$_2$CH—	—N(i-Pr)$_2$	30
Me—S(O)$_2$—CH$_2$CH$_2$—	—N(i-Pr)$_2$	27
O$_2$N-C$_6$H$_4$-CH$_2$CH$_2$—	—N(i-Pr)$_2$	28
(4-pyridyl)-CH$_2$CH$_2$—	—N(i-Pr)$_2$	31
(2-Me-C$_6$H$_4$)-CH$_2$—	—N(i-Pr)$_2$	32
(2-Cl-C$_6$H$_4$)-CH$_2$—	—N(i-Pr)$_2$	32
Ph—	—NEt$_2$	33
O$_2$N-C$_6$H$_4$—	—NEt$_2$	33
C$_6$Cl$_5$—	—NEt$_2$	33

to –78°C. The precipitate is isolated by filtration and dried under vacuum. The purity of these deoxyribonucleoside phosphoramidites is ca. 95% according to ^{31}P-NMR spectroscopy.

3.5. General Procedure for the Preparation of Deoxyribonucleoside Phosphoramidites from Bis-(N,N-Dialkylamino)alkoxyphosphines

N,N-Diisopropylammonium tetrazolide (0.5 mmol) and a 5'-*O*-DMTr-*N*-protected deoxyribonucleoside (**1a–d** Fig. 1, mmol) are dissolved in dry dichloromethane (5 mL). *Bis*-(*N,N*-diisopropylamino)-2-cyanoethoxyphosphine (1.1 mmol) is then added, and the reaction mixture is stirred under an inert atmosphere for ca. 5 h. Dichloromethane (20 mL) is added to the mixture and the solution is extracted twice with aqueous 2% sodium carbonate (25 mL) and once with brine (25 mL). The organic phase is dried over anhydrous sodium sulfate and filtered. The solvent is removed under reduced pressure affording a white foamy product that is then dissolved in anhydrous dichloromethane (5 mL) and precipitated in 300 mL of cold (–78°C) hexane under vigorous stirring. The deoxyribonucleoside phosphoramidites **11a–d** (Fig. 4) are isolated by filtration and dried under vacuum. Yield: 80–90%. These phosphoramidites can be stored for extended periods of time at –20°C under an inert atmosphere in a dessicator containing Drierite®. ^{31}P-NMR (CH$_2$Cl$_2$): δ in ppm; 146.0 (**11a**; Fig. 4); 146.3, 146.1 (**11b**; Fig. 4); 145.8 (**11c**; Fig. 4); 145.8, 145.4 (**11d**; Fig. 4) *(18)*.

The activation of deoxyribonucleoside phosphoramidites by 1*H*-tetrazole and their application to the solid-phase synthesis of oligodeoxyribonucleotides will not be described in detail. This standard technology is outlined in Note 2.

It should be emphasized that synthetic DNA segments free of contaminating shorter sequences are required for the ligation of multiple DNA duplexes dedicated to cloning experiments. It is known that synthetic DNA duplexes having sequences shorter than the expected length at the ligation junctions would, on insertion into a plasmid vector and bacterial transformation, lead to the production of bacterial colonies harboring deletion-mutant plasmids. The search for transformants carrying plasmids with the desired size insert could become difficult. Consequently, the meticulous purification of synthetic oligonucleotides is critical for these specific applications.

3.6. General Procedures for the Purification and Characterization of Oligonucleotides

3.6.1. Polyacrylamide Gel Electrophoresis (PAGE)

The application of this purification technique is described in the following protocol: The ammoniacal solution of a completely deprotected oligonucleotide obtained from a 0.2-μmol-scale synthesis is evaporated to dryness *in vacuo*. The oligomer is dissolved in water (1 mL) and extracted three times with ethyl ether saturated with water (1 mL) to remove deprotection side products, such as benzamide, isobutyramide, and 4,4'-dimethoxytritanol. Half the volume of the aqueous phase (500 μL) is evaporated to dryness under reduced pressure. A loading buffer (0.2% bromphenol blue:10× TBE buffer:water [8:1:1]) (55 μL) is then added to the residue, and after heating in boiling water for 3–5 min, the solution is quickly chilled on ice. The oligomer (up to 40 bases long) is loaded on a 20% polyacrylamide–$7M$ urea gel (1.5 mm × 20 cm [width] × 40 cm [length]) equilibrated with 1× TBE buffer (pH 8.3) and electrophoresed at ca. 600 V until the bromphenol blue dye reached the bottom of the gel. The purification of larger oligomers (40–60 mers) requires a 17% polyacrylamide–$7M$ urea gel for optimum results.

On completion of the electrophoresis, the apparatus is disassembled, and one side of the gel is covered with a thin transparent plastic sheet (Saran Wrap™) and laid over a silica gel TLC plate (20 × 20 cm) containing a fluorescent indicator. The oligonucleotide appears as a dark blue band on irradiation with UV light (254 nm). The desired band (usually the major slowest migrating band) is excised with a razor blade and transferred into a plastic tube in which the gel is crushed to a paste with a glass rod. Doubly distilled (dd) water (3 mL) is added, and the suspension is briefly vortexed. After standing overnight at ambient temperature, the suspension is centrifuged at low speed (3000 rpm) for 5 min, and the supernatant is collected. The remaining paste is vortexed with dd-water (1 mL) and then centrifuged at low speed. The supernatant is collected, combined with the previous one, and evaporated to dryness under reduced pressure. The residue is dissolved in a minimum volume of dd-water (ca. 250 μL) and loaded on the top of a 10-mL PD-10 Sephadex® G-25M column equilibrated with dd-water. The oligonucleotide is eluted by serial addition of dd-water (1 mL).

Eight fractions (1 mL each) are collected and analyzed by UV spectroscopy at 260 nm. The fractions containing the oligomer (usually fraction 3–6) are evaporated to dryness *in vacuo*. Oligodeoxyribonucleotides are best kept as dry pellets at –20°C. Polydeoxyribonucleotides purified according to this procedure are suitable for in vitro biochemical experiments.

3.6.2. Reverse-Phase High-Performance Liquid Chromatography (HPLC)

Relative to PAGE, reverse-phase HPLC allows the purification of larger amounts of synthetic oligonucleotides. The following chromatographic conditions have been useful in the purification and characterization of natural oligodeoxyribonucleotides and oligodeoxyribonucleoside phosphorothioates: The ammoniacal solution of a deprotected oligodeoxyribonucleotide (1 µmol scale synthesis) carrying a 5'-O-di-p-dimethoxytrityl (DMTr) group is evaporated to dryness under reduced pressure. HPLC-grade water (200 µL) is added to the crude oligomer, and 100 µL of the solution are injected on a reverse-phase PRP-1 column (10 mm OD × 270 mm). A linear gradient of 20% acetonitrile (MeCN)/0.1M triethylammonium acetate (TEAA) pH 7.0 to 30% MeCN/TEAA is pumped through the column at a flow rate of 3 mL/min for 10 min. The gradient is then isocratically held for 10 min at the same flow rate. The fractions containing the purified oligomer are pooled together and evaporated to dryness under *vacuo*. Aqueous 80% acetic acid is added to the oligomer, and, after 30 min at ambient temperature, the acid is removed under reduced pressure. Water (1 mL) is added to the residue, and the solution is extracted three times with ethyl ether saturated with water (1 mL). The aqueous phase is concentrated to ca. 250 µL and desalted on a PD-10 Sephadex® G-25M column (*vide supra*). The purity of the oligonucleotide can be checked by PAGE or reverse-phase HPLC (PRP-1 column; 30 min linear gradient of 1% MeCN/TEAA to 30% MeCN/TEAA at a flow rate of 2 mL/min).

3.6.3. Characterization of Oligonucleotides by Ezymatic Digestion

A purified oligodeoxyribonucleotide (0.4 A_{260} U) is incubated with snake venom phosphodiesterase *(Crotallus durissus,* 9×10^{-3} U) in 100 mM Tris-HCl, pH 9.0 (500 µL), for 3 h at 37°C. Bacterial alkaline phosphatase (*E. coli,* 0.4 U) is then added to the digest, and the mixture is incubated overnight at 37°C.

An aliquot (200 µL) of the hydrolysate is analyzed by reverse-phase HPLC using a PRP-1 column and a 36-min linear gradient of 1% MeCN/TEAA to 35% MeCN/TEAA at a flow rate of 3 mL/min. Typically, the peaks obtained from the enzymatic hydrolysis of an oligodeoxyribonucleotide have retention times of 5.4 min (dC), 8.5 min (dG), 8.6 min (dT), and 11.1 min (dA). The integrated peak areas provide the relative content of each deoxyribonucleoside. The extinction coefficients of 6100 (dC), 13,700 (dG), 6600 (dT), and 11,900 (dA) measured at 254 nm in 7.5% MeCN/TEAA are relevant to the analysis conditions.

3.6.4. Ion-Exchange Protocol

In order to be suitable for biochemical applications in cell cultures, purified oligonucleotides are preferably converted into their sodium salt form. Thus, an oligomer (up to 250 A_{260} U) dissolved in dd-water (500 mL) is loaded on the top of an AG® 50-X12 (sodium form) cation-exchange column (1 × 8 cm) equilibrated with dd-water (*see* Note 3). The oligonucleotide is eluted from the column with dd-water, and fractions (1 mL) are collected in microfuge tubes. The fractions containing the oligonucleotide are identified by UV spectroscopy, pooled together, and evaporated to dryness under vacuum. The oligomer is then desalted on a 10 mL PD-10 Sephadex®G-25M column as described in Section 3.6.1. The salt-free solution of the oligomer in dd-water is syringed through a sterile 3-mm Cameo® nylon-membrane (0.22 µ*M*) filter to a specific concentration in a sterile container. The solution is kept frozen at –20°C.

4. Discussion

The preparation of deoxyribonucleoside phosphoramidites from suitably protected deoxyribonucleosides and monochlorophosphoramidite or phosphorodiamidite derivatives, as reported herein, is facile and occurs in high isolated yields. Moreover, the efficacy of the deoxyribonucleoside phosphoramidites structurally related to **11a–d** (Fig. 4) on activation with 1*H*-tetrazole has been demonstrated by the synthesis of relatively large oligomers (150 mer) on nonporous silica microbeads *(34)* or on rigid nonswelling polystyrene beads *(35)*. Stepwise yields averaging 98–99% were reported. The dinucleotide phosphoramidites **16** and **17** (Fig. 6) have also been efficient in the solid-phase synthesis of polydeoxyribonucleotides *(36,37)*. The acti-

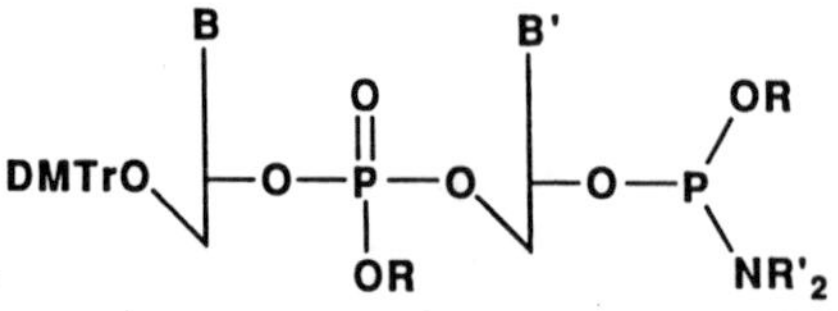

16 R= R'= Me;
B or B'= protected pyrimidine and purine nucleobases.
17 R= CH$_2$CH$_2$CN; R'= *i*-Pr; B= B'= T

Fig. 6.

vation of **16** with 1*H*-tetrazole enabled coupling reactions with a deoxyribonucleoside bound to CPG to occur as efficiently (ca. 99%) as with monomeric phosphoramidites *(36)*. Alternatively, the activation of **17** (Fig. 6) with 5-(*p*-nitrophenyl)-1*H*-tetrazole *(38)* led to the preparation of a large oligomer (101 mer) *(37)*.

The mechanism of activation of deoxyribonucleoside phosphoramidites by 1*H*-tetrazole has recently attracted considerable attention. It has been argued that the protonation of the phosphoramidite function by 1*H*-tetrazole was rapid and followed by the reversible and slower formation of a phosphorotetrazolide intermediate *(39)*. It is to be noted that relative to 5-(*p*-nitrophenyl)-1*H*-tetrazole *(38)*, 1-hydroxybenzotriazole *(24)*, *N*-methylanilinium trifluoroacetate *(40)*, *N*-methylanilinium trichloroacetate *(41)*, 5-trifluoromethyl-1*H*-tetrazole *(23)*, *N*-methylimidazole hydrochloride *(23)*, and *N*-methylimidazoletrifluoromethane sulfonate *(42)*, which have also been tested as activators, 1*H*-tetrazole still remained the most commonly used reagent for the activation of deoxyribonucleoside phosphoramidites.

Dahl and his colleagues *(43)* investigated various factors influencing the coupling rates of activated deoxyribonucleoside phosphoramidites. They reported that the coupling rates varied with the nature of the *O*-alkyl/aryl and *N*,*N*-dialkylamino functions of the phosphoramidites. Decreasing coupling rates were observed in the following order: *O*-methyl > *O*-(2-cyanoethyl) > *O*-(1-methyl-2-cyanoethyl) > *O*-(1,1-dimethyl-2-cyanoethyl) >> *O*-(*o*-chlorophenyl) and *N*,*N*-diethylamino > *N*,*N*-diisopropylamino > *N*-morpholino > *N*-methylanilino.

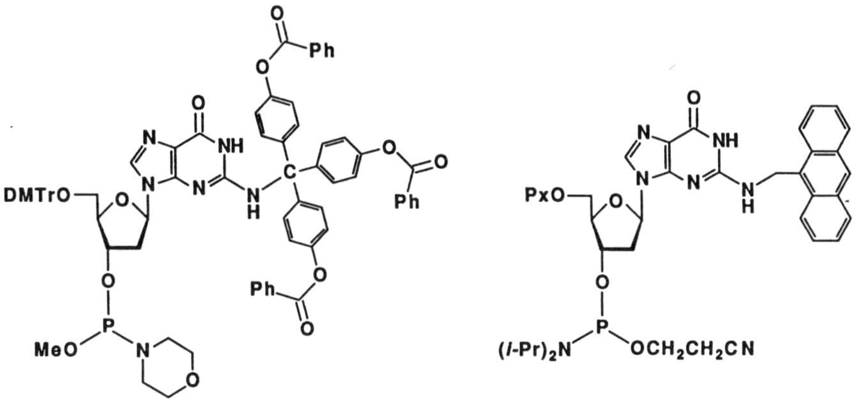

18

19

Px: 9-phenylxanthen-9-yl

Fig. 7.

The steric bulk of substituents at the exocyclic amino function of the guanine nucleobase has been shown to affect significantly the coupling rates and efficiency of the deoxyribonucleoside phosphoramidites **18** and **19** (Fig. 7) *(44,45)*. Specifically, the activation of **18** (Fig. 7) with 1*H*-tetrazole and a condensation time of 10 min generated coupling yields averaging 45% on a CPG support *(44)*. Under similar conditions, **19** (Fig. 7) led to coupling yields of 65–70%, but required a condensation time of at least 60 min *(45)*. These coupling yields were far below those obtained with the common deoxyribonucleoside phosphoramidites (98–99%) and, thus, stressed the importance of steric hindrance when designing nucleobase-protecting groups and/or modified bases toward the synthesis of oligonucleotide analogs.

In spite of the high efficiency of the phosphoramidite approach, the chain extension step did not occur in quantitative yields even under optimum conditions. Consequently, the desired oligomer was mixed with a population of shorter oligomers. With increasing chain lengths, the isolation of the desired n-mer oligonucleotide from the $(n-1)$-mer may be difficult. To alleviate this problem, the acetylation of unphosphitylated chains (capping step) was performed in a stepwise manner to terminate the elongation of these truncated oligomers. A solution of acetic anhydride, 2,6-lutidine, and *N*-methylimidazole in tetrahydrofuran has been effective as a capping formulation *(46)*. In addition to preventing the elongation of unphosphitylated oligomers, the capping reagent has also been effective in reducing the concentration of O^6-phosphitylated guanine residues generated during the coupling step *(47,48)*. The oxidation of such adducts would yield to the considerably more stable O^6-phosphorylated guanine derivatives and promote the subsequent ramification of the oligonucleotidic chain. It is, therefore, important to perform the capping reaction before the oxidation step to minimize this potential problem (*see* Note 4).

The oxidation of the internucleotide phosphite triester to the phosphotriester represents an important step in the automated synthesis of oligonucleotides by the phosphoramidite approach. An aqueous solution of iodine containing 2,6-lutidine *(49)* or pyridine *(50)* has been generally used for this task given its stability, mildness, and rapid reaction kinetics. This oxidation reaction occurred with an overall retention of configuration *(51)*. The substitution of enriched (^{17}O or ^{18}O) water for water in the oxidation formulation allowed the preparation of oligonucleotides having chiral phosphodiester linkages *(52)* as useful probes for studying the stereochemical course of hydrolyses catalyzed by phosphodiesterases *(53)*.

In specific applications, nonaqueous oxidizing reagents may advantageously offer an alternative to aqueous iodine for the oxidation of oligodeoxyribonucleoside phosphite triesters. Particularly, dinitrogen tetroxide *(54)*, *tert*-butyl hydroperoxide *(55)*, di-*tert*-butyl hydroperoxide *(55)*, cumene hydroperoxide *(55)*, hydrogen peroxide *(55)*, *bis*-trimethylsilyl peroxide in the presence of catalytic amounts of trimethylsilyl triflate *(55,56)*, *m*-chloroperbenzoic acid *(57,58)*, iodobenzene diacetate *(59)*, tetra-*n*-butylammonium periodate *(59)*, and molecular oxygen in the presence of 2,2'-azobis(2-methylpropionitrile) (AIBN) under thermal or photochemical conditions *(54)* have been effective.

Interestingly, the aqueous iodine oxidation of 2-cyano-1,1-dimethylethyl deoxyribodinucleoside phosphite or *o*-methylbenzyl-deoxyribodinucleoside phosphite prepared from deoxyribonucleoside phosphoramidite intermediates resulted in the formation of the corresponding phosphate diesters *(60)*. Furthermore, Caruthers et al. *(32)*

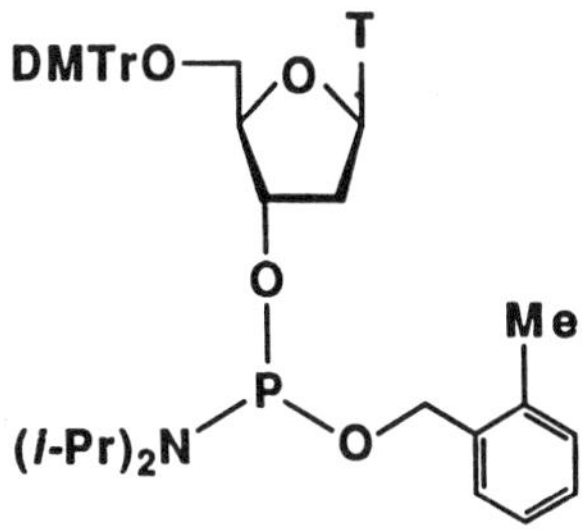

Phosphoramidite Approach

20

Fig. 8.

demonstrated that the absence of phosphate-protecting groups did not inhibit the subsequent elongation of the DNA chain. Indeed, the deoxyribonucleoside phosphoramidite **20** (Fig. 8) enabled the solid-phase synthesis of an oligothymidylic acid (20 mer) with an average stepwise yield of 96%. It was speculated that phosphate-phosphite mixed anhydrides resulting from the interaction of phosphate diesters with activated deoxyribonucleoside phosphoramidites could be cleaved by excess 1H-tetrazole to regenerate deoxyribonucleoside phospho-tetrazolide intermediates.

The versatility of deoxyribonucleoside phosphoramidites has been further illustrated by the synthesis of oligodeoxyribonucleotides in the absence of nucleobase-protecting groups. Recently, Gryaznof and Letsinger *(61)* described the solid-phase synthesis of a 20 mer from 5'-O-DMTr-2'-deoxyribonucleoside 3'-O-(N,N-diisopropylamino)methoxyphosphines having unprotected nucleobases. The success of the approach depended on a modified synthetic protocol involving a treatment with an equimolar solution (0.1M) of pyridine hydrochloride and aniline in acetonitrile after each coupling reaction, prior to the oxidation step, to cleave nucleobase adducts originating from the oligonucleotidic chain. This strategy should also facilitate the synthesis of oligonucleotides bearing base-sensitive functional groups on selected solid supports.

Of interest, the hydrolysis of the deoxyribonucleoside phosphorodiamidites **21** (Fig. 9) with acetic acid followed by pyridine water yielded the corresponding deoxyribonucleoside H-phosphonates in yields

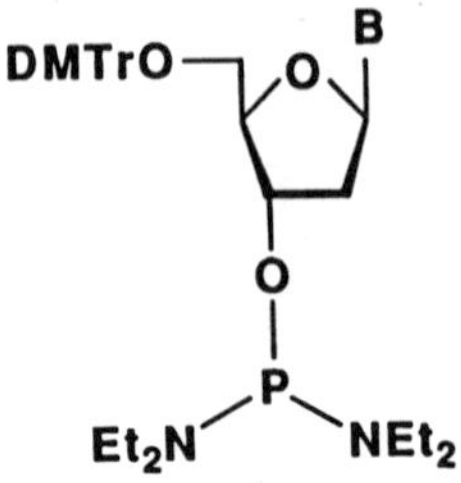

21

Fig. 9.

exceeding 85% *(62)*. The latter monomers have found application in the synthesis of oligonucleotides *(63,64)* and their analogs *(65)*. Alternatively, the phosphorodiamidites **21** (Fig. 9) generated *in situ* from deoxyribonucleosides, such as **1a–d**, *tris*-(diethylamino) phosphine, and *N,N*-diisopropylammonium tetrazolide, were used without purification in the solid-phase synthesis of a decanucleotide *(66)*. A hydrolytic step effected by 1*H*-tetrazole and water after each coupling reaction generated *H*-phosphonate internucleotide links, which could subsequently be converted into natural or modified phosphodiester linkages *(65)*. This procedure led to coupling yields averaging 97% *(66)*.

Monomeric deoxyribonucleoside phosphoramidites have been particularly useful in the incorporation of modified nucleobases into synthetic oligonucleotides to provide a better understanding of the dynamics *(67)* and thermal stability of DNA duplexes in solution *(68)*, and to probe protein–DNA interactions *(69,70)*. Additionally, a plethora of deoxyribonucleoside phosphoramidites having modified bases *(71–76)* or modified 5'-termini *(77,78)* have been introduced into oligonucleotides to facilitate the attachment of either fluorescent markers *(71,72,76)*, crosslinking agents *(73,74)*, or DNA cleaving agents *(75)* to these oligomers. These conjugates have found application as diagnostic probes *(71,76)*, DNA sequencing primers *(72,77)*, affinity ligands *(71,78)*, and as antisense molecules *(73,74)*.

Nucleosidic and nucleotidic phosphoramidite intermediates have also played an important role in biomedical investigations. For example, the mechanisms involved in the repair of abasic sites in DNA have

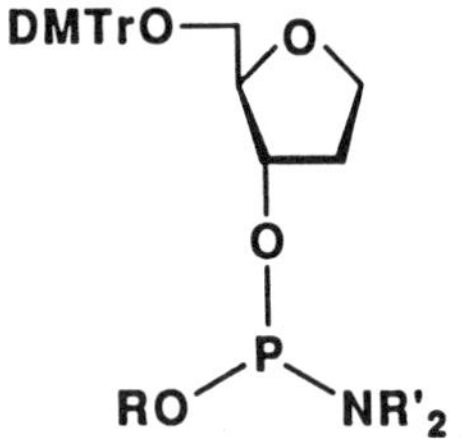

Fig. 10.

22 R= Me; R'= *i*-Pr
23 R= CH$_2$CH$_2$CN; NR'$_2$= morpholino

Fig. 11.

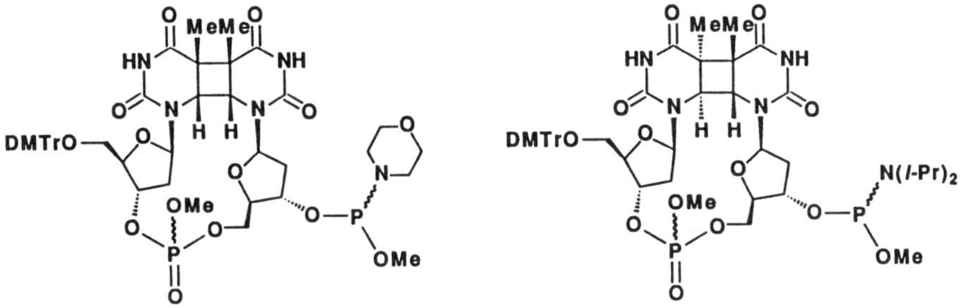

24

25

motivated Takeshita et al. *(79)* and Eritja et al. *(80)* to incorporate substituted tetrahydrofuran moieties into DNA to mimic the 2'-deoxyribose structure in abasic sites. The tetrahydrofuranyl phosphoramidites **22** and **23** (Fig. 10) have been developed for this purpose. Along similar lines, the dimeric phosphoramidites **24** and **25** (Fig. 11) have been applied to the construction of unique photolesion-containing DNA duplexes to instigate physical, enzymological, and mutagenesis experiments aimed at shedding light on DNA repair mechanisms induced on exposure of DNA to UV light *(81,82)*.

The occurrence of carcinogenic or mutagenic effects associated with polycyclic aromatic hydrocarbons and those pertaining to DNA lesions induced by γ-irradiation has been investigated by the synthesis

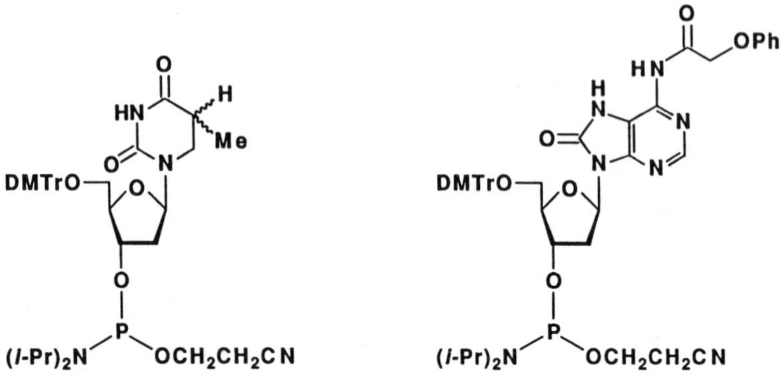

Fig. 12.

of oligonucleotides carrying well-known modifications at predetermined sites. The preparation of the deoxyribonucleoside phosphoramidites **19** (Fig. 7), **26** (Fig. 12), and **27** (Fig. 12) has allowed the incorporation of such defined modifications into oligonucleotides *(45,83,84)*.

Because of the scope of this chapter, much of the synthetic applications of phosphoramidite and nucleoside phosphoramidite derivatives have not been covered. Detailed accounts of such applications along with the recent advances in oligonucleotide synthesis by the phosphoramidite approach have recently been reported *(85,86)*.

5. Notes

1. Only a limited number of phosphorodichlorodites can be safely distilled; others may decompose or explode on heating *(20,21)*. Consequently, caution must be exercised when attempting the distillation of unknown phosphorodichloridites.
2. General methodology for the solid-phase synthesis of oligonucleotides.
   A. Acetonitrile wash.
   B. Cleavage of the 5'-*O*-di-*p*-methoxytrityl group: 3% (v/v) trichloro acetic acid in dichloromethane.
   C. Dry acetonitrile wash.
   D. Coupling step: $0.1M$ nucleosidic phosphoramidite and $0.4M$ 1*H*-tetrazole in dry acetonitrile. Depending on the nature of the nucleosidic phosphoramidite, a coupling time of 1–15 min may be required.
   E. Dry acetonitrile wash.

F. Capping step: A solution of acetic anhydride:2,6-lutidine:tetrahydrofuran (1:1:8) (v/v/v) and a $2M$ solution of 1-methylimidazole in tetrahydrofuran are equally mixed and reacted with the support.
G. Acetonitrile wash.
H. Oxidation step: $0.1M$ iodine in tetrahydrofuran:pyridine:water (90:5:5) (v/v/v).
I. Repeat the steps above until completion of the chain assembly.
J. Deprotection step: Specific reagents for the deprotection of certain phosphate-protecting groups may be necessary (see Tables 1 and 2 and references therein). Otherwise, concentrated ammonium hydroxide may be used at ambient temperature or at 55°C for 1–16 h, depending on the sensitivity of the nucleobases and nucleobase-protecting groups to alkaline conditions.

The quantity of each reagent and the time required for each step are adjusted to the scale at which the synthesis is carried out.
3. The cation-exchange resin AG® 50-X12 sodium form is prepared by sequentially washing a column (1 × 20 cm) filled with the resin (hydrogen form) with $1M$ hydrochloric acid (50 mL), water (50 mL), $0.1M$ sodium hydroxide until basic to pH paper and, finally, with dd-water to neutral pH.
4. It must be noted that the sulfurization reaction effected by $3H$-1,2-benzodithiol-3-one 1,1-dioxide during the solid-phase synthesis of oligodeoxyribonucleoside phosphorothioates has been performed before the capping step to minimize the unwanted oxidation of phosphite triester functions caused by the capping reagent (87). Occasionally, larger than expected oligodeoxyribonucleotide analogs (<5%) have been observed as a consequence of this change. These were easily separated from the desired oligomer by reverse-phase HPLC or PAGE.

References

1. Beaucage, S. L. and Caruthers, M. H. (1981) Deoxynucleoside phosphoramidites—A new class of key intermediates for deoxypolynucleotide synthesis. *Tetrahedron Lett.* **22**, 1859–1862.
2. Letsinger, R. L. and Lunsford, W. B. (1976) Synthesis of thymidine oligonucleotides by phosphite triester intermediates. *J. Am. Chem. Soc.* **98**, 3655–3661.
3. Caruthers, M. H., Beaucage, S. L., Becker, C., Efcavitch, W., Fisher, E. F., Galluppi, G., Goldman, R., deHaseth, P., Martin, F., Matteucci, M., and Stabinsky, Y. (1982) New methods for synthesizing deoxyoligonucleotides, in *Genetic Engineering: Principles and Methods*, vol. 4 (Setlow, J. K. and Hollaender, eds.), Plenum, New York, pp. 1–17.
4. Josephson, S., Lagerholm, E., and Palm, G. (1984) Automatic synthesis of oligodeoxynucleotides and mixed oligodeoxynucleotides using the phosphoamidite method. *Acta Chem. Scand.* **B38**, 539–545.

5. McBride, L. J. and Caruthers, M. H. (1983) An investigation of several deoxyribonucleoside phosphoramidites useful for synthesizing deoxyoligonucleotides. *Tetrahedron Lett.* **24**, 245–248.
6. Adams, S. P., Kavka, K. S., Wykes, E. J., Holder, S. B., and Galluppi, G. R. (1983) Hindered dialkylamino nucleoside phosphite reagents in the synthesis of two DNA 51 mers. *J. Am. Chem. Soc.* **105**, 661–663.
7. Dörper, T. and Winnacker, E.-L. (1983) Improvements in the phosphoramidite procedure for the synthesis of oligodeoxyribonucleotides. *Nucl. Acids Res.* **11**, 2575–2584.
8. Sinha, N. D., Biernat, J., and Köster, H. (1984) Polymer support oligonucleotide synthesis XVI: Synthesis of oligonucleotides using suitably protected deoxynucleoside-*N*-morpholinophosphoramidites on porous glass beads. *Nucleosides Nucleotides* **3**, 157–171.
9. Ott, J. and Eckstein, F. (1984) Filter disc supported oligonucleotide synthesis by the phosphite method. *Nucl. Acids Res.* **91**, 9137–9142.
10. Daub, G. W. and van Tamelen, E. E. (1977) Synthesis of oligoribonucleotides based on the facile cleavage of methyl phosphotriester intermediates. *J. Am. Chem. Soc.* **99**, 3526–3528.
11. Andrus, A. and Beaucage, S. L. (1988) 2-mercaptobenzothiazole—An improved reagent for the removal of methyl phosphate-protecting groups from oligodeoxynucleotide phosphotriesters. *Tetrahedron Lett.* **29**, 5479–5482.
12. Sinha, N. D., Biernat, J., and Köster, H. (1984) Polymer support oligonucleotide synthesis XVIII: use of β-cyanoethyl-*N,N*-dialkylamino-/*N*-morpholino phosphoramidite of deoxynucleosides for the synthesis of DNA fragments simplifying deprotection and isolation of the final product. *Nucl. Acids Res.* **12**, 4539–4557.
13. Zon, G., Gallo, K. A., Samson, C. J., Shao, K., Summers, M. F., and Byrd, R. A. (1985) Analytical studies of "mixed sequence" oligodeoxyribonucleotides synthesized by competitive coupling of either methyl- or β-cyanoethyl-*N,N*-diisopropylamino phosphoramidite reagents, including 2'-deoxyinosine. *Nucl. Acids Res.* **13**, 8181–8196.
14. Beaucage, S. L. (1984) A simple and efficient preparation of deoxynucleoside phosphoramidites *in situ*. *Tetrahedron Lett.* **25**, 375–378.
15. Moore, M. F. and Beaucage, S. L. (1985) Conceptual basis of the selective activation of *bis*-(dialkylamino)methoxyphosphines by weak acids and its application toward the preparation of deoxynucleoside phosphoramidites in situ. *J. Org. Chem.* **50**, 2019–2025.
16. Barone, A. D., Tang, J.-Y., and Caruthers, M. H. (1984) *In situ* activation of *bis*-dialkylaminophosphines—a new method for synthesizing deoxyoligonucleotides on polymer supports. *Nucl. Acids Res.* **12**, 4051–4061.
17. Lee, H.-J. and Moon, S.-H. (1984) *Bis*-(*N,N*-dialkylamino)-alkoxyphosphines as a new class of phosphite coupling agent for the synthesis of oligonucleotides. *Chem. Lett.* 1229–1232.
18. Caruthers, M. H., Barone, A. D., Beaucage, S. L., Dodds, D. R., Fisher, E. F., McBride, L. J., Matteucci, M., Stabinsky, Z., and Tang, J.-Y. (1987) Chemical

synthesis of deoxyoligonucleotides by the phosphoramidite method, in *Methods in Enzymology*, vol. 154 (Wu, R. and Grossman, L., eds.), Academic Press, San Diego, pp. 287–313.
19. Atkinson, T. and Smith, M. (1984) Solid-phase synthesis of oligodeoxyribonucleotides by the phosphite-triester method, in *Oligonucleotide Synthesis: A Practical Approach* (Gait M., ed.), IRL, Oxford, pp. 35–81.
20. Claesen, C. A. A., Segers, R. P. A. M., and Tesser, G. I. (1985) Ar(alk)ylsulfonyl ethyl groups as phosphorus-protecting functions. *Rec. Trav. Chim. Pays-Bas* **104,** 119–122.
21. Ogilvie, K. K., Theriault, N. Y., Seifert, J.-M., Pon, R. T., and Nemer, M. J. (1980) The chemical synthesis of oligoribonucleotides. IX. A comparison of protecting groups in the dichloridite procedure. *Can. J. Chem.* **58,** 2686–2693.
22. Marugg, J. E., Dreef, C. E., van der Marel, G. A., and van Boom, J. H. (1984) Use of 2-cyano-1,1-dimethylethyl as a protecting group in the synthesis of DNA via phosphite intermediates. *Rec. Trav. Chim. Pays-Bas* **103,** 97,98.
23. Hering, G., Stöcklein-Schneiderwind, R., Ugi, I., Pathak, T., Balgobin, N., and Chattopadhyaya, J. (1985) Preparation and properties of chloro-*N,N*-dialkylamino-2,2,2-trichloroethoxy-and chloro-*N,N*-dialkylamino-2,2,2- trichloro-1,1-dimethylethoxyphosphines and their deoxynucleoside phosphiteamidates. *Nucleosides Nucleotides* **4,** 169–171.
24. Claesen, C., Tesser, G. I., Dreef, C. E., Marugg, J. E., van der Marel, G. A., and van Boom, J. H. (1984) Use of 2-methylsulfonylethyl as a phosphorus protecting group in oligonucleotide synthesis via a phosphite triester approach. *Tetrahedron Lett.* **25,** 1307–1310.
25. Beiter, A. H. and Pfleiderer, W. (1984) Solution synthesis of protected di–2'-deoxynucleoside phosphotriesters via the phosphoramidite approach. *Tetrahedron Lett.* **25,** 1975–1978.
26. Schwarz, M. W. and Pfleiderer, W. (1984) Solution synthesis of fully protected thymidine dimer using various phosphoramidites. *Tetrahedron Lett.* **25,** 5513–5516.
27. Nielsen, J., Marugg, J. E., van Boom, J. H., Honnens, J., Taagaard, M., and Dahl, O. (1986) Thermal instability of some alkyl phosphorodiamidites. *J. Chem. Res.* (S), 26,27.
28. Hamamoto, S. and Takaku, H. (1986) New approach to the synthesis of deoxyribonucleoside phosphoramidite derivatives. *Chem. Lett.* 1401–1404.
29. Hayakawa, Y., Wakabayashi, S., Kato, H., and Noyori, R. (1990) The allylic protection method in solid-phase oligonucleotide synthesis. An efficient preparation of solid-anchored DNA oligomers. *J. Am. Chem. Soc.* **112,** 1691–1696.
30. Takaku, H., Watanabe, T., and Hamamoto, S. (1988) Use of 1,1,1,3,3,3-hexafluoro-2-propyl protecting group in the synthesis of DNA fragments via phosphoramidite intermediates. *Tetrahedron Lett.* **29,** 81–84.
31. Hamamoto, S., Shishido, Y., Furuta, M., Takaku, H., Kawashima, M., and Takaki, M. (1989) Use of the 2–(4–pyridyl)ethyl protecting group in the synthesis of DNA fragments via phosphoramidite intermediates. *Nucleosides Nucleotides* **8,** 317–326.

32. Caruthers, M. H., Kierzek, R., and Tang, J. Y. (1987) Synthesis of oligonucleotides using the phosphoramidite method, in *Biophosphates and Their Analog—Synthesis, Structure Metabolism and Activity* (Bruzik, K. S. and Stec, W. J., eds.), Elsevier, Amsterdam, pp. 3–21.
33. Eritja, R., Smirnov, V., and Caruthers, M. H. (1990) O-aryl phosphoramidites: Synthesis, reactivity and evaluation of their use for solid-phase synthesis of oligonucleotides. *Tetrahedron* **46**, 721–730.
34. Seliger, H., Kotschi, U., Scharpf, C., Martin, R., Eisenbeiss, F., Kinkel, J. N., and Unger, K. K. (1989) Polymer support synthesis XV. Behaviour of nonporous surface-coated silica gel microbeads in oligonucleotide synthesis. *J. Chrom.* **476**, 49–57.
35. Vu, H., McCollum, C., Lotys, C., and Andrus, A. (1990) New reagents and solid support for automated oligonucleotide synthesis. *Nucl. Acids Res. Symp. Ser.* **#22**, 63,64.
36. Kumar, G. and Poonian, M. S. (1984) Improvements in oligodeoxyribonucleotide synthesis: Methyl, N,N-dialkyl phosphoramidite dimer units for solid support phosphite methodology. *J. Org. Chem.* **49**, 4905–4912.
37. Wolter, A., Biernat, J., and Köster, H. (1986) Polymer support oligonucleotide synthesis XX: Synthesis of a henhectacosa deoxynucleotide by use of a dimeric phosphoramidite synthon. *Nucleosides Nucleotides* **5**, 65–77.
38. Froehler, B. and Matteucci, M. D. (1983) Substituted 5-phenyltetrazoles: Improved activators of deoxynucleoside phosphoramidites in deoxyoligonucleotide synthesis. *Tetrahedron Lett.* **24**, 3171–3174.
39. Berner, S., Mühlegger, K., and Seliger, H. (1989) Studies on the role of tetrazole in the activation of phosphoramidites. *Nucl. Acids Res.* **17**, 853–864.
40. Fourrey, J. L. and Varenne, J. (1984) Improved procedure for the preparation of deoxynucleoside phosphoramidites: Arylphosphoramidites as a new convenient intermediates for oligodeoxynucleotide synthesis. *Tetrahedron Lett.* **25**, 4511–4514.
41. Fourrey, J. L., Varenne, J., Fontaine, C., Guittet, E., and Yang, Z. W. (1987) A new method for the synthesis of branched ribonucleotides. *Tetrahedron Lett.* **28**, 1769–1772.
42. Hostomsky, Z., Smrt, J., Arnold, L., Tocik, Z., and Paces, V. (1987) Solid-phase assembly of cow colostrum trypsin inhibitor gene. *Nucl. Acids Res.* **15**, 4849–4856.
43. Dahl, B. H., Nielsen, J., and Dahl, O. (1987) Mechanistic studies on the phosphoramidite coupling reaction in oligonucleotide synthesis. I. Evidence for nucleophilic catalysis by tetrazole and rate variations with the phosphorus substituents. *Nucl. Acids Res.* **15**, 1729–1743.
44. Sekine, M., Masuda, N., and Hata, T. (1986) Synthesis of oligodeoxyribonucleotides involving a rapid procedure for removal of base-protecting groups by use of the 4,4',4'-tris(benzoyloxy)trityl (TBTr) group. *Bull. Chem. Soc. Jpn.* **59**, 1781–1789.
45. Casale, R. and McLaughlin, L. W. (1990) Synthesis and properties of an oligodeoxynucleotide containing a polycyclic aromatic hydrocarbon site specifically bound to the N^2 amino group of a 2'-deoxyguanosine residue. *J. Am. Chem. Soc.* **112**, 5264–5271.

46. Farrance, I. K., Eadie, J. S., and Ivarie, R. (1989) Improved chemistry for oligodeoxyribonucleotide synthesis substantially improves restriction enzyme cleavage of a synthetic 35mer. *Nucl. Acids Res.* **17,** 1231–1245.
47. Pon, R. T., Usman, N., Damha, M. J., and Ogilvie, K. K. (1986) Prevention of guanine modification and chain cleavage during the solid phase synthesis of oligonucleotides using phosphoramidite derivatives. *Nucl. Acids Res.* **14,** 6453–6470.
48. Eadie, J. S. and Davidson, D. S. (1987) Guanine modification during chemical DNA synthesis. *Nucl. Acids Res.* **15,** 8333–8349.
49. Letsinger, R. L., Finnan, J. L., Heavner, G. A., and Lunsford, W. B. (1975) Phosphite coupling procedure for generating internucleotide links. *J. Am. Chem. Soc.* **97,** 3278–79.
50. Usman, N., Pon, R. T., and Ogilvie, K. K. (1985) Preparation of ribonucleoside 3'-O-phosphoramidites and their application to the automated solid phase synthesis of oligonucleotides. *Tetrahedron Lett.* **26,** 4567–4570.
51. Cullis, P. M. (1984) The stereochemical course of iodine-water oxidation of dinucleoside phosphite triesters. *J. Chem. Soc. Chem. Commun.* 1510–1512.
52. Herdering, W., Kehne, A., and Seela, F. (1985) Phosphoramidites of chiral (R_p)-and (S_p)-configurated d(T[P–^{18}O]-A): Synthesis, configurational assignment, and use as dimer blocks in oligonucleotide synthesis. *Helv. Chim. Acta* **68,** 2119–2127.
53. Seela, F., Kehne, A., Herdering, W., and Kretschmer, U. (1987) Phosphoramidites of [^{18}O] chiral (Rp)-and (Sp)-configurated dimer-blocks and their use in automated oligonucleotide synthesis. *Nucleosides Nucleotides* **6,** 451–456.
54. Bentrude, W. G., Sopchik, A. E., and Gajda, T. (1989) Stereo-and regio-chemistries of the oxidations of 2-methoxy-5-*tert*-butyl-1,3,2-dioxaphosphorinanes and the cyclic methyl 3',5'-phosphite of thymidine by H_2O/I_2 and O_2/AIBN to P-chiral phosphates. ^{17}O NMR assigment of phosphorus configuration to the diastereomeric thymidine cyclic methyl 3',5'-monophosphates. *J. Am. Chem. Soc.* **111,** 3981–3987.
55. Hayakawa, Y., Uchiyama, M., and Noyori, R. (1986) Nonaqueous oxidation of nucleoside phosphites to the phosphates. *Tetrahedron Lett.* **27,** 4191–4194.
56. Hayakawa, Y., Uchiyama, M., and Noyori, R. (1986) Solid-phase synthesis of oligodeoxyribonucleotides using the *bis*(trimethylsilyl)peroxide oxidation of phosphites. *Tetrahedron Lett.* **27,** 4195–4196.
57. Ogilvie, K. K. and Nemer, M. J. (1981) Nonaqueous oxidation of phosphites to phosphates in nucleotide synthesis. *Tetrahedron Lett.* **22,** 2531–2532.
58. Tanaka, T. and Letsinger, R. L. (1982) Syringe method for the stepwise chemical synthesis of oligonucleotides. *Nucl. Acids Res.* **10,** 3249–3260.
59. Fourrey, J.-L. and Varenne, J. (1985) Introduction of a nonaqueous oxidation procedure in the phosphite triester route for oligonucleotide synthesis. *Tetrahedron Lett.* **26,** 1217–1220.
60. Nielsen, J. and Caruthers, M. H. (1988) Directed Arbuzov-type reactions of 2-cyano-1,1-dimethylethyl deoxynucleoside phosphites. *J. Am. Chem. Soc.* **110,** 6275–6276.
61. Gryaznov, S. M. and Letsinger, R. L. (1991) Synthesis of oligonucleotides via monomers with unprotected bases. *J. Am. Chem. Soc.* **113,** 5876–5877.

62. Marugg, J. E., Tromp, M., Kuyl-Yeheskiely, E., van der Marel, G. A., and van Boom, J. H. (1986) A convenient and general approach to the synthesis of properly protected d-nucleoside3'-hydrogenphosphonates via phosphite intermediates. *Tetrahedron Lett.* **27,** 2661–2664.
63. Garegg, P. J., Regberg, T., Stawinski, J., and Strömberg, R. (1985) Formation of internucleotidic bonds via phosphonate intermediates. *Chem. Scr.* **25,** 280–282.
64. Froehler, B. and Matteucci, M. D. (1986) Nucleoside H-phosphonates: Valuable intermediates in the synthesis of deoxyoligonucleotides. *Tetrahedron Lett.* **27,** 469–472.
65. Froehler, B. (1986) Deoxynucleoside H-phosphonate diester intermediates in the synthesis of internucleotide phosphate analogs. *Tetrahedron Lett.* **27,** 5575–5578.
66. Yamana, K., Nishijima, Y., Oka, A., Nakano, H., Sangen, O., Ozaki, H., and Shimidzu, T. (1989) A simple preparation of 5'-*O*-dimethoxytrityl deoxyribonucleoside 3'-*O*-phosphorbisdiethylamidites as useful intermediates in the synthesis of oligodeoxyribonucleotides and their phosphorodiethylamidate analogs on a solid support. *Tetrahedron* **45,** 4135–4140.
67. Spaltenstein, A., Robinson, B. H., and Hopkins, P. B. (1988) A rigid and nonperturbing probe for duplex DNA motion. *J. Am. Chem. Soc.* **110,** 1299–1301.
68. Eritja, R., Horowitz, D. M., Walker, P. A., Ziehler-Martin, J. P., Boosalis, M. S., Goodman, M. F., Itakura, K., and Kaplan, B. E. (1986) Synthesis and properties of oligonucleotides containing 2'-deoxynebularine and 2'-deoxyxanthosine. *Nucl. Acids Res.* **14,** 8135–8153.
69. Dubendorff, J. W., deHaseth, P. L., Rosendahl, M. S., and Caruthers, M. H. (1987) DNA functional groups required for the formation of open complexes between Escherichia coli RNA polymerase and the λ P_r promoter. *J. Biol. Chem.* **262,** 892–898.
70. Kupferschmitt, G., Schmidt, J., Schmidt, Th., Fera, B., Buck, F., and Rüterjans, H. (1987) ^{15}N labeling of oligodeoxynucleotides for NMR studies of DNA-ligand interactions. *Nucl. Acids Res.* **15,** 6225–6241.
71. Roget, A., Bazin, H., and Teoule, R. (1989) Synthesis and use of labelled nucleoside phosphoramidite building blocks bearing a reporter group: biotinyl, dinitrophenyl, pyrenyl and dansyl. *Nucl. Acids Res.* **17,** 7643–7651.
72. Brumbaugh, J. A., Middendorf, L. R., Grone, D. L., and Ruth, J. L. (1988) Continuous, on-line DNA sequencing using oligodeoxynucleotide primers with multiple fluorophores. *Proc. Natl. Acad. Sci. USA* **85,** 5610–5614.
73. Meyer, R. B., Tabone, J. C., Hurst, G. D., Smith, T. M., and Gamper, H. (1989) Efficient, specific, crosslinking and cleavage of DNA by stable, synthetic complementary oligodeoxynucleotides. *J. Am. Chem. Soc.* **111,** 8517–8519.
74. Pieles, U., Sproat, B. S., Neuner, P., and Cramer, F. (1989) Preparation of a novel psoralen containing deoxyadenosine building block for the facile solid phase synthesis of psoralen-modified oligonucleotides for a sequence specific crosslink to a given target sequence. *Nucl. Acids Res.* **17,** 8967–8978.
75. Dreyer, G. B. and Dervan, P. B. (1985) Sequence-specific cleavage of single-stranded DNA: Oligodeoxynucleotide-EDTA. Fe(II) *Proc. Natl. Acad. Sci. USA* **82,** 968–972.

76. Haralambidis, J., Chai, M., and Tregear, G. W. (1987) Preparation of base-modified nucleosides suitable for nonradioactive label attachment and their incorporation into synthetic oligodeoxyribonucleotides. *Nucl. Acids Res.* **15,** 4857–4876.
77. Smith, L. M., Kaiser, R. J., Sanders, J. Z., and Hood, L. E. (1987) The synthesis and use of fluorescent oligonucleotides in DNA sequence analysis, in *Methods in Enzymology,* vol. 155 (Wu, R., ed.), Academic Press, San Diego, pp. 260–301.
78. De Vos, M.-J., Cravador, A., Lenders, J.-P., Houard, S., and Bollen, A. (1990) Solid phase non isotopic labelling of oligodeoxynucleotides using 5'-protected aminoalkyl phosphoramidites: Application to the specific detection of human papilloma virus DNA. *Nucleosides Nucleotides* **9,** 259–273.
79. Takeshita, M., Chang, C.-N., Johnson, F., Will, S., and Grollman, A. P. (1987) Oligodeoxynucleotides containing synthetic abasic sites. *J. Biol. Chem.* **262,** 10,171–10,179.
80. Eritja, R., Walker, P. A., Randall, S. K., Goodman, M. F., and Kaplan, B. E. (1987) Synthesis of oligonucleotides containing the abasic site model compound 1,4-anhydro-2-deoxy-D-ribitol. *Nucleosides Nucleotides* **6,** 803–814.
81. Taylor, J.-S. and Brockie, I. R. (1988) Synthesis of a *trans-syn* thymine dimer buiding block. Solid phase synthesis of CGTAT[t,s]TATGC. *Nucl. Acids Res.* **16,** 5123–5136.
82. Taylor, J.-S., Brockie, I. R., and O'Day, C. L. (1987) A building block for the sequence-specific introduction of *cis-syn* thymine dimers into oligonucleotides. Solid-phase synthesis of TpT[c,s]pTpT. *J. Am. Chem. Soc.* **109,** 6735–6742.
83. Schulhof, J. C., Molko, D., and Teoule, R. (1988) Synthesis of DNA fragments containing 5,6-dihydrothymine, a major product of thymine gamma radiolysis. *Nucl. Acids Res.* **16,** 319–326.
84. Guy, A., Molko, D., and Téoule, R. (1989) Synthesis of DNA fragments bearing 7,8-dihydro-8-oxo-adenine: a gamma radiation product. *Nucl. Acids Res. Symp. Ser.* **#21,** 135,136.
85. Beaucage, S. L. and Iyer, R. P. (1992) Advances in the synthesis of oligonucleotides by the phosphoramidite approach. *Tetrahedron* **48,** 2223–2311.
86. Beaucage, S. L. and Iyer, R. P. (1993) Phosphoramidite derivatives and their synthetic applications. *Tetrahedron* **49,** in press.
87. Iyer, R. P., Phillips, L. R., Egan, W., Regan, J. B., and Beaucage, S. L. (1990) The automated synthesis of sulfur-containing oligodeoxyribonucleotides using 3H-1,2-benzodithiol-3-one 1,1-dioxide as a sulfur-transfer reagent. *J. Org. Chem.* **55,** 4693–4699.

CHAPTER 4

Oligodeoxynucleotide Synthesis

H-Phosphonate Approach

Brian C. Froehler

1. Introduction

Ribonucleoside H-phosphonates were introduced by Todd in 1957 to prepare diribonucleotide phosphate diesters *(1)*. The H-phosphonates were activated for condensation with diphenyl chlorophosphate, and this activator was later extended to the synthesis of deoxyribonucleotide dimers *(2)*. Activation of deoxynucleoside H-phosphonates with acyl chlorides allows for the synthesis of oligodeoxynucleotides greater than 100 bases in length *(3,4)*. Since these initial results, a number of laboratories have used acyl chlorides and nucleoside H-phosphonates in the synthesis of oligodeoxynucleotides *(5–9)*. This chemistry has also been extended to the synthesis of phosphate analogs of oligodeoxynucleotides *(10–16)*.

One of the advantages of the H-phosphonate method is the short cycle times (4 min or less). The features that allow for the rapid cycle time are the fast coupling reaction, an optional capping step, and oxidation taking place only once at the end of synthesis, rather than after each coupling (*see* Fig. 1). The nucleoside H-phosphonate monomers are easy to prepare from commercially available reagents (Fig. 2) and, additionally, mononucleoside H-phosphonates are stable to hydrolysis and oxidation, making these reagents easy to handle.

From: *Methods in Molecular Biology, Vol. 20: Protocols for Oligonucleotides and Analogs*
Edited by: S. Agrawal

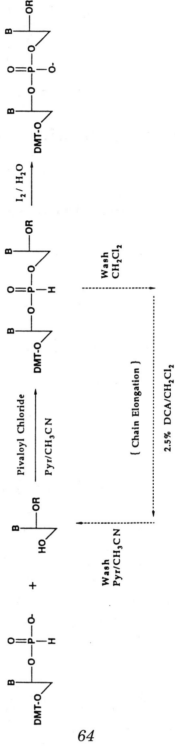

Fig. 1. Hydrogen-phosphonate synthesis scheme.

DNA Synthesis via H-Phosphonates

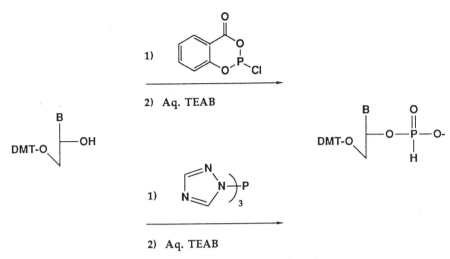

Fig. 2. Synthesis of nucleoside H-phosphonates.

Dinucleoside H-phosphonate dimers have been used as intermediates for the synthesis of phosphorothioates *(10)*, phosphoramidates *(10)*, phosphate triesters *(10)*, and methyl phosphonates *(15)* (Fig. 3). This method has been extended to the synthesis of polynucleoside phosphorothioates *(7,12,13)* and phosphoramidates *(7,10,11,16)*. Much of this chapter is devoted to the coupling and the oxidation reactions. The coupling step is very sensitive to overactivation by acyl chlorides, and thus, precautions must be taken to avoid the consequent side reactions. The oxidation reaction is important because of the fact that all linkages are oxidized at once, and care must be taken at this step to ensure complete oxidation, yet prevent the alkaline hydrolysis of the backbone. Often, poor quality of synthetic oligodeoxynucleotide is the result of hydrolysis of the backbone or incomplete oxidation, not inadequate chain elongation.

2. Materials

1. 2-Chloro-4H-1,3,2-benzo-dioxaphosphorin-4-one.
2. Diethyl phosphite.
3. 1,2,4-Triazole.
4. PCl$_3$.
5. Dichloroacetic acid (DCA).
6. Triethylamine.
7. 1,8-Diazabicyclo[5.4.0]undec-7-ene (DBU).
8. *N*-Methylmorpholine.

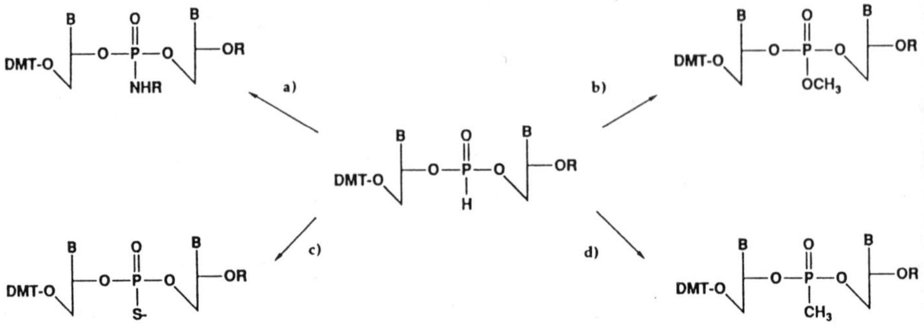

Fig. 3. Conversion of dinucleoside H-phosphonate diester to phosphate analogs: (a) RNH$_2$/ Pyridine/ CCl$_4$, (b) CH$_3$OH/ Et$_3$N/ CCl$_4$, (c) S$_8$/ Pyridine/ CS$_2$, and (d) BuLi/ CH$_3$I/ THF.

9. I$_2$.

All of the above are obtained from Aldrich, Milwaukee, WI, and used without further purification.
10. Trimethylacetyl chloride (pivaloyl chloride, Aldrich) is distilled and stored under Ar at room temperature.
11. Controlled Pore Glass LCAA is obtained from CPG, Inc., Fairfield, NJ, and functionalized by standard procedures *(17)*.

Acetonitrile and pyridine are obtained from Baxter, Hayward, CA (B&J, low water content) and dried over 3-Å molecular sieves (activated at 165°C/1 mm Hg/24 h) for at least 48 h. Anhydrous CH$_2$Cl$_2$ is dried over activated 4-Å molecular sieves and all other solvents are used directly without further purification. Dowex (50X2–100) resin is obtained from Aldrich and washed thoroughly with 3N HCl, H$_2$O, MeOH, and Et$_2$O, and dried under vacumm for 24 h before use. Triethylammonium bicarbonate (TEAB) and DBU bicarbonate are prepared by bubbling CO$_2$ into a stirred suspension of the corresponding volume of water and base until all base is in solution and the pH is at the desired value.

3. Methods

3.1. Synthesis of Nucleoside H-Phosphonates

3.1.1. 2-Chloro-4H-1,3,2-Benzo-Dioxaphosphorin-4-One Method

5'-DMT-protected nucleoside (1 mmol) is evaporated from anhydrous pyridine (1 × 20 mL), and dissolved into anhydrous pyridine (10 mL) and anhydrous CH$_2$Cl$_2$ (10 mL). In a 100-mL round-bottom flask is placed anhydrous pyridine (10 mL) and 10 mL of 1M PA solution (Note 1), the mixture cooled to 0°C (Note 2). The nucleoside solution is slowly added to the stirred PA solution under an inert atmosphere; the resulting mixture is removed from the ice bath and stirred for an additional 15 min. The reaction mixture is poured into 1M TEAB (100 mL, pH = 8.5), stirred vigorously until evolution of CO_2 ceases, transferred to a separatory funnel with 50 mL CH_2Cl_2, the phases are separated, the aqueous phase is extracted with CH_2Cl_2 (2 × 50 mL), and the combined organic phases are dried over Na_2SO_4 and evaporated to a foam. Purification is by silica gel chromatography with a 0.5% Et_3N/CH_2Cl_2 to 0.5% $Et_3N/10\%MeOH/CH_2Cl_2$ gradient. The product fractions are combined, evaporated to a foam, and evaporated from CH_3CN (2 × 50 mL; Note 3). The pure H-phosphonate is dissolved into CH_2Cl_2 (50 mL) and the solution washed with 0.2M DBU bicarbonate (pH = 8.7, 2 × 50 mL; Note 4), dried over Na_2SO_4, and evaporated to yield the DBU salt of the nucleoside H-phosphonate.

3.1.2. PCl_3 / Triazole Method

To a stirred solution of 1,2,4-triazole (1.2 g, 17 mmol) and triethylamine (7 mL, 50 mmol) in anhydrous CH_2Cl_2 (50 mL) is added PCl_3 (450 µL, 5 mmol) at room temperature under Ar. After 30 min, the reaction mixture is cooled to 0°C, and 5'-DMT-protected nucleoside (1 mmol, dried by evaporation from pyridine) in CH_2Cl_2 (15 mL) is added dropwise over 10–15 min, the mixture stirred for 15 min at 0°C, allowed to warm to ambient temperature, and poured into 1M TEAB (75 mL, pH = 8.5) with stirring. The mixture is transferred to a separatory funnel, the phases are separated, the aqueous phase is extracted with CH_2Cl_2 (1 × 50 mL), and the combined organic phases washed with 1M TEAB (1 × 75 mL), dried over Na_2SO_4, and evaporated. The product is isolated and purified as described above.

3.2. Synthesis of Ethyl H-Phosphonate

In a 2-L round-bottom flask is placed concentrated NH_4OH (675 mL) and diethyl phosphite (129 mL, 1 mol, *see* Note 5) is slowly added with stirring. The flask is capped and stirred at room temperature for 24 h, and evaporated to dryness. The solid is suspended into 500 mL of CH_3CN, evaporated to dryness, and dried by dessication overnight to yield solid ammonium ethyl H-phosphonate (mol wt = 127 g/mol).

A solution of triethylammonium ethyl H-phosphonate is prepared for the synthesizer as follows: 3.2 g of ammonium ethyl H-phosphonate is weighted into a 250-mL round-bottom flask, and to this is added CH_3CN (75 mL) and Et_3N (25 mL). The contents are stirred until all material is in solution (caution: NH_3 gas is evolved). The solution is evaporated to an oil, then evaporated from CH_3CN (2 × 100 mL; Note 6), and the resulting oil is dissolved into anhydrous pyridine/acetonitrile (1/1, 250 mL) to a final concentration of 100 mM and transferred to the appropriate bottle for machine synthesis.

3.3. Synthesis of Oligodeoxynucleotides

The synthetic cycle described below is performed on a Milligen/Biosearch 8750 DNA synthesizer. One of the important aspects of the synthetic cycle is to maintain short cycle times and avoid overactivation of the nucleoside H-phosphonate during the coupling reaction (see Section 4). The other important aspect is the method of oxidation; the solution must be basic enough to obtain complete oxidation of the polynucleoside H-phosphonate but not so basic as to lead to hydrolysis of the H-phosphonate backbone (see Section 4).

3.3.1. Reagent Preparation

The nucleoside H-phosphonates are evaporated to dryness twice with anhydrous CH_3CN and dissolved into anhydrous pyridine/CH_3CN (1/1) to a concentration of 40 mM. The activator solution (2% pivaloyl chloride [130 mM] in anhydrous prydine/CH_3CN [1/1]) is stable for 2 mo under Ar in an amber bottle. Wash A (CH_2Cl_2), Wash B (pyridine/CH_3CN), and the DCA solution are degassed with He prior to placement on the machine to minimize outgassing during synthesis, and wash B is maintained on activated 3-Å sieves. Teflon™ frits are used for the activator and anhydrous I_2 bottles, because stainless-steel frits are not inert to these solutions. Gas manifolds, with or without check valves, do not always prevent vapors from one bottle entering another bottle. To prevent pyridine vapors from entering the DCA/CH_2Cl_2 bottle, the DCA gas line is protected from pyridine with Dowex ion-exchange resin (H^+) using an in-line column (Omni fit, 10 × 150 mm; Note 7).

3.3.2. Synthetic Cycle

1. Wash A—CH_2Cl_2 (6 × 5 s).
2. Deblock—2.5% dichloroacetic acid/CH_2Cl_2 (6 × 10 s).
3. Wash A—CH_2Cl_2 (4 × 5 s).
4. Wash B—anhydrous pyridine/CH_3CN (6 × 5 s).
5. Coupling reaction—20 mM nucleoside H-phosphonate and 65 mM pivaloyl chloride in anhydrous pyridine/CH_3CN (60 s).
6. Wash B—anhydrous pyridine/CH_3CN (4 × 5 s).
7. Cap (optional)—50 mM ethyl H-phosphonate and 65 mM pivaloyl chloride in anhydrous pyridine/CH_3CN (60 s).
8. Wash B—anhydrous pyridine/CH_3CN (4 × 5 s).
9. Repeat step 1 until sequence complete.
10. Oxidation: 1 to 1 mixture of Ox. A and Ox. B (15 × 10 s) followed by 1 to 1 mixture of Ox. A and Ox. C (15 × 10 s).

3.3.3. Coupling Reaction

Nucleoside H-phosphonate and activator are briefly mixed in-line as the solution is delivered to the column. The premix time is kept to a minimum by moving the reagent quickly through the lines with a very brief wait step when the mixture reaches the column (≤5 s) and the overall coupling reaction, with delivery and wait times, is approx 1 min. Overactivation is avoided by using alternating small aliquots of nucleoside and activator rapidly. The nucleoside and activator are not allowed to premix for extended periods of time (>10 s) prior to exposing the solution to the machine-bound solid support *(see Section 4)*.

3.3.4. Capping

Ethyl H-phosphonate is used as the capping reagent when a cap step is desired. This step is essentially another coupling step and can also be replaced by a double-coupling step with the nucleoside H-phosphonate. The capping step is incorporated into the synthetic cycle after the coupling step and before the DMT deprotection step. The reaction is carried out in the same fashion as the coupling reaction *(see above)*. When ethyl H-phosphonate is used as the capping agent, the concentration (50 mM) is higher than that used for the nucleoside H-phosphonate (20 mM) during the coupling reaction.

3.3.5. Oxidation

Oxidation by I_2 is carried out by in-line mixing of a 1:1 mixture of a two-component system. THF is the solvent used, and the system is comprised of three bottles—Ox. A: $0.2M$ I_2 in THF, Ox. B: *N*-methyl morpholine/H_2O/THF (1/1/8), and Ox. C: Triethylamine/H_2O/THF (1/1/8). The advantage of this system is that oxidation of both short- and long-polynucleoside H-phosphonates to the corresponding phosphodiesters is very rapid, with minimal hydrolysis of the backbone *(see* Section 4). The disadvantage of this system is the need to use three bottles on the machine *(see* Note 8).

An alternative oxidation utilizes pyridine as the solvent and consists of only one bottle: $0.1M$ I_2 in H_2O/Pyridine (2/98). The primary advantage of this system is that it requires only one bottle on the machine; the disadvantage is that the oxidation is very slow (hours), and this oxidation mixture may not completely oxidize long oligodeoxynucleotides.

3.4. Alternative Plumbing Diagram

An alternate plumbing scheme can be used to aid in the automated synthesis of phosphodiesters and phosphorothioates (Fig. 4). The diagram shown is for the modification of a Milligen/Biosearch 8750. There are three valve trains, the A train, B train, and an oxidation train. The wash Ox. solution is from the same bottle as the wash B and is anhydrous pyridine/CH_3CN. The diester oxidation bottles are the same as those described above. The phophorothioate solutions are 2.5% S_8 in carbon disulfide/pyridine (1/1) and 10% triethylamine/pyridine (this reagent is not necessary for complete oxidation, but increases the rate of oxidation). This valve train configuration also contains an output for an optional capping reagent.

3.5. Recovery of Excess Nucleoside H-Phosphonate

Hydrolysis of the activated H-phosphonate leads to the starting nucleoside H-phosphonate, and thus, allows for the recovery and recycling of excess H-phosphonate from oligodeoxynucleotide synthesis. The activated H-phosphonate is quenched into $1M$ aq. TEAB (pH = 8.5), extracted into CH_2Cl_2, dried over Na_2SO_4, and evaporated. After column chromatography *(see above)*, the H-phosphonate can be reused

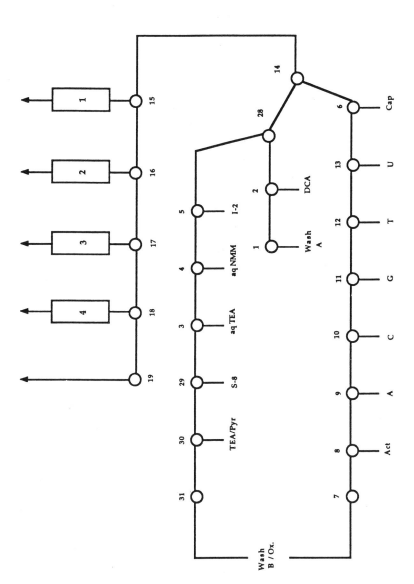

Fig. 4. Alternate plumbing diagram for the automated synthesis of oligonucleotide phosphodiesters and phosphorothioates. This scheme also incorporates an output for a capping reagent.

for synthesis. This is a valuable feature if expensive or modified nucleosides are used and offers an economical advantage for the large scale synthesis of oligodeoxynucleotides.

4. Discussion

4.1. Nucleoside H-Phosphonates

Synthesis of the nucleoside H-phosphonate monomers (Fig. 2) is readily accomplished by use of the phosphitylating agent introduced by Marugg et al. *(18)*. The reaction is rapid and quantitative with very little 3'-3' symmetrical dimer formation. A $1M$ solution of this reagent in CH_2Cl_2 can be kept for at least 6 mo if stored dessicated at $-20°C$, and is therefore a convenient reagent for the routine and rapid synthesis of nucleoside H-phosphonates. There are two disadvantages to this reagent: (1) hydrolysis of the intermediate phosphite leads to approx 5% of the starting 5'-DMT, 3'-OH nucleoside, which must be purified away from the product, and (2) the salicylic acid generated on hydrolysis can be difficult to separate from the product H-phosphonate. This reagent may be too reactive when modified nucleobases, with or without protecting groups, are used and can lead to many byproducts. An alternative phosphitylating agent, derived from PCl_3/triazole, is a less reactive reagent that is also useful in the synthesis of nucleoside H-phosphonates *(4)*. This reagent can be used with many reactive functionalities and also gives a high yield of product H-phosphonate, in general with no detectable 5'-DMT, 3'-OH nucleoside on aqueous workup. Additionally, purification is not complicated by salicylic acid. The drawback of this reagent is that it must be generated *in situ* prior to synthesis.

The fluorescent salicylic acid is generally the most difficult component of the reaction mixture to separate from the product H-phosphonate. Purification is accomplished by flash chromatography, using 0.5% Et_3N/CH_2Cl_2 to 0.5% Et_3N/10% $MeOH/CH_2Cl_2$ gradient. The product fractions are combined, concentrated, and evaporated from CH_3CN to remove triethylamine. It is important to remove all excess triethylamine from the product H-phosphonate after chromatography for two reasons: (1) conversion to the DBU salt will be incomplete and (2) strong base must not be present during the coupling reaction. When the nucleoside H-phosphonate contains base-sensitive protecting groups, such as FMOC, pyridine (1–2%) can be substituted for

Et$_3$N in the above system. CH$_3$CN/H$_2$O (9/1) works well as the solvent system for especially polar nucleoside H-phosphonates. CH$_3$CN/H$_2$O solvent system gives the best resolution from salicylic acid, but solubility problems can be encountered when loading the reaction mixture onto the column. TLC mobility of these compounds is very characteristic of the highly polar H-phosphonate anion. CH$_3$CN/H$_2$O (9/1) or 0.5% Et$_3$N/10% MeOH/CH$_2$Cl$_2$ solvent systems are well suited for development on silica gel TLC.

The most reliable method for characterization of the product H-phosphonate is ^{1}H and ^{31}P NMR. Hydrogen-phosphonates have a characteristic coupling constant of the P-H (J_{P-H} = 600–605 Hz) that can be seen in the ^{1}H-NMR spectrum as well as in the ^{1}H-coupled ^{31}P NMR spectrum. Furthermore, the ^{31}P NMR spectrum of a nucleoside 3'-H-phosphonate is split into another doublet due to the C-3' hydrogen (J_{P-H} = 8–12 Hz).

Nucleoside H-phosphonates are very stable, hygroscopic solids that can be stored under Ar at 0°C for more than 3 yr. In a pyridine/CH$_3$CN solution H-phosphonates are also very stable; when decomposition does takes place it is the slow loss of the 5'-DMT-protecting group *(6)*. The DBU salt is more stable than the TEA salt, with the loss of the DMT-protecting group occurring in weeks *(6)*. Nucleoside H-phosphonates are resistant to oxidation and are only oxidized by strong oxidants, such as MnO$_4$ *(19)*. Additionally, H-phosphonate monoesters are stable to hydrolysis under acidic and basic conditions. Since nucleoside H-phosphonates are resistant to autooxidation, and are not sensitive to hydrolysis, they offer distinct advantages over phosphoramidite reagents for the synthesis of DNA.

4.2. Oligodeoxynucleotides

Synthesis of oligodeoxynucleotides via H-phosphonates consists of a DMT-deprotection reaction, a coupling reaction, and an optional capping step with intermittent wash steps (Fig. 1). One single oxidation at the end of the synthesis is used to convert the H-phosphonate linkages to phosphodiester linkages. In general, the shorter the cycle time, the better. Long cycle times can lead to the degradation of the polymer-bound polynucleoside H-phosphonate backbone via hydrolysis of the diester.

4.2.1. DMT Deprotection

The DMT-protecting group is removed under standard conditions, 2.5% DCA/CH_2Cl_2 for 1 min. To shorten the overall cycle time, wash A is changed from CH_3CN to CH_2Cl_2; CH_3CN is not used as the wash solvent because the initial rate of deprotection is reduced with CH_3CN present. The DCA bottle is protected from pyridine vapors, introduced via the gas manifold, by the use of a strongly acidic ion-exchange resin (Dowex 50X2-100) contained within an in-line column.

4.2.2. Coupling Reaction

Deoxynucleoside H-phosphonates are activated by acid chlorides *(3)*, generally pivaloyl chloride *(4)*, to yield a very reactive mixed anhydride intermediate (Fig. 5) *(6,20–22)*. The coupling of this intermediate to the solid-support-bound 5'-hydroxy of the growing oligonucleotide chain is usually complete in 20–30 s, although the more sterically hindered 2'-protected ribonucleoside H-phosphonates may take longer. Overactivation of the H-phosphonate leads to a number of undesirable byproducts, and therefore, coupling reactions in excess of 1 min are generally more harmful than helpful. Pyridine acts both as a base and a nucleophilic catalyst for this reaction. The rate of coupling is greatly reduced in the absence of pyridine or a similar base. Recently, quinoline has been suggested as a replacement for pyridine; quinoline is less basic than pyridine and, thus, side reactions are reduced *(22)*. Use of strong bases, or strongly basic nucleophilic catalysts, during the coupling reaction accelerates these side reactions and is avoided at all times *(6,22)*. The activator is used in two- to fivefold excess over nucleoside H-phosphonate and anhydrous pyridine/CH_3CN seems to be the best solvent system for this reaction *(6)*. Prior to coupling, the solid support is washed thoroughly with pyridine/CH_3CN, and after coupling to remove excess activated H-phosphonate monomer. If necessary, double coupling can be employed, although it is often not needed.

Figure 5 shows some of the pathways, intermediates, and products derived from overactivation of the nucleoside H-phosphonate. The desired reaction course is reaction A followed by B to generate the product H-phosphonate diester **(3)**. Reaction C, to form the acyl phosphonate diester **(4)**, is significant only in the presence of strong bases (such as triethylamine) or basic nucleophilic catalysts (such as *N*-methyl imidazole or *N,N*-dimethylaminopyridine) *(6,20,23)*, and therefore, these

DNA Synthesis via H-Phosphonates

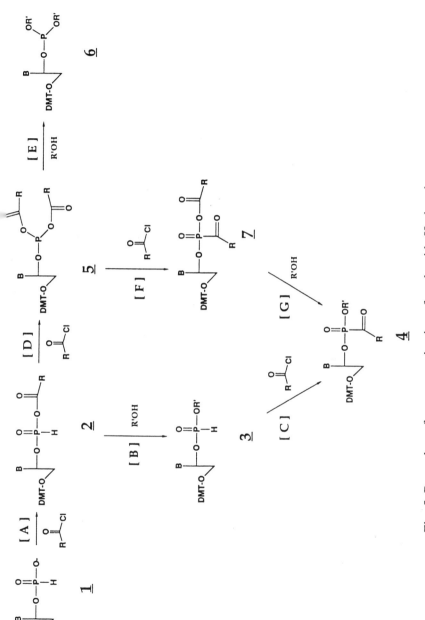

Fig. 5. Byproducts from overactivation of nucleoside H-phosphonates; R = $(CH_3)_3C-$, R' = protected nucleoside or CH_3CH_2-.

reagents are avoided in the coupling reaction. Reaction D, to form the *bis*-acylphosphite **(5)**, slowly takes place in the presence of pyridine and requires approx 5 min for complete formation *(6)*. Therefore, it is minimized by brief mixing and coupling times. Reaction F is very slow under pyridine catalysis *(22,24)*. The reaction of any of these intermediates **(2, 5,** and **7)** with a primary or secondary alcohol (reactions B, E, and G) is very fast. Once the desired product diester **(3)** is formed, strong base or nucleophilic catalysis is required during coupling to modify the linkage further *(6,20)*. The acyl phosphonate diester **(4)**, if formed, is not oxidized under the conditions used for DNA synthesis; therefore, on concentrated ammonium hydroxide treatment, the acyl function is cleaved, generating the H-phosphonate diester **(3)** *(23)*, which is further hydrolyzed under these conditions leading to chain cleavage. The trivalent product **6** is oxidized by the aqueous I_2 solution and leads to a branched phosphate triester.

4.2.3. Stability of Oligodeoxynucleoside H-Phosphonate

During synthesis, the internucleoside H-phosphonate diester has no protecting group at phosphorus. The H-phosphonate diester is stable to the condition of DNA synthesis, as long as the synthesis is rapid and anhydrous *(4)*. H-phosphonate diesters are susceptible to hydrolysis under mildly basic conditions; therefore, prolonged coupling reactions and pyridine/CH_3CN washes should be avoided. DCA deprotection can slowly cleave the backbone to a measurable extent and should also be kept to a minimum. Often when a synthesis does not go well, the tendency is to increase reaction and wash times, although this may be more harmful than helpful. Many times the reason for impure product oligonucleotide is incomplete oxidation or hydrolysis during oxidation, not incomplete coupling. Another potential problem is water contamination in one or more of the reagents leading to the slow hydrolysis of the backbone during synthesis.

4.2.4. Capping

The capping step in H-phosphonate synthesis is essentially another coupling reaction with an aliphatic H-phosphonate, such as cyanoethyl *(25)* or iso-propyl *(9)* H-phosphonate. This step is optional and not always necessary. The preparation of ethyl H-phosphonate and its use in capping is described above.

4.2.5. Oxidation

Oxidation of the polynucleoside H-phosphonate can be the most troublesome reaction. All linkages are oxidized at the same time; therefore, any decrease in yield of oxidation is amplified by the number of linkages present and can be a significant problem with long oligodeoxynucleotides. The aqueous I_2 oxidation of dialkyl H-phosphonates is subject to general base catalysis *(26)*, and it has been previously shown that the addition of strong bases, or nucleophilic catalysts, increases the rate of oxidation *(4)*. The greatest difficulty encountered on oxidation of a polynucleoside H-phosphonate is the alkaline hydrolysis of the H-phosphonate diester linkage, leading to chain cleavage. This is most readily overcome by the use of a weak base (pyridine or *N*-methyl morpholine) in the initial aqueous I_2 oxidation, followed by oxidation in the presence of a stronger base (triethylamine). Without the second oxidation step, the reaction time for complete oxidation is very long, and this step is required for the oxidation of long oligodeoxynucleotides *(4)*. The other problem commonly encountered is loss of activity by the I_2 oxidation solution. Aqueous I_2 disproportionates to IO_3^- and I^-, and an alkaline IO_3^- solution will not oxidize H-phosphonate diesters *(26)*; therefore, it is necessary to prepare separate solutions of the I_2 reagent and aqueous base. Stainless-steel frits are not inert to an anhydrous THF solution of I_2, and Teflon™ frits should be used with this reagent. I_2 in aqueous pyridine has been used to oxidize nucleoside H-phosphonate diesters *(27)*, but the reaction is very slow and may not be useful for the synthesis of long oligodeoxynucleotides. Aqueous CCl_4 has also been used successfully to oxidize polynucleoside H-phosphonates to the corresponding diester in the presence of I_2-sensitive functionalities *(28)*.

One of the advantages of H-phosphonate chemistry is the ease with which phosphate analogs are prepared (Fig. 3). Phosphorothioates are readily prepared by oxidation with S_8 in pyridine/CS_2. This is a very clean reaction leading to virtually 100% substitution by sulfur on phosphate. This reaction has been extended to the synthesis of oligophosphorothioates *(7,12,13)*, as well as uniformly labeled ^{35}S phosphorothioates *(29,30)*. The synthesis of phosphorothioates has been discussed in greater detail in Chapter 8.

Phosphoramidates are prepared in high yield by oxidation with CCl_4 in the presence of a primary or secondary amine. This procedure has led to the successful synthesis of oligodeoxynucleotides with a high degree of substitution at phosphorus by phosphoramidates *(7,10, 11,16)*.

4.2.6. Recovery of H-Phosphonate

The activated nucleoside H-phosphonate, when quenched into aqueous TEAB, generates the starting nucleoside H-phosphonate. A recent report describes simultaneous synthesis of multiple oligodeoxynucleotides with the recovery of excess nucleoside H-phosphonate *(31)*. This allows for the recovery of valuable or expensive nucleoside derivatives, and is an attractive attribute with the advent of large-scale DNA synthesis.

5. Notes

1. $1M$ PA is 2-chloro-4H-1,3,2-benzo-dioxaphosphorin-4-one in CH_2Cl_2 and is stable for at least 6 mo stored dessicated at -20 °C.
2. In general, a higher yield and cleaner product are obtained with deoxyguanosine if the reaction is carried out at -20 °C.
3. The CH_3CN evaporation is necessary to remove excess Et_3N, otherwise conversion to the DBU salt is incomplete.
4. The DBU bicarbonate wash performs two functions: (a) converts the salt form from Et_3N to DBU and (b) removes dissolved silica from the column chromatography. This procedure should not be performed when the nucleoside contains base-sensitive protecting groups, such as FMOC.
5. Addition of diethyl phosphite should be very slow, since the reaction is exothermic.
6. Evaporation from CH_3CN is necessary to remove excess Et_3N.
7. Contamination of the DCA bottle with pyridine vapors is most readily detected when the DCA bottle is opened: If a white cloud forms, pyridine is contaminating the system.
8. Stainless-steel frits are not inert to anhydrous I_2, and Teflon™ frits (common on peptide synthesizers) are used for this bottle.

References

1. Hall, R. H., Todd, S. A., and Webb, R. F. (1957) Nucleotides. Part XLI. Mixed anhydrides as intermediates in the synthesis of dinucleoside phosphates. *J. Chem. Soc.* **3291–3296**.
2. Garegg, P. J., Regberg, T., Stawinski, J., and Strömberg, R. (1985) Formation of internucleotidic bonds via phosphonate intermediates. *Chemica Scripta* **25**, 280–282.

3. Froehler, B. C. and Matteucci, M. D. (1986) Nucleoside H-phosphonates: valuable intermediates in the synthesis of deoxyoligonucleotides. *Tetrahedron Lett.* **27,** 469–472.
4. Froehler, B. C., Ng, P. G., and Matteucci, M. D. (1986) Synthesis of DNA via deoxynucleoside H-phosphonate intermediates. *Nucl. Acids Res.* **14,** 5399–5407.
5. Garegg, P. J., Lindh, I., Regberg, T., Stawinski, J., and Strömberg, R. (1986) Nucleoside H-phosphonates. III. Chemical synthesis of oligodeoxyribonucleotides by the hydrogen-phosphonate approach. *Tetrahedron Lett.* **27,** 4051–4054.
6. Froehler, B. C. and Matteucci, M. D. (1987) The use of nucleoside H-phosphonates in the synthesis of deoxyoligonucleotides. *Nucleosides Nucleotides* **6,** 287–291.
7. Agrawal, S., Goodchild, J., Civeira, M., Thornton, A., Sarin, P., and Zamecnik, P. (1988) Oligodeoxynucleoside phosphoramidates and phosphorothioates as inhibitors of human immunodeficiency virus. *Proc. Natl. Acad. Sci. USA* **85,** 7079–7083.
8. Vasser, M., Ng, P., Jhurani, P., and Bischofberger, N. (1990) Error rates in oligonucleotides synthesized by the H-phosphonate method. *Nucl. Acids Res.* **18,** 3089.
9. Andrus, A., Efcavitch, J. W., McBride, L. J., and Giusti, B. (1988) Novel activating and capping reagents for improved hydrogen-phosphonate DNA synthesis. *Tetrahedron Lett.* **29,** 861–864.
10. Froehler, B. C. (1986) Deoxynucleoside H-phosphonate diester intermediates in the synthesis of internucleotide phosphate analogues. *Tetrahedron Lett.* **27,** 5575–5578.
11. Froehler, B. C., Ng, P., and Matteucci, M. (1988) Phosphoramidate analogues of DNA: synthesis and thermal stability of heteroduplexes. *Nucl. Acids Res.* **16,** 4831–4839.
12. Agrawal, S. and Tang, J.-Y. (1990) Efficient synthesis of oligoribonucleotide and its phosphorothioate analogue using H-phosphonate approach. *Tetrahedron Lett.* **31,** 7541–7544.
13. Andrus, A. and Zon, G. (1988) Phosphorothioate DNA: synthesis via improved hydrogen-phosphonate chemistry. *Nucl. Acids Res.* **Symposium Series No. 20,** 121,122.
14. Letsinger, R. L., Zhang, G., Sun, D. K., Ikeuchi, T., and Sarin, P. S. (1989) Cholesteryl-conjugated oligonucleotides: synthesis, properties, and activity as inhibitors of replication of human immunodeficiency virus in cell culture. *Proc. Natl. Acad. Sci. USA* **86,** 6553–6556.
15. Seela, F. and Kretschmer, U. (1991) Diastereomerically pure R_p and S_p dinucleoside H-phosphonates: The stereochemical course of their conversion into P-methylphosphonates, phosphorothioates, and [^{18}O] chiral phosphates. *J. Org. Chem.* **56,** 3861–3869.
16. Letsinger, R. L., Singman, C. N., Histand, G., and Salunkhe, M. (1988) Cationic oligonucleotides. *J. Am. Chem. Soc.* **110,** 4470–4471.

17. Damha, M. J., Giannaris, P. A., and Zabarylo, S. V. (1990) An improved procedure for derivatization of controlled-pore glass beads for solid-phase oligonucleotide synthesis. *Nucl. Acids Res.* **18,** 3813–3821.
18. Marugg, J. E., Tromp, M., Kuyl-Yeheskiely, E., van der Marel, G. A., and van Boom, J. H. (1986) A Convenient and General approach to the synthesis of properly protected d-nucleoside-3'-hydrogenphosphonates via phosphite intermediates. *Tetrahedron Lett.* **27,** 2661–2664.
19. Brown, D. M. and Hammond, P. R. (1960) Studies on phosphorylation. Part XX. The oxidation of mono- and di-alkyl phosphites. *J. Chem. Soc.* 4229,4230.
20. Garegg, P. J., Regberg, T., Stawinski, J., and Strömberg, R. (1987) Studies on the synthesis of oligonucleotides via the hydrogenphosphonate approach. *Nucleosides Nucleotides* **6,** 283–286.
21. Garegg, P. J., Regberg, T., Stawinski, J., and Strömberg, R. (1987) Nucleoside H-phosphonates. V. The mechanism of hydrogenphosphonate diester formation using acyl chlorides as coupling agents in oligonucleotide synthesis by the hydrogenphosphonate approach. *Nucleosides Nucleotides* **6,** 655–662.
22. Efimov, V. A., Dubey, I. Y., and Chakhmakhcheva, O. G. (1990) NMR Study and improvement of H-phosphonate oligonucleotide synthesis. *Nucleosides Nucleotides* **9,** 473–477.
23. Kume, A., Fujii, M., Sekine, M., and Hata, T. (1984) Acylphophonates. 4. Synthesis of dithymidine phosphonate: a new method for generation of phosphonate function via aroylphosphonate intermediates. *J. Org. Chem.* **49,** 2139–2143.
24. de Vroom, E., Spierenburg, M., Dreef, C., van der Marel, G., and van Boom, J. (1987) A general procedure to convert H-phosphonate mono- or diesters of nucleic acids into valuable phosphate di- or triesters. *Recl. Trav. Chim. Pays-Bas.* **106,** 65,66.
25. Gaffney, B. L. and Jones, R. A. (1988) Large-scale oligonucleotide synthesis by the H-phosphonate method. *Tetrahedron Lett.* **29,** 2619–2622.
26. Lewis, E. S. and Spears, Jr., L. G. (1985) Ionization of the pH bond in diethyl phosphonate. *J. Am. Chem. Soc.* **107,** 3918–3921.
27. Garegg, P. J., Regberg, T., Stawinski, J., and Strömberg, R. (1987) Nucleoside phosphonates: Part 7. Studies on the oxidation of nucleoside phosphonate esters. *J. Chem. Soc. Perkin Trans.* **1,** 1269–1273.
28. Sinha, N. D. and Cook, R. M. (1988) The preparation and application of functionalized synthetic oligonucleotides: III. Use of H-phosphonate derivatives of protected amino-hexanol and mercapto-propanol or -hexanol. *Nucl. Acids Res.* **16,** 2659–2669.
29. Stein, C., Iverson, P., Subasinghe, C., Cohen, J., Stec, W., and Zon, G. (1990) Preparation of ^{35}S-Labeled polyphosphorothioate oligodeoxyribonucleotides by use of hydrogen phosphonate chemistry. *Analytical Biochemistry* **188,** 11–16.
30. Shaw, J-P., Kent, K., Bird, J., Fishback J., and Froehler, B. (1991) Modified deoxyoligonucleotides stable to exonuclease degradation in serum. *Nucl. Acids Res.* **19,** 747–750.
31. Seliger, H. and Rösch, R. (1990) Laboratory methods: simultaneous synthesis of multiple oligonucleotides using nucleoside-H-phosphonate intermediates. *DNA and Cell Biology* **9,** 691–696.

CHAPTER 5

Oligoribonucleotide Synthesis

The Silyl-Phosphoramidite Method

Masad J. Damha and Kelvin K. Ogilvie

1. Introduction

In 1968, one of us began a quest to develop a general method for the synthesis of RNA sequences. The major problems associated with the chemical assembly of ribonucleotide chains were already well established. RNA sequences are very sensitive to chemical and enzymatic degradation. Consequently, all procedures involved in the assembly of RNA chains must respect the delicacy of the assembled chain. Further, the assembly of RNA sequences is complicated by the presence of a 2'-hydroxyl group in the monomeric ribonucleosides that form the starting materials for the chemical assembly of RNA chains. The 2'-hydroxyl must be blocked (or "protected") with a protecting group that remains stable throughout the many steps of a chain assembly yet that is labile enough to be removed at the end of chain assembly without leading to cleavage of the assembled chain. Finally, it was clear even in the late 1960s that in order to be convincing the method must be capable of producing an RNA chain of the length of a tRNA in order to overcome the inherent skepticism of chemical synthesis of RNA.

Our first approach was to attempt to mask the 2'-hydroxyl in the form of an "anhydro-" or "cyclonucleoside" *(1)*. This approach failed in our hands, but convinced us that the ideal way to mask the 2'-hydroxyl would be with a protecting group that was stable to acid and base, but

From: *Methods in Molecular Biology, Vol. 20: Protocols for Oligonucleotides and Analogs*
Edited by: S. Agrawal Copyright ©1993 Humana Press Inc., Totowa, NJ

that could be removed under neutral conditions. Although this was a tall order, we had become intrigued with silyl ethers from our attempts to render anhydronucleosides sufficiently volatile for mass-spectrometric characterization. A search of the literature showed that the solvolysis of silyl ethers had been extensively studied and that bulky alkyl silyl ethers were remarkably stable *(2)*. Thus began a long development of silyl ethers as 2'-hydroxyl-protecting groups for oligoribonucleotide synthesis *(3–5)*. This led ultimately to the development of a method amenable to both solution and automated solid-phase synthesis *(6)*, and the first chemical synthesis of RNA molecules of the size and character of tRNAs *(7,8)*.

In this chapter, we will attempt to provide general procedures and recipes for the synthesis of oligoribonucleotides. We will also attempt to point out areas where precautions are absolutely necessary. Finally, we will not attempt to review the whole literature of RNA synthesis, but would refer the reader to an excellent review by Michael Gait and colleagues *(9)*.

1.1. General Precautions

We wish to draw attention immediately to the need to standardize definitions. For example, when we refer to dry reagents, we do not mean casually dry, such as simply storing reagent-grade solvents over molecular sieves. We mean dry according to the descriptions to follow. We also recommend that silylated nucleosides be prepared on a millimolar scale (or larger), not a micromolar scale. Purchased reagents, such as silylating reagents, should be of a high quality, and if in doubt, they should be purified (e.g., redistilled or sublimed). Purchased fully protected silylated phosphoramidites should be carefully checked for purity and rechromatographed if any doubt exists.

Deprotection conditions are crucial to the utimate success of the overall procedure, and the protocols described below should be followed carefully. In particular, the conditions used for *N*-deprotection should be ethanol/concentrated NH_4OH or saturated methanolic ammonia, as described in Section 3.3.2., and not the concentrated NH_4OH conditions used to deprotect DNA oligomers. Finally, care must be taken in handling deprotected ribonucleotides because of the ever-present ribonucleases (in the air, hands, sweat glands, and breath). In spite of these areas of caution, RNA sequences can now be prepared routinely, and the novice should have no hesitation in setting forth.

1.2. Silyl Migration

This aspect of the silyl-phosphoramidite method has plagued the literature on RNA synthesis for some time. Silyl groups will migrate under certain conditions, and we have described this in detail *(5,10)*. There is, however, no need for silyl groups to migrate, and if you follow the steps outlined here, migration will not enter the picture. Hopefully, the unsubstantiated criticism of the method will have abated with the detailed characterization that we have carried out *(6–8,11)*, and the number of independent laboratories that have repeated and extended our procedures *(9,12–26)*.

1.3. N-Protecting Groups

The benzoyl group is a perfectly acceptable protecting group for exocyclic amino groups on A, C, and G *(6,7)*. The phenoxyacetyl group makes a very good and easily removed protecting group for A and G *(15,27)*, and permits milder deprotection of assembled chains. The isobutyryl group is the most widely used protecting group for guanosine, and we describe below the preparation of a silylated *N*-isobutyrylguanosine.

1.4. O^6-Protection

In general, O^6-protection of guanosine is not essential for the preparation of short oligomers (2–24 mers), and longer sequences containing very few guanosines. However, our experience *(6,7)* suggests that for oligomers rich in guanosine and for long sequences, O^6-protection is advisable *(28,29)*. At the very least, O^6-protected guanosine renders oligomers more soluble in the TBAF reagent.

1.5. Phosphate Protection

Both the cyanoethyl and methyl groups have been used in protecting the phosphate internucleotide linkage, and we find them to be virtually interchangeable *(6,7,12–22,27)*. The cyanoethyl group has the advantage of ease of removal, but the methyl group is more stable. In our experience, the average coupling efficiencies achieved with methyl phosphoramidites are slightly higher (97–99%) when compared with the yields obtained with the cyanoethyl protected ribophosphoramidites (95–97%). However, for the synthesis of short oligomers (e.g., 6–18 mers), we find comparable overall yields following deprotection regardless of which group is used.

1.6. Silylated Ribonucleoside Phosphoramidites

The general procedures for the preparation of silylated *N*-protected nucleosides have been described in detail, and include the traditional *N*-acyl-protecting groups *(6,30)* as well as the phenoxyacetylated nucleosides *(27)*. The procedures for the preparation of phosphoramidites from silylated nucleosides are also fully documented *(6,27)*. Standard protocols for silylation and phosphitylation reactions are provided in Sections 3.1.2. and 3.1.3., respectively.

2. Materials

2.1. Silylation and Phosphitylation of Nucleosides

1. *tert*-Butyldimethylsilyl chloride (Aldrich, WI).
2. Silver nitrate: Grind finely before use.
3. Imidazole: Recrystallize from THF and dry *in vacuo* over P_2O_5 prior to use.
4. 4-Dimethylaminopyridine: Recrystallize from THF and dry *in vacuo* over P_2O_5 prior to use.
5. Anhydrous tetrahydrofuran (THF): Dry THF overnight at room temperature over KOH. Decant THF, and reflux with sodium and benzophenone under nitrogen until a purple color persists. Distill prior to use removing THF from a collection bulb by syringe via a septum port.
6. Anhydrous pyridine: Reflux (16 h), and distill from phthalic anhydride, redistill from calcium hydride, and store over activated (at 400°C) molecular sieves (Type 4A).
7. *N,N*-dimethylformamide: Dry by fractional distillation from calcium hydride, and store over activated (at 400°C) molecular sieves (Type 4A).
8. *N*-methylimidazole: Dry by fractional distillation from calcium hydride, and store over activated (at 400°C) molecular sieves (Type 4A).
9. *N,N*-diisopropyl(2-cyanoethyl)phosphonamidic chloride and *N,N*-diisopropylmethylphosphonamidic chloride: Store at −20°C in 15 mL glass Hypo Vials (Pierce) sealed with a Viton serum cap (Chromatographic Specialties, Ontario, Canada).
10. Silica gel column and thin-layer chromatography: (A) Merck Kieselgel 60 (230–400 mesh) for gravity and flash columns. (B) Merck Kieselgel 60F 254 silica gel analytical sheets (0.2 mm × 20 cm × 20 cm) (Merck #5735) for the analysis of silylation and phosphitylation reactions. **Caution: Eye and skin protection required**. Nucleosides and nucleotide derivatives are detected using a UV light source (output ca. 254 nm).

2.2. Solid-Phase RNA Synthesis

1. Ribonucleoside monomers: Silylated nucleosides and their phosphoramidite derivatives can be prepared as described in Sections 3.1.2. and 3.1.3., or can be obtained commercially. The following ribophosphoramidites are currently available from Dalton Chemical Laboratories (DCL), Inc. (York University, Toronto, Canada), Peninsula Laboratories, Inc. (California, USA), and Milligen-Biosearch, CA:

    5'-DMT-2'-TBDMS-uridine-3'-N,N'-diisopropyl(2-cyanoethyl) phosphoramidite
    N-Bz-5'-DMT-2'-TBDMS-cytidine-3'-N,N'-diisopropyl(2-cyanoethyl)phosphoramidite
    N-Bz-5'-DMT-2'-TBDMS-adenosine-3'-N,N'-diisopropyl(2-cyanoethyl) phosphoramidite
    N-iBu-5'-DMT-2'-TBDMS-guanosine-3'-N,N'-diisopropyl(2-cyanoethyl)phosphoramidite

2. Ribonucleoside-derivatized long-chain alkylamine controlled pore glass (LCAA-CPG): Ribonucleoside-functionalized LCAA-CPG can be prepared according to procedures described in detail in Chapter 19 or can be obtained commercially. The following supports are currently available from DCL and Peninsula Laboratories:

    N-Bz-5'-DMT-2'(3')-Ac-A-LCAA-CPG;
    N-iBu-5'-DMT-2'(3')-Ac-GLCAA-CPG;
    N-Bz-5'-DMT-2'(3')-Ac-C-LCAA-CPG;
    5'-DMT-2'(3')-U-Ac-LCAA-CPG.

3. Ancillary reagents: These reagents are the same as those used for DNA synthesis and can be obtained from Applied Biosystems or other suppliers.
    (a) Anydrous acetonitrile: Dry acetonitrile initially from phosphorus pentoxide (**Caution: corrosive;** 5 g/L of CH_3CN), and distill at atmospheric pressure after an overnight reflux. Reflux continuously under nitrogen over CaH_2, and distill just prior to use. We recommend the use of anhydrous acetonitrile for both amidite dissolution and intermediate washing steps.
    (b) Detritylation: 3% trichloroacetic acid/1,2-dichloroethane (or dichloromethane) for deblocking DMT and MMT groups.
    (c) Activation: $0.5M$ tetrazole in acetonitrile.
    (d) Capping: 10% acetic anhydride/10% 2,6-lutidine/THF and 16% N-methylimidazole/THF (Ac_2O/N-MeIm).
    (e) Oxidation: $0.1M$ iodine in THF:pyridine:water (75:20:2, v/v/v).

2.3. Deprotection of Oligoribonucleotides

1. Demethylation: Thiophenol:triethylamine:dioxane (1:2:2, v/v/v). **Caution: Toxic stench. Wear gloves. Store and use in fumehood. Discard excess reagent by pouring it into bleach.**
2. Oligomer cleavage/decyanoethylation/N-deacylation:
   (a) Aqueous ammonia:ethanol (3:1, v/v): add ethanol (10 mL) to concentrated NH_4OH solution (30 mL) prior to use. Mix thoroughly and use immediately. **Warning:** Concentrated (29–35%) NH_4OH should be stored tightly sealed in a refrigerator and opened only briefly before use.
   (b) Anhydrous saturated methanolic (or ethanolic) ammonia: Prepare this solution just prior to use by bubbling ammonia gas through a stirred anhydrous methanol (or ethanol) solution at 0°C until saturation is achieved (15–20 min). Use immediately.
3. Desilylation: $1M$ tetra-n-butylammonium fluoride (TBAF) in THF (Aldrich).

2.4. Purification and Desalting of Oligoribonucleotides

It is critical that sterile equipment, reagents, and handling techniques be used in handling free oligoribonucleotides. All water, Sephadex, silanized glassware, and plasticware must be autoclaved in the presence of diethylpyrocarbonate (DEP, *31*) as follows:

1. Deionized sterile water: Sterilize double-distilled water by treatment with DEP as a 1% solution followed by autoclaving at 120°C for 20 min. When cool, add sodium azide (0.001%) to inhibit microbial growth. Water sterilized in this way can be used safely for a 2-wk period. Whenever in doubt, discard water and sterilize a new batch.
2. Glassware and plasticware: Glassware used in the deprotection of nucleotides (pipets, glass wool, and so on) should be silanized prior to sterilization using Sigmacote® (Sigma, St. Louis, MO). Small glassware and plasticware are sterilized by autoclaving at 120°C for 20 min. Large pieces of glassware (e.g., Sephadex and cation-exchange glass columns) are sterilized by washing with double-distilled water (1% DEP), and drying in an oven (110°C).
3. Ion exchange and size-exclusion resins:
   a. Dowex 50W-W8 (Na^+ form, 20–50 mesh) (J. T. Baker Chem. Co., NJ): Transfer Dowex-Na^+ exchange resin (100 mL) into a sintered-glass funnel. Wash resin first with $0.1N$ HCl (500 mL), then with

0.1N NaOH (500 mL), and finally with double-distilled water until washings are neutral and colorless. Transfer resin into an Erlenmeyer flask containing double-distilled water (1% DEP), cover the rim of flask with aluminum paper, and autoclave (120°C, 20 min). After cooling, stopper flask and store beads at room temperature under sterile water. Prior to use, pour resin solution into a 10-mL syringe barrel equipped with a sterile silanized glass-wool plug and wash beads thoroughly with 10–20 mL sterile water until the washings are colorless. Do not allow the column to go dry during the entire procedure.

b. Sephadex G-25 F (Pharmacia, Quebec, Canada): Transfer Sephadex (25 g) to an Erlenmeyer flask containing double-distilled water (1% DEP, 250 mL), and cover neck of flask with aluminum paper. Allow beads to swell (6 h), and autoclave (120°C, 20 min). When cool, add sodium azide (0.001%) to inhibit microbial growth, stopper the flask, and store at room temperature under sterile water. Prior to use, pour bead solution into a 10-mL syringe barrel equipped with a sterile silanized glass-wool plug without allowing the column to go dry. Sephadex G-25 F is also available in prepacked Sephadex NAP columns (Pharmacia).

4. Thin-layer chromatography (TLC). **Caution: Eye and skin protection required**. Cellulose F TLC plates (0.1 mm × 20 cm × 20 cm) (Merck, Darmstadt, Germany, #5565) or cellulose plates (0.16 mm × 20 cm × 20 cm) (Eastman Kodak Co. #13254). Develop plates with n-propanol/conc. NH_4OH:water (55:10:35). Oligoribonucleotides are detected using a UV light source (output ca 254 nm).

5. Polyacrylamide gel electrophoresis (PAGE). **Caution: Acrylamide, *bis*-acrylamide and formamide are quite toxic**. Keep all PAGE reagents refrigerated.

a. 50% Acrylamide/2.5% *bis*-acrylamide in double-distilled water.

b. Urea.

c. 10X TBE Buffer: 0.9M Tris-base borate, 93 mM EDTA (pH 8.3).

d. Loading buffer: Deionized formamide in 10X TBE buffer (8:2 or 80%). Formamide is deionized by stirring with mixed bed AG 501–X8 resin (0.3 g/10 mL) for 1–2 h.

e. Dye solution: 0.5% xylene cyanol/0.5% bromophenol blue in 80% formamide loading buffer.

f. Tetramethylethylene diamine (TEMED). Use as such.

g. Ammonium persulfate: Use as a 10% ammonium persulfate solution. Prepare just prior to use.

6. HPLC: Filter buffer and methanol through a 0.45-micron filter, and degas before use.
   a. Reversed-phase buffer $0.02M$ KH_2PO_4 (pH = 5.5, adjusted with KOH). Make up buffer using deionized water that has been distilled twice from glass.
   b. Methanol (HPLC grade).
   c. Column: Whatman Partisil ODS-2 (4.6 × 250 mm; 10-micron particles).
7. C_{18} SEP-PAK® desalting.
   a. C_{18} SEP-PAK cartridges (Waters Associates).
   b. 50-mM Triethylammonium acetate (TEAA): Make up $0.1M$ stock solution by adding to sterile water (500 mL) freshly distilled triethylamine (6.7 mL), and acetic acid (2.8 mL, analytical grade). After mixing, the pH is adjusted to 7.0 by addition of either triethylamine or acetic acid as required. Store at 5°C. Dilute to 50 mM before use. A $2M$ stock TEAA solution can also be obtained from Applied Biosystems.
   c. Methanol (HPLC grade).

2.5. Enzymatic Digestions of Oligoribonucleotides

1. List of enzymes:
   a. Snake venom phosphodiesterase (SVPDE, from *Crotalus durissus*) (Boehringer Mannheim): The enzyme is obtained as a solution (2 mg/mL) in 50% (v/v) glycerol, pH ca. 6.0.
   b. Nuclease Pl (from *Pencillium citrinum*) (Boehringer Mannheim): dissolve the lyophilized enzyme in 30 mM NH_4OAc (pH 5.3) (1 mg enzyme/mL or 300 U/mL).
   c. Alkaline phosphatase (AP, from calf-intestine) (Boehringer Mannheim): The enzyme is obtained as a suspension in $3.2M$ $(NH_4)_2SO_4$/0.1 mM $ZnCl_2$, pH ca. 7.0.
   d. Phosphodiesterase (CSPDE, from calf spleen) (Boehringer Mannheim): The enzyme is obtained as a suspension (2 mg/mL) in $(NH_4)_2SO_4$, pH ca. 6.0.
   e. RNase T2 (from *Aspergillus oryzal*) (Sigma): Dissolve the lyophilized powder in double-distilled sterile water (600 U enzyme/mL).
2. Incubation buffers: Make up buffers using water that has been distilled twice from glass and autoclaved. Filter through a 0.45 micron filter and store at −20°C.
   a. SVPDE/AP: 50 mM Tris-HCl/10 mM M_gCl_2 (pH 8.0).
   b. Nuclease Pl/AP: $0.1M$ Tris-HCl/1 mM $ZnCl_2$ (pH 7.2).
   c. CSPDE: $0.5M$ NH_4OAc (pH 6.5).
   d. RNase T2: 10 mM NaOAc (pH 4.5).

3. Methods

3.1. General Procedures for the Preparation of Ribonucleosides Phosphoramidite

3.1.1. N-Protection and 5'-O-Tritylation

Nucleosides are first protected at the exocyclic amino groups with either the benzoyl (A, G, and C) *(32)*, isobutyryl (G) *(33,34)*, or phenoxyacetyl (A and G) *(27)* groups through the "transient" procedure developed by Jones and coworkers for protection of deoxynucleosides *(35)*. Following N-protection the nucleosides are tritylated (DMT or MMT) at the 5'-position by the standard pyridine *(27)* or pyridine-silver nitrate *(30)* procedures. Detailed NMR, UV, and TLC data of 5'-tritylated N-protected ribonucleosides have appeared in the literature *(6,27,29,30,36-38)*.

3.1.2. Silylation of Nucleosides (see "Notes" Section 5.1.)

The *t*-butyldimethylsilyl (TBDMS) group is used for protection of U, C, A, and G residues having the following protecting group combinations: 5'-MMT-U, N-Bz-5'-MMT-C, N-Bz-(or PhOAc)-5'-MMT-A, and N-iBu-5'-MMT-G. The triisopropylsilyl (TIPS) group is used for protection of N-Bz(or PhOAc)-5'-MMT-G in preference to the TBDMS group because of the ease of separation of the resulting 2'-TIPS and 3'-TIPS isomers on a silica gel gravity column. In general, the 2'-isomer (TBDMS or TIPS protection) travels faster than the 3'-isomer on silica gel gravity columns and TLC.

The isomerization of the 3'-silyl isomer to a ca. 1:1 equilibrium mixture of 2- and 3'-silyl isomers can be effected in basic solutions, such as 9:1 pyridine:water (TBDMS protection) or 1% NH$_4$OH: ethanol (TIPS/N-Bz combination). This provides a method to convert (recycle) more of the "unwanted" 3'-isomer into the more useful 2'-isomer.

Either the THF-pyridine-AgNO$_3$ *(30)* or the DMF-imidazole *(37,38)* system can be employed for the silylation of nucleosides (Schemes 1 and 2). The THF-pyridine-AgNO$_3$ system provides higher conversion yields with highly selective 2'-silylation in most cases except for N-protected guanosines. The DMF-imidazole system gives the best results for TBDMS or TPS protection of N-protected guanosines *(6,27)*. These silylation procedures are illustrated here for the synthesis of N-Bz-5'-MMT-2'-TBDMS-A (**2c**), and N-iBu-5'-MMT-2'-TBDMS-G (**2e**).

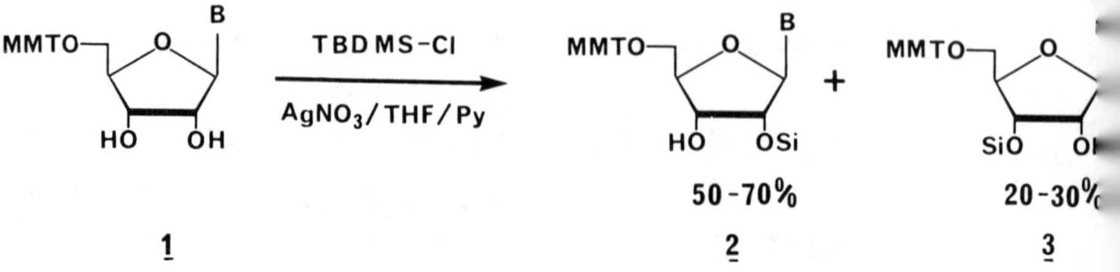

Scheme 1. Selective 2'-silylation of ribonucleosides.

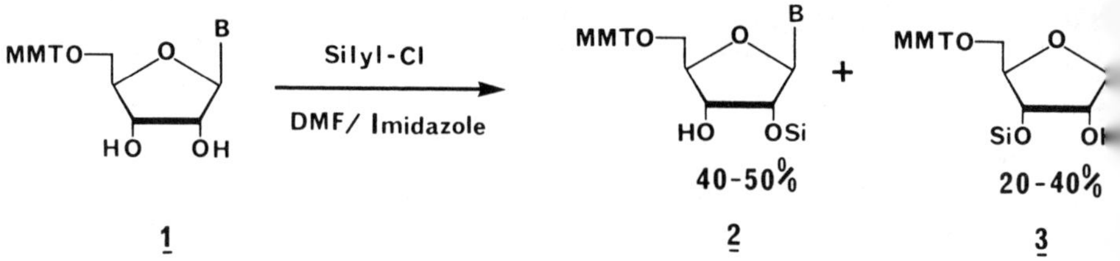

	Base	Reference
a.	Ur	(37)
b.	Cy^{Bz}	(38)
c.	Ad^{Bz}	(34)
d.	Ad^{PhOAc}	(27)

	Base	Silyl-Cl	Reference
e.	Gu^{iBu}	TBDMS-Cl	(sect. 3.1.2)
f.	Gu^{Bz}	TIPS-Cl	(6)
g.	Gu^{PhOAc}	TIPS-Cl	(27)

Scheme 2. Silylation of 5'-MMT-N2-acylated guanosines.

3.1.2.1. THE THF-PYRIDINE-SILVER NITRATE PROCEDURE

N-Bz-5'-MMT-A (**1d**) (6.44 g, 10 mmol) is dissolved in dry THF (100 mL). Dry pyridine (3.3 mL, 40 mmol) and AgNO$_3$ (2.2 g, 13 mmol) are added, and stirring is continued for 5 min until almost all AgNO$_3$ dissolves. TBDMS-Cl (2.0 g, 13 mmol) is added all at once, and the resulting cloudy-milky solution is stirred at room temperature. After 3 h, TLC analysis (1:3 ether:CH$_2$Cl$_2$) indicates the presence of the 2'-TBDMS

(R_f = 0.66), and 3'-TBDMS (R_f = 0.50) isomers, and unreacted starting material (R_f = 0.32) in ca. 2:1:1 ratio. More AgNO$_3$ (0.5 g, 3 mmol), and TBDMS-Cl (0.45g, 3 mmol) is added and the reaction allowed to proceed for another 2 h (total 5 h). The solution is filtered off into a 5% NaHCO$_3$ solution (100 mL) to prevent detritylation and extracted with CH$_2$Cl$_2$ (2× 200 mL). The combined organic solutions are dried over anhydrous Na$_2$SO$_4$, filtered, and evaporated to yield a white foam. This material is purified by column chromatography on silica gel (20 g/ g of product) by slow elution with ether:CH$_2$Cl$_2$ (gradient 1:7 to 1:3). Trace amounts of N-Bz-5'-MMT-2',3'-diTBDMS-A elute first from the column, followed by pure N-Bz-5'-MMT-2'-TBDMS-A (**2c**), mixtures of (**2c**) and (**3c**), and pure N-Bz-5'-MMT-3'-TBDMS-A (**3c**). The pooled fractions containing the mixtures of (**2c**) and (**3c**) are evaporated and repurified. The 2'-isomer is obtained in 54% yield (4.2 g), and the 3'-isomer in 26% yield (2.0 g).

This general procedure is employed for the silylation of nucleosides **1a–d**. The desired 2'-isomers are separated from the 3'-isomers by silica gel column chromatography using the following solvent systems: **2a/3a**, apply mixture to the column as a 20% CH$_2$Cl$_2$ solution and elute using ether/hexanes (gradient 1:5 to 1:1; gravity column); **2b/3b** ether:CH$_2$Cl$_2$ (gradient 1:9 to 1:2; gravity column); **2d/3d**, EtOAc: hexanes (4:6, flash column). Chromatographic and spectroscopic data of these derivatives have been fully documented in the references listed in Scheme 1.

3.1.2.2. The DMF/Imidazole Procedure

To a solution of N-iBu-5'-MMT-G (**1e**) (6.3 g, 10.0 mmol) in anhydrous DMF (50 mL) are added imidazole (1.8 g, 26.0 mmol) and TBDMS-Cl (2.0 g, 13 mmol), and the solution is stirred overnight at room temperature. The consumption of starting material is confirmed by TLC (EtOAc:CH$_2$Cl$_2$, 1:2), and the reaction stopped by addition of 5% sodium bicarbonate (5 mL). After evaporation of most of the DMF, the residue is taken up in CH$_2$Cl$_2$ (125 mL), washed with saturated brine (2 × 150 mL), and concentrated to a foam. This is purified by column chromatography on silica gel (6 cm diameter, 250 g or 25 g/g of product) using EtOAc:CH$_2$Cl$_2$ (gradient 1:7 to 1:2) to yield 10% (0.85 g) of N-iBu-5'-MMT-2',3'-diTBDMS-G, 38% (2.8 g) of N-iBu-5'-MMT-2'-TBDMS-G (**2e**), and 28% (2.0 g) of N-iBu-5'-MMT-3'-TBDMS-G (**3e**) as white foams.

¹H-NMR of N-iBu-5'-MMT-2'-TBDMS-G (**2e**) (CDCl₃, 400 MHz): selected resonances in ppm, 7.77 (s, 1, H8), S.70 (d, 1, H1'), 5.31 (dd, 1, H2'), 4.33 (bd, 1, H3'), 0.83 (s, 9, t-Bu-Si), 0.02 s, 3, Me-Si), -0.19 (s, 3, Me-Si). N-iBu-5'-MMT-3'-TBDMS-G (**3e**) (CDCl₃, 400 MHz): ppm, 7.82 (s, 1, H8), 5.69 (d, 1, H1'), 4.85 (m, 1, H2'), 4.44 (dd, 1, H3'), 0.84 (s, 9, *t*-Bu-Si), 0.01 (s, 3, Me-Si),-0.09 (s, 3, Me-Si).

N-Bz-5'-MMT-G (**1f**) and *N*-PhOAc-5'-MMT-G (**1g**) derivatives are silylated using the same procedure, except that TIPS-Cl is used instead of TBDMS-Cl *(6,27)*. The desired 2'-TIPS isomers are separated from the 3'-TIPS isomers by silica gel column chromatography using the following solvent systems: **2f/3f**, CHCl₃:CH₂Cl₂/NEt3 (9:0.5:0.5, gravity column); **2g/3g** EtOAc:CH₂Cl₂:NEt₃ (20:75:5, flash column). Chromatographic and spectroscopic data of these derivatives have been fully documented in the references listed in Scheme 2.

3.1.3 Phosphitylation of Nucleosides (see "Notes" Section 5.2.)

The general procedure can be illustrated by the synthesis of N-Bz-5'-MMT-2'-TBDMS-A-3'-*N,N*-diisopropyl(2-cyanoethyl)phosporamidite (**4c**) (Scheme 3). The protocol is similar to that described previously by Atkinson and Smith *(39)* for the preparation of deoxynucleoside-3'-phosphoramidites.

3.1.3.1. PHOSPHITYLATION PROCEDURE

To a stirred THF (5 mL) solution of N-Bz-5'-MMT-2'-TBDMS-A (**2c**) (1.0 g, 1.40 mmol), 4-dimethylaminopyridine (20 mg, 0.15 mmol), and diisopropylethylamine (1.0 mL, 5.75 mmol) is added dropwise, over 20 s, *N,N*-diisopropyl(2-cyanoethyl)phosphonamidic chloride (375 μL, 1.68 mmol). A white precipitate (diisopropylethylammonium hydrochloride) appears after 20 min, indicative of the desired reaction. The complete consumption of starting material is confirmed by TLC (ether:CH₂Cl₂, 1:2) after 4 h. EtOAc (150 mL, prewashed with 5% NaHCO₃) is added, and the resulting solution washed with saturated brine (5 × 100 mL). The aqueous washes are back-extracted with EtOAc (50 mL), and the combined EtOAc phases are dried over anhydrous Na₂SO₄, filtered, and evaporated under vacuum to yield a white foam. The crude product is purified by column chromatography on silica gel (3 cm diameter, 20 g) by rapid elution with CH₂Cl₂:Hexanes:NEt₃ 50:44:6. After evaporation of the pooled fractions, the product is coevaporated first with 95% EtOH (2 × 30 mL)

Oligoribonucleotide Synthesis

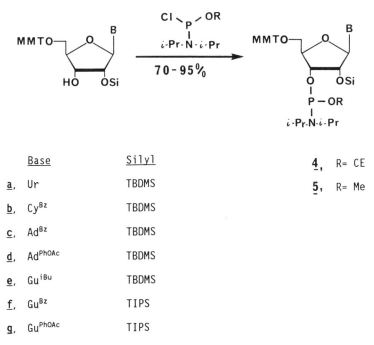

Scheme 3. Preparation of ribonucleoside 3'-phosphoramidites.

to remove traces of triethylamine followed by ether (3 × 20 mL) providing the pure product **4c** as a white foam (1.20 g, 88% yield). The corresponding methylphosphoramidite derivative **5c** is prepared in 84% yield using the same procedure. Uridine (**4a, 5a**), and cytidine (**4b, 5b**) methyl and 2-cyanoethyl phosphoramidites are prepared in 80–95% yield using the same procedure.

Guanosine phosphoramidites **4e, 4f, 5f**, and **5g** are prepared using the standard procedure, except that 2.5–3.0 equivalents of phosphitylating reagent and 8.5–10 equivalents of diisopropylethylamine are used, no 4-dimethylamino pyridine is used, and the phosphitylation reactions are allowed to proceed for 24 h. Sometimes, the products are obtained as viscous oils after column chromatography because of contamination of N,N-$(iPr_2)PO(H)OR$ (R = CE or Me). In these cases, the amidites are repurified by precipitating 20% CH_2Cl_2 solutions of the amidites into cold (dry ice/isopropyl alcohol bath), rapidly stirred hexanes. Chromatographic and ^{31}P-NMR data of these derivatives are reported in Table 1.

Table 1
^{31}P-NMR Data and Purification Solvents of Ribophosphoramidites

Ribophosphoramidite	Purification solvent,[a] CH$_2$Cl$_2$:hexanes/NEt$_3$	^{31}P-NMR, ppm CDCl$_3$[b]	R_f
Cyanoethylamidites			
4a	50/46/4	150.5, 150.1	0.75, 0.68[c]
4b	40/58/2	149.9	0.68, 0.50[c]
4c	50/44/6	151.7, 149.7	0.83, 0.70[c]
4e	70/24/6	151.8, 149.9	0.49[c]
4f	[f]	153.0, 148.8	0.61, 0.55[c]
Methylamidites			
5a	50/46/4	150.4, 150.3	0.79[c]
5b	40/58/2	150.0, 148.7	0.81, 0.75[c]
5c	40/58/2	151.3, 149.5	0.90, 0.83[c]
5d	50/47/3	151.9, 150.1	0.48, 0.32[d]
5f	55/35/10	153.7, 149.0	0.69, 0.61[c]
5g	75/20/5[g]	152.3, 150.4	0.62[e]

[a]Fast elution (ca. 15 drops/5 s) on a gravity silica gel column (20 g silica/g of amidite).
[b]Referenced to external 85% H$_3$PO$_4$. ^{1}H-NMR, UV, and chromatographic data of compounds, **4a**, **5a–d**, and **5f** and **g** have appeared in the literature (6,27).
[c]CH$_2$Cl$_2$:EtOAc, 2:1.
[d]CH$_2$Cl$_2$:EtOAc, 8:2.
[e]CH$_2$Cl$_2$:EtOAc, 1:1.
[f]Derivative **4f** decomposes during chromatography on a silica gel column; **4f** can be purified by precipitation into cold (dry ice/isopropyl alcohol bath) vigorously stirred hexanes.
[g]CH$_2$Cl$_2$:EtOAc:NEt$_3$.

^{1}H-NMR of N-iBu-5'-MMT-G amidite **4e** (CD$_3$CN, 400 MHz): selected resonances in ppm, diastereoisomer I, 7.86 (s, 1, H8), 5.86 (d, 1, H1'), 4.90 (dd, 1, H2'), 4.36 (m, 1, H3'), 0.77 (s, 9, t-Bu-Si), –0.02 (s, 3, Me-Si), –0.19 (s, 3, Me-Si). Diastereoisomer II, 7.87 (s, 1, H8), 5.82 (d, 1, H1'), 5.15 (dd, 1, H2'), 4.28 (m, 1, H3'), 0.76 (s, 9, t-Bu-Si), –0.01 (s, 3, Me-Si), –0.18 (s, 3, MeSi).

3.2. Automated Solid-Phase Synthesis of Oligoribonucleotides

All of our solid-phase syntheses are carried out on an Applied Biosystems 381A synthesizer as described below. Solutions of ribonucleosides phosphoramidite are made up at 0.15M concentration by transferring via syringe the appropriate amount of anhydrous acetonitrile into 0.25-g bottles. For the commercially available amidites (Section 2.2.) the amount of acetonitrile required is: A and G, 1.7 mL, C 1.8 mL, and U 1.9 mL.

Prepacked 0.2- and 1-µmol columns containing about 30–40 µmol/g nucleoside-derivatized supports are available from DCL, Inc. and Peninsula Laboratories, Inc. One can also pack the column by transferring the appropriate amount of support into the empty Teflon™ body (Applied Biosystems), and crimping the Teflon™ end fittings with filter into place using aluminum Hypo-Vial seals (Pierce). An ISCO Retriever II fraction collector was attached to the instrument to collect the released trityl cation.

3.2.1. The Synthesis Cycle

In a typical 0.2-µmol scale synthesis, the support is first treated with the capping reagents (Ac_2O/N-MeIm). This step blocks off sites on the support's surface that are unmasked during CPG storage and removes any trace of water at the start of the synthesis *(40,41)*. This may be easily performed on the DNA synthesizer by use of an automatic or manual capping cycle. We recommend a 45-s delivery "cap to column" step followed immediately by a 300-s "wait" step, during which time the reagents are allowed to remain in contact with the support. Finally, a 60-s acetonitrile washing step leaves the support ready for oligonucleotide assembly.

Assembly of oligoribonucleotides is carried out using the following minor modifications of the recommended Applied Biosystems DNA synthesis cycle:

1. Phosphoramidite coupling: "wait" step, 600 s (as opposed to 30 s for DNA synthesis).
2. Ac_2O/N-MeIm capping: "cap to column" step, 20 s (as opposed to 10 s); "wait" step, 30 s (this extra step is added by editing the standard cycle program). This prolonged capping treatment efficiently reverses guanine modification, which takes place during the amidite coupling step *(29)*.
3. Oxidation: "iodine to column" step, 30 s (as opposed to 12 s); "wait" step, 20 s (as opposed to 30 s).
4. Detritylation: "TCA to column" step, 120 s for DMT deblocking and 160 s for MMT deblocking.

Each synthesis cycle takes about 20 min to complete. Quantitation of the trityl cation released during each cycle from the 5'-terminus of the growing oligonucleotide chain is used as a preliminary monitor of synthesis performance. Comparison of each synthesis cycle to the previous one usually indicates 95–97% average yield per cycle with cyanoethyl-protected phosphoramidites and 97–99% with methyl-protected phosphoramidites.

3.2.2. The Trityl Assay for Monitoring Synthesis Performance

In a typical 0.2-µmol scale synthesis, each TCA eluate (collected in glass test tubes) is diluted to 10 mL with 5% TCA/1,2-dichloroethane and the absorbance at 504 nm (DMT, E 76 mL cm^{-1} µmol^{-1}) or 478 nm (MMT, E 56 mL cm^{-1} µmol^{-1}) measured. The value obtained for each TCA eluate is compared as a percentage to that of the previous TCA eluate.

It should be noted that evaluation of synthesis performance by PAGE and HPLC is more informative than the trityl assay. On humid summer days, in particular, a significant percentage of the released trityl solutions is hydrolyzed in the collection tubes, creating the illusion of a less-than-optimal synthesis. When the oligomers of these syntheses are deblocked and analyzed (PAGE, HPLC), far better yields are obtained than would be indicated to be present by the trityl cation measurements.

3.3. Deblocking of Synthetic Oligoribonucleotide Sequences

Deprotection of RNA oligomers synthesized by the silyl-phosphoramidite method is straightforward and is simply an extension of the methodologies currently used for the preparation of DNA oligomers. The difference lies in the need for an additional deblocking step to remove the silyl group (TBDMS and TIPS) at 2'-hydroxyl positions.

First, the methyl groups are removed from the phosphate triester internucleotide linkages by using a triethylammonium thiophenoxide treatment. The next step is to treat with an ammonia/alcohol solution that concomitantly cleaves the oligoribonucleotide from the support and removes exocyclic amino-protecting groups. Finally, the oligomer

Oligoribonucleotide Synthesis

is treated with TBAF for removal of the silyl protection groups. Oligomers with cyanoethyl phosphate-protecting groups do not require thiophenoxide treatment, since they are removed in the same step that cleaves the oligomer from the support and removes *N*-acyl-protecting groups (Section 3.3.2.).

3.3.1. Deprotection of Phosphates

3.3.1.1. Cyanoethyl Deprotection

Proceed to cleavage from LCAA-CPG and *N*-deacylation, which simultaneously removes the cyanoethyl group (Section 3.3.2.).

3.3.1.2. Methyl Deprotection

This step is performed manually after removal of the column from the synthesizer. The methyl-protecting groups on the phosphates are removed by treatment with thiophenol:triethylamine:dioxane (1:2:2, vol/vol/vol; *see* "Materials" Section 2.3., Step 1), for 1 h. The method of introducing the thiophenol reagent into the column is identical to that described below in Section 3.3.2. for introducing ammonia solutions. The reagent is then expelled from the column, and the support washed extensively with ethanol. The RNA oligomer remains covalently bound to the support during this procedure.

3.3.2. Cleavage and N-Deacylation

The removal of *N*-isobutyryl or benzoyl-protecting groups from assembled oligoribonucleotide chains is the rate-limiting step during deprotection, traditionally requiring severe NH_4OH conditions. We *(27)* and others *(24)* have found that these conditions led to some cleavage of the assembled chain, resulting in the formation of shorter, default sequences. The cleavage is more severe in the aqueous ammonium hydroxide conditions routinely employed for the deacylation of oligodeoxynucleotides (i.e., 29% NH_4OH, 55°C, 16 h), and greatly reduced (but still detectable) in 29% NH_4OH:ethanol (3:1 v/v, 55°C, 16 h). The problem can be virtually eliminated by using anhydrous-saturated ethanolic ammonia (55°C, 16 h) *(20)* or 29% NH_4OH:ethanol (3:1, v/v, room temperature, 2–3 d). Under these conditions, N-iBu groups on G residues are efficiently removed, and chain cleavage is minimized. We prefer to use the more convenient

29% NH₄OH:ethanol (3:1, v/v, 55°C, 16 h) system for deprotection of 2–12 mers, but recommend anhydrous-saturated ethanolic ammonia (55°C, 16 h, or 29% NH₄OH:ethanol (3:1, v/v, room temperature, 2–3 d) for the deprotection of longer oligomers.

Oligomers assembled with *N*-PhOAc-G, *N*-PhOAc-A, and N-Bz-C amidites can be deacylated under much milder conditions using anhydrous-saturated methanolic ammonia *(27)* at room temperature for 16–18 h without any detectable chain cleavage. This improved deacylation condition not only eliminates the small amount of unwanted chain cleavage, but also allows the synthesis of oligoribonucleotide analogs containing base-sensitive moieties *(25)*.

A minor modification of the manual deprotection protocol recommended by Applied Biosystems *(42)* follows: The column obtained from Section 3.3.1. is fitted onto a 1-mL sterile polypropylene syringe (Luer). The other end of the column is fitted with a male-to-male Luer connector that is, in turn, fitted with a sterile syringe needle. The ammonia reagent is drawn to fill the column (ca. 0.5 mL) minimizing the volume in the syringe. The syringe needle is then inserted into a rubber stopper, and the reaction allowed to proceed at room temperature. After 30 min, the syringe plunger is withdrawn to the 0.5–1.0 mL mark, as the needle is removed from the stopper and the solution discharged into a 16-mL polypropylene test tube. This treatment is repeated two more times for 30 min each, collecting the ammoniacal eluates in the same tube. Finally, the support is washed by passing more ammonia reagent through the column and expelling it into the tube until the total volume of the solution is ca. 8 mL. The RNA oligomer is now in the tube. The tube is sealed and incubated at the desired temperature (*vide supra*). On completion, the sample is frozen (dry ice/ethanol bath), and lyophilized to dryness using a Speed-Vac Concentrator (Savant Instruments).

Alternatively, the ammonia treatment can be carried out by simply opening the synthesis column and carefully transferring the solid support into a 16-mL polypropylene test tube. The appropriate ammonia reagent is added (8 mL), the tube is sealed, and the mixture incubated at the desired temperature. On completion, the solution is filtered, then frozen (dry ice/ethanol bath), and finally, lyophilized to dryness. (Note: here, filtration is necessary to remove the CPG glass beads that interfere with the subsequent desilylation step.)

3.3.3. Desilylation

Removal of the TBDMS-protecting groups is carried out by treatment of the residue obtained from the deacylation step (Section 3.3.2.) with $1.0M$ tetrabutylammonium fluoride (TBAF)/THF solution. The reaction is carried out at room temperature for 16 h using 50 equivalents of TBAF per TBDMS group. For example, a decaribonucleotide chain prepared on a 0.2-µmol scale requires 100 µmol TBAF or 100 uL of the solution. If prepared on the 1-µmol scale, 0.50 mL TBAF solution is needed. The reaction is quenched with sterile water (5.0 mL), and the amount of recovered crude product quantitated in A_{260} U (the amount of material that will produce an absorbance of 1.0 at 260 nm, when dissolved in 1.0 mL of water, in a 1-cm cell). Following quantitation, the solution can be safely stored frozen at $-20°C$. Lyophilize to a white powder just prior to purification.

3.4. Purification of Oligoribonucleotides (see "Notes" Section 5.3.)

Generally, oligoribonucleotides are purified and desalted through one or two efficient chromatographic procedures. These include:

1. Preparative polyacrylamide-gel electrophoresis (PAGE) (Table 2) *(43)*.
2. Size-exclusion chromatography on Sephadex G-25F (desalting).
3. Reversed-phase chromatography on C_{18} SEP PAK cartridges (desalting) *(39,42)*.
4. Ion-exchange chromatography (Dowex-Na$^+$).
5. Thin-layer chromatography (TLC) on cellulose.
6. High-pressure liquid chromatography (HPLC) *(44)*.

In our experience, the sequential use of preparative PAGE, followed by either size-exclusion chromatography on Sephadex G-25F or reversed-phase purification on SEP-PAK cartridges (SEP-PAK giving generally better results) is the most convenient and economical combination for the purification of up to 1 µmol of material. Material obtained in this way is sufficiently pure for thermal denaturation and most biological studies (splicing and ribozyme work, and so forth). If large amounts of ribonucleotides (1–5 mg) are required for biophysical studies, such as NMR, they are best purified by the sequential use of Dowex-Na$^+$ exchange chromatography, TLC on cellulose (or semipreparative HPLC, refs. *9,44)*, and the Sephadex G-25F (or C_{18} SEP-PAK) system.

Table 2
Recipes for the Preparation
of Polyacrylamide/7M Urea (Denaturing) Gels[a] (43)

	Final acrylamide concentration, %[b]		
	12	16	24
Acrylamide[c] (mL)	12.0	16.0	24.0
10× TBE buffer (mL)	5.0	5.0	5.0
Urea (g)	21.0	21.0	21.0
dd-Water	[d]	[d]	[d]

[a]Using a 125-mL vacuum filtration flask with a rubber stopper, the gel mixture is degassed under vacuum for about 10 min until bubbling subsides. TEMED (12 µL) is then added followed by freshly prepared 10% ammonium persulfate (100 µL). After mixing thoroughly, the gel mixture is quickly poured between siliconized glass plates, and the slot former (10 mm and 10 cm wide for analytical and preparative gels, respectively) is positioned. The gel polymerizes within 1 h. We recommend the standard 0.75 and 1.5 mm-thick gels for analytical and preparative gels, respectively. Nondenaturing gels are prepared in the same way, with the exception that urea is omitted from the above recipe.

[b]This depends on the length of the oligoribonucleotide. We recommend 12% for 30–80 mers, 16% for 20–30 mers, and 24% for 2–20 mers.

[c]Stock solution: 50% acrylamide/2.5% bis-acrylamide in double-distilled water.

[d]Dilute with double-distilled water to a final vol of 50.0 mL.

3.4.1. Polyacrylamide Gel Electrophoresis (PAGE)

PAGE is a relatively fast and economical method for the separation of the desired oligoribonucleotide product from shorter fragments. Oligoribonucleotides containing 2–20 nucleotides are run on 24% polyacrylamide gels at constant current (10 mA for the first 15 min followed by 20 mA to completion [ca. 4–5 h]) using $0.09M$ Tris-borate-EDTA buffer (pH 8.3) (Section 2.4.). Oligomers containing 20–30 nucleotide residues are run on 16% gels, and 30–80 mers are run on 12% gels. Both denaturing ($7M$ urea), and nonenaturing (no urea) gels give good results.

We recommend checking the purity of the crude oligomer by analytical PAGE prior to purification by preparative gel electrophoresis as follows: To the crude product obtained after the TBAF treatment (Section 3.3.3.) are added 10 µL of loading buffer, and the resultant loaded onto the gel (0.75 mm × 20 cm × 20 cm) in a single lane (12 lanes/gel). Load 0.2–0.4 A_{260} U/lane on nondenaturing gels and 0.4–0.8 A_{260} U/lane on denaturing gels. Following electrophoresis, the gels are wrapped in commercially available plastic wrap (Saran Wrap™),

Oligoribonucleotide Synthesis

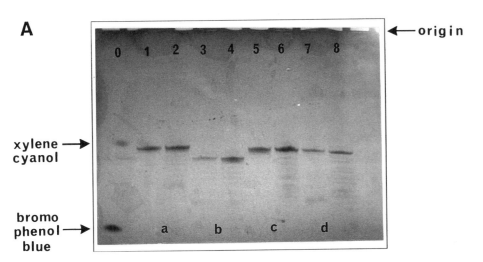

Fig. 1A. 24% Analytical polyacrylamide/7M urea gel electrophoresis of pure (lanes 1, 3, 5, 7), and crude (lanes 2, 4, 6, 8) oligoribonucleotides. **a:** GUG UGA GCG AGU G; **b:** GUG AGC GAG UGU G; **c:** AUA AAG ACA CAC A; **d:** AGA AGG GAG AGG G.

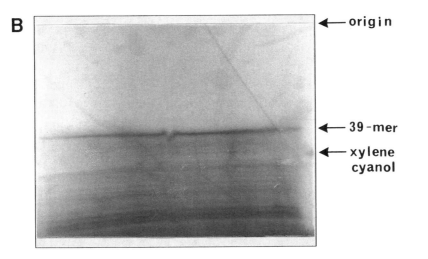

Fig. 1B. 12% Preparative polyacrylamide/7M urea gel electrophoresis of crude 5'-GAA AUA CGC AUA UCA GUG AGG AUU CGU CCG AGA UUG UGU-3' (39 mer).

placed over a fluorescent TLC plate illuminated with a UV lamp (254 nm), and analyzed **(Caution: Eye protection required)**. The illuminated gels can be photographed using Polaroid PolaPan™ 4 × 5 cm Instant Sheet Film (#52, medium contrast, ISO 400/27°) through a Kodak Wratten gelatin filter (#58 green, cat. no. 1495860). The desired product should be the major and least mobile band present. Below the desired band, one observes low, but detectable levels of shorter failure sequences. Typical results are shown in Figs. 1A and 1B.

Once the presence of the desired sequence is established according to its electrophoretic mobility, the sequence is purified by preparative PAGE using 1.5 mm thick gels and a single 11 cm-wide lane. Up to 30 A_{260} U of crude material may be loaded per gel. After electrophoresis, the least mobile band is excised using a scalpel. The gel piece is placed in a sterile 16-mL polypropylene test tube and incubated overnight in sterile water (5 mL) at 37°C. The tube is then vortexed and centrifuged, and the supernatant carefully transferred to a polypropylene tube. The gel pieces are washed with more water (3 mL), and the combined washings are lyophilized and quantitated. The sample is now ready for desalting (Sections 3.4.2. or 3.4.3.).

3.4.2. Desalting—Size-Exclusion Chromatography on Sephadex G-25F

The material obtained from Section 3.4.1. (or 3.4.5.) is dissolved in water (1 mL), applied to a Sephadex G-25F column (4 × 2 cm, Section 2.4.), and eluted with water (ca. 15 mL). The eluent is collected in ten 1.5-mL fractions, and each of the fractions is lyophilized to dryness. It is possible to see which fractions contain salts or urea by monitoring the presence of a white residue (flake) after lyophilization. The pure desalted oligomer is usually in the first three fractions (UV quantitation), which contain a small, barely visible pellet. Sometimes the next three fractions contain a mixture of the oligomer and various salts that can be combined and desalted once again. The last remaining fractions usually contain salts and are discarded. In some cases where the Sephadex system fails to desalt the oligomer adequately (e.g., oligomers 2–10 residues in length), or if extremely pure oligomers are needed, C_{18} SEP-PAK cartridge purification (3.4.3.) can be used following this procedure.

3.4.3. Desalting—Reversed-Phase Chromatography on C_{18} SEP-PAK® Cartridges

The cartridge is attached to a 10-mL sydnge and flushed with methanol (10 mL) followed by water (10 mL). The RNA sample (2–10 A_{260} U) obtained from Section 3.4.1. (or 3.4.6.) is dissolved in 50 mM aqueous triethylammonium acetate (TEAA) and loaded on the cartridge collecting the eluant (A). Next, the cartridge is flushed with 50 mM TEAA (3 mL), collecting again the eluant (B). The oligomer is eluted from the cartridge with 7:3 50 mM TEAA:methanol (10 mL) collecting ten 1-mL fractions (C). The oligomer elutes in the first two TEAA:methanol fractions (C) (UV quantitation at A_{260}).

Solutions A and B should have zero absorbance at 260 nm. If not, these fractions are combined, lyophilized, and repurified.

3.4.4. Dowex-Na⁺ Exchange Chromatography

This step is carried out immediately following deblocking of silyl groups by the TBAF reagent. After quenching the TBAF reaction with water (Section 3.3.3.), the oligonucleotide sample is applied directly to a Dowex-Na⁺ column (4 × 2 cm, Section 2.4.) and eluted using sterile water (10 mL). The eluant, collected in a single polypropylene test tube, is then lyophilized to dryness. This procedure efficiently removes tetrabutylammonium ions, which interfere with NMR analyses, and provides the oligomer as a sodium salt.

3.4.5. Thin-Layer Chromatography on Cellulose

The crude oligoribonucleotide (100–200 A_{260} U) obtained from Section 3.4.4. is dissolved in sterile water (0.25 mL), and applied, on one plate, as a uniform 15-cm band at the origin. After development in 55:10:35 n-propanol:conc. NH$_4$OH:water for 6–8 h, the slowest migrating major band (detected by UV) is scraped off and suspended in sterile water (5 mL in a 16-mL polypropylene tube). The mixture is stirred on an automatic vortex and then centrifuged, and the supernatant is gently collected. The extraction step is repeated two more times, and the combined supernatants are lyophilized in a Speed-Vac concentrator. The oligomer should be further purified by size-exclusion (Section 3.3.2.) or reversed-phase (Section 3.3.3.) chromatography to remove residual cellulose and impurities that leach from the TLC plates.

3.4.6. HPLC

We recommend this technique as an analytical check on the purity of the oligoribonucleotide following deprotection and at different stages of the purification steps. This method of separation is also useful as an additional purification step after an initial separation (e.g., PAGE followed by C_{18} SEP-PAK) or if ultrapure oligomer samples are needed. HPLC analysis of the enzymatic digests of oligomers is also a powerful method to obtain information concerning purity and base composition. Slight variations of the reversed-phase system developed by McLaughlin and Piel can be used for these purposes *(44)*.

3.4.6.1. HPLC CONDITIONS

Column: Whatman Partisil ODS-2 4.6 × 250 mm
Mobile phase:
A: 20 mM KH_2PO_4 (pH 5.5)
B: methanol
Linear gradient: 0–50% B in 30 min
Flow: 2.0 mL/min
Temperature: 22°C

Under these conditions, the order of elution of nucleosides and nucleoside 3'-monophosphates is Cp (1.7 min), Up (2.6 min), C (6.4 min), Gp (7.8 min), and U (8.0 min), I (11.7 min), G (12.1 min), Ap (12.4 min), and A (17.3 min). Peak areas and the previously reported extinction coefficient values (in mL cm^{-1} µmol^{-1}) at 254 nm *(45)* were used to calculate concentration of monomers: Cp (5.8), Up (9.1), C (6.7), Gp (13.0), and U (9.0), I (10.1), G (13.4), Ap (14.5), and A (14.0).

3.4.6.2. ANALYTICAL SEPARATIONS OF OLIGORIBONUCLEOTIDES

Dissolve 0.5–3.0 A_{260} units of oligoribonucleotide sample (obtained after Section 3.3.3., or after purification by PAGE and C_{18} SEP-PAK) in 20 µL of water. Inject sample on the analytical 4.6 × 250 mm ODS-2 column, and run using the conditions described above. The oligomer, which generally elutes as a single major peak in the range of 15–22 min, is collected in a sterile polypropylene tube. The solution is then lyophilized and the residue desalted on SEP-PAK cartridges (Section 3.4.3.). A typical analysis of a crude and sample of an 11 mer is shown in Fig. 2.

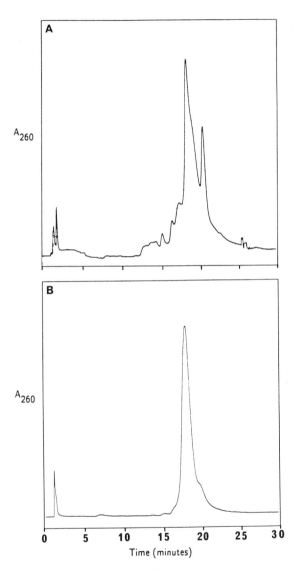

Fig. 2. **A.** A typical reversed-phase HPLC chromatogram of a crude oligoribonucleotide (AUGAUCAUACA) prepared using conventionally protected ribophosphoramidites (Section 2.). **B.** The chromatogram of the same oligomer after purification by the sequential use of PAGE and Sephadex G25-F (Section 3.4.). HPLC conditions: Section 3.4.6.

3.4.6.3. Enzymatic Digestion of Oligoribonucleotides

The oligoribonucleotides are digested using the following mixtures at 37°C for 2 h *(see* Section 2.4.).

Snake venom phosphodiesterase (SVPDE)/alkaline phosphatase (AP):
0.3 A_{260} U oligoribonucleotide
50 µL of 50 mM Tris-HCl/10 mM MgCl$_2$ (pH 8.0)
1 µL SVPDE (Section 2.4.)
1 µL AP (Section 2.4.)

Nuclease Pl/alkaline phosphatase (AP):
0.3 A_{260} U oligoribonucleotide
17 µL of 0.1M Tris-HCl/1 mM ZnCl$_2$ (pH 7.2)
3 µL nucl. Pl (Section 2.4.)
1 µL AP (Section 2.4.)

These degradations generate a mixture of U, C, I, G, and A nucleosides that is directly analyzed on the ODS-2 column under the conditions described above. Inosine (I) results from the deamination of adenosine by adenosine deaminase, which is a contaminant in the enzyme preparations. If the incubation is allowed to proceed for 2 h, only U, C, I, and G can be detected. A typical analysis of total digestion of a oligoribonucleotide is shown in Fig. 3.

Calf spleen phosphodiesterase (CSPDE):
0.3 A_{260} U oligoribonucleotide
50 µL of 0.5M NH$_4$OAc (pH 6.5)
1 µL CSPDE (Section 2.4.)

RNase T2:
0.3 A_{260} U oligoribonucleotide
50 µL of 10 mM NaOAc (pH 4.5)
1 µL RNase T2 (Section 2.4.)

Both CSPDE and RNase T2 cleave oligomers to give a mixture of the four ribonucleoside 3'-monophosphates and a nucleoside (the 3'-terminal residue). An aliquot (25 µL) of this mixture can then be analyzed on the ODS-2 column using the mobile phase described above.

4. Discussion

The above procedures for the synthesis of RNA have previously been illustrated by the synthesis of several RNA sequences, including a 77–nucleotide-long tRNA$^f_{Met}$ species *(7,8)* and various ham-

Oligoribonucleotide Synthesis

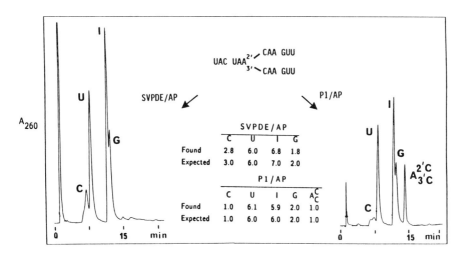

Fig. 3. Reversed-phase HPLC analysis of the mixtures generated by treatment of branched oligoribonucleotide IV (Table 3) with SVPDE/AP and nuclease Pl/AP. Inosine results from the deamination of adenosine by adenosine deaminase, which is a contaminant in the enzyme preparations. The peak appearing at ca. 1 min is a nonnucleotidic material. HPLC conditions: Section 3.4.6.

merhead ribozymes (46). Here we describe the synthesis using conventionally protected, commercially available ribophosphoramidites (DCL, Section 2.2.) of various oligoribonucleotides for use in protein/RNA recognition and RNA splicing studies (Table 3). All oligoribonucleotides were assembled using the abovementioned protocols. Oligomer IV, a 39 mer, contains a sequence corresponding to positions 15–53 of yeast U4 small nuclear RNA (47). Oligomer V, a branched 18 mer, possesses a base sequence similar to that of RNA lariats generated in the splicing of *S. cerevisiae* precursor mRNA. The vicinal (2'–5'), and (3'–5') phosphodiester bonds located at the branch point of this molecule were introduced in one step by reaction of polymer-bound 6-mer 5'-CAA GUU-(LCAA-CPG, 37 μmol/g) with a 0.03M solution of N-Bz-5'-MMT-A-2',3'-*bis*-([2-cyanoethyl]-N,N-diisopropylphosphoramidite) (48,49). Chain extension of the resulting branched oligomer in the usual 5'-direction afforded the "Y"-like oligomer V. The RNA sequences were deprotected in 29% NH$_4$OH:ethanol (3:1 v/v; oligomers I–III: 55°C, 16 h; oligomers IV and V: rt, 2 d) followed by TBAF (rt, 16 h), as described above. The

Table 3
Synthesis of Oligoribonucleotides

No.	Oligomer, 5'→3'	Length of oligomer	Scale of synthesis, μmol	Average coupling yield, %	Isolated yield,[a] A_{260} U Crude	Isolated yield,[a] A_{260} U Pure
	Linear					
I	UAG UA	5	1.0	94	39	11.9
II	GUG AGC GAG UGU G	13	0.2	97	18	9.6
III	AUA AAG ACA CAC A	13	0.2	96	16	6.5
IV	GAA AUA CGC AUA UCA GUG AGG AUU CGU CCG AGA UUG UGU	39	0.2	97	81	2.2
	Branched					
V	UAC UAA 2'—CAA GUU 3'—CAA GUU	18	1.0	97	61	12

[a]After purification by PAGE and Sephadex G25-F.

chain lengths were then confirmed by PAGE (Fig. 1A,B) and HPLC (Fig. 2). Appropriate bands were cut out, under UV shadow, extracted with water at 37°C overnight, and desalted by Sephadex G-25F or C_{18} SEP PAK chromatography. Isolated yield of crude and purified oligoribonucleotides is reported in Table 3.

As a check on the purity, structure, and base composition of the oligoribonucleotides, a small sample (0.2 A_{260} units) was subjected to enzymatic degradation by either SVPDE/AP, nuclease P1/AP, CSPDE, or RNase T2, and the resulting digests were analyzed by HPLC on a reversed-phase column. Integration of the peaks corresponding to the nucleotide and/or nucleoside products gave the expected ratio in all cases (Fig. 3).

The procedures described in this chapter are completely general. Although improvements will certainly continue to be made, the silylphosphoramidite method of RNA synthesis allows, and is being used for, the synthesis of RNA sequences having a wide range of applications including: ribozymes *(12,13,20–22,25,46,50)*, tRNAs *(7,8)*, branched RNA *(48,49)*, RNA with modified internucleotide linkages (e.g., 2'–5' RNA *[26,51,52]*, RNA-thioates *[26]*), antisense sequences *(23,51)*, and sequences used to elucidate further the structure and function of RNA *(16,17)*. The availability of reagents from commercial sources makes the method available to anyone desiring an RNA sequence.

5. Notes

5.1. Silylation of Nucleosides

The silylation reaction is extremely moisture sensitive. Therefore, all solvents and reagents must be anhydrous. Reaction vessels (glass round-bottom flasks) should be oven-dried, flushed with argon, and equipped with a stopper. Ribonucleosides, imidazole, and finely ground $AgNO_3$ must be dried overnight before use over P_2O_5 *in vacuo*. Allow TBDMS-Cl and TIPS-Cl reagents (stored at −20°C) to warm to room temperature in a desiccator before use.

5.2. Phosphitylation of Nucleosides

Like silylation reactions, the phosphitylation of nucleosides is extremely moisture sensitive. Therefore, all solvents, reagents, and apparatus (syringes, needles, and reaction vessel) must be anhydrous. Ribonucleosides and 4-dimethylamino pyridine must be dried before use for 24 h at room temperature over P_2O_5 *in vacuo*. Phosphitylation reactions are carried out at room temperature in oven-dried (110–120°C), argon-purged, 15-mL Hypo-Vials (Pierce) closed by a rubber septum. The phosphitylating reagent must be allowed to warm to room temperature before transferring it via syringe to the reaction vessel. All other liquids are also injected via syringe. A $CaCl_2$-filled syringe barrel with a needle attached to the rubber septum is used to equalize the pressure in the reaction vessel during injections of reagents.

5.3. Purification and Handling of Oligoribonucleotides

Once all the protecting groups have been removed, it is critical that sterile equipment, reagents, and handling techniques be used in handling the free RNA oligomers (Section 2.). *During all steps of the deprotection, gloves must be worn to prevent enzymatic degradation by ribonucleases present on one's hands.* If these procedures are rigorously adhered to, the accidental degradation of the full deprotected oligomers can be completely avoided.

Acknowledgments

We gratefully acknowledge financial support from the Natural Sciences and Engineering Council of Canada (NSERC). We are indebted to K. Ganeshan and P. A. Giannaris (U. of Toronto) for their assistance in the synthesis and purification of oligoribonucleotides and in the preparation of this manuscript.

References

1. Ogilvie, K. K. and Iwacha, D. (1974) Synthesis of nucleotides from O2,2'-cyclouridine. *Can. J. Chem.* **52,** 1787–1797.
2. Bazant, B. and Chvalowsky, V. (1965) *Chemistry of Organosilicon Compounds,* vol. I, Academic, New York.
3. Ogilvie, K. K., Sadana, K. L., Thompson, E. A., Quilliam, M. A., and Westmore, J. B. (1974) The use of silyl groups in protecting the hydroxyl functions of ribonucleosides. *Tetrahedron Lett.* **15,** 2861–2864.
4. Ogilvie, K. K., Beaucage, S. L., Entwistle, D. W., Thompson, E. A., Quilliam, M. A., and Westmore, J. B. (1976) Alkylsilyl groups in nucleoside and nucleotide chemistry. *J. Carbohydrates Nucleosides Nucleotides* **3,** 197–227.
5. Ogilvie, K. K. (1983) *Proceedings of the 5th International Round Table on Nucleosides, Nucleotides and Their Biological Applications* (Rideout, J. L. Henry, D. W., and Beacham L. M., III, eds.), Academic, London, pp. 209–256.

6. Usman, N., Ogilvie, K. K., Jiang, M.-Y., and Cedergren, R. J. (1987) The automated chemical synthesis of long oligoribonucleotides using 2'-O-silylribonucleoside-3'-O-phosphoramidites on a controlled pore glass support: the synthesis of a 43-nucleotide sequence similar to the 3'-half molecule of an E. coli formyl-methionine tRNA. *J. Am. Chem. Soc.* **109,** 7845–7854.
7. Ogilvie, K. K., Usman, N., Nicoghosian, K., and Cedergren, R. J. (1988) Total chemical synthesis of a 77-nucleotide long RNA sequence having methionine acceptance activity. *Proc. Natl. Acad. Sci. USA* **85,** 5764–5768.
8. Bratty, J., Wu, T., Nicoghosian, K., Ogilvie, K. K., Perrault, J.-P., Keith, G., and Cedergren, R. (1990) Characterization of a chemically synthesized RNA having the sequence of the yeast initiator tRNA(Met) *FEBS Lett.* **269,** 60–64.
9. Gait, M. J., Pritchard, C., and Slim, G. (1991) *Oligonucleotides and Their Analogs: A Practical Approach* (Gait, M. J., ed.), Oxford University Press Oxford, England, pp. 25–48.
10. Ogilvie, K. K. and Entwistle, D. W. (1981) Isomerization of *t*-butyldimethylsilyl protecting group in ribonucleosides. *Carbohydrate Res.* **89,** 203–210.
11. Wu, T. and Ogilvie, K. K. (1990) A study on the alkylsilyl protecting groups in oligonucleotide synthesis. *J. Org. Chem.* **55,** 4717–4724.
12. Slim, G. and Gait, M. J. (1991) Configurationally defined phosphorothioate-containing oligoribonucleotides in the study of the mechanism of cleavage of hammerhead ribozymes. *Nucl. Acids Res.* **19,** 1183–1188.
13. Pieken, W. A., Olsen, D. B., Benseler, F., Aurup, H., and Eckstein, F. (1991) Kinetic characterization of ribonuclease-resistant 2'-modified hammerhead ribozymes. *Science* **252,** 314–317.
14. Stawinski, J., Stromberg, R., Thelin, M., and Westman, E. (1988) Studies on the *t*-butyldimethylsilyl group as 2'-O-protection in oligoribonucleotide synthesis via the H-phosphonate approach. *Nucl. Acids Res.* **16,** 9285–9298.
15. Chaix, C., Molko, D., and Teoule, R. (1989) The use of labile base protecting groups in oligoribonucleotide synthesis. *Tetrahedron Lett.* **30,** 71–74.
16. Talbot S. J., Goodman, S., Bates, S. R. E., Fishwick, C. W. G., and Stockley, P. G. (1990) Use of synthetic oligoribonucleotides to probe RNA-protein interactions in the MS2 translational operator complex. *Nucl. Acids Res.* **18,** 3521–3528.
17. Chou, S.-H., Flynn, P., and Reid, B. (1989) Solid-phase synthesis and high-resolution NMR studies of two synthetic double-helical RNA dodecamers: r(CGCGAAUUGCG) and r(CGCGUAUACGCG). *Biochemistry* **28,** 2422–2435.
18. Milecki, J., Dembek, P., and Antkowiak, W. Z. (1989) On the application of *t*-butyldimethylsilyl group in chemical RNA synthesis. ^{31}P NMR study of 2'-O-*t*-BDMS group migration during nucleoside 3'-OH phosphorylation and phosphitylation reactions. *Nucleosides and Nucleotides* **8,** 463–474.
19. Wang, Y. Y., Lyttle, M. H., and Borer, P. N. (1990) Enzymatic and NMR analysis of oligoribonucleotides synthesized with 2'-*tert*-butyldimethylsilyl protected cyanoethylphosphoramidite monomers. *Nucl. Acids Res.* **18,** 3347–3352.

20. Scaringe, S. A., Francklyn, C., and Usman, N. (1990) Chemical synthesis of biologically active oligoribonucleotides using 2-cyanoethyl-protected ribonucleoside phosphoramidites. *Nucl. Acids Res.* **18**, 5433–5441.
21. Doudna, J. A., Szostak, J. W., Rich, A., and Usman, N. (1990) Chemical synthesis of oligoribonucleotides containing 2-aminopurine: substrates for the investigation of ribozyme function. *J. Org. Chem.* **55**, 5547–5549.
22. Pyle, A. M. and Cech, T. R. (1991) Ribozyme recognition of RNA by tertiary interactions with specific ribose 2'-OH groups. *Nature* **350**, 628–630.
23. Agrawal, S. and Tang, J.-Y. (1990) Efficient synthesis of oligoribonucleotide and its phosphorothioate analogue using H-phosphonate approach. *Tetrahedron Letts.* **31**, 7541–7544.
24. Stawinski J., Stromberg, R., Thelin, M., and Westman, E. (1988) Studies on the *t*-butyldimethylsilyl group as 2'-*O*-protection in oligoribonucleotide synthesis via the H-phosphonate approach. *Nucl. Acids Res.* **16**, 9285–9298.
25. Slim, G., Pritchard, C., Biala, E., Asseline, U., and Gait, M. J. (1991) Synthesis of site-specifically modified oligoribonucleotides for studies of the recognition of TAR RNA by HIV-1 tat protein and studies of hammerhead ribozymes. International Symposium on Synthetic Oligonucleotides: Problems and Frontiers of Practical Application, June 23–30, Moscow, USSR. *Nucl. Acids Res.* Symp. Series **24**, 55–58.
26. Charubala, R., Pfleiderer, W., Suhadolnik, R. J., and SoboL R. W. (1991) Chemical synthesis and biological activity of 2'-5'-phosphorothioate tetramer cores. *Nucleosides and Nucleotides* **10**, 383–388.
27. Wu, T., Ogilvie, K. K., and Pon, R. T. (1989) Prevention of chain cleavage in the chemical synthesis of 2'-silylated oligoribonucleotides. *Nucl. Acids Res.* **17**, 3501–3517.
28. Pon, R. T., Damha, M. J., and Ogilvie, K. K. (1985) Modification of guanine bases by nucleoside phosphoramidite reagents during the solid-phase synthesis of oligonucleotides. *Nucl. Acids Res.* **13**, 6447–6465.
29. Pon, R. T., Usman, N., Damha, M. J., and Ogilvie, K. K. (1986) Prevention of guanine modification and chain cleavage during the solid-phase synthesis of oligonucleotides using phosphoramidite reagents. *Nucl. Acids Res.* **16**, 6453–6470.
30. Hakimelahi, G. H., Proba, Z. A., and Ogilvie, K. K. (1982) New catalysts and procedures for the dimethoxytritylation and selective silylation of ribonucleosides. *Can. J. Chem.* **60**, 1106–1113.
31. Ehrenberg, L. and Fedorcsak, F. (1976) Diethylpyrocarbonate in nucleic acid research. *Prog. Nucl. Acids Res.* **16**, 189–262.
32. McLaughlin, L. W., Piel, N., and Hellman, T. (1985) Preparation of protected ribonucleosides suitable for chemical oligoribonucleotide synthesis. *Synthesis* 322,323.
33. Chu, C. K., Bhadti, V. S., Doboszewskin, B., Gu, Z. P., Kosugi, Y., Pullaiah, K. C., and Roey, P. V. (1989) General synthesis of 2',3'-dideoxynucleosides and 2',3'-didehydro-2',3'-dideoxynucleosides. *J. Org. Chem.* **54**, 2217–2225.
34. Ohtsuka, E., Nakagawa, E., Tanaka, T., Markham, A. F., and Ikehara, M. (1978) Studies on transfer ribonucleic acids and related compounds. XXI. Synthesis and properties of guanine rich fragments from *E. coli* tRNA$_f^{Met}$ 5'-end. *Chem. Pharm. Bull.* **26**, 2998–3006.

35. Ti, G. S., Gaffney, B. L., and Jones, R. A. (1982) Transient protection: efficient one-flask synthesis of protected deoxynucleosides. *J. Am. Chem. Soc.* **104,** 1316–1319.
36. Smith, M., Rammler, D. H., Goldberg, I. H., and Khorana, H. G. (1962) Studies on polynucleotides. XIV. Specific synthesis of the C3'-C5' inter-ribonucleotide linkage. Syntheses of uridylyl-*(*3'-5')-uridine and uridylyl-*(*3'-5')-adenosine. *J. Am. Chem. Soc.* **84,** 430–440.
37. Ogilvie, K. K., Beaucage, S. L., Schifman, A. L., Theriault, N. Y., and Sadana K. (1978) The synthesis of oligoribonucleotides. II. The use of silyl-protecting groups in nucleoside and nucleotide chemistry VII. *Can. J. Chem.* **56,** 2768–2780.
38. Ogilvie, K. K., Schifman, A. L., and Penney, C. (1979) The synthesis of oligoribonucleotides III. The use of sllyl protecting groups in nucleoside and nucleotide chemistry VIII. *Can. J. Chem.* **57,** 2230–2238.
39. Atkinson, T. and Smith, T. (1984) Solid-phase synthesis of oligodeoxyribonucleotides by the phosphite-triester method, in *Oligonucleotide Synthesis: A Practical Approach* (Gait, M. J., ed.). IRL Press, Oxford, pp. 35–81.
40. Pon, R. T., Usman, N., and Ogilvie, K. K. (1988) Derivatization of controlled pore glass beads of solid-phase oligonucleotide synthesis. *BioTechniques* **6,** 768–775.
41. Damha, M. J., Giannaris, P. A., and Zabarylo, S. V. (1990) An improved procedure for derivatization of controlled-pore glass beads for solid-phase oligonucleotide synthesis. *Nucl. Acids Res.* **18,** 3813–3821.
42. *Applied Biosystems User Bulletin,* Issue No. 13, April 1, 1987.
43. Rickwood, D. and Hames, B. D. (ed.) (1987) *Gel Electrophoresis of Nucleic Acids: A Practical Approach.* IRL Press, Oxford.
44. McLaughlin, L. W. and Piel, N. (1991) Chromatographic purification of synthetic oligonucleotides, in *Oligonucleotides and Their Analogues: A Practical Approach* (Gait, M. J., ed.). IRL Press, Oxford, pp. 117–133.
45. Pon, R. T. (1984*),* Ph.D. Thesis, McGill University, Montreal, P. Q., Canada.
46. Perreault, J.-P., Wu, T., Cousineau, B., Ogilvie, K. K., and Cedergren, R. (1990) Mixed deoxyribo- and ribooligonucleotides with catalytic activity. *Nature* **344,** 565–567.
47. Siliciano, P. G., Brow, D. A., Roiha, H., and Guthrie, C. (1987) An essential snRNA from *S. cerevisiae* has properties predicted for U4, including interaction with a U6-like snRNA. *Cell (Cambridge Mass.),* **50,** 585–592.
48. Damha, M. J. and Zabarylo, S. (1989) Automated solid-phase synthesis of branched oligonucleotides. *Tetrahedron Lett.* **30,** 6295–6298.
49. Damha, M. J., Hudson, R. H. E., Ganeshan, K., Zabarylo, S. V., and Guo, Y. (1991) Synthesis and properties of branched and hyperbranched oligonucleotides. International Symposium on Synthetic Oligonucleotides: Problems and Frontiers of Practical Application, June 23–30, Moscow, USSR. *Nucl. Acids Res.* **Symp. Series, 24,** 203,204.
50. Yang, J., Perreault, J. P., Labuda, D., Usman, N., and Cedergren, R. (1990)

Mixed DNA/RNA polymers are cleaved by the hammerhead ribozyme. *Biochemistry* **29,** 11,156–11,160.
51. Damha, M. J., Giannaris, P. A., and Khan, N. (1991) 2'-5'-Linked oligoribonucleotides form stable complexes with complementary RNA and DNA. International Symposium on Synthetic Oligonucleotides: Problems and Frontiers of Practical Application, June 23–30, Moscow, USSR. *Nucl. Acids Res.* **Symp. Series, 24,** 290.
52. Kierzek, R., He, L., and Turner, D. H. (1992) Association of 2'-5' oligoribonucleotides. *Nucl. Acids. Res.,* **20,** 1685–1690.

CHAPTER 6

Synthesis of 2'-*O*-Alkyloligoribonucleotides

Brian S. Sproat

1. Introduction

2'-*O*-Alkyloligoribonucleotides possess properties that render them ideally suited for studying RNA processing and for the affinity chromatography of RNA–protein complexes. Particularly useful compounds are those in which alkyl is methyl *(1–7)*, ethyl *(8)*, allyl *(9)*, or butyl *(10)*, with most studies being carried out with the methyl and allyl analogs.

In particular, 2'-*O*-alkyloligoribonucleotides are highly stable chemically, much more so than either oligodeoxyribonucleotides or oligoribonucleotides, and are moreover totally resistant to degradation by either RNA or DNA specific nucleases *(7, 9)*. A 2'-*O*-methyloligoribonucleotide–RNA duplex has a higher T_m than the corresponding oligodeoxyribonucleotide–RNA one *(2)*, and most importantly, the former duplex is not a substrate for RNase H *(6)*, an enzyme that degrades the RNA component of RNA–DNA heteroduplexes *(11)*. This means that one can perform antisense experiments with crude nuclear extracts containing large amounts of RNase H using these analogs, without any damage to the hybridized RNA.

When the alkyl group is branched, for example, 3,3-dimethylallyl, then the analogs, although very stable to degradation, are not very useful *(9)* and form only weak hybrids with RNA. This is probably because of a combination of steric effects and solvation problems caused by the lipophilic dimethylallyl residue. Analogs with large alkyl groups are also likely to suffer from solubility problems, and it is our experience that analogs in which alkyl is higher than butyl will not be very useful.

We recently described new synthetic routes for the preparation of purine 2'-*O*-methylriboside-3'-*O*-phosphoramidites *(12)* using highly versatile reaction intermediates, and then went on to prepare the 2'-*O*-allylribonucleotide monomers using a facile allylation procedure *(13,14).* Since we have recently described the synthesis and applications of 2'-*O*-methyloligoribonucleotides in considerable detail *(15),* this chapter will be devoted largely to 2'-*O*-allyloligoribonucleotides, with the emphasis placed on monomer synthesis. The advantages of 2'-*O*-allyloligoribonucleotides over their 2'-*O*-methyl counterparts have been previously discussed *(9);* the major advantage as far as the reader is concerned is the reduced nonspecific interactions, an important consideration for many antisense applications.

The application of such 2'-*O*-alkyloligoribonucleotides has proved important for in vitro studies of RNA processing *(16–21),* and for the in vitro *(22)* and in vivo *(23)* localization of the small nuclear ribonucleic acids. Many other interesting applications can be foreseen for these useful analogs.

2. Materials

2.1. Solvents

The following laboratory-grade solvents are required:

1. Chloroform (care, toxic).
2. Diethyl ether.
3. Dimethylsulfoxide.
4. Ethanol.

In addition, the following anhydrous and/or highly purified solvents are required for use in reactions in the absence of moisture and for preparative liquid chromatography:

1. Acetonitrile (care, toxic).
2. Benzene (care, highly toxic).
3. Dichloromethane (care, toxic).
4. 1,2-Dichloroethane (care, toxic).

5. *N,N*-Diisopropylethylamine (care, irritant).
6. *N,N*-Dimethylformamide.
7. Ethyl acetate.
8. Methanol (care, toxic).
9. Petrol, bp 40–60°C.
10. Pyridine (care, highly toxic).
11. Tetrahydrofuran.
12. Toluene.
13. Triethylamine (care, irritant).

2.2. Chemicals

The following high-purity reagents are required:

1. Anhydrous ammonia (great care required).
2. Anhydrous stannous chloride (toxic).
3. Benzoyl chloride.
4. 1,4-Bis(diphenylphosphino)butane.
5. Calf intestinal adenosine deaminase.
6. Chlorotrimethylsilane.
7. 2-Cyanoethoxy *N,N*-diisopropylaminochlorophosphine.
8. Deuterochloroform.
9. 1,4-Diazabicyclo[2.2.2]octane.
10. 2,6-Dichlorophenol.
11. 1,3-Dichloro-1,1,3,3-tetraisopropyldisiloxane.
12. 4,4'-Dimethoxytrityl chloride.
13. 4-Dimethylaminopyridine.
14. *N,N*-Dimethylformamide dimethylacetal.
15. Dimethylsulfoxide-d_6.
16. Dowex 50 W × 4–200 resin.
17. 2-Mesitylenesulfonyl chloride.
18. 2-Nitrobenzaldoxime.
19. Pivaloyl chloride.
20. Pyridine-d_5.
21. Sodium azide (highly toxic).
22. Tetrabutylammonium fluoride (1.1M solution in tetrahydrofuran).
23. 1,1,3,3-Tetramethylguanidine.
24. Thiophenol (toxic, stench).
25. *p*-Toluenesulfonic acid monohydrate.
26. Tris(dibenzylideneacetone)dipalladium(0).
27. Uridine.

2.3. Reagents to Be Synthesized

Prepare allyl ethyl carbonate, required for the allylation reactions, according to the detailed published procedure *(13)*. Prepare the condensing agent 5-(4-nitrophenyl)-1H-tetrazole as previously described *(15)*. Synthesize 3',5'-O-(tetraisopropyldisiloxane-1,3-diyl)-6-(2,6-dichlorophenoxy)purine riboside and 3',5'-O-(tetraisopropyldisiloxane-1,3-diyl)-2-chloro-6-(2,6-dichlorophenoxy)purine riboside, required as starting materials for the synthesis of the 2'-O-allyl A and G monomers, respectively, according to the detailed published procedures *(15)*.

2.4. Chromatographic Materials

Perform ascending mode thin-layer chromatography (TLC) on aluminum-foil-supported silica gel containing a 254-nm fluor. In general, purify nucleoside derivatives by preparative liquid chromatography on silica cartridges (30 × 5 cm, 300 g silica) using a Waters system 500 A liquid chromatograph (Waters Millipore, Milford, USA) and refractive index detection (for isocratic runs), or by short-column chromatography on silica gel 60.

2.5. Materials for Pressure Reactions

Perform reactions requiring anhydrous ammonia at high-pressure and elevated temperature in a Teflon™-lined, stainless-steel laboratory reactor equipped with stirring, heating, and temperature measurement and control. The necessary items can be obtained from Berghof Labortechnik GmbH, Harretstrasse 1, 7412 Eningen u.A., Germany:

1. High-pressure laboratory reactor HR700.
2. Heating mantle RHM790 for step motor.
3. Stirring drive BRS756.
4. Temperature controller BTR841.
5. Temperature display BTA840.

3. Methods

3.1. Synthesis of the 2'-O-Allyluridine Monomer (Fig. 1)

3.1.1. 3',5'-O-(Tetraisopropyldisiloxane-1,3-Diyl)-4-O-(2,6-Dichlorophenyl)Uridine (**1**); Mol Wt 631.70

Dissolve dry uridine (146.5 g, 600 mmol) in anhydrous pyridine (1.5 L), and cool the solution in ice. Add a solution of 1,3-dichloro-1,1,3,3-tetraisopropyldisiloxane (210 g, 669 mmol) in dichloromethane

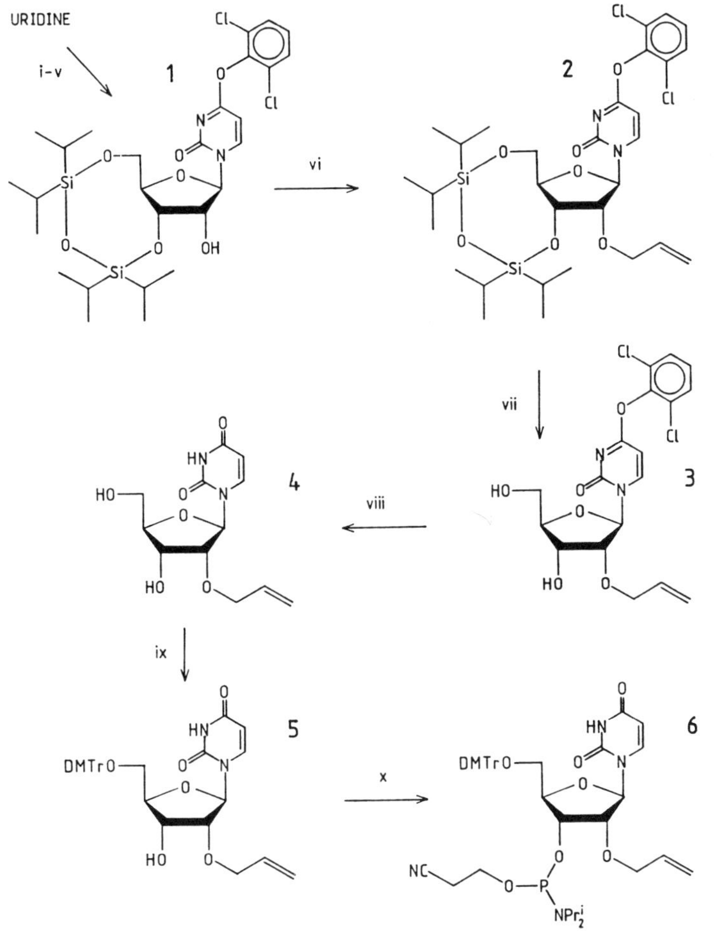

2'-O-Alkyloligoribonucleotides

Fig. 1. Reaction scheme for the synthesis of the 2'-O-allyluridine building block. Reagents: i, 1,3-dichloro-1,1,3,3-tetraisopropyldisiloxane in pyridine; ii, chlorotrimethylsilane, and triethylamine in 1,2-dichloroethane; iii, 2-mesitylenesulfonyl chloride, triethylamine, and 4-dimethylaminopyridine in dichloromethane; iv, 2,6-dichlorophenol, 1,4-diazabicyclo[2.2.2]octane and triethylamine; v, p-toluene sulfonic acid monohydrate in THF/dichloromethane; vi, allyl ethyl carbonate, 1,4-bis(diphenylphosphino)butane and tris(dibenzylideneacetone)dipalladium(0) in tetrahydrofuran; vii, tetrabutylammonium fluoride in tetrahydrofuran; viii, 2-nitrobenzaldoxime and 1,1,3,3-tetramethylguanidine in acetonitrile; ix, 4,4'-dimethoxytrityl chloride and triethylamine in pyridine; x, 2-cyanoethoxy N,N-diisopropylaminochlorophosphine and N,N-diisopropylethylamine in 1,2-dichloroethane.

(100 mL) during 15 min with stirring and exclusion of moisture. On completion of the addition, stir the mixture for 3 h at room temperature, at which time silica gel TLC in ethanol/chloroform (1:9 by vol) should show complete reaction with a spot of R_f 0.58. Quench the reaction with methanol (50 mL), and remove solvent *in vacuo*. Dissolve the residue in dichloromethane (2 L), and wash the solution with 1M aqueous sodium bicarbonate (2 × 2 L). Dry the organic phase (Na_2SO_4), filter it, and evaporate it *in vacuo*. Dry the residue by addition and evaporation of toluene (2 × 250 mL) *in vacuo* to leave a white foam.

Dissolve the 3',5'-*O*-(tetraisopropyldisiloxane-1,3-diyl)uridine from above in anhydrous 1,2-dichloroethane (2 L); add triethylamine (420 mL, 3 mol), and chlorotrimethylsilane (225 mL, 1.8 mol) with stirring and exclusion of moisture. After 30 min, silica gel TLC in petrol/ethyl acetate (2:1 by vol) should show complete reaction with a new spot of R_f 0.39. Pour the reaction mixture into vigorously stirred 1M aqueous sodium bicarbonate (5 L). Separate the organic phase, dry it (Na_2SO_4), filter, and remove solvent *in vacuo*. Dry the foam by two evaporations of dry toluene (250 mL).

Dissolve the 2'-*O*-trimethylsilyl-3',5'-*O*-(tetraisopropyldisiloxane-1,3-diyl)uridine in anhydrous dichloromethane (3 L), and add triethylamine (420 mL, 3 mol), 2-mesitylenesulfonyl chloride (195 g, 900 mmol), and 4-dimethylaminopyridine (18 g, 150 mmol) with stirring and exclusion of moisture. Silica gel TLC of the deep-red solution in petrol/ethyl acetate (2:1 by vol) should show complete reaction after 30 min with a new spot of R_f 0.63. Add 1,4-diazabicyclo[2.2.2]octane (13.5 g, 120 mmol) and 2,6-dichlorophenol (196 g, 1.2 mol), and stir the reaction mixture for 2 h at room temperature. TLC in petrol/ethyl acetate (2:1 by vol) should show complete reaction with a spot of R_f 0.56. Pour the mixture into stirred 1M aqueous sodium bicarbonate (5 L). Separate the organic layer, dry it (Na_2SO_4), filter, and remove solvent *in vacuo* to leave the 2'-*O*-trimethylsilyl ether of compound **1** as a syrup.

Dissolve the above syrup in dichloromethane (3 L). Add a solution of *p*-toluenesulfonic acid monohydrate (285 g, 1.5 mol) in tetrahydrofuran (1 L) with stirring, and after 2.5 min, quench the acid by addition of triethylamine (280 mL, 2 mol). Pour the reaction mixture into vigorously stirred $1M$ aqueous sodium bicarbonate (5 L). Separate the organic layer, dry it (Na_2SO_4), filter, and remove solvent *in vacuo*. Silica gel TLC of the crude product in petrol/ethyl acetate (1:1 by vol) should show a spot of R_f 0.59 owing to 2,6-dichlorophenyl 2-mesitylenesulfonate and a spot of R_f 0.41 owing to the desired product. Purify the crude product in 16 aliquots by preparative liquid chromatography using a total of eight cartridges with petrol/ethyl acetate (2:1 by vol) as eluant. Evaporate solvent to obtain pure product as a solid cream-colored foam (256 g, 67.5%) of R_f 0.23 on TLC in petrol/ethyl acetate (2:1 by vol). ^{13}C-NMR spectrum ($CDCl_3$) δ: 169.82 (C-4), 154.70 (C-2), 144.94 (C-6), 144.77 (phenyl C-1), 128.92 (phenyl C-2 and C-6), 128.71 (phenyl C-3 and C-5), 127.11 (phenyl C-4), 94.05 (C-5), 92.22 (C-1'), 82.01 (C-4'), 74.88 (C-2'), 68.89 (C-3'), 60.35 (C-5'), 17.40–16.85 (isopropyl CH_3s), 13.34, 12.91, 12.83, and 12.48 ppm (isopropyl CHs).

3.1.2. 3',5'-O-(Tetraisopropyldisiloxane-1,3-Diyl)-2'-O-Allyl-4-O-(2,6-Dichlorophenyl)Uridine (2); Mol Wt 671.77

Suspend tris(dibenzylideneacetone)dipalladium(0) (1.83 g, 2 mmol), and 1,4-bis(diphenylphosphino)butane (3.41 g, 8 mmol) in dry tetrahydrofuran (400 mL) under argon. Add a solution of compound **1** (126.3 g, 200 mmol) and allyl ethyl carbonate (52 g, 400 mmol) in dry tetrahydrofuran (600 mL), and reflux the mixture for 30 min. Silica gel TLC in petrol/ethyl acetate (2:1 by vol) should show complete reaction with a new spot of R_f 0.48. When cool, filter the mixture, and remove solvent *in vacuo*. Purify the crude product in four aliquots by preparative liquid chromatography on two silica cartridges using 3% ethyl acetate in dichloromethane as eluant. Remove solvent *in vacuo* from the pure product fractions to leave a solid cream-colored foam (110 g, 81.9%) of R_f 0.51 on TLC in petrol/ethyl acetate (2:1 by vol). ^{13}C-NMR spectrum ($CDCl_3$) δ: 169.61 (C-4), 154.49 (C-2), 144.64 (phenyl C-1), 144.37 (C-6), 134.29 (allyl CH), 128.75 (phenyl C-2 and C-6), 128.51 (phenyl C-3 and C-5), 126.93 (phenyl C-4), 116.85 (allyl =CH_2), 93.50 (C-5), 89.94 (C-1'), 81.64 (C-2'), 80.40 (C-4'), 70.90 (O—CH_2 of allyl), 67.49 (C-3'), 59.34 (C-5'), 17.24, 17.10, 16.79, and 16.63 (isopropyl CH_3s), 13.21, 12.83, 12.70, and 12.30 ppm (isopropyl CHs).

3.1.3. 2'-O-Allyl-4-O-(2,6-Dichlorophenyl)Uridine (3); Mol Wt 429.25

Dissolve compound **2** (55 g, 81.9 mmol) in dry tetrahydrofuran (200 mL), and add 1.1M tetrabutylammonium fluoride in tetrahydrofuran (180 mL) with stirring. Silica gel TLC in ethanol/chloroform (5:95 by vol) should show complete reaction after 5 min with a new spot of R_f 0.22. Quench the reaction mixture with pyridine/methanol/water (500 mL, 3:1:1 by vol), and pour the solution into stirred pyridinium form Dowex 50 W × 4–200 resin (300 g) suspended in pyridine/methanol/water (500 mL, 3:1:1 by vol). Stir the mixture for 20 min, filter off the resin, and wash with the above solvent mixture (3 × 500 mL). Combine the filtrate and washings, and evaporate to dryness *in vacuo*. Dry the residual glass by evaporation of toluene. Purify the crude product in three aliquots by preparative liquid chromatography on two silica cartridges using 6% ethanol in chloroform as eluant. Remove solvent *in vacuo*, and remove residual ethanol and traces of pyridine by addition and evaporation of toluene *in vacuo* at 75°C to obtain 2'-O-allyl-4-O-(2,6-dichlorophenyl)uridine as a solid white foam (29.1 g, 82.9%) of R_f 0.57 on TLC in ethanol/chloroform (1:4 by vol). ^{13}C-NMR spectrum (CDCl$_3$) δ: 169.74 (C-4), 155.28 (C-2), 146.20 (C-6), 144.42 (phenyl C-1), 133.54 (allyl CH), 128.67 (phenyl C-2 and C-6), 128.57 (phenyl C-3 and C-5), 127.13 (phenyl C-4), 117.98 (allyl =CH$_2$), 94.41 (C-5), 89.60 (C-1'), 84.54 (C-4'), 81.01 (C-2'), 71.10 (allyl CH$_2$O), 67.43 (C-3'), and 59.55 ppm (C-5').

3.1.4. 2'-O-Allyluridine (4); Mol Wt 284.27

Dissolve compound **3** (29.1 g, 67.9 mmol) in dry acetonitrile (200 mL), add a solution of 2-nitrobenzaldoxime *(28.2 g, 169.8 mmol)* and 1,1,3,3-tetramethylguanidine (17.6 g, 152.8 mmol) in dry acetonitrile (200 mL), and stir the mixture at room temperature for 18 h. Silica gel TLC in ethanol/chloroform (1:4 by vol) should show complete reaction with the product spot at R_f 0.37. Remove solvent *in vacuo* to leave a red oil. Dissolve the residue in dichloromethane (1 L), and extract the desired product into water (1 L). Wash the aqueous phase with dichloromethane (1 L) followed by diethyl ether (1 L). Stir the yellow aqueous phase with pyridinium form Dowex 50 W × 4–200 resin (250 g) for 5 min, filter off the resin, and wash the opalescent filtrate with dichloromethane (2 × 500 mL) followed by ether (1 L), to remove 2,6-dichlorophenol. The desired product remains solely in the aqueous phase. Remove solvent *in vacuo*, and remove residual water by evaporation of methanol followed by tetrahydrofuran. Crystallize the crude 2'-*O*-allyluridine from methanol, filter it off, wash with ether, and then dry it to obtain a white solid (18.3 g, 94%) of R_f 0.39 on TLC in ethanol/chloroform (1:4 by vol). ^{13}C-NMR spectrum (pyridine-d_5) δ: 164.48 (C-4), 151.72 (C-2), 140.76 (C-6), 135.11 (CH of allyl), 116.96 (allyl =CH_2), 102.17 (C-5), 88.26 (C-1'), 85.72 (C-4'), 82.57 (C-2'), 71.38 (allyl CH_2O), 69.41 (C-3'), and 60.75 ppm (C-5').

3.1.5. 5'-O-Dimethoxytrityl-2'-O-Allyluridine (5); Mol Wt 586.64

Dry 2'-*O*-allyluridine (18.3 g, 64.4 mmol) by evaporation of pyridine (2 × 150 mL) *in vacuo*. Add anhydrous pyridine (300 mL), triethylamine (14 mL, 100 mmol), and 4,4'-dimethoxytrityl chloride (27.1 g, 80 mmol) with stirring and exclusion of moisture. Silica gel TLC in triethylamine/ethanol/chloroform (1:10:89 by vol) should show complete reaction after 1 h with a new spot of R_f 0.57. Quench the reaction by addition of methanol (10 mL), and remove solvent *in vacuo*. Dissolve the residual syrup in ethyl acetate (1 L), and wash the solution with 1*M* aqueous sodium bicarbonate (2 × 1 L). Separate the organic phase, dry it (Na_2SO_4), filter, and remove solvent *in vacuo*. Remove residual pyridine by evaporation of toluene *in vacuo* at room temperature. Purify the crude product in three aliquots by preparative liquid chromatography on two silica cartridges, using ethyl acetate/ dichloromethane (6:1 by vol) containing 1% triethylamine as eluant. Pool the pure product fractions, and remove solvent *in vacuo* at room temperature to obtain the title compound as a solid cream-colored foam (37.2 g, 98.4%) of R_f 0.34 on TLC in ethyl acetate/dichloromethane (6:1 by vol) containing 2% triethylamine.

3.1.6. 5'-O-Dimethoxytrityl-2'-O-Allyluridine-3'-O-(2-Cyanoethyl N,N-Diisopropylphosphoramidite) (6); Mol Wt 786.86

Dry compound **5** (11.73 g, 20 mmol) by evaporation of acetonitrile (100 mL) *in vacuo*, dissolve the residual foam in dry 1,2-dichloroethane (100 mL) containing *N,N*-diisopropylethylamine (7 mL, 40 mmol), and cool the solution in ice under argon. Add 2-cyanoethoxy N,N-diisopropylaminochlorophosphine (6.66 mL, 30 mmol) dropwise with stirring during 2 min. Keep the mixture 5 min at 0°C, and then stir for 1 h at room temperature. TLC in triethylamine/dichloromethane (5:95 by vol) should show complete reaction with a product spot of R_f 0.45. Add dichloromethane (100 mL), and wash the solution with $1M$ aqueous sodium bicarbonate (250 mL) followed by saturated brine (250 mL). Dry the organic layer (Na_2SO_4), filter, and remove solvent *in vacuo*. Purify the crude product by preparative liquid chromatography on two silica cartridges using dichloromethane/petrol (5:2 by vol) containing 7% triethylamine as eluant. Monitor the product fractions by ^{31}P NMR spectroscopy to check for H-phosphonate impurities. Pool the pure phosphonate-free product fractions (the front of the peak), and remove solvent *in vacuo* at room temperature. Lyophilize from benzene to afford the title compound as a solid white foam (10.4 g, 66%) of R_f 0.42 and 0.39 on TLC in ethyl acetate/dichloromethane (4:1 by vol) containing 5% triethylamine. ^{31}P NMR spectrum (CH_2Cl_2, concentric external D_2O lock) δ: + 146.84 and 146.33 ppm.

3.2. Synthesis of the 2'-O-Allylcytidine Monomer (Fig. 2)

3.2.1. 3',5'-O-(Tetraisopropyldisiloxane-1,3-Diyl)-N^4-Benzoyl-2'-O-Allylcytidine (7); Mol Wt 629.91

Treat compound **2** (55 g, 81.9 mmol) in dry tetrahydrofuran (250 mL) with dry ammonia (100 g) in a Teflon™-lined, stainless-steel bomb at 70°C for 72 h. Silica gel TLC in ethanol/chloroform (1:9 by vol) should show complete reaction with a new spot at R_f 0.32. Open the cooled bomb carefully in a well-ventilated fume cupboard. Remove solvent *in vacuo,* and dry the residue by evaporation of pyridine (2 × 250 mL). Dissolve the residue in dry pyridine (500 mL), and cool the solution in an ice bath. Add benzoyl chloride (38 mL, 328 mmol) during 5 min with stirring and exclusion of moisture. Continue stirring for 1.5 h at room temperature at which time TLC should show complete reaction (spot R_f 0.71 in 10% ethanol/chloroform). Cool the mixture in ice, and quench the reaction by addition of water (20 mL) followed after 5 min by 25% ammonia solution (80 mL). Stir for an additional 20 min at room temperature, and then remove solvent *in vacuo*. Dissolve the residual gum in ethyl acetate (1 L), and wash the

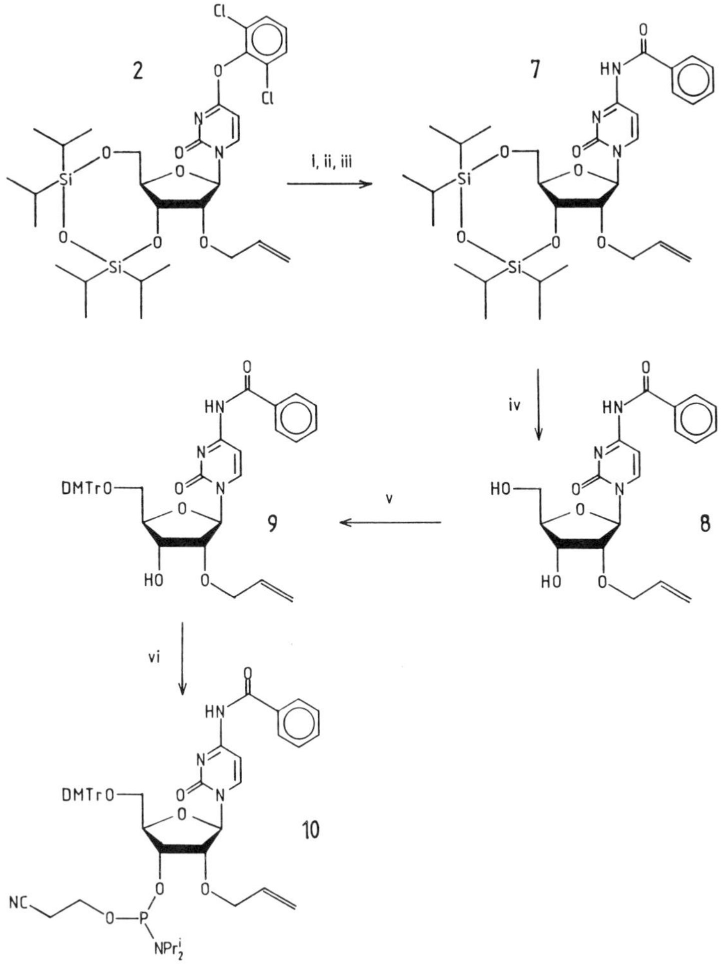

Fig. 2. Reaction scheme for the synthesis of the 2'-O-allylcytidine building block. Reagents: i, anhydrous ammonia in tetrahydrofuran; ii, benzoyl chloride in pyridine; iii, dilute aqueous ammonia/pyridine; iv, tetrabutylammonium fluoride in tetrahydrofuran; v, 4,4'-dimethoxytrityl chloride and triethylamine in pyridine; vi, 2-cyanoethoxy N,N-diisopropylaminochlorophosphine and N,N-diisopropylethylamine in 1,2-dichloroethane.

solution with 1M sodium bicarbonate (2 × 1 L). Dry the organic phase (Na_2SO_4), filter, and remove solvent *in vacuo*. Purify the crude product in three aliquots by preparative liquid chromatography on two silica cartridges using petrol/ethyl acetate (1:1 by vol) as eluant. Remove solvent *in vacuo* to obtain the title compound as a solid white foam (46.6 g, 90.3%) of R_f 0.26 on TLC in petrol/ethyl acetate (1:1 by vol). ^{13}C-NMR spectrum ($CDCl_3$) δ: 166.84 (benzoyl C=O), 162.65 (C-4), 154.47 (C-2), 144.13 (C-6), 134.31 (allyl CH), 133.08 (phenyl C-1), 132.81 (phenyl C-4), 128.65 (phenyl C-3 and C-5), 127.66 (phenyl C-2 and C-6), 117.03 (allyl =CH_2), 96.22 (C-5), 89.80 (C-1'), 81.75 (C-2'), 80.63 (C-4'), 71.01 (allyl CH_2O), 67.55 (C-3'), 59.34 (C-5'), 17.36–16.69 (isopropyl CH_3s), 13.27, 13.01, 12.75, and 12.36 ppm (isopropyl CHs).

3.2.2. N^4-Benzoyl-2'-O-Allylcytidine (8); Mol Wt 387.40

Desilylate compound **7** (46.6 g, 74 mmol), and work up the reaction according to the procedure for preparing compound **3** above. Purify the crude product in three aliquots by preparative liquid chromatography on two silica cartridges using methanol/dichloromethane (5:95 by vol) as eluant. Evaporate the solvent to leave a solid. Wash this with pentane, and dry it to give N^4-benzoyl-2'-*O*-allylcytidine as an off-white powder (21.7 g, 75.7%) of R_f 0.55 on TLC in ethanol/chloroform (1:4 by vol). ^{13}C-NMR spectrum (DMSO-d_6) δ: 167.11 (benzoyl C=O), 162.94 (C-4), 154.34 (C-2), 144.60 (C-6), 134.42 (allyl CH), 133.02 (phenyl C-1), 132.20 (phenyl C-4), 128.12 (phenyl C-3 and C-5), 127.98 (phenyl C-2 and C-6), 116.41 (allyl =CH_2), 96.07 (C-5), 88.44 (C-1'), 83.88 (C-4'), 81.49 (C-2'), 70.24 (allyl CH_2O), 67.03 (C-3'), and 58.96 ppm (C-5').

3.2.3. 5'-O-Dimethoxytrityl-N^4-Benzoyl-2'-O-Allylcytidine (9); Mol Wt 689.77

Dimethoxytritylate compound **8** (21.5 g, 55.4 mmol) according to the procedure for preparing compound **5** above. Purify the crude product in three aliquots by preparative liquid chromatography on two silica cartridges using ethyl acetate/dichloromethane (6:1 by vol) containing 1% triethylamine as eluant. Product R_f 0.34 on TLC in this solvent mixture. You will obtain the title compound as a solid, generally pale yellow foam (37.3 g, 97.6%) of R_f 0.60 on TLC in triethylamine/ethanol/chloroform (1:10:89 by vol).

3.2.4. 5'-O-Dimethoxytrityl-N^4-Benzoyl-2'-O-Allylcytidine-3'-O-(2-Cyanoethyl N,N-Diisopropylphosphoramidite) (10); Mol Wt 889.99

Phosphitylate compound **9** (10.34 g, 15 mmol) according to the procedure for preparing compound **6** above. Purify the crude product by preparative liquid chromatography on a single silica cartridge using petrol/dichloromethane/triethylamine (5:4:1 by vol) as eluant. Remove solvent *in vacuo* and lyophilize from benzene to obtain the title compound as a solid white foam (12.3 g, 92.1%) of R_f 0.48 and 0.43 on TLC in ethyl acetate/dichloromethane (4:1 by vol) containing 5% triethylamine. ^{31}P-NMR spectrum (CH$_2$Cl$_2$, concentric external D$_2$O lock) δ: +146.86 and 146.04 ppm.

3.3. Synthesis of the 2'-O-Allyladenosine Monomer (Fig. 3)

3.3.1. 3',5'-O-(Tetraisopropyldisiloxane-1,3-Diyl)-2'-O-Allyl-6-(2,6-Dichlorophenoxy)Purine Riboside (12); Mol Wt 695.80

Allylate 3',5'-*O*-(tetraisopropyldisiloxane-1,3-diyl)-6-(2,6-dichlorophenoxy)purine riboside, **11**, (131.1 g, 200 mmol) as described above for preparing compound **2**. Purify the crude product in four aliquots by preparative liquid chromatography on two silica cartridges using 2% ethyl acetate in dichloromethane as eluant. Remove solvent *in vacuo* to obtain the pure title compound as a solid cream-colored foam (110 g, 79%) of R_f 0.53 on TLC in petrol/ethyl acetate (2:1 by vol). ^{13}C-NMR spectrum (CDCl$_3$) δ: 158.29 (C-6), 152.38 (C-4), 151.82 (C-2), 145.45 (phenyl C-1), 141.92 (C-8), 134.26 (allyl CH), 129.50 (phenyl C-2 and C-6), 128.80 (phenyl C-3 and C-5), 127.08 (phenyl C-4), 121.84 (C-5), 117.44 (allyl =CH$_2$), 88.96 (C-1'), 81.71 and 81.61 (C-2' and C-4'), 71.97 (allyl CH$_2$O), 69.50 (C-3'), 59.88 (C-5'), 17.48–16.93 (isopropyl CH$_3$s), 13.48, 12.97, and 12.72 ppm (isopropyl CHs).

3.3.2. 3',5'-O-(Tetraisopropyldisiloxane-1,3-Diyl)-N^6-Pivaloyl-2'-O-Allyladenosine (13); Mol Wt 633.94

Treat compound **12** (55 g, 79 mmol) in dry tetrahydrofuran (250 mL) with dry ammonia (100 g) in a Teflon™-lined, stainless-steel bomb for 7 d at 95°C. Silica gel TLC in ethanol/chloroform (1:9 by vol) should show complete reaction with a new spot of R_f 0.43. Remove

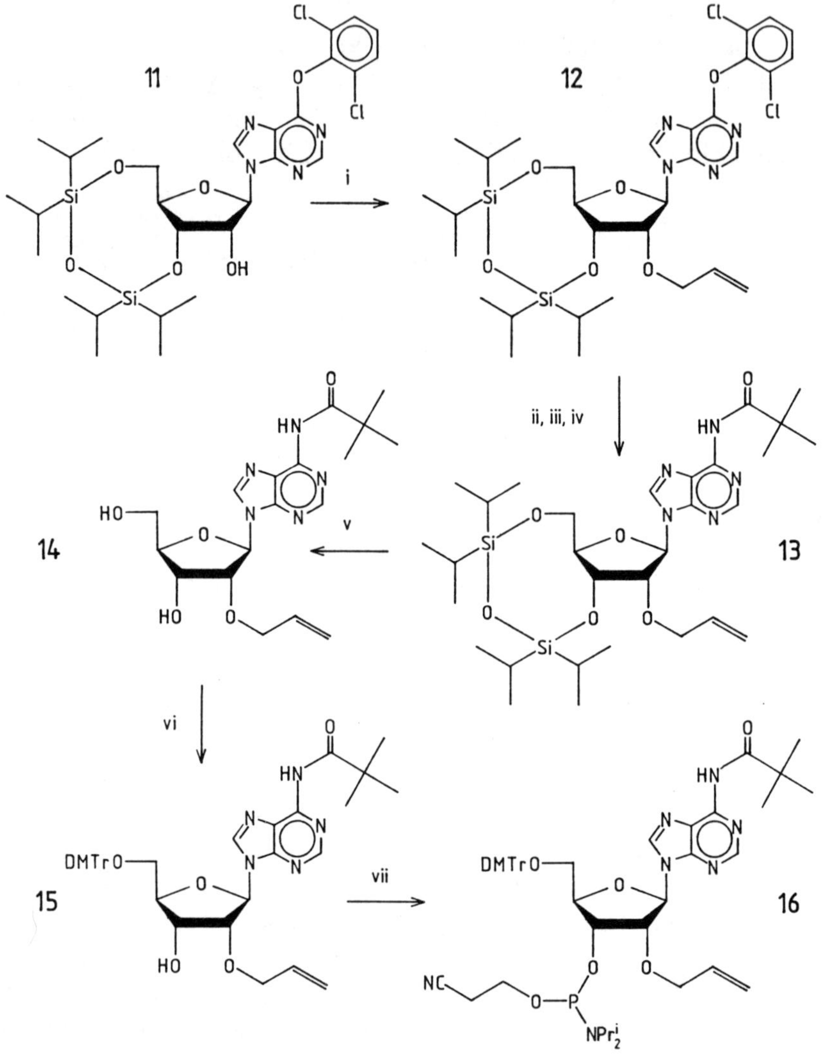

Fig. 3. Reaction scheme for the synthesis of the 2'-O-allyladenosine building block. Reagents: i, allyl ethyl carbonate, 1,4-bis(diphenylphosphino)butane and tris(dibenzylideneacetone)dipalladium(0) in tetrahydrofuran; ii, anhydrous ammonia/tetrahydrofuran; iii, pivaloyl chloride in pyridine; iv, dilute aqueous ammonia/pyridine; v, tetrabutylammonium fluoride in tetrahydrofuran, vi, 4,4'-dimethoxytrityl chloride and triethylamine in pyridine; vii, 2-cyanoethoxy N,N-diisopropylaminochlorophosphine and N,N-diisopropylethylamine in 1,2-dichloroethane.

solvent *in vacuo*, and dry the residue by evaporation of pyridine (500 mL). Dissolve the residue in anhydrous pyridine (500 mL), cool in an ice bath, and add pivaloyl chloride (60 mL, 480 mmol). Stir the mixture overnight at room temperature with exclusion of moisture. Silica gel TLC should show complete reaction with a spot of R_f 0.7 in ethanol/chloroform (1:9 by vol). Quench, work up, and purify as described for preparing compound **7** above. The title compound will be obtained as a solid, generally cream-colored foam (32.46 g, 64.8%) of R_f 0.29 on TLC in petrol/ethyl acetate (1:1 by vol), and R_f 0.65 in ethanol/chloroform (1:9 by vol). ^{13}C-NMR spectrum (CDCl$_3$) δ: 175.21 (pivaloyl C=O), 151.97 (C-2), 150.28 (C-6), 149.16 (C-4), 140.86 (C-8), 133.75 (allyl CH), 123.24 (C-5), 116.97 (allyl =CH$_2$), 88.42 (C-1'), 81.08 (C-2'), 80.41 (C-4'), 71.40 (allyl CH$_2$O), 69.09 (C-3'), 59.47 (C-5'), 39.97 (quaternary C of pivaloyl), 26.94 (pivaloyl CH$_3$s), 16.99–16.45 (isopropyl CH$_3$s), 12.93, 12.48, 12.39, and 12.21 ppm (isopropyl CHs).

3.3.3. N^6-Pivaloyl-2'-O-Allyladenosine (**14**); Mol Wt 391.43

Desilylate compound **13** (27.7 g, 43.7 mmol), and then work up the reaction according to the procedure for synthesizing compound **3**. Purify the crude product in two aliquots by preparative liquid chromatography on two silica cartridges using 6% ethanol in chloroform as eluant to obtain N^6-pivaloyl-2'-*O*-allyladenosine as a white solid (15.85 g, 92.7%) of R_f 0.52 on TLC in ethanol/chloroform (1:4 by vol). ^{13}C-NMR spectrum (DMSO-d$_6$) δ: 176.25 (pivaloyl C=O), 151.82 (C-6), 151.51 (C-2), 150.49 (C-4), 142.76 (C-8), 134.59 (allyl CH), 125.84 (C-5), 116.81 (allyl =CH$_2$), 86.18 (C-1'), 85.95 (C-4'), 80.55 (C-2'), 70.36 (allyl CH$_2$O), 68.89 (C-3'), 61.11 (C-5'), quaternary C of pivaloyl under DMSO signal, and 26.88 ppm (pivaloyl CH$_3$s).

3.3.4. 5'-O-Dimethoxytrityl-N^6-Pivaloyl-2'-O-Allyladenosine (**15**); Mol Wt 693.80

Dimethoxytritylate compound **14** (15.5 g, 39.6 mmol) according to the procedure for preparing compound **5** above. Purify the crude product in two aliquots by preparative liquid chromatography on two silica cartridges using ethyl acetate/dichloromethane (1:1 by vol) containing 2% triethylamine as eluant. You will obtain the title compound as a solid, generally pale cream-colored foam (26.3 g, 95.6%) of R_f 0.59 on TLC in triethylamine/ethanol/chloroform (1:10:89 by vol), and R_f 0.25 in triethylamine/ethyl acetate/dichloromethane (2:49:49 by vol).

3.3.5. 5'-O-Dimethoxytrityl-N^6-Pivaloyl-2'-O-Allyladenosine-3'-O-(2-Cyanoethyl N,N-Diisopropylphosphoramidite) (16); Mol Wt 894.02

Phosphitylate compound **15** (10.41 g, 15 mmol) according to the preparation of compound **6** above. Purify the crude product by preparative liquid chromatography on a single silica cartridge using petrol/dichloromethane (3:2 by vol) containing 10% triethylamine as eluant. Remove solvent *in vacuo* and lyophilize from benzene to obtain the title compound as a solid white foam (12.2 g, 91%) of R_f 0.53 and 0.49 on TLC in ethyl acetate/dichloromethane (4:1 by vol) containing 5% triethylamine. ^{31}P-NMR spectrum (CH_2Cl_2, concentric external D_2O lock) δ: +147.01 and 146.73 ppm.

3.4. Synthesis of the 2'-O-Allylguanosine Monomer (Fig. 4)

3.4.1. 3',5'-O-(Tetraisopropyldisiloxane-1,3-Diyl)-2'-O-Allyl-2-Chloro-6-(2,6-Dichlorophenoxy) Purine Riboside (18); Mol Wt 730.24

Allylate 3',5'-*O*-(tetraisopropyldisiloxane-1,3-diyl)-2-chloro-6-(2,6-dichlorophenoxy)purine riboside, compound **17** (14.49 g, 21 mmol) according to the synthesis of compound **2** above. Purify the crude product by preparative liquid chromatography on a single silica cartridge using petrol/ethyl acetate (9:2 by vol) as eluant, to obtain the title compound as a pale yellow foam (13.7 g, 89.4%) of R_f 0.31 on TLC in petrol/ethyl acetate (4:1 by vol). ^{13}C-NMR spectrum (CDCl$_3$) δ: 158.20 (C-6), 153.19 (C-2), 152.64 (C-4), 144.82 (phenyl C-1), 142.26 (C-8), 133.93 (allyl CH), 128.95 (phenyl C-2 and C-6), 128.65 (phenyl C-3 and C-5), 127.19 (phenyl C-4), 120.63 (C-5), 117.66 (allyl =CH$_2$), 88.77 (C-1'), 81.50 (C-2'), 80.48 (C-4'), 71.71 (allyl CH$_2$O), 69.17 (C-3'), 59.68 (C-5'), 17.32–16.76 (isopropyl CH$_3$s), 13.28, 12.81, and 12.47 ppm (isopropyl CHs).

3.4.2. 3',5'-O-(Tetraisopropyldisiloxane-1,3-Diyl)-2'-O-Allyl-2,6-Diazidopurine Riboside (19); Mol Wt 616.84

Stir compound **18** (13.45 g, 18.4 mmol) and sodium azide (3 g, 46 mmol) in dry *N,N*-dimethylformamide (400 mL) under argon for 6 h at 55°C. Allow the solution to cool overnight, and remove solvent *in vacuo*. Dissolve the residual dark oil in dichloromethane (500 mL), wash the solution with water (500 mL), dry it (Na$_2$SO$_4$), filter, and remove solvent *in vacuo*. Purify the residue by column chromatography on silica gel (220 g) using petrol/ethyl acetate (5:1 by vol) as eluant. Evaporate pure product fractions to obtain a pale yellow oil (8.95 g, 79%) of R_f 0.16 on TLC in petrol/ethyl acetate (4:1 by vol). ^{13}C-NMR spectrum (CDCl$_3$) δ: 155.40 (C-6), 153.04 (C-2), 151.94 (C-4), 141.07 (C-8), 133.73 (allyl CH), 121.47 (C-5), 116.78 (allyl =CH$_2$), 88.09 (C-1'), 81.18 (C-2'), 80.52 (C-4'), 71.14 (allyl CH$_2$O), 68.53 (C-3'), 59.33 (C-5'), 16.96–16.36 (isopropyl CH$_3$s), 12.95, 12.51, and 12.18 ppm (isopropyl CHs).

3.4.3. 3',5'-O-(Tetraisopropyldisiloxane-1,3-Diyl)-2'-O-Allyl-2,6-Diaminopurine Riboside (20); Mol Wt 564.84

Dissolve anhydrous stannous chloride (8.25 g, 43.5 mmol) with stirring in dry acetonitrile (290 mL), and then add thiophenol (17.9 mL, 174 mmol) and triethylamine (18.15 mL, 130.5 mmol). Dissolve compound **19** (8.95 g, 14.5 mmol) in dry acetonitrile (50 mL), and add with stirring to the above yellow solution. TLC in 5% ethanol/chloroform should show complete reaction after 15 min. Remove solvent *in vacuo*, and dissolve the yellow residue in dichloromethane (300 mL). Wash the solution with 1*M* aqueous sodium hydroxide (300 mL), and back-wash the aqueous phase with dichloromethane (2 × 200 mL). Combine the colorless organic layers, dry (Na$_2$SO$_4$), filter, and remove solvent *in vacuo*. TLC in 5% ethanol/dichloromethane should show a single spot of R_f 0.16. Purify the crude product by column chromatography on silica gel (200 g), eluting with a gradient of ethanol from 0 to 5% in dichloromethane. Evaporate solvent to leave the title compound as a solid white foam (7.66 g, 93%). ^{13}C-NMR spectrum (CDCl$_3$) δ: 159.87 (C-6), 155.80 (C-2), 150.30 (C-4), 134.80 (C-8), 133.96 (allyl CH), 116.90 (allyl =CH$_2$), 113.83 (C-5), 87.62 (C-1'), 80.83 (C-2'), 80.66 (C-4'), 71.27 (allyl CH$_2$O), 68.78 (C-3'), 59.57 (C-5'), 17.04–16.46 (isopropyl CH$_3$s), 13.04, 12.55, and 12.22 ppm (isopropyl CHs).

3.4.4. 2'-O-Allyl-2,6-Diaminopurine Riboside (21); Mol Wt 322.33

Desilylate compound **20** (4.55 g, 8.05 mmol) according to the procedure for synthesizing compound **3** above. Purify the crude product by column chromatography on silica gel (100 g) eluting with a gradient of ethanol from 5 to 15% in dichloromethane, to obtain pure title compound as a solid white foam (2.5 g, 96%) of R_f 0.45 on TLC in ethanol/dichloromethane (1:4 by vol). ^{13}C-NMR spectrum (DMSO-d$_6$) δ: 160.20 (C-6), 156.43 (C-2), 151.39 (C-4), 136.57 (C-8), 134.73 (allyl CH), 116.92 (allyl =CH$_2$), 113.67 (C-5), 86.50 (C-1'), 85.81 (C-4'), 80.47 (C-2'), 70.43 (allyl CH$_2$O), 69.46 (C-3'), and 61.86 ppm (C-5').

3.4.5. N^2-Dimethylaminomethylidene-2'-O-Allylguanosine (22); Mol Wt 378.39

Dissolve compound **21** (2.5 g, 7.75 mmol) in a mixture of dimethylsulfoxide (66 mL) and 0.1M aqueous sodium phosphate (160 mL, pH 7.5). Add crude adenosine deaminase (150 mg), and shake the solution gently at 37°C for 72 h. Silica gel TLC in ethanol/dichloromethane (1:4 by vol) should show complete reaction with a new spot at R_f 0.25 because of 2'-O-allylguanosine. Remove solvent *in vacuo*, and dry the residue by several evaporations of methanol. Suspend the residue in methanol (100 mL), add *N,N*-dimethylformamide dimethyl acetal (5 mL, 38 mmol), and stir the mixture overnight at room temperature. TLC in ethanol/dichloromethane (1:4 by vol) should show complete reaction with a new spot of R_f 0.30. Evaporate the reaction mixture to dryness *in vacuo*, and dry further by evaporation of toluene. Purify by column chromatography on silica gel (100 g) eluting with a gradient of ethanol from 5 to 20% in dichloromethane to afford the title compound as a solid white foam (2.90 g, 99%). ^{13}C-NMR spectrum (CDCl$_3$) δ: 157.96 (C-6 and amidine CH), 156.79 (C-2), 149.11 (C-4), 139.94 (C-8), 133.38 (allyl CH), 120.45 (C-5), 117.59 (allyl =CH$_2$), 87.46 (C-1'), 86.08 (C-4'), 80.53 (C-2'), 71.28 (allyl CH$_2$O), 67.77 (C-3'), 61.67 (C-5'), 41.03, and 34.72 ppm (N-CH$_3$s).

3.4.6. 5'-O-Dimethoxytrityl-N^2-Dimethylaminomethylidene-2'-O-Allylguanosine (23); Mol Wt 680.76

Dimethoxytritylate compound **22** (2.90 g, 7.6 mmol) according to the procedure for preparing compound **5**. Purify the crude product by column chromatography on silica gel (100 g). Elute initially with dichloromethane/ethyl acetate (1:1 by vol) containing 0.5% triethylamine, and then with a gradient of ethanol from 0 to 8% in ethyl acetate/triethylamine (199:1 by vol). Remove solvent from pure product fractions *in vacuo* to obtain the title compound as a solid white foam (5.14 g, 99%) of R_f 0.15 on TLC in triethylamine/ethanol/dichloromethane (1:5:94 by vol).

3.4.7. 5'-O-Dimethoxytrityl-N^2-Dimethylaminomethylidene-2'-O-Allylguanosine-3'-O-(2-Cyanoethyl N,N-Diisopropylphosphoramidite) (24); Mol Wt 880.99

Phosphitylate compound **23** (4.94 g, 7.25 mmol) according to the procedure for preparing compound **6**. Work up as described; however, quench the reaction with ethanol (2 mL). Purify the crude product by column chromatography on silica gel (120 g) eluting with a gradient of ethanol from 0 to 2% in dichloromethane/triethylamine (98:2 by vol) to obtain the title compound as a solid white foam (5.95 g, 93%) after lyophilizing from benzene. Product R_f 0.66 on TLC in dichloromethane/ethanol/triethylamine (98:10:2 by vol). ^{31}P-NMR spectrum (CH_2Cl_2, external concentric D_2O lock) δ: +147.19 and 147.02 ppm.

3.5. Synthesis of Supports

Convert the 5'-O-dimethoxytrityl base-protected 2'-O-allylribonucleosides (generally 2 mmol) into their 3'-O-succinates, and then turn these into the reactive 4-nitrophenyl esters. Subsequently react these with aminopropyl controlled pore glass (500 Å pore diameter) to give the functionalized CPG supports for solid-phase synthesis. Use the identical procedure as described in detail for the 2'-O-methyl compounds *(15)*.

3.6. 2'-O-Allyloligoribonucleotide Synthesis, Deprotection, and Purification

3.6.1. Synthesis

1. Weigh out the requisite amounts of phosphoramidite monomers viz. compounds **6, 10, 16,** and **24** and the condensing agent, 5-(4-nitrophenyl)-1H-tetrazole for the syntheses to be performed in bottles appropriate for the solid-phase synthesizer. Dry the materials (open bottles plus contents) thoroughly, preferably overnight in high vacuum over separate containers of potassium hydroxide and phosphorus pentoxide. This ensures removal of residual triethylamine, moisture, and solvents, such as alcohol, that will impair coupling efficiency.
2. Release the vacuum carefully with dry argon, dissolve each monomer in turn, and finally the coupling agent in the requisite volume of anhydrous acetonitrile (<30 ppm water) to give a 0.1*M* solution. Attach the bottles immediately to the synthesizer, keeping atmospheric contact to a minimum. The condensing agent may need gentle warming with a hair drier to achieve total solution.
3. Oligomers will be synthesized in the normal 3' to 5' direction, so place the required amount (depends on scale) of the appropriate functionalized

controlled pore glass support (this corresponds to the 3'-base) into an empty cartridge or column appropriate for your synthesizer. Attach the column or cartridge to the machine, ready to start.

4. To synthesize 2'-O-allyloligoribonucleotides is very simple; choose a standard β-cyanoethylphosphoramidite DNA cycle, and just alter the coupling wait time to 8 min. All the other steps can be the same. Of course, the coupling is performed with 5-(4-nitrophenyl)-1H-tetrazole and not with tetrazole. Incorporate single or multiple biotinylations at the 5' and/or 3'-ends of the oligomer, during the solid-phase synthesis using the previously described biotin phosphoramidite *(15,24)*. Choose trityl on or off depending on whether you prefer to use reversed-phase HPLC or PAGE to purify the oligomer.

3.6.2. Deprotection

1. Cleave the partially protected oligomer from the carrier on the machine with 25% ammonia solution as for a DNA oligomer.
2. Transfer the solution to a sealed vial and keep at 60°C for 8–10 h (generally overnight) to remove the heterocyclic protection.
3. When cool, open the vial, and evaporate the solution to dryness *in vacuo* at room temperature.

3.6.3. Purification

1. Purify trityl on 2'-O-allyloligoribonucleotides by reversed-phase HPLC on a μ-Bondapak C_{18} column or equivalent using a gradient of acetonitrile in $0.1M$ aqueous triethylammonium acetate, pH 7.0. The product peak will elute at a concentration of acetonitrile of about 35–40%. Lyophilize the product fraction on a Speedvac, and remove the dimethoxytrityl group in the usual way with 80% acetic acid. In contrast to oligodeoxyribonucleotides, depurination during synthesis of 2'-O-allyloligoribonucleotides does not occur, so that the DMTr-oligomer peak is homogeneous even for long syntheses. If you rerun your fully deprotected 2'-O-allyloligoribonucleotide on reversed-phase HPLC, it will elute at an acetonitrile concentration in the range of 25–30% because of the lipophilic allyl groups.
2. Purify trityl off oligomers by denaturing polyacrylamide gel electrophoresis as for an oligodeoxyribonucleotide.

4. Discussion

In order to obtain high-yield specific 2'-O-alkylation of ribonucleosides, use the 3',5'-O-(tetraisopropyldisiloxane-1,3-diyl)-protecting group *(25)* in combination with appropriate protection of the heterocycle to avoid *N*-alkylation *(12)*. The highly versatile intermediates **1**, **11**, and **17** are readily 2'-*O*-alkylated and then, depending on the reaction conditions employed, can be turned into a large number of 2'-*O*-alkylribonucleoside derivatives. Nyilas and Chattopadhyaya *(26)* have shown that 3',5'-*O*-(tetraisopropyldisiloxane-1,3-diyl)-2'-*O*-methyl-4-*O*-(2-nitrophenyl)uridine reacts with oximate reagent *(27,28)* to give the siloxane-protected 2'-*O*-methyluridine in high yield, and with ammonia to give the siloxane-protected 2'-*O*-methylcytidine. This forms the basis of intermediate **1**, which can be converted into uridine or cytidine, and of intermediate **11**, which can be converted into adenosine or inosine by the same principle.

Figure 1 illustrates the synthesis of the 2'-*O*-allyluridine monomer, **6**. Compound **1** is synthesized from uridine in five steps, viz. siloxane-protection of the 3'- and 5'-hydroxyl groups followed by trimethylsilylation of the 2'-hydroxyl group. The lactam moiety of the heterocycle is then enolized and turned into the 4-(2-mesitylene sulfonate). Subsequent nucleophilic substitution by 2,6-dichlorophenoxide followed by acidic cleavage of the 2'-*O*-TMS ether and preparative LC purification yields the desired compound **1**. The recently described alcohol allylation procedure using allyl ethyl carbonate and a Pd(0) catalyst *(29)* proves highly successful for the rapid 2'-OH allylation of compound **1**. Conversion of **1** to **2** is complete within 30 min in refluxing THF, and the product is readily isolated in over 80% yield. Desilylation of compound **2** gives 2'-*O*-allyl-4-*O*-(2,6-dichlorophenyl) uridine, **3**, which is subsequently treated with the tetramethylguanidinium salt of 2-nitrobenzaldoxime in dry acetonitrile to give 2'-*O*-allyluridine, **4**, as a white solid in 94% yield after workup. Subsequent dimethoxytritylation and phosphitylation *(30)* afford the desired 2'-*O*-allyluridine monomer **6**.

Figure 2 illustrates the synthesis of the 2'-*O*-allylcytidine monomer, **10**. Treatment of intermediate **2** with ammonia/THF in a bomb at 70°C for 72 h results in nucleophilic displacement of 2,6-dichlorophenoxide and the formation of 3',5'-*O*-(tetraisopropyldisiloxane-1,3-diyl)-2'-*O*-allylcytidine. This material is not isolated, but directly *N*-benzoylated to give compound **7** in 90% yield. Subsequent desilylation, dimethoxytritylation, and phosphitylation lead to the desired 2'-*O*-allylcytidine monomer, **10**.

Figure 3 illustrates the synthesis of the 2'-*O*-allyladenosine monomer, **16**. The starting material, **11**, is prepared from 6-chloropurine riboside in two steps, as previously described *(12,15)*. Allylation as described above affords compound **12** in 79% yield. Displacement of the 2,6-dichlorophenoxide by ammonia is rather slow and takes 7 d at 95°C in a bomb; however, the conversion of compound **12** to the intermediate 3',5'-*O*-(tetraisopropyldisiloxane-1,3-diyl)-2'-*O*-allyladenosine is very clean. Pivaloylation affords compound **13** in moderate yield. Subsequent desilylation, dimethoxytritylation, and phospitylation each go in over 90% yield to give the desired 2'-*O*-allyladenosine monomer, **16**.

Figure 4 illustrates the synthesis of the 2'-*O*-allylguanosine monomer, **24**. The starting material **17** is prepared in five steps from 2-amino-6-chloropurine riboside as previously described *(12,15)*. Allylation of **17** affords **18** in 89% yield. Subsequent reaction with 2.5 equivalents of sodium azide in DMF displaces the 2,6-dichlorophenoxide and chloride moieties to give compound **19**. Attempted reduction of **19** to **20** with hydrogen/Lindlar catalyst poisoned with quinoline causes partial or complete reduction of the allyl group to propyl. The stannous tris(thiophenolate) anion recently reported in the literature *(31,32)* cleanly reduces compound **19** to compound **20** in 93% isolated yield in 15 min at room temperature. Desilylation of compound **20** affords 2'-*O*-allyl-2,6-diaminopurine riboside, which is cleanly deaminated by adenosine deaminase during 2–3 d to give 2'-*O*-allylguanosine. Robins et al. previously observed that 2'-*O*-methylguanosine could be obtained by adenosinedeaminase-catalyzed deamination of 2'-*O*-methyl-2,6-diaminopurine riboside *(33)*. The exocyclic amino group of the 2'-*O*-allylguanosine is directly protected using dimethylformamide dimethyl acetal *(34)* in dry methanol to give compound **22** in excellent yield. Dimethoxytritylation and phosphitylation afford the 2'-*O*-allylguanosine monomer, **24**.

Two useful derivatives whose syntheses are not described here are the 2'-*O*-allylinosine monomer *(13,14)*, and the 2'-*O*-allyl-2-aminoadenosine monomer *(13,14)*. The 2'-*O*-allylinosine monomer is obtained from compound **12** by desilylation, oximate reaction (requires 24 h reflux in acetonitrile), and dimethoxytritylation followed by phosphitylation. The 2'-*O*-allyl-2-aminoadenosine monomer is obtained from compound **20** by protection of the exocyclic amino groups followed by desilylation, dimethoxytritylation, and finally phosphitylation.

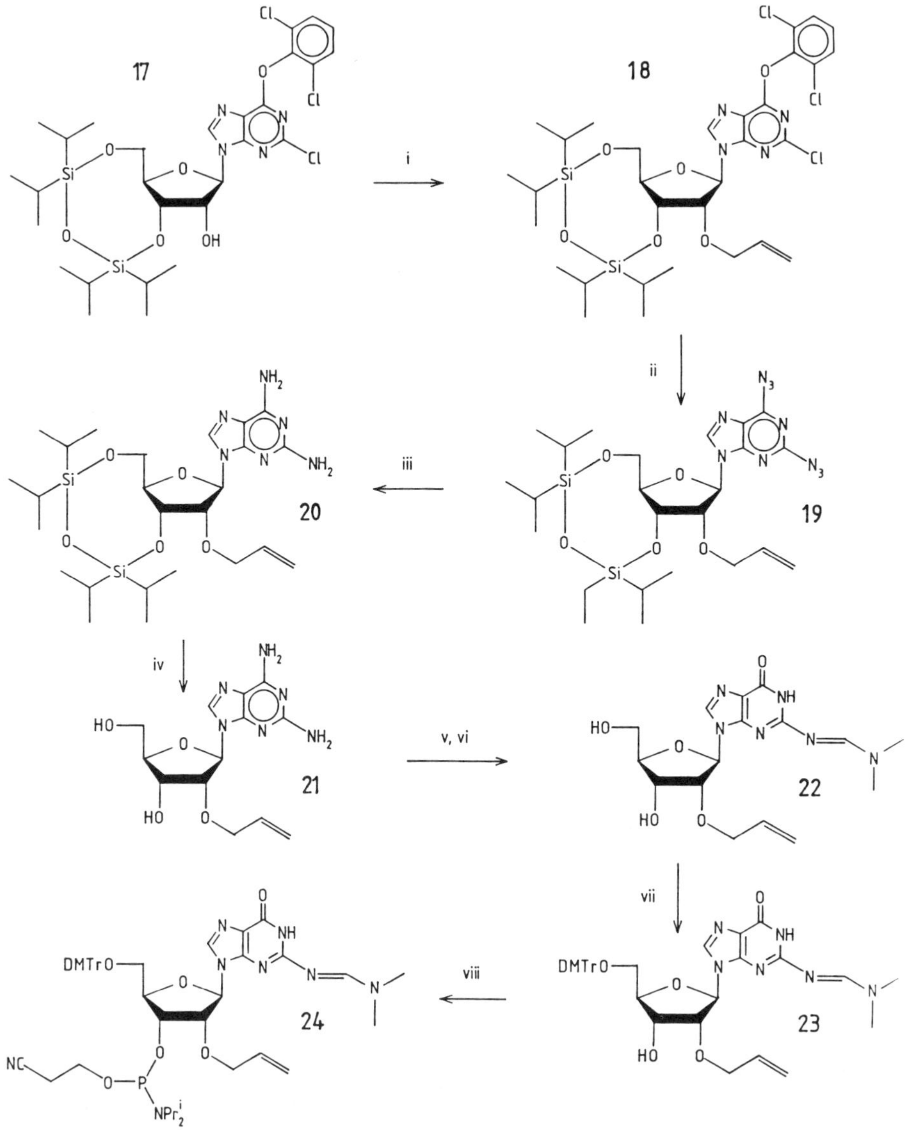

2'-O-Alkyloligoribonucleotides

Fig. 4. Reaction scheme for the preparation of the 2'-O-allyguanosine building block. Reagents: i, allyl ethyl carbonate, 1,4-bis(diphenylphosphino)butane and tris (dibenzylideneacetone)dipalladium(0) in tetrahydrofuran; ii, sodium azide in N,N-dimethylformamide; iii, stannous chloride, thiophenol, and triethylamine in acetonitrile; iv, tetrabutylammonium fluoride in tetrahydrofuran; v, adenosine deaminase in aqueous phosphate buffer pH 7.4/dimethylsulfoxide; vi, N,N-dimethylformamide dimethyl acetal in methanol; vii, 4,4'-dimethoxytrityl chloride and triethylamine in pyridine; viii, 2-cyanoethoxy N,N-diisopropylaminochlorophosphine and N,N-diisopropylethylamine in 1,2-dichloroethane.

Procedures for performing affinity selection or depletion of RNAs or RNA/protein complexes with biotinylated antisense oligos are described in detail elsewhere *(15,20)*. It should be mentioned that incorporation of 2'-*O*-allylinosine in place of 2'-*O*-allylguanosine is advisable for isolating targeted complexes with minimal background. However, in order to obtain extracts that are quantitatively depleted of a particular complex, it is essential to use biotinylated antisense probes containing 2'-*O*-allylguanosine. Moreover, in some cases, the replacement of 2'-*O*-allyladenosine by 2'-*O*-allyl-2-aminoadenosine has been found to be an additional requirement for quantitative depletion *(21)*. A multitude of applications of 2'-*O*-alkyloligoribonucleotides are to be found in the references in the introduction.

5. Notes

1. The majority of solvents and reagents used in Section 3 provide some sort of health risk. Many of them are toxic and/or corrosive or irritant. It is therefore, absolutely essential that the syntheses described here and the preparative LC are only performed by qualified chemists in well-ventilated fume hoods. In addition, the large quantities of petrol and ethyl acetate required for preparative liquid chromatography present a potential fire hazard. Adequate precautions must be taken. Always wear safety glasses, a laboratory coat, and if necessary, a face shield. Two of the synthetic steps require the use of ammonia under pressure at elevated temperature, so make sure there is a gas mask available should anything go wrong. Finally, never work alone; in case of an emergency, immediate help may be needed from a colleague.
2. The phosphoramidite monomers should be stored desiccated under argon at –20°C. The author has kept chemicals this way for several years without loss of quality.
3. The 5-(4-nitrophenyl)-1H-tetrazole solution used as coupling agent for the assembly of 2'-*O*-alkyloligoribonucleotides is almost saturated at ambient temperature, so it is advisable to use a fresh bottle of reagent every 1 or 2 d, and to back-flush the delivery line to avoid potential blockage caused by crystallization.
4. 2'-*O*-Allyloligoribonucleotides cannot be ethanol precipitated, so desalting must be performed by dialysis or gel filtration.

Acknowledgments

The synthesis of the 2'-*O*-allylguanosine monomer and Fig. 4 have been reproduced from the article (ref. *13*) by Sproat et al. in *Nucl. Acids Res.* (1991), **19,** pp. 733–738 by kind permission of Oxford University Press.

Figures 1, 2, and 3 have been reprinted from the article (ref. *14*) by Sproat et al. in *Nucleosides & Nucleotides* (1991), **10,** pp. 25–36 by courtesy of Marcel Dekker, Inc.

In addition, I would like to express my sincere thanks to all my coworkers who have been involved in this work during the past four years, and to Julia Pickles for typing the manuscript.

References

1. Inoue, H., Hayase, Y., Asaka, M., Imura, A., Iwai, S., Miura, K., and Ohtsuka, E. (1985) Synthesis and properties of novel nucleic acid probes. *Nucl. Acids Res.* **Symposium Series No. 16,** 165–168.
2. Inoue, H., Hayase, Y., Imura, A., Iwai, S., Miura, K., and Ohtsuka, E. (1987) Synthesis and hybridization studies on two complementary nona(2'-O-methyl)ribonucleotides. *Nucl. Acids Res.* **15,** 6131–6148.
3. Mukai, S., Shibahara, S., and Morikawa, H. (1988) A new method for the unidirectional deletion of DNA with Bal 31 nuclease using 2'-O-MeRNA-DNA chimeric adaptors. *Nucl. Acids Res.* **Symposium Series No. 19,** 117–120.
4. Inoue, H., Hayase, Y., Iwai, S., and Ohtsuka, E. (1988) Sequence-specific cleavage of RNA using chimeric DNA splints and RNase H. *Nucl. Acids Res.* **Symposium Series No. 19,** 135–138.
5. Shibahara, S., Mukai, S., Nishihara, T., Inoue, H., Ohtsuka, E., and Morisawa, H. (1987) Site-directed cleavage of RNA. *Nucl. Acids Res.* **15,** 4403–4415.
6. Inoue, H., Hayase, Y., Iwai, S., and Ohtsuka, E. (1987) Sequence-dependent hydrolysis of RNA using modified oligonucleotide splints and RNase H. *FEBS Lett.* **215,** 327–330.
7. Sproat, B. S., Lamond, A. I., Beijer, B., Neuner, P., and Ryder, U. (1989) Highly efficient chemical synthesis of 2'-O-methyloligoribonucleotides and tetrabiotinylated derivatives; novel probes that are resistant to degradation by RNA or DNA specific nucleases. *Nucl. Acids Res.* **17,** 3373–3386.
8. Cotten, M., Oberhauser, B., Brunar, H., Holzner, A., Issakides, G., Noe, C. R., Schaffner, G., Wagner, E., and Birnstiel, M. L. (1991) 2'-O-Methyl, 2'-O-ethyl oligoribonucleotides and phosphorothioate oligodeoxyribonucleotides as inhibitors of the *in vitro* U7 snRNP-dependent mRNA processing event. *Nucl. Acids Res.* **19,** 2629–2635.
9. Iribarren, A. M., Sproat, B. S., Neuner, P., Sulston, I., Ryder, U., and Lamond, A. I. (1990) 2'-O-Alkyl oligoribonucleotides as antisense probes. *Proc. Natl. Acad. Sci. USA* **87,** 7747–7751.
10. Sproat, B. S., Lamond, A. I., Guimil Garcia, R., Beijer, B., Pieles, U., Douglas, M., Bohmann, K., Carmo-Fonseco, M., Weston, S., and O'Loughlin, S. (1991) 2'-O-Alkyloligoribonucleotides, synthesis and applications in molecular biology. *Nucl. Acids Res.* **Symposium Series No. 24,** 59–62.
11. Berkower, I., Leis, J., and Hurwitz, J. (1973) Isolation and characterization of an endonuclease from *Escherichia coli* specific for ribonucleic acid in ribonucleic acid deoxyribonucleic acid hybrid structures. *J. Biol. Chem.* **248,** 5914–5921.

12. Sproat, B. S., Beijer, B., and Iribarren, A. (1990) New synthetic routes to protected purine 2'-O-methylriboside-3'-O-phosphoramidites using a novel alkylation procedure. *Nucl. Acids Res.* **18,** 41–49.
13. Sproat, B. S., Iribarren, A., Guimil Garcia, R., and Beijer, B. (1991) New synthetic routes to synthons suitable for 2'-O-allyloligoribonucleotide assembly. *Nucl. Acids Res.* **19,** 733–738.
14. Sproat, B. S., Iribarren, A., Beijer, B., Pieles, U., and Lamond, A. I. (1991) 2'-O-Alkyloligoribonucleotides: Synthesis and applications in studying RNA splicing. *Nucleosides Nucleotides* **10,** 25–36.
15. Sproat, B. S. and Lamond, A. I. (1991) 2'-O-Methyloligoribonucleotides: synthesis and applications, in *Oligonucleotides and Analogs: A Practical Approach* (Eckstein, F., ed.) IRL Press at Oxford University Press, Oxford, pp. 49–86.
16. Lamond, A. I., Sproat, B., Ryder, U., and Hamm, J. (1989) Probing the structure and function of U2 snRNP with antisense oligonucleotides made of 2'-OMe RNA. *Cell* **58,** 383–390.
17. Blencowe, B. J., Sproat, B. S., Ryder, U., Barabino, S., and Lamond, A. I. (1989) Antisense probing of the human U4/U6 snRNP with biotinylated 2'-OMe RNA oligonucleotides. *Cell* **59,** 531–539.
18. Barabino, S. M. L., Sproat, B. S., Ryder, U., Blencowe, B. J., and Lamond, A. I. (1989) Mapping U2 snRNP-pre-mRNA interactions using biotinylated oligonucleotides made of 2'-OMe RNA. *EMBO J.* **8,** 4171–4178.
19. Barabino, S. M. L., Blencowe, B. J., Ryder, U., Sproat, B. S., and Lamond, A. I. (1990) Targeted snRNP depletion reveals an additional role for mammalian U1 snRNP in spliceosome assembly. *Cell* **63,** 293–302.
20. Ryder, U., Sproat, B. S., and Lamond, A. I. (1990) Sequence-specific affinity selection of mammalian splicing complexes. *Nucl. Acids Res.* **18,** 7373–7379.
21. Lamm, G. M., Blencowe, B. J., Sproat, B. S., Iribarren, A. M., Ryder, U., and Lamond, A. I. (1991) Antisense probes containing 2-aminoadenosine allow efficient depletion of U5 snRNP from HeLa splicing extracts. *Nucl. Acids Res.* **19,** 3193–3198.
22. Carmo-Fonseca, M., Tollervey, D., Pepperkok, R., Barabino, S. M. L., Merdes, A., Brunner, C., Zamore, P. D., Green, M. R., Hurt, E., and Lamond, A. I. (1991) Mammalian nuclei contain foci which are highly enriched in components of the pre-mRNA splicing machinery. *EMBO J.* **10,** 195–206.
23. Carmo-Fonseca, M., Pepperkok, R., Sproat, B. S., Ansorge, W., Swanson, M. S., and Lamond, A. I. (1991) *In vivo* detection of snRNP-rich organelles in the nuclei of mammalian cells. *EMBO J.* **10,** 1863–1873.
24. Pieles, U., Sproat, B. S., and Lamm, G. M. (1990) A protected biotin containing deoxycytidine building block for solid-phase synthesis of biotinylated oligonucleotides. *Nucl. Acids Res.* **18,** 4355–4360.
25. Markiewicz, W. T. (1979) Tetraisopropyldisiloxane-1,3-diyl, a group for simultaneous protection of 3'- and 5'-hydroxy functions of nucleosides. *J. Chem. Res. (S),* 24,25.

26. Nyilas, A. and Chattopadhyaya, J. (1986) Synthesis of $O^{2'}$-methyluridine, $O^{2'}$-methylcytidine, N^4-$O^{2'}$-dimethylcytidine and $N^4,N^4,O^{2'}$-trimethylcytidine from a common intermediate. *Acta Chem. Scand.* **B40,** 826–830.
27. Reese, C. B. and Skone, P. A. (1984) The protection of thymine and guanine residues in oligodeoxyribonucleotide synthesis. *J. Chem. Soc. Perkin Trans. I,* 1263–1271.
28. Jones, S. S., Reese, C. B., Sibanda, S., and Ubasawa, A. (1981) The protection of uracil and guanine residues in oligonucleotide synthesis. *Tetrahedron Lett.* **22,** 4755–4758.
29. Lakhmiri, R., Lhoste, P., and Sinou, D. (1989) Allyl ethyl carbonate/palladium (0), a new system for the one step conversion of alcohols into allyl ethers under neutral conditions. *Tetrahedron Lett.* **30,** 4669–4672.
30. Sinha, N. D., Biernat, J., McManus, J., and Köster, H. (1984) Polymer support oligonucleotide synthesis XVIII; use of β-cyanoethyl-N,N-dialkyl-amino-/N-morpholino phosphoramidite of deoxynucleosides for the synthesis of DNA fragments simplifying deprotection and isolation of the final product. *Nucl. Acids Res.* **12,** 4539–4557.
31. Bartra, M., Urpí, F., and Vilarrasa, J. (1987) New synthetic tricks. [Et$_3$NH][Sn(SPh)$_3$] and Bu$_2$SnH$_2$, two useful reagents for the reduction of azides to amines. *Tetrahedron Lett.* **28,** 5941–5944.
32. Bartra, M., Romea, P., Urpí, F., and Vilarrasa, J. (1990) A fast procedure for the reduction of azides and nitro compounds based on the reducing ability of Sn(SR)$_3^-$ species. *Tetrahedron* **46,** 587–594.
33. Robins, M. J., Hansske, F., and Bernier, S. E. (1981) Nucleic acid related compounds. 36. Synthesis of the 2'-O-methyl and 3'-O-methyl ethers of guanosine and 2-aminoadenosine and correlation of O'-methylnucleoside ^{13}C-NMR spectral shifts. *Can. J. Chem.* **59,** 3360–3364.
34. McBride, L. J., Kierzek, R., Beaucage, S. L., and Caruthers, M. H. (1986) Amidine protecting groups for oligonucleotide synthesis. *J. Amer. Chem. Soc.* **108,** 2040–2048.

CHAPTER 7

An Improved Method for the Synthesis and Deprotection of Methylphosphonate Oligonucleotides

Richard I. Hogrefe, Mark A. Reynolds, Morteza M. Vaghefi, Kevin M. Young, Timothy A. Riley, Robert E. Klem, and Lyle J. Arnold, Jr.

1. Introduction

The value of oligonucleotides as therapeutic agents has become increasingly apparent over the past decade *(1–3)*. In order to be useful as therapeutic agents, oligonucleotides must possess a number of properties. These include:

1. Resistance to enzyme degradation;
2. The ability to enter target cells;
3. A lack of interference with normal DNA and RNA processing enzymes; and
4. The ability to bind to a target and alter its expression in a sequence-specific manner.

The nonionic methylphosphonates fulfill these requirements. Studies have been carried out on methylphosphonate oligonucleotides for more than 20 years originating with the pioneering work of Ts'o and Miller *(4,5)*. Methylphosphonates have shown sequence-specific inhibition of translation, are taken into cells readily, and have a minimum of toxicity *(4–12)*. In addition, methylphosphonates have been shown to form triple strands with DNA targets *(13)*.

From: *Methods in Molecular Biology, Vol. 20: Protocols for Oligonucleotides and Analogs*
Edited by: S. Agrawal Copyright ©1993 Humana Press Inc., Totowa, NJ

Historically, methylphosphonate oligonucleotide synthesis has been inefficient, costly, and difficult to conduct above a 1-μmol scale. We have devised methods that significantly reduce cost, increase coupling efficiency, improve recovery, and are applicable to larger-scale synthesis. We routinely carry out 2000-μmol scale syntheses with product yields in excess of 2 g, corresponding to an overall product yield of ≈20%. In this chapter, we describe the use of these methods for the reliable synthesis, deprotection, and purification of methylphosphonate oligonucleotides at 1- and 15-μmol scales.

1.1. General Considerations

1.1.1. Synthesis

The phosphonamidite reagents needed to synthesize methylphosphonate oligonucleotides are commercially available. The standard procedures used for the synthesis of phosphodiester oligonucleotides work poorly for methylphosphonate synthesis, particularly at >1-μmol scale. There are two primary factors that account for the poor efficiency of methylphosphonate synthesis. First, NMR studies have shown that the methylphosphonite P(III) diester intermediate is more susceptible to hydrolysis by tetrazole, the capping reagent, and water compared to the phosphite diester intermediate (*see* Fig. 1 for a schematic of the synthesis cycle and degradative side reactions). For this reason, it is necessary to:

1. Minimize the time methylphosphonate linkages exist in the trivalent phosphonite form;
2. Reduce the water in the oxidizing agent to minimize backbone hydrolysis (for larger-scale syntheses, >1 μmol); and
3. Oxidize prior to capping.

Fig. 1. *(opposite page)* Methylphosphonate synthesis method and undesired side reaction. The crosshatched structure indicates the DNA synthesis support, and the circles depict previous nucleotide additions. Water is a major factor in several undesired side reactions; it hydrolyzes the activated monomer (**I**) and, either alone or in combination with tetrazole, hydrolyzes the methyphosphonate intermediate (**II**) to give a mixture of shorter fragments. Water may be introduced by wet solvents or reagents. In particular, the presence of excess water in the oxidizer reagent can cause hydrolysis of the methylphosphonite intermediate (**II**) before oxidation occurs. Furthermore, the capping catalyst *N*-methylimidazole (NMI) can catalyze hydrolysis at any pentavalent methylphosphonate linkage (**III**) to generate a mixture of side products of varying length. These side reactions can be avoided by using dry solvents and reagents, a low-water oxidizer, and DMAP as the capping catalyst.

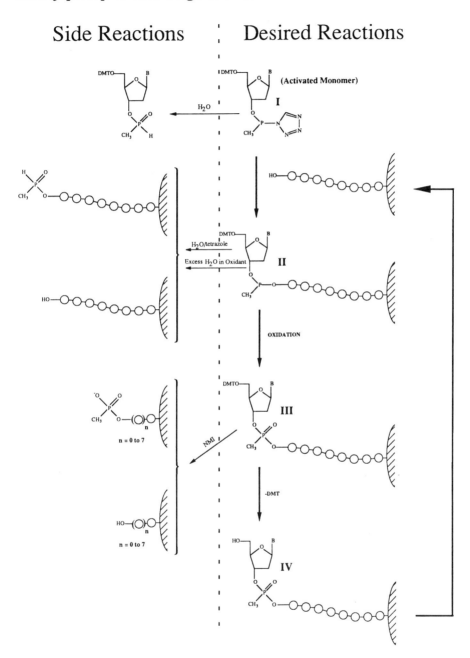

Second, the oxidized pentavalent methylphosphonate P(V) linkage is susceptible to hydrolysis by *N*-methylimidazole (NMI), which is the most commonly used capping reagent. This is solved by using dimethylaminopyridine (DMAP) (*see* Fig. 1), which does not readily cleave the backbone.

The general scheme for synthesizing methylphosphonate oligonucleotides, which incorporates these modifications, follows:

1. Wash with acetonitrile.
2. Detritylate.
3. Wash well with acetonitrile to dry column.
4. Couple using subroutine.
   a. Add monomer and activator (1:4 ratio; *see* Section 3.1.2.1 for number of equivalents over oligonucleotide).
   b. Couple (same amount of time as standard amidites).
   c. Oxidize immediately, with no prewash, using a low-water-content oxidant.
   d. Wash until oxidant is rinsed away.
5. Cap using acetic anhydride and DMAP.
6. Wash well with acetonitrile.
7. Begin cycle again.

Standard times recommended by the manufacturer can be used for detritylation, coupling, and capping. The oxidation step must be modified, however. It must precede the capping step and must be done as soon after the coupling as possible. The improved programs eliminate the wash step between the coupling and oxidation reactions. When using multiple-column instruments, the oxidation should be incorporated into the coupling subroutine. Section 2.2. details programming parameters.

One other synthesis change was to use *N*-isobutyryl-protected deoxycytidine, instead of the more commonly used *N*-benzoyl-protecting group. This change is necessary because the improved deprotection conditions described below generate transamination side products with benzoylated cytosine, but not with isobutyryl-protected cytosine *(14)*. The isobutyryl-protected dC monomer and the corresponding isobutyryl-protected dC DNA synthesis support can be obtained from JBL Scientific (San Luis Obispo, CA).

In order to compare standard and improved methylphosphonate synthesis conditions, an all-methylphosphonate oligonucleotide (5'-ACTAGTGCATCGACTGA) was synthesized on a 15-µmol scale.

Methylphosphonate Oligonucleotides

Table 1
Standard vs Improved 15-μmol Scale Synthesis

Syn no.	Program	Capping Oxidizer	Capping Agent	Coupling Efficiency	crude yield OD	Prod. int. area by HPLC	
(1)	STD[b]	STD[b]	STD[b]	85.7%	250	None	Obs.
(2)	IMP[c]	STD[b]	IMP[c]	87.4%	336	>5%	
(3)	IMP[c]	STD[b]	STD[b]	88.2%	232	None	Obs.
(4)	STD[b]	STD[b]	IMP[c]	91.3%	257	None	Obs.
(5)	STD[b]	IMP[c]	STD[b]	93.3%	1272	20%	
(6)	STD[b]	IMP[c]	IMP[c]	93.1%	1154	31%	
(7)	IMP[c]	IMP[c]	IMP[c]	95.2%	1351	42%	
(8)[a]	IMP[c]	IMP[c]	IMP[c]	93.1%	1143	44%	

[a]This synthesis was done on Gentrix™ polymeric support; all others were done on CPG.
[b]STD: Standard conditions are: program—Milligen 15-μM programs; oxidizer—2.0% water content oxidizer; capping agent—NMI.
[c]IMP: Improved conditions are: program—Genta 15-μM programs; oxidizer—0.25% water content oxidizer; capping agent—DMAP (see Section 3.1.).

The oligonucleotide was synthesized under eight different conditions, where one or more of the critical parameters (oxidant mixture, cap B reagent, solid support, and synthesis program) were varied. The product oligonucleotides were all deprotected using the new "one-pot" procedure described below. Table 1 lists the conditions of each synthesis along with average coupling efficiency (ACE) and crude yield. Also shown (Figs. 2–6) are the chromatograms that result from reverse-phase HPLC analysis of representative crude product mixtures.

This comparison clearly demonstrates that the new synthesis procedures give superior products and yields. The procedure using standard conditions gave no discernible product. (Standard conditions include the Milligen 15-μmol synthesis programs, Milligen oxidizer, and NMI in cap B) (Synthesis Nos. 1–4; Figs. 2 and 4). In all cases, a conventional oxidant mixture resulted in very poor syntheses. NMI in cap B instead of DMAP resulted in backbone cleavage, although some product was evident (Synthesis No. 5; Fig. 5).

The average trityl yields, although indicative of the coupling efficiencies, were not accurate predictors of oligomer purity as shown by HPLC analyses. The coupling efficiencies of the failed syntheses varied considerably from coupling to coupling.

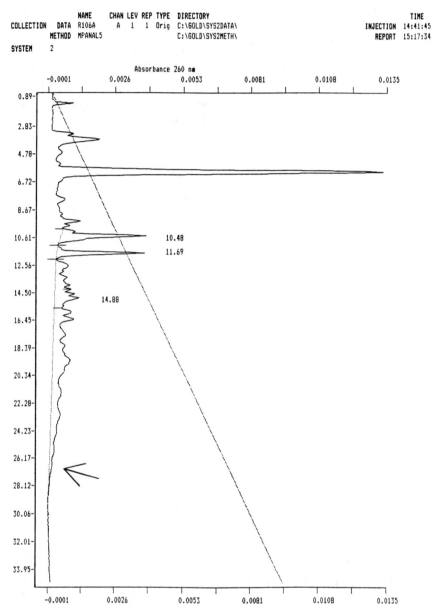

Fig. 2. Crude product mixture (synthesis no. 1) resulting from 15-µmol scale synthesis using all standard synthesis procedures and "one-pot" deprotection. Standard conditions include manufacturer's supplied 15-µmol programs using ~9 Eq of monomer/coupling, commercially available oxidant containing 2% water, and N-methylimidazole (NMI) capping reagent. The product is shown by the arrow. Chromatography was done on a Whatman RAC-II C-18 analytical column (100 × 4.6 mm, 5 mol) using a gradient of 0–70% B over 35 min. Flow—1.5 mL/min. A—0.05M triethylammonium acetate, pH 7.0–7.2, B—50% acetonitrile in A.

Methylphosphonate Oligonucleotides

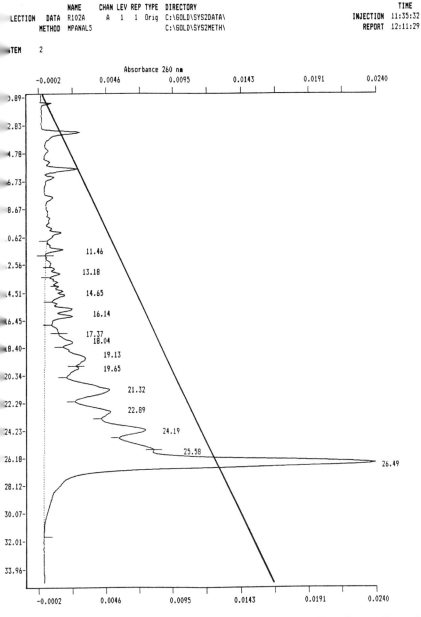

Fig. 3. Crude product mixture (synthesis no. 7) resulting from 15-µmol scale synthesis using all of the improved procedures and "one-pot" deprotection. Improved conditions include new program, low-water oxidation reagent, and dimethylaminopyridine (DMAP) capping reagent (*see* Section 3.1.). Chromatography was done as described in Fig. 2. Integrated product area = 42%.

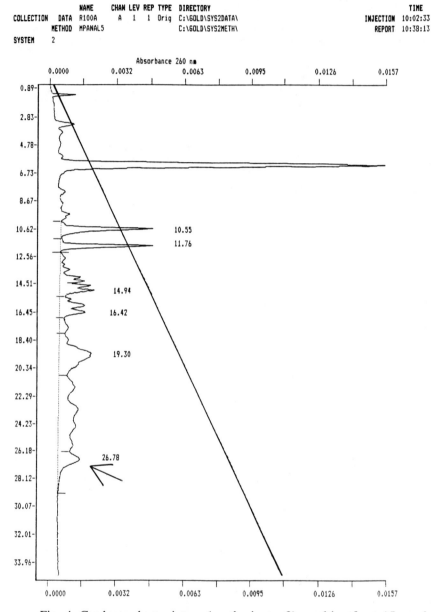

Fig. 4. Crude product mixture (synthesis no. 2) resulting from 15-µmol scale synthesis using the conditions described in Fig. 3, but replacing the low-water oxidant with the standard 2% water oxidant. The product is indicated by the arrow. The comparison to Fig. 3 dramatically shows the importance of using low water for good synthesis of methylphosphonate oligonucleotides at larger scales. Chromatography was done on a Whatman RAC-II C-18 analytical column (100 × 4.6 mm, 5 mol) using a gradient of 0–70% B over 35 min, as described in Fig. 2.

Methylphosphonate Oligonucleotides

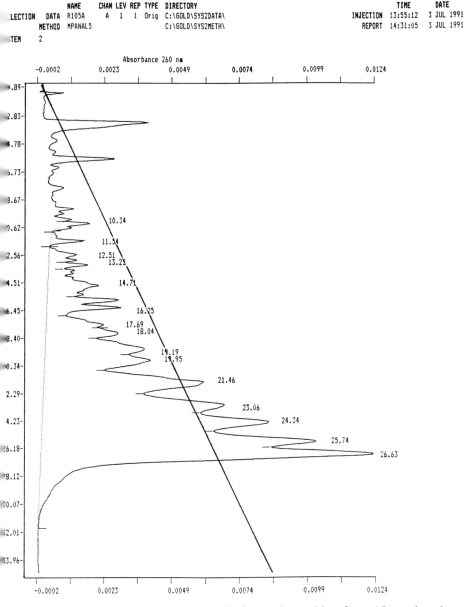

Fig. 5. Crude product mixture (synthesis no. 5) resulting from 15-μmol scale synthesis using the conditions described in Fig. 2, but replacing the 2% water oxidant with the low-water oxidant mixture. Although more product is available (integrated product area = 20% vs nearly no product), the deleterious effect of the more basic NMI is evident. Chromatography was done on a Whatman RAC-II C-18 analytical column (100 × 4.6 mm, 5 mol) using a gradient of 0–70% B over 35 min, as described in Fig. 2.

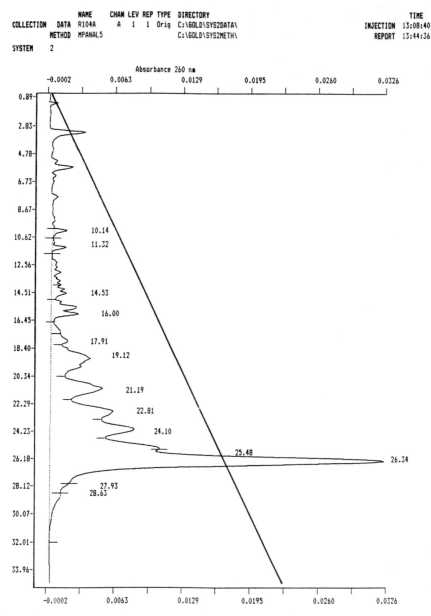

Fig. 6. Crude product mixture (synthesis no. 8) resulting from 15-μmol scale synthesis using the conditions described in Fig. 3, except that the synthesis was done on the polymeric support available from JBL Scientific, San Luis Obispo, CA, instead of CPG. The product is comparable (integrated product area = 44%) to that in Fig. 3. Chromatography was done on a Whatman RAC-II C-18 analytical column (100 × 4.6 mm, 5 mol) using a gradient of 0–70% B over 35 min, as described in Fig. 2. Standard CPG was used in all other syntheses except this one.

1.1.2. Deprotection

The deprotection of methylphosphonate oligonucleotides must be modified, because the standard concentrated ammonium hydroxide treatment used to deprotect phosphodiester oligonucleotides degrades the methylphosphonate backbone. To avoid degradation, the more gentle deprotecting reagent ethylenediamine/95% ethanol (EtOH) (1:1) has been used *(6,15)*. Under these conditions, however, the commonly used N-4 benzoyl protected dC is susceptible to transamination with ethylenediamine replacing the C-4 amine *(14,15)*. This has necessitated removal of the benzoyl-protecting group prior to treatment with ethylenediamine. The benzoyl group can be removed with either an overnight treatment with hydrazine hydrate in acetic acid and pyridine *(16–18)* or a 2-h room temperature treatment with concentrated ammonium hydroxide *(19,20)*. The deprotection is then concluded by treatment with ethylenediamine/95% EtOH (1:1) at room temperature for 4–6 h.

The hydrazine hydrate method was found to give transaminated products, whereas the ammonium hydroxide treatment results in low yields *(14)*. Furthermore, the half-life of cleavage of the benzoyl group from dC with ammonium hydroxide at room temperature is 2 h *(21)*. Consequently, the benzoyl group is not completely removed by a 2-h ammonium hydroxide treatment, which may also lead to transaminated side products. If longer ammonium hydroxide treatments are used, unacceptable levels of backbone cleavage can occur.

To resolve the issue of transamination, we changed the protecting group used with dC from benzoyl to isobutyryl. This isobutyryl group can be cleanly removed with EDA alone, eliminating transamination side products *(14)*. The deprotection procedure was then modified to give quantitative recovery of methylphosphonate oligonucleotides, while minimizing base modifications and backbone cleavage. (We found that conventional deprotection methods give only about 50% recovery of oligonucleotide from the support.)

This chapter describes a simple "one-pot" procedure that utilizes a short preincubation step with dilute ammonium hydroxide, followed by treatment with ethylenediamine. The reaction is stopped by dilution and neutralization to yield a solution ready for reverse-phase HPLC purification or other purification method. The purpose of the ammonium hydroxide step in this method is to hydrolyze minor modifications on dG residues that can occur when DMAP is used as the capping catalyst *(14)*.

1.1.3. Purification

The preferred method of purification is reverse-phase chromatography (22), since ion-exchange chromatography is not possible with neutral oligonucleotides. However, if the methylphosphonate contains one or more phosphodiester linkages, both ion exchange (i.e., DEAE-Sephadex) and polyacrylamide gel electrophoresis can be used.

A variation useful for loading the large sample volumes obtained from the neutralization step of the "one-pot" method on a reverse-phase HPLC column is described in this chapter. Large-sample volumes can be loaded on a reverse-phase column if the organic content of the solution is low enough to adsorb the compound at the top of the column. For methylphosphonate oligonucleotides, the organic content should be 10% or less. We use an inexpensive high-pressure pump (Eldex Laboratories, Inc., San Carlos, CA) to load the sample onto the column. This procedure ensures against the possibility of damage to the more expensive HPLC pumps because of precipitation of the sample. We have successfully loaded up to 2 L of solution on columns containing in excess of 500 mg of crude oligonucleotide in this manner. After loading the sample, the column is reattached to the HPLC, and the product eluted.

2. Methods

The synthesis procedures described below were developed for the Milligen/Biosearch 8600/8700 series instruments, although they can be modified to utilize any standard DNA synthesizer.

2.1. Reagents and Materials

1. DMT-Nucleoside Methylphosphonamidites; dC protected with isobutyryl (JBL Scientific, San Luis Obispo, CA; cat. nos: DMT-MAP-Thymidine—1800A; DMT-*N*-ibu-MAP-deoxyguanosine—1810A; DMT-*N*-bz-MAP-deoxyadenosine—1830A; DMT-*N*-ibu-MAP-deoxycytidine—2170A).
2. Anhydrous acetonitrile (≤30 ppm water) for wash steps.
3. Anhydrous acetonitrile (≤20 ppm water) for methylphosphonamidite solutions.
4. Dichloroacetic acid (2.5%) in dichloromethane (≤30 ppm water).
5. Acetic anhydride (40%) in anhydrous acetonitrile (Cap A).
6. Dimethylaminopyridine (DMAP) (0.625%) in anhydrous pyridine (Cap B) (≤50 ppm water).

7. I$_2$ (0.1M)in tetrahydrofuran:2,6-lutidine:water (74.75:25:0.25) (available from JBL Scientific, San Luis Obispo, CA; cat. no. 2440A).
8. Tetrazole (0.45M)in anhydrous acetonitrile (≤30 ppm water).
9. Support-bound nucleosides; dC protected with isobutyryl (JBL Scientific, San Luis Obispo, CA; cat. no: Gentrix™ Support A—2400A; Gentrix™ Support C—2410A; Gentrix™ Support G—2420A; Gentrix™ Support T—2430A).
10. Diester oxidizer. (Standard mixture available from several sources. Used to couple a 5'-diester linkage to the methylphosphonate oligonucleotide—optional.)
11. Anhydrous dichloromethane (≤20 ppm water).

2.2. Programs

Modified programs for the Milligen/Biosearch 8600/8700 series DNA synthesizer are provided in Figs. 7 and 8. These programming changes result in a reduction in the number of equivalents of phosphonamidite reagents and tetrazole for both the 1- and 15-µmol scales as shown in Table 2. We have also moved the oxidation reaction so that it occurs within the coupling cycle. This was done to allow multiple-column syntheses on the 8700/8750. Since the methylphosphonite intermediate is labile, it is important to oxidize it as rapidly as possible. When standard programming is used for a multicolumn synthesis on a Milligen/Biosearch instrument, the phosphonite intermediate sits on the column until all four columns have undergone the coupling subroutine step. This time delay leads to partial hydrolysis of the phosphonite intermediate, thereby reducing overall yields. It is necessary to oxidize prior to capping, since the capping mixture is also capable of hydrolyzing this intermediate. The only other change from standard programs is a reduction in flow rate to 5.6 mL/min through the U vial, which lowers phosphoramidite use. All other instrument parameters are the same as the manufacturer's specifications.

2.3. Preparation of Synthesis Reagents

All reagents should be prepared with clean, dry solvents—HPLC grade or redistilled. All reagents are commercially available, except Cap B, which is prepared by adding 6.25 g of dimethylaminopyridine (DMAP)/L of anhydrous pyridine. The oxidant for the methylphosphonate couplings is critical at scales above 1 µmol. The low-water oxidizer described here offers substantially higher coupling efficien-

```
                    Main Program                    Coupling Subroutine

P#-S#  Length    Function    Outputs        P#-S#  Length    Function    Outputs
15- 1   1.0     FLUSH B      7,14,19        3- 1    3.0      AUX 1      * 8,14,15
15- 2   2.0     FLUSH A      1,19                                         16,17,18
15- 3  10.0     WASH A     * 1,15,16,17     3- 2    0.5      COUPLE     *14,15,16
                             18                                           17,18
15- 4   0.5     DEBLOCK    * 2,19           3- 3    0.5      AUX 1      * 8,14,15
15- 5  10.0     DEBLOCK      2,15,16,17                                   16,17,18
                             18,20,22,24
                             26             3- 4             REPEAT     Step 2 2X
15- 6   5.0     WAIT       *15,16,17,18     3- 5    0.6      WASH B     * 7,14,15
15- 7           REPEAT     Step 5 6X                                      16,17,18
15- 8  15.0     WASH A     * 1,15,16,17     3- 6    6.0      WAIT       *14,15,16
                             18,20,22,24                                  17,18
                             26             3- 7             REPEAT     Step 5 4X
15- 9   1.0     AUX 3      * 1, 2, 3, 4
15-10   3.0     FLUSH B      7,14,19        3- 8    3.0      WASH B     * 7,14,15
15-11   5.0     WASH B     * 7,14,15,16                                   16,17,18
                             17,18          3- 9    1.0      FLUSH B      7,14,19
15-12  20.0     WASH A     * 1,15,16,17     3-10   10.0      WASH A     *1,15,16
                             18                                           17,18
15-13           SUB CPL                     3-11   15.0      OXIDIZE    * 5,15,16
15-14   6.0     WASH A     * 1,15,16,17                                   17,18
                             18
15-15  15.0     WASH A     * 1,15,16,17     3-12   10.0      WAIT
                             18             3-13   10.0      FLUSH A      1,19
15-16   2.0     CAP B      * 4,15,16,17     3-14   20.0      WASH A     *1,15,16
                             18                                           17,18
15-17   0.5     CAP A      * 3,15,16,17     3-15             SUB   0 RETURN
15-18   0.5     CAP B      * 4,15,16,17
                             18
15-19           REPEAT     Step 17    5X
15-20   0.5     WASH A     * 1,15,16,17
                             18
15-21   2.0     WAIT       *15,16,17,18
15-22           REPEAT     Step 20    4X
15-23  10.0     WASH A     * 1,15,16,17
                             18
15-24   5.0     WAIT       *15,16,17,18
15-25           GO WH BA     3
15-26  20.0     WASH A     * 1,15,16,17
                             18
15-27   5.0     WAIT       *15,16,17,18
15-28  10.0     WASH A     * 1,15,16,17
                             18
```

1 µM flow rates — Train A: 6.6 ml/min. (Wash A)
                  Train B: 4.7 ml/min. (U Vial)

Fig. 7. Improved program for the synthesis of methylphosphonate oligonucleotide at the 1-µmol scale, for the Milligen/Bioresearch 8600/8700 series instrument. The monomer excess using this program is eight times the support-bound monomer.

cies at the 15-µmol scale (*see* Table 1 for comparison data). The oxidant must be changed back to the standard mixtures when the 5'-diester coupling is done, if one is desired. Apparently, the β-cyanoethyl phosphoramidites require more water for quantitative oxidation under normal synthesis conditions.

Methylphosphonate Oligonucleotides

	Main Program				Coupling Subroutine		
P#-S#	Length	Function	Outputs	P#-S#	Length	Function	Outputs
1- 1	2.0	FLUSH A	1,19	3- 1	3.0	AUX 1	* 8,14,15
1- 2	2.0	FLUSH B	7,14,19				16,17,18
1- 3	15.0	WASH A	* 1,15,16,17 18	3- 2	1.0	COUPLE	*14,15,16 17,18
1- 4	2.0	DEBLOCK	* 2,19	3- 3	1.0	AUX 1	* 8,14,15
1- 5	10.0	DEBLOCK	* 2,15,16,17 18,20,22,24 26	3- 4		REPEAT Step 2	16,17,18 4X
1- 6	7.0	WAIT	*20,22,24,26	3- 5	0.5	WASH B	* 7,14,15 16,17,18
1- 7		REPEAT	Step 5 7X				
1- 8	10.0	WASH A	* 1,15,16,17 18,20,22,24 26	3- 6	15.0	WAIT	17,18
				3- 7		REPEAT	Step 5 5X
1- 9	1.0	AUX 3	* 1, 2, 3, 4				
1-10	30.0	WASH A	* 1,15,16,17 18	3- 8	10.0	WASH B	* 7,14,15 16,17,18
1-11	5.0	WAIT	* 15,16,17,18	3- 9	1.0	FLUSH B	7,14,19
1-12		REPEAT Step 10	4X	3-10	30.0	OXIDIZE	* 5,15,16 17,18
1-13		SUB CPL	18	3-11	30.0	WASH A	* 1,15,16 17,18
1-14	60.0	WASH A	* 1,15,16,17 18	3-12	7.0	WAIT	*15,16,17 18
1-15	4.0	CAP B	* 4,15,16,17 18	3-13		REPEAT Step 11	2X
1-16	1.0	CAP A	* 3,15,16,17 18	3-14	2.0	FLUSH A	1,19 17,18
1-17	1.0	CAP B	* 4,15,16,17 18	3-15		SUB 0 RETURN	
1-18	1.0	WAIT	* 15,16,17,18				
1-19		REPEAT	Step 16 9X				
1-20	5.0	WAIT	*15,16,17,18				
1-21	0.5	WASH A	* 1,15,16,17 18				
1-22		REPEAT Step 20	3X				
1-23	90.0	WASH A	* 1,15,16,17 18				
1-24		GO WH BA	4				
1-25		GO WH BA	3				
1-26	90.0	WASH A	* 1,15,16,17 18				

15 µM flow rates – Train A: 9.8 ml/min. (Wash A)
Train B: 5.6 ml/min. (U Vial)

Fig. 8. Improved program for synthesis of methylphosphonate oligonucleotide at the 15-µmol scale, for the Milligen/Bioresearch 8600/8700 series instrument. The monomer excess using this program is three times the support-bound monomer.

Three of the monomers (DMT-MAP-thymidine, DMT-*N*-bz-MAP-deoxyadenosine, and DMT-*N*-ibu-MAP-deoxycytidine) are dissolved at a concentration of $0.1M$ in dry acetonitrile, whereas DMT-*N*-ibu-MAP-deoxyguanosine is dissolved in acetonitrile:dichloromethane (1:1) (*see* Table 3). The solvents used for amidite solutions must be ≤20 ppm in water for good coupling efficiencies.

Table 2
Equivalents of Monomer Used per Coupling

Scale	Method	Eq.
1 µM	Standard	30
	Improved	11
15 µM	Standard	9
	Improved	3

Table 3
Monomer Solvation Table

Monomer	Mol wt	0.5 g	1.0 g	2.0 g
DMT-dA-Bz-MAP	803	6.2 mL	12.4 mL	24.8 mL
DMT-dC-Ibu-MAP	740	6.8 mL	13.6 mL	27.2 mL
DMT-dG-Ibu-MAP[a]	785	6.4 mL	12.7 mL	25.4 mL
DMT-T-MAP	690	7.3 mL	14.5 mL	29.0 mL

[a]Acetonitrile:dichloromethane (1:1); all others use acetonitrile only.

Solvents are stored under argon and transferred by dry syringe immediately before use. One convenient way to dry the syringe is to wash it at least four times with dry acetonitrile.

3. Oligonucleotide Synthesis, Deprotection, and Purification

3.1. Synthesis

Once the instrument is loaded with reagents, synthesis is initiated as usual. We generally remove the final dimethoxytrityl (DMT) group automatically. After the synthesis is complete, remove and dry the column with a stream of N_2 or argon gas.

3.2. Deprotection

3.2.1. Reagents and Materials

1. Absolute EtOH:acetonitrile:ammonium hydroxide (45:45:10); store at 4°C.
2. Ethylenediamine (EDA): distilled and stored under an inert atmosphere.
3. Water (HPLC grade).
4. Acetonitrile (HPLC grade).
5. Acetonitrile:100 mM triethylammonium bicarbonate (TEAB) (1:1).
6. TEAB (25 mM).

Methylphosphonate Oligonucleotides

7. Acetonitrile:water (1:1).
8. Waters C-18 125A reverse-phase bulk packing material (cat. no. 20594) or equivalent.
9. Waters C-18 Sep-Pak reverse-phase cartridges (cat. no. 51910).
10. Column for bulk C-18: 2.5 cm diameter, at least 5 cm in height.

3.2.2. Procedure

1. Remove the dried support from the synthesis column, and place it in an appropriately sized screw-cap vessel. For 1-μmol scales, a 1-dram glass sample vial works well, whereas a 13 × 100 mm screw-top test tube can be used for 15-μmol scale syntheses. Both vessels are available from the major distributors (e.g., Fisher Scientific, Pittsburgh, PA, Baxter Scientific Products, Megan Park, IL).
2. Add the ethanol:acetonitrile:ammonium hydroxide (45:45:10) solution (*see* Table 4 for the appropriate volume).
3. Vortex and allow to stand at room temperature for 30 min.
4. Add 1 vol of ethylenediamine, vortex well, and attach to a rotating mixer. If a vortex mixer is not available, vortex well every hour for a total of 6 h.
5. Decant the solution, and rinse the beads twice with acetonitrile:water (1:1). Add the rinsing to the deprotection solution. Dilute the solution with water. Neutralize the solution with 6N HCl in 10% acetonitrile:water (approx 4 mL for the 1-μmol scale and about 8 mL for 15-μmol scale). Monitor neutralization using pH paper.
6. At this time, the sample is ready for reverse-phase HPLC purification. If another purification (such as DEAE) is desired, the sample can be desalted using the following technique:

    15-μmol scale—Pour a column of the C-18 packing approx 3 × 2.5 cm in diameter. A Kontes Flex Column™ with a 60-mL syringe attached to the luer fitting on the cap for positive pressure works well. Preswell and wash the column sequentially with 40 mL each of acetonitrile, acetonitrile:100 mM triethylammonium bicarbonate (TEAB) (1:1), and 25 mM TEAB, successively. Load the sample, and wash with 40 mL of water, saving the effluent in case the sample does not adhere to the column. Elute the sample with 2 × 20 mL aliquots of acetonitrile:water (1:1). Read the absorbance at 260 nm to determine if the sample is in one or both aliquots. If the 260-nm recovery is low, check the effluent. If the absorbance is high, dilute the volume with an equal volume of water, and try desalting again. The column must be reequilibrated as described above before reloading the sample.

Table 4
Reagent Volumes for Deprotection (in mL)

Scale, μM	Soln. I[a]	EDA[b]	Water[c]	6N HCl[d]
1	1	1	30	4
15	2	2	60	8
30	3	3	90	12
45	4	4	120	16
60	5	5	150	20

[a]Soln. I—(45:45:10) acetonitrile:EtOH (abs):ammonium hydroxide.
[b]EDA—ethylenediamine; neat; freshly distilled; stored under inert gas
[c]Water—distilled; for dilution.
[d]6N HCl—Contains 10% by vol acetonitrile. Approximate vol; follow tritration with pH paper. The vol of the (1:1) acetonitrile:water washes (2) are equal to the vol of solution I used.

1-μmol scale—The procedure is similar to that above, except the Waters C-18 Sep-Pak (0.4-g size) is used, and the volumes of the equilibration volumes are reduced to 10 mL each, the water wash is reduced to 20 mL, and the elution volume to 5 mL.

3.3. Purification

3.3.1. Reagents and Materials

1. Buffer A: 0.05*M* triethylammonium acetate (TEAA).
2. Buffer B: 50% acetonitrile in 0.05*M* TEAA.
3. Whatman RAC II ODS-3 semipreparative HPLC column; 5 mm, 9.0 × 100 mm (or equivalent).
4. Whatman RAC II ODS-3 analytical HPLC column; 5 mm, 4.6 × 100 mm (or equivalent).

3.3.2. Procedure

3.3.2.1. Filtration

Prior to HPLC purification, the neutralized methylphosphonate oligonucleotide mixture should be filtered. The absorbance at 260 nm should be checked both before and after filtration, since some methylphosphonate oligonucleotides will precipitate after neutralization of the deprotection solution, depending on sequence. If precipitation occurs, the oligonucleotide should be collected and resuspended. Spin down the solution, decant the supernatant (checking absorbance), and resuspend the oligonucleotide pellet in acetonitrile:water (1:1). If the sample does not appear to have precipitated, but a loss is noticed after filtration, rinse the filter with acetonitrile:water (1:1), and check the absorbance of the rinse solution.

3.3.2.2. GRADIENT

The column should be washed and equilibrated prior to use with 100% buffer A. Program the following gradient into the HPLC: 0–20% B for 5 min, followed by 20–60% B for 40 min at a flow of 3 mL/min.

3.3.2.3. COLUMN LOADING

Sample injection depends on two parameters; the available equipment and the percent of acetonitrile required to maintain solubility The easiest way to load large volumes of sample onto a reverse-phase HPLC column is to load the sample off the HPLC using a small high-pressure pump, such as the Eldex Model B-100-S. (We do not recommend using the HPLC pumps for this purpose because of the possibility of precipitation in the pump.) If the sample is in ≤10 acetonitrile, it will stick at the top of the column. If such a pump is unavailable, purchase a large loop, preferably 5 or 10 mL.

Loading can be done in one of three ways:

1. Sample in 10% acetonitrile and an auxiliary pump is available. Load the solution using the pump being careful not to pump air into the column. Wash the system with 10% acetonitrile:water to complete loading of the methylphosphonate and to start washing the salts off the column. Connect the column to the HPLC and begin the gradient.
2. Sample in 10% acetonitrile, but no pump is available. By running 0% B (no acetonitrile), multiple injections may be done until the sample is completely loaded. Allow sufficient time for the sample to get through the loop prior to the next injection. Check your HPLC manual or ask technical support how to do this without starting the gradient.
3. Dissolution of sample requires > 10% acetonitrile. This purification is still possible using multiple injections through a large injection loop. Inject only 10% of the loop volume at a time. Add the sample to the loop fairly rapidly to cause mixing in the loop, and then quickly inject onto the column. Do not allow the sample to sit in the loop since precipitation may occur. Allow extra time between injections to ensure that the sample is bound well to the column.

3.3.2.4. Elution

Run the gradient and collect 0.5-min fractions. Read the absorbance of the fractions in the product region, and analyze by reverse-phase HPLC. Use the Whatman RAC II C-18 column with the same gradient as above, but with a flow of 1.0 mL/min. Pool together the fractions with the desired purity. Dry the pooled fractions, and reconstitute the pellet in acetonitrile:water (1:1) for aliquoting. A typical recovery is 0.5 –2 mg of isolated product/1 µmol of starting support.

3.3.2.5. Lyophilization

The sample should be lyophilized with acetonitrile:water (1:1) or ethanol:water (1:1) three times prior to use to ensure complete removal of TEAA.

4. Summary

The methylphosphonate oligonucleotide synthesis methods described here give the desired products in good yield. Superior amounts of product are achieved by modifying both the DNA synthesis program and the reagent to compensate for the unstable methylphosphonite intermediate. Deprotection conditions have also been altered to maximize the recovery of oligonucleotide from DNA synthesis supports and to minimize the amount of base modification. Mass-spectrometry analysis of our oligonucleotides has verified their purity and confirmed the absence of modified bases.

When compared to standard DNA synthesis methods, this procedure uses only about one-third the usual amount of monomer. Using these procedures, it should be possible to synthesize reliably methylphosphonate oligonucleotides at 1- and 15-µmol scales.

Acknowledgments

We would like to acknowledge Paul Miller, Paul Ts'o, Tom Adams, Bob Wang, and Lloyd Snyder for their insightful discussions; Eric McCampbell and Eric Hesselberth for their technical assistance; and finally Kathleen M. Sweeting and Cynthia Reindal for preparation of the manuscript.

References

1. Stein, C. A. and Cohen, J. S. (1988) Oligodeoxynucleotides as inhibitors of gene expression: A Review. *Cancer Res.* **48,** 2659–2668.
2. Tidd, D. M. (1991) Synthetic oligonucleotides as therapeutic agents. *Br. J. Cancer* **63,** 6–8.

3. Riorden, M. L. and Martin, J. C. (1991) Oligonucleotide-based therapeutics. *Nature* **350,** 442,443.
4. Miller, P. S. (1991) Oligonucleoside methylphosphonates as anti-sense reagents. *Biotechnology* **9,** 358–362.
5. Miller, P. S. (1989) Non-ionic antisense oligonucleotides, in *Oligodeoxyribonucleotides: Antisense Inhibitors of Gene Expression, vol. 12, Topics in Molecular and Structural Biology* (Cohen, J. S., ed.) MacMillin, London, pp. 79–95.
6. Blake, K. R., Murakami, A., Spitz, S. A., Glave, S. A., Reddy, M. P., Ts'o, P. O. P., and Miller, P. S. (1985) Hybridization arrest of globin synthesis in rabbit reticulocyte lysates and cells by oligodeoxyribonucleoside methylphosphonate. *Biochemistry* **24,** 6139–6142.
7. Chang, E. H., Zu, Y., Shinozuka, R., Zon, G., Wilson, W. D., and Strekowska, A. (1989) Comparative inhibition of *ras* P21 protein synthesis with phosphorus-modified antisense oligonucleotides. *Anti-Cancer Drug Design* **4,** 221–232.
8. Temsamani, J., Agrawal, S., and Pederson, T. (1991) Biotinylated antisense methylphosphonate oligonucleotide. *J. Biol. Chem.* **266,** 468–472.
9. Zamecnik, P. C., Goodchild, J., Taguchi, Y., and Sarin, P. S. (1986) Inhibition of replication and expression of human T-cell lymphotropic virus type III in cultured cells by exogenous synthetic oligonucleotides complementary to viral RNA. *Proc. Natl. Acad. Sci. USA* **83,** 4143–4146.
10. Miller, P. S., McFarland, K. B., Jayaraman, K., and Ts'o, P. O. P. (1981) Biochemical and biological effects of non-ionic nucleic acid methylphosphonates. *Biochemistry* **20,** 1874–1880.
11. Marcus-Sekura, C. J., Woerner, A. M., Shinozuka, K., Zon, G., and Quinnan, G. V. (1987) Comparative inhibition of chloramphenicol acetyltransferase gene expression by antisense oligonucleotide analogues having alkyl phosphotriester, methylphosphonate and phosphorothioate linkage. *Nucl. Acids Res.* **15,** 5749–5752.
12. Vasanthakumar, G. and Ahmed, N. K. (1989) Modulation of drug resistance in a daunorubicin resistant subline with oligonucleoside methylphosphonates. *Cancer Commun.* **1,** 225–232.
13. Callahan, D. E., Trapane, T. L., Miller, P. S., Ts'o, P. O. P., and Kan, L.-S. (1991) Comparative circular dichroism and fluorescence studies of oligodeoxyribonucleotide and oligodeoxyribonucleoside methylphosphonate strands in duplex and triplex formation. *Biochemistry* **30,** 1650–1655.
14. Hogrefe, R. I., Vaghefi, M. M., Reynold, M. A., Young, K., and Arnold, L. J. (1992) Deprotection of methylphosphonate oligonucleotides using a novel one-pot procedure. *Nucl. Acids Res.* submitted.
15. Miller, P. S., Reddy, M. P., Murakami, A., Blake, K. R., Lin, S.-B., and Agris, C. H. (1986) Solid phase syntheses of oligodeoxyribonucleoside methylphosphonate. *Biochemistry* **25,** 5092–5095.
16. Letsinger, R. L., Miller, P. S., and Gram, G. W. (1968) Nucleotide chemistry. XII. Selective *N*-debenzoylation of *N,O*-polybenzoylnucleoside. *Tetrahedron Lett.* **22,** 2621–2634.
17. Barnett, R. W. and Letsinger, R. L. (1981) Debenzoylation of *N*-benzoylnucleoside derivatives with ethylenediamine/phenol. *Tetrahedron Lett.* **22,** 991–994.

18. Maher, L. J. and Dolnick, B. J. (1988) Comparative hybrid arrest by tandem antisense oligodeoxyribonucleotides or oligodeoxyribonucleoside methylphosphonate in a cell-free system. *Nucl. Acid Res.* **16,** 3341–3358.
19. Agrawal, S. and Goodchild, J. (1987) Oligodeoxynucleoside methylphosphonate: Synthesis and enzymatic degradation. *Tetrahedron Lett.* **28,** 3539–3542.
20. Sarin, P. S., Agrawal, S., Civeira, M. P., Goodchild, J. Ikeuchi, T., and Zamecnik, P. C. (1988) Inhibition of acquired immunodeficiency syndrome virus by oligonucleoside methylphosphonates. *Proc. Natl. Acad. Sci. USA* **85,** 7448–7451.
21. Schulhof, J. C., Molko, D., and Teoule, R. (1987) The final deprotection step in oligonucleotide synthesis reduced to a mild and rapid ammonia treatment by using labile base-protection groups. *Nucl. Acid Res.* **15,** 397–416.
22. Ebright, Y., Tous, G. I., Tsao, J., Fausnaugh, J., and Stein, S. (1988) Chromatographic purification of non-ionic methylphosphonate oligodeoxyribonucleoside. *J. Liquid Chromatogr.* **11,** 2005–2017.

CHAPTER 8

Oligonucleoside Phosphorothioates

Gerald Zon

1. Introduction

There are three constitutionally isomeric forms of internucleoside phosphorothioate linkages. Replacement of either the 3'- or 5'-oxygen in a phosphodiester linkage, **1**, results in the respective thiolo structures **2** or **3**, whereas substitution of a nonbridging oxygen with sulfur gives a thiono structure **4**. Oligonucleoside thiolo phosphorothioates **2** and **3**, which have an achiral phosphorus moiety and are electronically similar to **1**, are currently not synthesized by automated methods *(1,2)* as easily as their thiono counterpart *(3)*, **4**. Thiono form **4** also differs from **2** and **3** by having a chiral phosphorus *(R_p, S_p)* center and a negative charge localized on sulfur **5** *(4,5)*.

Eckstein pioneered both chemical *(6,7)* and biochemical *(8,9)* syntheses of thiono-type oligonucleoside phosphorothioates, **4/5**, and also determined that these modified linkages in either DNA or RNA could exhibit resistance to cleavage by nucleases *(6–9)*. As detailed elsewhere *(3,10–12)*, this work led over a 20 year period to numerous applications of "oligonucleotide phosphorothioate analogs," which is a term now generally meant to imply either DNA or RNA modified with one or more thiono-type linkages, **4/5**. The chemobiological applications of these phosphorothioates have dealt with enzyme biochemistry, interferon induction, autolytic processing of RNA (ribozymes) *(13)*, interactions with proteins, oligonucleotidedirected mutagenesis, and structural dynamics.

$$\begin{array}{ccc}
\text{O} & \text{O} & \text{O} \\
\| & \| & \| \\
3'\text{O}-\text{P}-\text{O}5' & 3'\text{S}-\text{P}-\text{O}5' & 3'\text{O}-\text{P}-\text{S}5' \\
| & | & | \\
\text{O}^- & \text{O}^- & \text{O}^- \\
\mathbf{1} & \mathbf{2} & \mathbf{3}
\end{array}$$

$$\begin{array}{ccc}
\text{S} & & \text{S}^- \\
\| & & | \\
3'\text{O}-\text{P}-\text{O}5' & \longleftrightarrow & 3'\text{O}-\text{P}-\text{O}5' \\
| & & \| \\
\text{O}^- & & \text{O} \\
\mathbf{4} & & \mathbf{5}
\end{array}$$

Structures 1–5.

Recently, "antisense" inhibition of genetic expression by oligonucleotide phosphorothioate analogs has emerged as an exciting new use for these compounds that has attracted considerable attention. Antisense and related studies of oligonucleoside phosphorothioates, such as those listed in Table 1, have sparked interest in more efficient and scaleable synthetic methods, purification procedures, labeling techniques, and new analytical methods—all of which are important for investigations of these biologically active compounds in cell culture and, especially, in vivo as potential therapeutic agents.

In view of the availability of published alternative procedures, as well as related information contained in companion chapters of this book, the synthetic method described here will be limited to the latest and most flexible high-yielding route to the phosphorothioates using conventional phosphoramidite chemistry *(42)*, in conjunction with stepwise sulfurization by tetraethylthiuram disulfide (TETD) *(43)*. For similar reasons, the recommended phosphorothioate purification methodology given here is limited to isolation of 5'-dimethoxytrityl (DMT) derivatives either by reversed-phase manual procedures or HPLC *(44)*. Both routes for purification have proven to be effective

Table 1
Antisense and Related Investigations
of Phosphorothioate-Modified Oligonucleotides

Viral/protein target	Reference(s)
Human immunodeficiency virus	*14–23*
Influenza virus types A and C	*24*
Hepatitis B virus	*25*
Herpes simplex virus types 1 and 2	*26–28*
Vesicular stomatitis virus	*29*
Human papilloma virus type 16	*30*
Interleukin 1β	*31*
Rabbit β-globin	*32*
Xenopus Vg1 protein	*33*
Human U1 snRNP	*34*
Human β-globin	*35*
Xenopus histone H4	*36*
Xenopus An2 and cyclin	*37*
Primer recognition proteins	*38*
Chloramphenicol acetyltransferase	*39,40*
ras	*41*

and relatively easy. Details for uniform ^{35}S-labeling have been reported earlier *(45)* and are also given in a companion chapter; consequently, an alternative nonradioisotopic fluorescent-labeling procedure is described herein. Finally, two new analytical methods are presented here: enzyme/HPLC-based composition analysis, which is analogous to that used for native DNA *(46),* and a procedure for quantifying sulfur content by use of strong anion-exchange HPLC, which is an unprecedented and advantageous application of HPLC relative to the traditional ^{31}P-NMR method.

2. Materials

1. Controlled pore glass (CPG) support.
2. β-Cyanoethyl diisopropyl phosphoramidite (CED), 0.1M in anhydrous acetonitrile (CH_3CN).
3. Tetraethylthiuram disulfide (TETD), 0.5M in anhydrous CH_3CN (ABI).
4. TCA/CH_2Cl_2 (3% w/w) 3% trichloroacetic acid/dichloromethane.
5. Acetonitrile (CH_3CN).
6. Tetrazole (0.5M) in anhydrous CH_3CN.
7. *N*-methyl imidazole (NMI) 2.4M.
8. Acetic anhydride (1M) (Ac_2O)/2,6-lutidine.

9. Polypropylene syringe.
10. Oligonucleotide purification cartridge (OPC™ Applied Biosystems, Inc., Foster City, CA).
11. Triethylamine acetate (TEAA) (2.0M).
12. Concentrated ammonia.
13. Deionized water.
14. Trifluoroacetic acid (2%).
15. Polystyrene-matrix HPLC column.
16. Acetic acid (80%).
17. Ethyl acetate (EtOAc).
18. Sodium chloride (NaCl) (1M).
19. Ethanol (EtOH).
20. Eppendorf tube (1.5 mL).
21. Sodium acetate (NaOAc) (3M).
22. LC-18DB HPLC column (Supelco, Bellefonte, PA).
23. Snake venom phosphodiesterase (SVP) 0.2 U/mg, Sigma, St. Louis, MO (cat. #P6761).
24. Nuclease P1 280 U/mg, Sigma, St. Louis, MO (cat. #8630).
25. Bacterial alkaline phosphatase (BAP) 110 U/mL, Sigma, St. Louis, MO (cat. #P4377).
26. Tris-Mg^{2+} 0.1M (0.02M $MgCl_2 \cdot 6H_2O$).
27. Tris (0.033M), pH 7.11.
28. Aminolink 2™ (Applied Biosystems, Inc., Foster City, CA).
29. $NaCO_3/NaHCO_3$ buffer (0.5M, pH 9).
30. Dimethyl sulfoxide (DMSO).
31. N-Hydroxy succimimide (NHS) ester of 5- (and 6-) carboxyfluoroscein.
32. PL-SAX HPLC column (Polymer Labs, Amherst, MA).
33. $(NH_4)_2HPO_4$ (50 mM), pH 8.2.
34. $NH_4H_2PO_4$ (50 mM), pH 6.7/1.5M KBr, 20% (v/v) CH_3CN.

3. Methods

3.1. Automated Synthesis Using Phosphoramidite Chemistry and TETD

The original approach developed *(47,48)* for automated synthesis of phosphorothioate-containing DNA using phosphoramidite chemistry involved reaction of the internucleoside phosphite linkage with elemental sulfur (S_8) in place of I_2/H_2O. This stepwise sulfurization strategy was followed by development *(15,49,50)* of a one-step procedure wherein full-length oligonucleoside H-phosphonate was reacted with S_8 just before cleavage from the solid support. Disadvantages of

Oligonucleoside Phosphorothioates

Structures 6–9.

the H-phosphonate route include the inability to sulfurize site specifically; use of an unconventional set of monomers and ancillary reagents/solvents, which include pyridine; and, perhaps most important, relatively low coupling yields leading to less product. These factors, and the desirability of eliminating CS_2 as a solvent for S_8, have led to recent reports *(43,51-53)* of several organic compounds, **6-9**, as alternatives to S_8 that may be used with high-yielding phosphoramidite (or other chemistry) for oligonucleotide synthesis. TETD **8** offers the advantages of being a relatively low-cost chemical commodity that is nonhazardous (used also as a pharmaceutical), and is both very soluble (0.5M) and indefinitely stable in acetonitrile—the prevailing solvent for DNA synthesis.

The chemistry cycle for solid-phase synthesis using phosphoramidite monomers and TETD is depicted in Fig. 1 and detailed further in Table 2. Except for use of capping after sulfurization, the cycle is essentially identical to that employed for making native DNA. The plumbing diagram for a synthesizer (ABI Model 390Z) used to conduct 50- to 200-µmol scales is shown in Fig. 2. This device for obtaining gram amounts of product features the availability of a glass-fritted activation vessel (AV) for preparation of the reactive tetrazolide (or other) species. The reactive intermediate is then automatically transferred

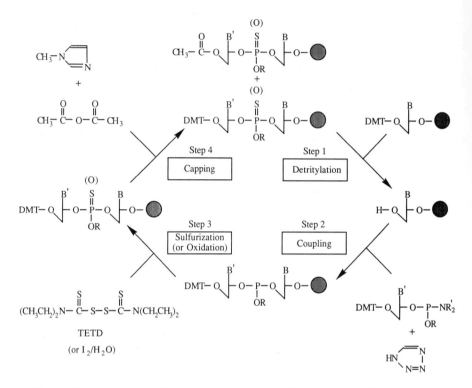

Fig. 1. The first cycle of automated sythesis of PS-modified DNA begins with detritylation (Step 1) of nucleoside attached to a solid support via the 3'-position. This is followed by coupling (Step 2) of the 5'-HO nucleoside with an incoming phosphoramidite reagent (R = CH$_2$CH$_2$CN, R' = CH [CH$_3$]$_2$) activated with 1H-tetrazole. The resultant 3', 5'-internucleoside phosphite linkage is subjected to sulfurization (Step 3) with TETD to give a thiono phosphotriester linkage; alternatively, oxidation can be carried with I$_2$/H$_2$O to give a conventional phosphotriester linkage. Capping (Step 4) unreacted 5'-HO nucleoside is achieved with acetic anhydride and N-methyl imidazole. This four-step synthesis cycle is repeated until the full-length sequence is constructed.

by means of argon gas pressure to a reaction vessel (RV), where it is vortexed with the CPG solid support at speeds of 1000–1500 rpm to ensure intimate mixing during coupling. Stepwise coupling efficiency is generally ≥98%, as measured by the DMT cation assay, and the overall sulfurization efficiency is generally ≥99%, as measured by ^{31}P NMR (Fig. 3). TETD *(54)* as well as **6** *(55)* have each been used with the phosphoramidite method to synthesize phosphorothioate-modified RNA.

Table 2
Automated Synthesis on 1- and 200-μmol Scales
Using Controlled Pore Glass (CPG) Support,
β-Cyanoethyl Diisopropyl Phosphoramidite (CED)
Chemistry, and Tetraethylthiuram Disulfide (TETD)

Step	1-μmol scale	200-μmol scale
TCA/CH_2Cl_2 (3% w/w)	60 s	19.3 min
CH_3CN wash	30 s	3.8 min
$0.1M$ CED + $0.5M$ tetrazole	21 s	4.7 min
		(6 Eq CED)
Wait	15 s	—
CH_3CN wash	—	1.0 min
$0.5M$ TETD	12 s	15.7 min
Wait	900 s	—
CH_3CN wash	40 s	5.5 min
$2.4M$ NMI + $1M$ Ac_2O/2,6-lutidine	12 s	1.0 min
Wait	10 s	—
CH_3CN wash	40 s	3.7 min

3.2. Purification by Reversed-Phase Isolation of 5'-DMT Oligonucleoside Phosphorothioates

As shown in Fig. 1, the synthetic cycle involves acetylation ("capping") of 5'-OH groups that failed to couple with the incoming phosphoramidite. The resultant 5'-acetylated "failure" sequences of chain length 1 to n-1 are cleaved from the support and deacetylated by reaction with ammonia, together with release and deprotection of the full-length 5'-DMT n-mer product. To the extent that the resultant crude material contains this n-mer product as a single 5'-DMT component, reversed-phase "affinity" purification can be highly efficient, yet relatively easy and scaleable (56). More realistically, however, the crude material will also contain shorter lengths of 5'-DMT oligomers derived from: spurious chain-growth on either the CPG or on base moieties; scission of apurinic sites; and incomplete detritylation or capping within one or more cycles of synthesis. For this reason, a simple step-gradient (batch-elution) process will generally afford more heterogeneous 5'-DMT isolate, relative to a slow-gradient (fractionation) procedure in which various product "cuts" are taken, analyzed for homogeneity, and pooled accordingly (5). On the other hand, because of the high level of refinement of the current chemistry cycle, batch elution nev-

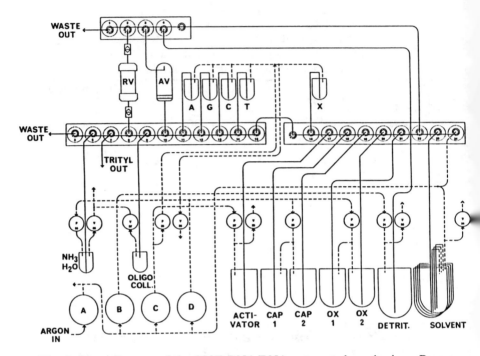

Fig. 2. Flow diagram of the 390Z DNA/RNA automated synthesizer. Preprogrammed computer-controlled flow delivery of reagents/solvents involves opening sets of individual valves (housed in three valve blocks) with concomitant argon pressurization of vessels via four pressure regulators, A–D. Abbreviations used: RV = vortexed reaction vessel (ca. 40-mL vol) for chain assembly; AV = activiation vessel for reaction of monomer with activator prior to delivery to RV, if desired; A, G, C, T, X = monomers; Cap 1 and Cap 2 = binary capping reagents; Ox 1 and Ox 2 = binary oxidants, such as H_2O/I_2, to introduce phosphodiester linkages, or tetraethylthiuram disulfide (TETD), to introduce phosphorothioate linkages, in any desired proportion/location in the oligonucleotide "backbone"; Oligo Coll. = oligonucleotide collection vessel; Detrit. = detritylation reagent.

ertheless can provide relatively high-quality final product, by comparison to the crude material. Batch elution, therefore, lends itself to very simple manual procedures using inexpensive disposable cartridges.

3.3. Purification of Phosphorothioate-Modified DNA Using Reversed-Phase HPLC/Precipitation

1. Trial separation: Depending on the HPLC column dimensions and UV detector (260 nm) sensitivity, inject ~1% of the crude 5'-DMT material at ~3 mL/min for a 10 × 250 mm column or ~12 mL/min for a 25 × 250 mm column:

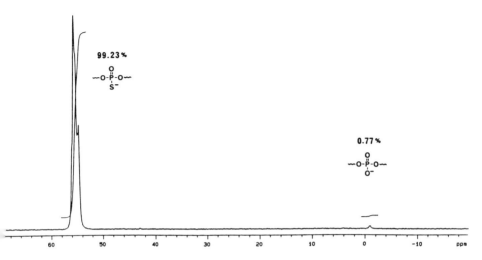

Fig. 3. ^{31}P-NMR spectrum of 5'd (TCACAGTCTGATCTCGAT)3' all-phosphorothioate derived from 100-µmol scale synthesis with TETD.

Time, min	% CH$_3$CN	% TEAA
0	20	80
10	20	80
40	50	50
50	50	50

The product elutes at a CH$_3$CN:TEAA ratio of ~40:60 and is often seen as a "doublet" because of the stereochemistry of the phosphorothioate linkage.

2. Preparative collection: Depending on the capacity of the column, inject a suitable portion of the sample using an appropriate longer wavelength (e.g., 280–300 nm) for detection. Collect the central region of the peak, discarding its leading and trailing edges. A "doublet" is sometimes seen as a result of oligomer chirality; fractionation is not necessary.
3. Concentrate the collected fraction *in vacuo* using a rotary evaporator and gentle heating (~45°C). **Caution: Avoid excess heating, which can lead to fragmentation of the product.**
4. To the residue, add 80% acetic acid to cause dissolution (a pale orange color indicates DMT cation formulation). After 30 min, remove the acetic acid *in vacuo* using a rotary evaporator either without heating or with slight warming.
5. Dissolve the residue in H$_2$O, extract twice with an equal vol of EtOAc, and reconcentrate *in vacuo*.

6. Precipitation: Dissolve the residue in $1M$ NaCl using ~1 mL/500 A_{260}. Cool in an ice bath, and add 3 vol of cold EtOH while vortexing. Collect the precipitate (centrifuge or filter) and repeat the precipitation twice more. A larger proportion of EtOH (4–5 vol), can be used if the recovery by precipitation is low.
7. Lyophilization: The final precipitate is diluted to an appropriate vol with H_2O, filtered, and lyophilized.

The original procedure developed (57) for an "oligonucleotide purification cartridge" (OPC™) format is given in Section 3.4., and is applicable to either unmodified or phosphorothioate DNA. The OPC™ procedure is particularly well suited for isolation of ^{35}S-labeled product (45). For larger amounts of analog DNA, the reversed-phase HPLC/precipitation method in Section 3.3. has proven to be efficient, reliable, and easy to implement. A typical HPLC profile is shown in Fig. 4 and was obtained using a polystyrene column charged with 100 µmol of crude 5'-DMT phosphorothioate-modified dC_{28} (S-dC_{28}). Capillary electrophoresis (Fig. 4 inset) was used to assess the purity of the resultant detritylated 5'-HO final product, which exhibits interesting sequence-independent antiHIV activity in a cytopathic-effect inhibition assay (14,20).

3.4. Manual Purification of Phosphorothioate-Modified DNA Using the OPC™ Method

1. After completion of 5'-DMT (trityl-on) synthesis, cleave the oligonucleotide from the support, and deprotect following normal protocols for the synthesis method utilized.
2. Connect an all-polypropylene syringe, an OPC™ cartridge, and male-to-male luer tip. Make sure all fittings are snug. The OPC™ cartridge may be immobilized with a laboratory clamp.
3. Flush the cartridge with 5 mL HPLC-grade acetonitrile followed by 5 mL 2.0M triethylamine acetate. Remove the syringe from the OPC™ cartridge before removing the plunger; then reinsert the syringe barrel prior to the next addition.
4. Dilute an aliquot containing ~20 A_{260} U of the crude, deprotected oligonucleotide still in concentrated ammonia with one-third volume of deionized water. The final volume of the solution should be 1–4 mL. **Important:** Keep the flow rate at 1–2 drop/s for all subsequent reagent additions.

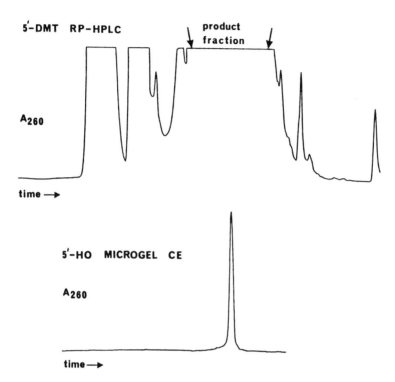

Fig. 4. **(Top)** HPLC trace of 100 µmol of crude 5'-DMT S-dC$_{28}$ eluted from a reversed-phase (polystyrene) column showing the start and stop points for the product fraction collected. **(Bottom)** Analytical trace obtained by use of a Microgel™ (Applied Biosystem, Inc., Foster City, CA) column for capillary electrophoresis of the final 5'-HO S-dC$_{28}$ product derived from the above HPLC fraction.

5. Place this solution (Step 4) in the syringe, and slowly push it through the cartridge. Save the eluted fraction, place it in the syringe, and gently push it through the cartridge. Again, this will load 5–10 A U of the crude oligonucleotide (depending on length, sequence, and synthesis quality) onto the cartridge.
6. Slowly wash the cartridge with 3 × 5 mL 1.5M ammonium hydroxide.
7. Flush the cartridge with 2 × 5 mL deionized water.
8. Detritylate the OPC™-bound oligonucleotide with 5 mL of the 2% trifluoroacetic acid solution. Gently push ~1 mL through the cartridge, wait 5 min, and then flush the remaining TFA solution through the cartridge.
9. Flush the cartridge with 2 × 5 mL deionized water.
10. Elute the purified, detritylated oligonucleotide by slowly washing the cartridge with 1mL of the 35% acetonitrile solution.

3.5. Base Composition Analysis of Phosphorothioate-Modified Oligonucleotides by HPLC of an Enzyme Digest

Exhaustive digestion of synthetic DNA using nucleases and phosphatases to give nucleotides and, ultimately, nucleosides allows the use of HPLC to measure base composition quantitatively *(46)*. At the same time, adventitious base modification can be monitored *(58)*. This procedure is readily extended to oligonucleotides having one or several phosphorothioate linkages *(47,59)*; however, a procedure for an all-phosphorothioate analog of antisense length has apparently not been reported. A new protocol for achieving complete digestion of polyphosphorothioate-modified DNA is detailed in Section 3.6. This method offers the significant advantage of using an internal nonDNA standard (inosine, I) to provide absolute mole counting and, thereby, direct evidence for complete digestion as well as a molecular formula rather than ratios of bases. This protocol can also be used to determine accurately absolute oligomer concentrations in place of approximate concentrations derived from calculated molecular extinction coefficients. Figure 5 provides representative results for a 27-mer all-phosphorothioate S-*rev*, which exhibits sequence-specific inhibition of HIV_{IIIB} expression in chronically infected H9 cells *(20)*. The deoxyuridine found in such digestions results from deamination of deoxycytidine by an enzymatic activity present in the commercially available nuclease used.

3.6. Enzyme Digest Assay for Phosphorothioate-Modified Oligonucleotides

1. Add ~1-2 nmol of oligomer and 20 nmol dI to a 1.5-mL Eppendorf tube, and dissolve in 100 µL H_2O.
2. Add 10 µL of each stock SVP and P, solution, and 5 µL BAP stock. Mix and incubate at 37°C for 20-40 h. Enzyme controls are prepared by treating 100 µL dI solution with stock enzymes and incubating similarly.
3. Digestions are stopped with 10 µL 3M NaOAc followed by precipitation of protein by addition of 500 µL EtOH, vortexing, and standing on dry ice for 0.5 h.
4. The samples are centrifuged, and supernatant is removed and taken to dryness. The resultant dried residue is dissolved in 150 µL H_2O, and 50- to 70-µL portions are analyzed by HPLC. Peak areas are converted to moles by use of updated calibration curves obtained with stock solutions containing known amounts of dNs and dI.

Oligonucleoside Phosphorothioates

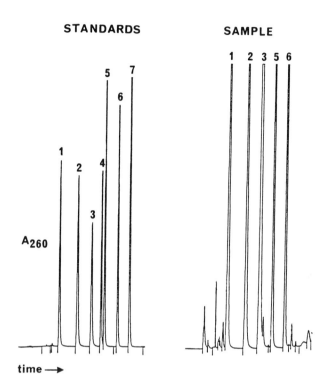

Fig. 5. (**Left**) HPLC trace of a mixture of standards used for the base composition assay: 1 = dC, 2 = dU, 3 = I, 4 = dI, 5 = dG, 6 = T, 7 = dA. (**Right**) HPLC trace of a typical mixture obtained from digestion of a representative oligonucleoside phosphorothioate sample. The peak identities are referred to by the aforementioned numbers; these peaks accounted for ca. 99% of the total integrated signal intensity.

5. HPLC conditions: Supelco LC-18DB column 0.46 × 25 cm at 45°C, flow rate = 1 µL/min. Solvent A = 1 mM TEAA, solvent B = CH_3CN, and elution is performed using a linear gradient of 1.5% B for 3 min, followed by increase to 14.5% B over 15 min, and then 10 min reequilibration at 1.5 mL/min to starting conditions.
6. Proportions of each dN are expressed according to the following general formula:

[dN (nmol) × n-mer length = mol fraction dN/mol n-mer ΣdN_n (nmol)](1)

For this compound, which has the formula $dA_0\, dG_5\, dC_{12}\, dT_{10}$, the found composition was $dA_0\, dG_{4.92}\, dC_{11.92}\, dT_{10.18}$, and the recovery of total dNs was 99.0% (average of six determinations).

3.7. Fluorescent Labeling of Oligonucleoside Phosphorothioates

Direct attachment of fluorescent "reporter groups" to one or more internucleoside phosphorothioate linkages by S-alkylation *(60–62)* results in a thiolo phosphotriester derivative. Although this type of chemical linkage may be useful for some applications *(60–62)*, its hydrolytic lability may be a concern in certain other situations, such as studies of antisense oligonucleotide cellular uptake and distribution. For such work, it is preferable to use more stable linkages; for example, aminoalkyl phosphoramidate or aminoalkyl phosphodiester. The latter type of linkage, which is extensively employed in fluorescent primer-based sequencing *(63)*, can be readily incorporated into oligonucleoside phosphorothioates at the 5'-terminus using phosphoramidite and TETD chemistry (Fig. 1) in conjunction with an aminoalkyl phosphoramidite reagent, Aminolink 2™ (Applied Biosystems, Inc.), $CF_3C(O)NH(CH_2)_6OP(OCH_3)N[CH(CH_3)_2]_2$. Fluorescently labeled products derived from the type of protocol given in Section 3.8., or suitable variants thereof, have been recently used by several research groups to investigate cellular pharmacokinetics and distribution *(64–66)*.

3.8. Fluorescent Labeling of 5'-Aminoalkyl Oligonucleoside Phosphorothioates

1. Synthesize the title amino compound according to Fig. 1 and use of Aminolink 2™ (or comparable reagent) in the final cycle.
2. To a solution of the crude amino compound (~5 µmol) in $NaCO_3$/$NaHCO_3$ buffer (0.5M, pH 9, 4 mL) add, while mixing, a DMSO solution (2 mL) of the NHS ester of 5- (and 6-) carboxyfluoroscein (200 mg), and let stand at room temperature overnight in the dark.
3. Divide into portions (6 × 1 mL), and to each add, while mixing, 5 vol (6 × 5 mL) of EtOH, and let stand on ice for ~30 min. Combine the resultant precipitates after centrifugation and removal of the supernatant.
4. Apply the combined precipitates to a polystyrene (or other reversed-phase) HPLC column (25 × 250 mm), and elute (13.5 mL/min) with 95:5 TEAA:CH_3CN for 10 min followed by a linear gradient of increasing CH_3CN (1%/min) for an additional 45 min.
5. A "center cut" of the main peak detected at 260 nm yields product (~2–3 µmol) that can be freed to TEAA:CH_3CN by repeated precipitation from

ag. NaCl:EtOH (cf. Section 3.4.). Then analyze for purity by PAGE and, if desired, HPLC using dual UV/fluorescence detection (e.g., 480 nm excitation, 520 nm emission).

3.9. Measurement of Oligonucleoside Phosphorothioate Content by Strong Anion-Exchange (SAX) HPLC

Of primary concern to any strategy for synthesis of oligonucleoside phosphorothioates is the efficiency of sulfur incorporation. ^{31}P-NMR spectroscopy, such as that used in conjunction with Fig. 3, has been the only analytical method available to date for quantitative measurement of the molar ratio of phosphodiester to phosphorothioate linkages (PO:PS). Interestingly, in the course of investigating the ion-exchange characteristics of phosphorothioate-modified oligodeoxyribonucleotides, Bergot has recently discovered and developed an alternative analytical procedure that uses HPLC *(67)*. There are significant advantages of this novel SAX-HPLC method over ^{31}P NMR: sensitivity, with regard to both dynamic range (low PO:PS) and amount of sample, accuracy, precision, and speed.

The procedure is outlined in Section 3.10. and is predicated on the perhaps surprising observation that, under the specified conditions, an *n*-mer oligonucleoside phosphorothioate is eluted from an SAX matrix in proportion to the PO:PS content, i.e., the higher the PO:PS ratio, the faster the elution—completely (or largely) independent of the location of the PO linkage (s) within the *n*-mer chain *(67)*. The SAX-HPLC profile for a resolution standard is shown in Fig. 6. This 21-mer phosphorothioate was prepared such that the PO:PS content of 3.5:96.5 would lead to accentuated PO-defect species II–IV.

3.10. Determination of Oligonucleoside Phosphorothioate Content by SAX-HPLC

Column: 15 × 0.46 cm PL-SAX (Polymer Labs)
Buffer: A: 50 mM (NH$_4$)$_2$HPO$_4$, pH 8.2
Buffer: B: 50 mM NH$_4$H$_2$PO$_4$, pH 6.7/1.5M KBr, 20% (v/v) CH$_3$CN
Gradient: 0–100% B, 40 min (linear), hold at 100% B 5 min
Flow rate: 1 mL/min
Temperature: 50°C

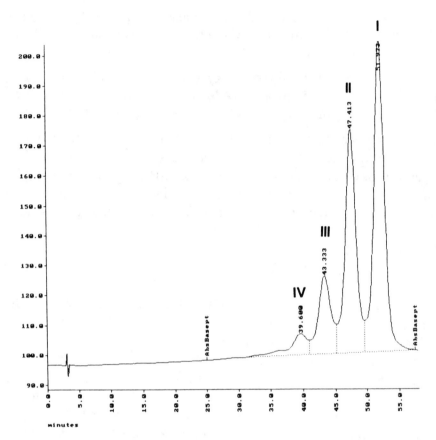

Fig. 6. Typical SAX HPLC trace obtained for a 21-mer oligonucleoside phosphorothioate resolution standard having PO/PS = 3.5/96.5, as measured by ^{31}P NMR. Peak I = 0 PO/20 PS, peak II = 1 PO/19 PS, peak III = 2 PO/18 PS, peak IV = 3 PO/17 PS.

Caution: Buffer B is very aggressive toward stainless-steel and pump seals. It is essential that the HPLC system be thoroughly purged with H$_2$O at conclusion of the assay period.

Quantitative assessment of total PS (and, by difference, PO) content is calculated using the general formula:

$$\%PS = (\text{peak height } \%/100)^{1/n-1} \times 100 \qquad (2)$$

n = number of dN residues in the oligomer; peak height = last eluted all-PS species (peak I in Fig. 6).

4. Discussion

As a final addendum to the commentary in Section 3., it is worthwhile to discuss briefly two subjects of continual interest in the area of potential pharmaceutical applications of oligonucleoside phosphorothioates: "scale-up *vis-à-vis* solid-phase or solution synthesis" and "stereochemistry."

4.1. Comparison of Solid-Phase and Solution Synthesis

Generalized solid-phase synthesis methods encompassing Fig. 1 have been analyzed *(68)* in terms of both current costs and future potential costs applicable to hypothetical automated processes up-scaled to ~10 mmol, which is 50× the 200-µmol scale mentioned above. A more recent and much more expanded analysis *(69)* has indicated that, at bulk (multikilogram) purchase prices of phosphoramidite monomers, the cost of 1 g of a typical HPLC-purified 21-mer product would include ~$750 for monomer. This monomer cost component and the costs of other necessary raw materials, which are approximately three times the monomer cost, are independent of the scale of synthesis. Consequently, increasing output to afford 100 g of purified product by either "multiplexing" smaller synthesis scales or using a single batch entails a total raw material cost of $300,000 at $3000/g. As previously noted *(69)*, total labor costs in current solid-phase synthesis and purification methodologies represent well over 75% of the product cost of goods (COGS), where COGS = total material costs + total burdened labor cost. Future improvements envisaged will adjust the contributing percentages for total raw material and total labor to approx 50:50.

With regard to maximal reduction of raw material costs, one can assume the use of a hypothetical *in situ* phosphitylation strategy with suitably derivatized (protected) nucleosides to achieve a minimum monomer cost per gram of 21 mer of $180 (as opposed to the current $750). On the other hand, the added *in situ* phosphitylation step in each cycle of solid-phase synthesis would necessarily add to the cost of chemicals and solvents. This and other factors indicate that a minimal raw materials cost for purified 21 mer is ~$200/g.

In the current manufacturing process, only ~15% of the total labor cost derives from synthesis, since the majority of work is associated with HPLC purification, postpurification processing, multiple in-process quality control analyses, final product analyses, and comprehensive documentation, as needed in the context of current Good Manufacturing Practices (GMP). Further scale-up of solid-phase synthesis could reduce synthesis labor to only ~5% of the total synthesis cost, which would therefore be almost entirely a function of the raw materials consumed, i.e., ~$200/g of a hypothetical purified 21 mer.

In the conventional solution approach using convergent synthesis *(69)*, fully protected dimer (2 mer) blocks are formed from monomers and then purified by existing chromatography methods. This process is repeated using 2 mers to make 4 mers, and so forth, thereby converging on the full-length product. The apparent utility of convergent synthesis lies in the hypothetical ability to minimize the labor of oligonucleotide synthesis by synthesizing the 2-mer, 4-mer, and so on, blocks in a very large single batch, by comparison to solid phase, thereby diluting labor costs over a large amount of material cost. Were a batch size to become large enough throughout the convergent synthesis of an oligonucleotide, the overall labor component for synthesis could be greatly reduced.

It has been additionally thought that, by purifying each fragment, the final product would be either easier to purify, of higher purity, or both, relative to the solid-phase approach. However, this assumption seems unrealistic for the near term, given the difficulty of completely deprotecting a full-length phosphotriester intermediate without causing chain scission or having other side reactions take place leading to contaminant n-1, n-2, and so forth, fragments of DNA that are not readily removed, if at all, as is already known from solid-phase methodology.

Unfortunately, it is impossible to do real cost accounting for scenarios involving convergent synthesis of phosphorothioate DNA because of the absence of a report that describes the preparation of a product in the n-mer size range of interest, i.e., $n = 20$–25. On the other hand, one can use the best fragment coupling approach to unmodified DNA and the accompanying yields that have been reported to date in order to calculate yield of crude n-mer before purification, keeping in mind that, to a first approximation, all of the postsynthesis costs (primarily labor) for convergent solution of solid phase would be roughly similar. Applying this method of analysis to a hypothetical 21 mer to be derived from a fixed amount of derivatized nucleoside starting material, it has been estimated that the current state of solid-phase synthesis already has a fourfold yield advantage over solution synthesis in terms of material cost *(69)*.

A much larger distinction between solid-phase and solution strategies is seen by analysis of the labor associated with synthesis. For solid phase, as discussed above, this cost is already known to be insignificant relative to material costs. By contrast, convergent solution synthesis involves stages that are progressively more difficult and costly to refine and scale up through systematic process development. Moreover, these stages involve progressively greater risk exposure, i.e., cost of failed batches, that impacts development expense and COGS. If one realistically balances such up-front development cost and actual operating risk with batch size, and then assumes economically acceptable batch sizes on the order of 30–50 L, the total labor for producing the final product remains large compared to solid phase *(69)*.

In conclusion, it would appear that the long-term further development of synthesis scale-up for manufacturing phosphorothioate DNA pharmaceuticals has greater potential in a solid-phase strategy compared to a convergent solution strategy. This advantage stems from much lower development cost and lower operating cost at batch sizes reflecting realistic concern for expenses, in both money and time, because of failed batches, i.e., risk exposure.

4.2. Stereochemistry of Oligonucleoside Phosphorothioates

It is generally recognized that incorporating R_p and S_p internucleoside linkages into DNA/RNA leads to the existence of 2^n-possible diastereomeric forms of a given sequence having n-centers of such chirality. Although the principles of stereochemistry dictate that diastereomers must have different physical and chemical properties, these same principles teach that neither the magnitude nor direction of such differences can be determined without recourse to either theoretical calculations or experimental observation.

Structure 10.

Up to this time, it has been possible to obtain chemically diastereomerically pure (or enriched) oligonucleoside phosphorothioates in cases with only one, two, or three phosphorothioate linkages *(47,70)*. This circumstance has therefore precluded potentially important evaluation of whether or not stereorandom and diastereomerically pure oligonucleoside phosphorothioates have significantly different properties with regard to degradation in vivo, distribution, cellular uptake, binding to RNA, antisense potency, and so forth.

Whereas one report *(70)* of a stereospecific coupling route to oligonucleoside phosphorothioates indicates little hope for practical application to constructs of antisense length, Stec et al. *(71)* have described a clever and promising new strategy to this end in Chapter 14. The new method involves synthesis and chromatographic separation of diastereomerically pure 5'-*O*-DMT nucleoside 3'-*O* (2-thio-1.3.2-oxathiaphospholane **10** for stereospecific (>99%) coupling with support-bound 5'-HO under conditions of basic catalysis. Hopefully, this method will soon afford diastereomerically enriched oligonucleoside phosphorothioates to enable evaluation of their potential utility.

References

1. Cosstick, R. and Vyle, J. S. (1989) Solid phase synthesis of oligonucleotides containing 3'-thiothymidine. *Tetrahedron Lett.* **30,** 4693–4696.
2. Mag, M., Lüking, S., and Engels, J. W. (1991) Synthesis and selective cleavage of an oligodeoxynucleotide containing a bridged internucleotide 5'-phosphorothioate linkage. *Nucl. Acids Res.* **19,** 1437–1441.
3. Zon, G. and Stec, W. J. (1991) Phosphorothioate oligonucleotides, in *Oligonucleotides and Their Analogues* (Eckstein, F., ed.) IRL Press, London, pp. 87–108.

4. Liang, C. and Allen, L. C. (1987) Sulfur does not form double bonds in phosphorothioate anions. *J. Am. Chem. Soc.* **109**, 6449–6543.
5. Baraniak, J. and Frey, P. A. (1988) Effect of ion pairing on bond order and charge localization in alkyl phosphorothioates. *J. Am. Chem. Soc.* **110**, 4059–4060.
6. Eckstein, F. (1967) A dinucleoside phosphorothioate. *Tetrahedron Lett.* 1157–1160.
7. Eckstein, F. (1967) Diuridin-3', 5'-thiophosphat. *Tetrahedron Lett.* 3495–3499.
8. Matzura, H. and Eckstein, F. (1968) A polyribonucleotide containing alternating —> P=O and —> P=S linkages. *Eur. J. Biochem.* **3**, 448–452.
9. Eckstein, F. and Gindl, H. (1970) Polyribonucleotides containing a phosphorothioate backbone. *Eur. J. Biochem.* **13**, 558–564.
10. Eckstein, F. (1985) Nucleoside phosphorothioates. *Ann. Rev. Biochem.* **54**, 367–402.
11. Gish, G. and Eckstein, F. (1988) DNA and RNA sequence determination based on phosphorothioate chemistry. *Science* **240**, 1520–1521.
12. Eckstein, F. and Gish, G. (1989) Phosphorothioates in molecular biology. *Trends in Biochem. Sci.* **14**, 97–100.
13. Herschlag, D., Piccirilli, J. A., and Cech, T. R. (1991) Ribozyme-catalyzed and nonenzymatic reactions of phosphate diesters: rate effects upon substitution of sulfur for a nonbridging phosphoryl oxygen atom. *Biochemistry* **30**, 4844–4854.
14. Matsukura, M., Shinozuka, K., Zon, G., Mitsuya, H., Reitz, M., Cohen, J. S., and Broder, S. (1987) Phosphorothioate analogs of oligodeoxynucleotides: inhibitors of replication and cytopathic effects of human immunodeficiency virus. *Proc. Natl. Acad. Sci. USA* **84**, 7706–7710.
15. Matsukura, M., Zon, G., Shinozuka, K., Stein, C. A., Mitsuya, H., and Broder, S. (1988) Synthesis of phosphorothioate analogues of oligodeoxyribonucleotides and their antiviral activity against human immunodeficiency virus (HIV). *Gene* **72**, 343–347.
16. Agrawal, S., Goodchild, J., Civeira, M. P., Thornton, A. H., and Zamecnik, P. C. (1988) Oligodeoxynucleoside phosphoramidates and phosphorothioates as inhibitors of human immunodeficiency virus. *Proc. Natl. Acad. Sci. USA* **85**, 7079–7083.
17. Agrawal, S., Ikeuchi, T., Sun, D., Sarin, P. S., Konopk, A., and Zamecnik, P. C. (1989) Inhibition of human immunodeficiency virus in early infected and chronically infected cells by antisense oligodeoxynucleotides and their phosphorothioate analogues. *Proc. Natl. Acad. Sci. USA* **86**, 7790–7794.
18. Letsinger, R. L., Zhang, G., Sun, D. K., Ikeuchi, T., and Sarin, P. S. (1989) Cholesteryl-conjugated oligonucleotides: Synthesis, properties, and activity as inhibitors of replication of human immunodeficiency virus in cell culture. *Proc. Natl. Acad. Sci. USA* **86**, 6553–6556.
19. Stein, C. A., Matsukura, M., Subasinghe, C., Broder, S., and Cohen, J. S. (1989) Phosphorothioate oligodeoxynucleotides are potent sequence nonspecific inhibitors of *de novo* infection by HIV. *Aids Res. Human Retroviruses* **5**, 639–646.
20. Matsukura, M., Zon, G., Shinozuka, K., Robert-Guroff, M., Shimada, T., Stein, C. A., Mitsuya, H., Wong-Staal, F., Cohen, J. S., and Broder, S. (1989) Regulation of viral expression of human immunodeficiency virus *in vitro* by an antisense phosphorothioate oligodeoxynucleotide against *rev* (*art/trs*) in chronically infected cells. *Proc. Natl. Acad. Sci. USA* **86**, 4244–4248.

21. Shibahara, S., Mukai, S., Morisawa, H., Nakashima, H., Kobayashi, S., and Yamamoto, N. (1989) Inhibition of human immunodeficiency virus (HIV-1) replication by synthetic oligo-RNA derivatives. *Nucl. Acids Res.* **17**, 239–252.
22. Majumdar, C., Stein, C. A., Cohen, J. S., and Broder, S. (1989) Stepwise mechanism of HIV reverse transcriptase: primer function of phosphorothioate oligodeoxynucleotide. *Biochemistry* **28**, 1340–1346.
23. Iyer, R. P., Uznanski, B., Boal, J., Storm, C., Egan, W., Matsukura, M., Broder, S., Zon, G., Wilk, A., Koziolkewicz, M., and Stec, W. J. (1990) Abasic oligodeoxyribonucleoside phosphorothioates: Synthesis and evaluation as anti-HIV-1 agents. *Nucl. Acids Res.* **18**, 2855.
24. Leiter, J. M. E., Agrawal, S., Palese, P., and Zamecnik, P. C. (1990) Inhibition of influenza virus replication by phosphorothioate oligodeoxynucleotides. *Proc. Natl. Acad. Sci. USA* **87**, 3430–3434.
25. Goodarzi, G., Gross, S. C., Tewari, A., and Watabe, K. (1990) Antisense oligodeoxyribonucleotides inhibit the expression of the gene for hepatitis B virus surface antigen. *J. Gen. Virol.* **71**, 3021–3025.
26. Ceruzzi, M. and Draper, K. (1989) The intracellular and extracellular fate of oligodeoxyribonucleotides in tissue culture systems. *Nucleosides & Nucleotides* **8**, 815–818.
27. Gao, W., Stein, C. A., Cohen, J. S., Dutschman, G.E., and Cheng, Y. C. (1989) Effect of phosphorothioate homo-oligodeoxynucleotides on herpes simplex virus type 2-induced DNA polymerase. *J. Biol. Chem.* **264**, 11,521–11,526.
28. Gao, W. Y., Hanes, R. N., Vazquez-Padua, M. A., Stein, C. A., Cohen, J. S., and Cheng, Y. C. (1990) Inhibition of herpes simplex virus type 2 growth by phosphorothioate oligodeoxynucleotides. *Antimicrobial Agents and Chemotherapy* **34**, 808–812.
29. Shea, R. G., Marsters, J. C., and Bischofberger, N. (1990) Synthesis, hybridization properties and antiviral activity of lipid-oligodeoxynucleotide conjugates. *Nucl. Acids Res.* **18**, 3777–3783.
30. Storey, A., Oates, D., Banks, L., Crawford, L., and Crook, T. (1991) Antisense phosphorothioate oligonucleotides have both specific and nonspecific effects on cells containing human papillomavirus type 16. *Nucl. Acids Res.* **19**, 4109–4114.
31. Manson, J., Brown, T., and Duff, G. (1990) Modulation of interleukin 1β gene expression using antisense phosphorothioate oligonucleotides. *Lymphokine Res.* **9**, 35–42.
32. Cazenave, C., Stein, C. A., Loreau, N., Thuong, N. T., Neckers, L. M., Subasinghe, C., Hélène, C., Cohen, J. S., and Toulmé, J. J. (1989) Comparative inhibition of rabbit globin mRNA translation by modified antisense oligodeoxynucleotides. *Nucl. Acids Res.* **17**, 4255–4273.
33. Woolf, T. M., Jennings, C. G. B., Rebagliati, M., and Melton, D. A. (1990) The stability, toxicity and effectiveness of unmodified and phosphorothioate antisense oligodeoxynucleotides in *Xenopus oocytes* and embryos. *Nucl. Acids Res.* **18**, 1763–1769.
34. Agrawal, S., Mayrand, S. H., Zamecnik, P. C., and Pederson, T. (1990) Site-specific excision from RNA by RNase H and mixed-phosphate-backbone oligodeoxynucleotides. *Proc. Natl. Acad. Sci. USA* **87**, 1401–1405.

35. Furdon, P. J., Dominski, Z., and Kole, R. (1989) RNase II cleavage of RNA hybridized to oligonucleotides containing methylphosphonate, phosphorothioate and phosphodiester bonds. *Nucl. Acids Res.* **17**, 9193–9204.
36. Baker, C., Holland, D., Edge, M., and Colman, A. (1990) Effects of oligo sequence and chemistry on the efficiency of oligodeoxyribonucleotide-mediated mRNA cleavage. *Nucl. Acids Res.* **18**, 3537–3543.
37. Dagle, J. M., Walder, J. A., and Weeks, D. L. (1990) Targeted degradation of mRNA in *Xenopus oocytes* and embryos directed by modified oligonucleotides: Studies of An2 and cyclin in embryogenesis. *Nucl. Acids Res.* **18**, 4751–4757.
38. Kumble, K. D., Iversen, P. L., and Vishwanatha, J. K. (1992) The role of primer recognition in DNA replication: inhibition of cellular proliferation by antisense oligodeoxyribonucleotides. *J. Cell Sci.* **101**, 35–41.
39. Marcus-Sekura, C. J., Woerner, A.M., Shinozuka, K., Zon, G., and Quinnan, G. V. (1987) Comparative inhibition of chloramphenicol acetyltransferase gene expression by antisense oligonucleotide analogues having alkyl phosphotriester, methylphosphonate and phosphorothioate linkages. *Nucl. Acids Res.* **15**, 5749–5763.
40. Young, S. and Wagner, R. W. (1991) Hybridization and dissociation rates of phosphodiester or modified oligodeoxynucleotides with RNA at near-physiological conditions. *Nucl. Acids Res.* **19**, 2463–2474.
41. Chang, E. H., Yu, Z., Shinozuka, K., Zon, G., Wilson, W. D., and Strekowska, A. (1989) Comparative inhibition of *ras* p21 protein synthesis with phosphorus-modified antisense oligonucleotides. *Anti-Cancer Drug Design* **4**, 221–232.
42. Efcavitch, J. W. (1988) Automated system for the optimized chemical synthesis of oligodeoxyribonucleotides, in *Macromolecular Sequencing and Synthesis Selected Methods and Applications* (Schlesinger, D. H., ed.) Liss, New York, pp. 221–234.
43. Vu, H. and Hirschbein, B. L. (1991) Internucleotide phosphite sulfurization with tetraethylthiuram disulfide phosphorothioate oligonucleotide synthesis via phosphoramidite chemistry. *Tetrahedron Lett.* **32**, 3005–3008.
44. Zon, G. (1990) Purification of synthetic oligodeoxyribonucleotides, in *High Performance Liquid Chromatography in Biotechnology* (Hancock, W. S., ed.) Wiley, New York, pp. 301–397.
45. Stein, C. A., Iversen, P. L., Subasinghe, C., Cohen, J. S., Stec, W. J., and Zon, G. (1990) Preparation of ^{35}S-labeled polyphosphorothioate oligodeoxyribonucleotides by use of hydrogen phosphonate chemistry. *Anal. Biochem.* **188**, 11–16.
46. Eadie, J. S., McBride, L. J., Efcavitch, W., Hoff, L. B., and Cathcart, R. (1987) High-performance liquid chromatographic analysis of oligodeoxyribonucleotide base composition. *Anal. Biochem.* **165**, 442–447.
47. Stec, W. J., Zon, G., Egan, W., and Stec, B. (1984) Automated solid-phase synthesis, separation, and stereochemistry of phosphorothioate analogues of oligodeoxyribonucleotides. *J. Am. Chem. Soc.* **106**, 6077–6079.
48. Ott, J. and Eckstein, F. (1987) Protection of oligonucleotide primers against degradation by DNA polymerase I. *Biochemistry* **26**, 8237.

49. Froehler, B. C. (1986) Deoxynucleoside H-phosphonate diester intermediates in the synthesis of internucleotide phosphate analogues, *Tetrahedron Lett.* **27**, 5575–5578.
50. Andrus, A., Efcavitch, J. W., McBride, L. J., and Giusti, B. (1988) Novel activating and capping reagents for improved hydrogen-phosphonate DNA synthesis. *Tetrahedron Lett.* **29**, 861–864.
51. Iyer, R. P., Phillips, L. R., Egan, W., Regan, J. B., and Beaucage, S. L. (1990) The automated synthesis of sulfur-containing oligodeoxyribonucleotides using 3H-1,2-benzodithiol-3-one 1,1-dioxide as a sulfur-transfer reagent. *J. Org. Chem.* **55**, 4693–4699.
52. Kamer, P. C. J., Roeien, H.C. P. F., van den Elst, H., van der Marel, G. A., and van Boom, J. H. (1989) An efficient approach toward the synthesis of phosphorothioate diesters via the Schönberg reaction. *Tetrahedron Lett.* **30**, 6757–6760.
53. Stawinski, J. and Thelin, M. (1991) Nucleoside H-phosphonates. 13. Studies on 3H-1,2-benzodithiol-3-one derivatives as sulfurizing reagents for H-phosphonate and H-phosphonothioate diesters. *J. Org. Chem.* **56**, 5169–5175.
54. Slim, G. and Gait, M. J. (1991) Configurationally defined phosphorothioate-containing oligoribonucleotides in the study of the mechanism of cleavage of hammerhead ribozymes. *Nucl. Acids Res.* **19**, 1183–1188.
55. Morvan, F., Rayner, B., and Imbach, J. L. (1990) Modified oligonucleotides: IV. Solid phase synthesis and preliminary evaluation of phosphorothioate RNA as potential antisense agents. *Tetrahedron Lett.* **31**, 7149–7152.
56. Zon, G. and Thompson, J.A. (1986) A review of high-performance liquid chromatography in nucleic acids research. *BioChromatography* **1**, 22–32.
57. McBride, L. J., McCollum, C., Davidson, S., Efcavitch, J. W., Andrus, A., and Lombardi, S. J. (1988) A new, reliable cartridge for the rapid purification of synthetic DNA, *BioTechniques* **6**, 362–367.
58. Eadie, J. S. and Davidson, D. S. (1987) Guanine modification during chemical DNA synthesis. *Nucl. Acids Res.* **15**, 8333–8349.
59. Stec, W. J. and Zon, G. (1984) Synthesis, separation, and stereochemistry of diastereomeric oligodeoxyribonucleotides having a 5'-terminal internucleotide phosphorothioate linkage. *Tetrahedron Lett.* **25**, 5275–5278.
60. Hodges, R. R., Conway, N. E., and McLaughlin, L. W. (1989) "Post-assay" covalent labelling of phosphorothioate-containing nucleic acids with multiple fluorescent markers. *Biochemistry* **28**, 261–267.
61. Fidanza, J. A. and McLaughlin, L. W. (1989) Introduction of reporter groups at specific sites in DNA containing phosphorothioate diesters. *J. Am. Chem. Soc.* **111**, 9117–9119.
62. Agrawal, S. and Zamecnik, P. C. (1990) Site specific functionalization of oligonucleotides for attaching two different reporter groups. *Nucl. Acids Res.* **18**, 5419–5423.
63. Connell, C., Fung, S., Heiner, C., Bridgham, J., Chakerian, V., Heron, E., Jones, B., Menchen, S., Mordan, W., Raff, J., Recknor, M., Smith, L., Springer, J., Woo, S., and Hunkapiller, M. (1987) Automated DNA sequence analysis. *BioTechniques* **5**, 342–347.

64. Chin, D. J., Green, G. A., Zon, G., Szoka, F.C., and Straubinger, R. M. (1990) Rapid nuclear accumulation of injected oligodeoxyribonucleotides. *The New Biologist* **2**, 1091–1100.
65. Iversen, P. L., Zhu, S., Meyer, A., and Zon, G. (1992) Cellular uptake and subcellular distribution of phosphorothioate oligonucleotides into cultured cells. *Antisense Res. Dev.* **2**.
66. Marti, G., Egan, W., Noguchi, P., Zon, G., Matsukura, M., and Broder, S. (1992) Oligodeoxyribonucleotide phosphorothioate fluxes and localization in hematopoietic cells. *Antisense Res. Dev.* **2**, 27–39.
67. Bergot, B. J. and Egan, W. (1992) Separation of synthetic phosphorothioate oligodeoxynucleotides from their oxygenated (phosphodiester) defect species by strong-anion exchange high-performance liquid chromatography. *J. Chromatogr.* **599**, 34–42.
68. Geiser, T. (1990) Large-scale economic synthesis of antisense phosphorothioate analogues of DNA for preclinical investigations. *Ann. NY Acad. Sci.* **616**, 173–183.
69. Zon, G. and Geiser, T. (1991) Phosphorothioate oligonucleotides: Chemistry, purification, analysis, scale-up and future directions. *Anticancer Drug Design,* **6**, 539–568.
70. Lesnikowski, Z. J. and Jaworska, M. M. (1989) Studies on stereospecific formation of p-chiral internucleotide linkage. Synthesis of (R_P,R_P)-and (S_P,S_P)-thymidylyl (3',5')thymidylyl (3',5')thymidine di(O,O,-phosphorothioate using 2-nitro-benzyl group as a new S-protection. *Tetrahedron Lett.* **30**, 3821–3824.
71. Stec, W. J., Grajkowski, A., Koziolkiewicz, M., and Uznanski, B. (1991) Novel route to oligo(deoxyribonucleoside phosphorothioates. Stereocontrolled synthesis of P-chiral oligo(deoxyribonucleoside phosphorothioates) *Nucl. Acids Res.* **19**, 5883–5888.

CHAPTER 9

Synthesis and Purification of Phosphorodithioate DNA

W. T. Wiesler, W. S. Marshall, and M. H. Caruthers

1. Introduction

Oligonucleotide analogs bearing modified phosphodiester linkages have been the focus of considerable interest in the antisense field, with methylphosphonates and phosphorothioates being the most extensively studied derivatives to date *(1–3)*. These modifications, whereby a single nonbridging oxygen atom is replaced with either a methyl group or sulfur atom, generate certain very desirable oligomer characteristics *(4–7)*, such as resistance toward nucleases, retention of the ability to form duplexes with natural DNA or RNA, and, relative to the phosphorothioate derivative, an ability to stimulate RNase H activity. Unfortunately, and in contrast to natural DNA, the phosphorus center is rendered chiral by these substitutions. Thus, for a deoxyoligonucleotide with n phosphodiester linkages, there can be 2^n stereoisomers with perhaps only one being the most active. Recent attempts to overcome this problem have focused on developing chemistries that either are stereoselective and generate oligomers enriched in one diastereomer *(8,9)* or are designed to synthesize new, phosphorus achiral analogs. In addition to being phosphorus achiral, perhaps new analogs may also have unique and attractive properties relative to the antisense area.

One derivative, called phosphorodithioate DNA (or dithioate DNA), which is in the latter category, is a relatively new analog having both nonbridging oxygen atoms replaced with sulfur *(10)*. Like natural DNA, it is achiral at phosphorus. Additionally, recent research has demonstrated that this analog is completely resistant to nuclease degradation *(11–13)*, forms complexes having somewhat reduced stability with normal DNA, activates endogenous RNase H in HeLa cells, binds with very high affinity to reverse transcriptases, and can be rapidly labeled postsynthetically in aqueous solution with fluorescent and spin-label probes *(14,15)*. Because of these encouraging results, there have been numerous synthetic efforts directed toward the development of this chemistry *(16)*, with perhaps the most currently satisfactory method using a silica support and deoxynucleoside phosphorothioamidites as synthons *(15–18)*. Here a new sulfur-protecting group for the deoxynucleoside phosphorothioamidite synthons is described. Using this new derivative, completely compatible methodologies have been developed for synthesizing phosphorodithioate DNA or oligomers having both dithioate and natural internucleotide linkages.

2. Materials

2.1. General Procedures

Proton nuclear magnetic resonance spectra (^{1}H-NMR) are recorded on either a Bruker WM-250 or a Varian VXR-300S in deuterated chloroform with tetramethylsilane as internal standard. Phosphorus nuclear magnetic resonance spectra (^{31}P-NMR) are recorded on either a JEOL FX90Q or a Varian VXR-300S in either acetonitrile (2'-deoxynucleoside 3'-phosphorothioamidites) or D_2O (deoxyoligonucleotides, 40,000 transients minimum) with 85% phosphoric acid as external standard. Thin-layer chromatography (TLC) is on aluminum-backed sheets (silica gel 60F, 0.2 mm, E. Merck, Darmstadt, Germany). Preparative chromatography is by flash-column chromatography on silica gel 60, 230–400 mesh (Macherey Nagel, Dueren, Germany). Solutions are concentrated *in vacuo* at 40°C or lower using an aspirator or an oil vacuum pump. Solids are dried at rt in a desiccator over phosphorus pentoxide or potassium hydroxide pellets. 5'-*O*-dimethoxytrityl-2'-deoxythymidine, 5'-*O*-dimethoxytrityl-4-*N*-benzoyl-2'-deoxycytidine, 5'-*O*-dimethoxytrityl-6-*N*-benzoyl-2'-deoxyadenosine, 5'-*O*-dimethoxytrityl-2-*N*-isobutyryl-2'-deoxyguanosine are from Cruachem (Sterling, VA).

2.2. Solvents and Reagents

Pyridine and dichloromethane are freshly distilled over calcium hydride. Triethylamine is distilled over toluenesulfonyl chloride and then calcium hydride. Anhydrous diethylether is used directly. Acetonitrile is distilled over phosphorus pentoxide and then calcium hydride. Tetrahydrofuran is distilled over sodium metal in the presence of benzophenone. 1-H-Tetrazole (tetrazole; Aldrich, Milwaukee, WI) is sublimed before use. The preparation or purification of other reagents and solvents has been presented elsewhere *(17)*. DNA synthesis is performed on an Applied Biosystems 380A automated DNA synthesizer.

2.3. Quantitation of Deoxyoligonucleotides

The concentrations of all oligonucleotide stock solutions are determined by their UV absorbance. Values of molar absorbtivity coefficients for the various bases at neutral pH are taken from *CRC Press Handbook of Biochemistry and Molecular Biology*. For heteropolymers, concentrations are determined at 260 nm by summing the absorbtivity coefficients for all bases in the oligomer and using Beers Law. Homopolymers are similarly quantitated at their respective wavelength of maximum absorbance.

2.4. Synthesis of Ethanedithiol Monobenzoate

Ethanedithiol (1.06 mol, 100 g) is placed in a 2-L three-necked flask containing anhydrous diethylether (400 mL), and anhydrous pyridine (100 mL), and the solution cooled to 0°C. Benzoyl chloride (1 Eq, 123 mL) is then added dropwise over 2 h with mechanical stirring. After an additional 1 h at 0°C, pyridine hydrochloride is removed by filtration from the soluble reaction product and washed with diethylether. The filtrates are combined, concentrated to an oil, dissolved in hot methanol (500 mL), and placed at –5°C overnight. The white crystals of *bis*-thiobenzoate are removed by filtration and washed with cold methanol. The combined filtrates are then concentrated to approx 150 mL of a yellow oil, and excess ethanedithiol removed under high vacuum while heating to 60°C. The ethanedithiol is collected in a –78°C trap. The desired product is collected (0.2 mm Hg, 110°C) using a microdistillation apparatus while the reaction mixture is at 160°C in an oil bath. The absence of ethanedithiol contamination of the product is confirmed by ^{1}H-NMR—^{1}H-NMR ($CDCl_3$) δ 7.95 (d, 2H), 7.57 (t, 1H), 7.44 (m, 2H), 3.30 (t, 2H), 2.78 (m, 2H), 1.70 (t, 1H).

2.5. Synthesis of Tris(Pyrrolidino)Phosphine

Phosphorus trichloride (370 mmol, 50.4 g), and anhydrous diethylether (400 mL) are placed in a 1-L flask fitted with a 250-mL addition funnel containing trimethylsilylpyrrolidine (3.3 Eq, 173.5 g). After cooling to −10°C (ice/brine) under argon with stirring, trimethylsilylpyrrolidine is added dropwise over 1.5 h, and the reaction is allowed to proceed for an additional hour. The salts are then removed by filtration from the soluble reaction product and washed with anhydrous diethylether. The filtrates are combined and concentrated using a rotary evaporator to yield approx 100 mL of crude phosphine. Distillation at reduced pressure (0.25 mm Hg) affords an initial fraction (40–103°C, 5–10 g) that is discarded. The product is then obtained in 86% yield as a clear, colorless oil (104–110°C, 76.3 g)—^{31}P-NMR δ 104.6 (THF); 102.8 (CDCl$_3$).

2.6. A General Procedure for the Synthesis of 2'-Deoxynucleoside-3'-yl S-(β-Thiobenzoylethyl) Pyrrolidinophosphorothioamidites

A suitably protected 5'-O-dimethoxytrityl-2'-deoxynucleoside (2 mmol) is dissolved in anhydrous dichloromethane (30 mL) containing approx 5 g of 3 Å molecular sieves. Tris(pyrrolidino)phosphine (460 mL, 2.0 mmol) and tetrazole in seven aliquots (7 × 0.2 mL of 0.5M tetrazole in anhydrous acetonitrile) at 2-min intervals are first added to the reaction mixture. This is followed by addition of trimethylsilylimidazole (30 µL, 0.1 Eq), and, after 5 min, first tetrazole (10.8 mL of 0.5M tetrazole in anhydrous acetonitrile) and then immediately ethanedithiol monobenzoate (440 µL, 2.6 mmol). The reaction is allowed to proceed for 105 s (thymidine), 120 s (deoxycytidine and deoxyadenosine), or 150 s (deoxyguanosine), and then quenched by pouring it into dichloromethane (75 mL) containing triethylamine (5 mL). The reaction mixture is washed successively with saturated sodium bicarbonate (100 mL), 10% sodium carbonate (2 × 100 mL), and brine (100 mL). The organic layer is dried over sodium sulfate for 15 min, reisolated free of salt by filtration, buffered against acids by addition of triethylamine (5 mL), and concentrated on a rotary evaporator. The resulting syrup is dissolved in toluene (25 mL) containing triethylamine (5 mL), and the product is isolated by precipitation into vigorously stirred heptane (900 mL). After decanting most of the heptane, the white precipitate is collected by filtration and dried in vacuum to give the desired product in 75–80% yield—^{31}P-NMR δ 162.7, 159.2 (T); 162.1, 159.2 (C^{Bz}); 160.3, 159.2 (A^{Bz}); 159.7, 158.6 (G^{ib}).

3. Methods

3.1. Preparation of 2'-Deoxynucleoside 3'-Phosphorothioamidites

Much of our effort has been directed toward the development of procedures that yield 2'-deoxynucleoside 3'-phosphorothioamidites in high yield and free of impurities. Since these monomers cannot be purified via the column chromatography procedures that are used for the standard 2'-deoxynucleoside 3'-phosphoramidites, reaction conditions are required whereby these monomers can be obtained free of unwanted and reactive byproducts. Using conditions similar to those previously developed in this laboratory (17,18), a one-pot, two-step synthesis has been devised (Scheme 1). Phosphitylation is achieved with tris(pyrrolidino)phosphine under tetrazole catalysis to give a *bis*-(pyrrolidino)phosphite intermediate. Without isolation, this intermediate is converted to the phosphorothioamidite product by treatment with monobenzoylethanedithiol and additional tetrazole. After an aqueous work-up to remove tetrazole and tetrazolide salts, the 2'-deoxynucleoside 3'-phosphorothioamidite is isolated by precipitation from heptane in very good yields (90% at 90–95% purity by ^{31}P-NMR). The precipitation step removes excess thiol and phosphines, while yielding the synthons as stable powders.

A potential problem that had to be solved was neutralization of the two equivalents of pyrrolidine that are generated during this reaction sequence. This is because pyrrolidine is very effective at removing the benzoyl-protecting group from the product. In order to avoid the formation of this byproduct, a large excess of tetrazole is used to increase the reaction rate and to buffer the basicity of pyrrolidine by the formation of the pyrrolidinium tetrazolide salt. Another byproduct is the 2'-dideoxynucleoside phosphoramidite, which forms by reaction of 2'-deoxynucleoside with the *bis*-(pyrrolidino)phosphite intermediate. This byproduct, even at levels of 1% or less, generates branched oligomers on the polymer support. These branched side products are easily detected as oligonucleotides that migrate more slowly than the product by polyacrylamide gel electrophoresis. We have recently found that a midsynthesis treatment with trimethylsilylimidazole effectively caps any unphosphitylated 2'-deoxynucleoside and, thus, prevents the subsequent formation of 2'-dideoxynucleoside phosphoramidites during the second step of the phosphorothioamidite synthesis.

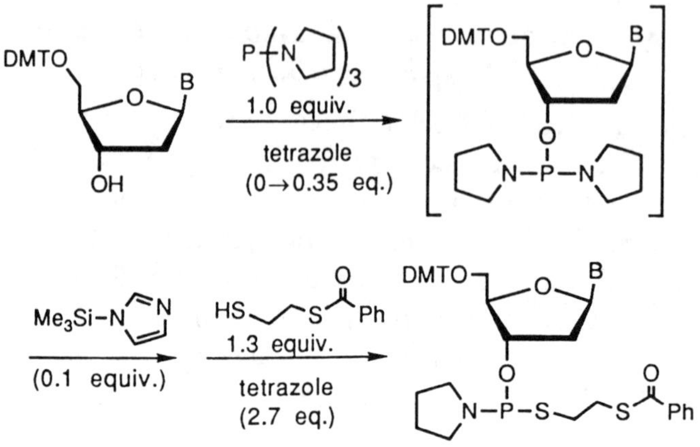

Scheme 1. Synthesis of deoxynucleoside phosphorothioamidites.

1. B = T
2. B = C^{Bz}
3. B = A^{Bz}
4. B = G^{ib}

3.2. Synthesis of Phosphorodithioate DNA

The solid-phase synthesis of phosphorodithioate DNA oligomers is similar to conventional procedures for preparing deoxyoligonucleotides (19), but there are a few important modifications (Scheme 2, Table 1). Following detritylation of a 2'-deoxynucleoside linked to controlled pore glass, the 2'-deoxynucleoside 3'-phosphorothioamidite synthons are coupled to the growing oligomers on the support by conventional tetrazole activation. The resulting thiophosphite triester is oxidized with elemental sulfur to the desired phosphorodithioate triester. Unreacted support-linked 2'-deoxynucleoside is then capped by acylation with acetic anhydride, which completes the cycle for addition of one nucleotide. By using a pyrrolidine amide leaving group in the phosphorothioamidite synthon, the coupling rates are comparable to those observed with conventional 2'-deoxynucleoside 3'-diisopropylphosphoramidites. Earlier work clearly demonstrates that this change in amide leaving groups is essential in order to retain rapid reaction rates (18). For the diisopropylphosphorothioamidites, tetrazole activation is very slow, and more acidic catalysts are required. Under these more acidic conditions, yields of condensations are less, and many side products are generated.

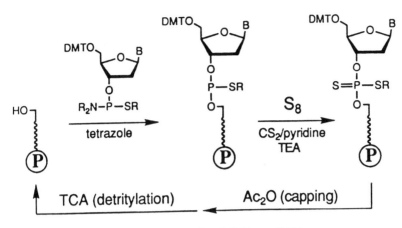

Scheme 2. Synthesis of dithioate DNA.

On completion of the automated synthesis, removal of the sulfur-protecting group converts the phosphorodithioate triester to the corresponding diester linkage (Scheme 3). Two different sulfur-protecting groups have been used in previous approaches with good, but limited success. β-cyanoethyl, which is widely used in conventional DNA synthesis for oxygen protection, can also serve to block sulfur. Although its removal conveniently proceeds under the ammonolysis treatment for deprotection of purines and pyrimidines, nonspecific hydrolysis at phosphorus competes with the desired β-elimination to product. As a consequence, high levels (8–10%) of phosphoromonothioate are incorporated in deoxyoligonucleotides (^{31}P-NMR). Efforts to reduce this contamination by varying deprotection solvents and conditions have been unsuccessful.

Phosphorothioate contamination is reduced to a much more acceptable level (2–4%) by using 2,4-dichlorobenzyl protection of sulfur *(17)*. This group can be removed from the phosphorodithiotriester linkage by treatment with a thiophenolate solution for 5 h. There are several reasons why this group is unsatisfactory. One is that thiophenol is an unattractive reagent for routine use in DNA synthesis. Another is that the 2,4-dichlorobenzyl group renders the phosphorothioamidite synthons relatively unstable toward oxidation in solvents. As a consequence, the solutions of 2'-deoxynucleoside 3'-phosphorothioamidites on the DNA synthesizer must be changed more frequently than desired.

Table 1
Chemical Steps for Synthesis of Dithioate DNA on a Solid Support

Step	Reagent or solvent[a]	Purpose	Time, min
(i)	a. Trichloroacetic acid in CH_2Cl_2 (3%, w/v)	Detritylation	0.50
	b. CH_2Cl_2	Wash	0.50
	c. Acetonitrile	Wash	0.50
	d. Dry acetonitrile	Wash	0.50
(ii)	a. Activated nucleotide in acetonitrile[b]	Add nucleotide	0.75
	b. Repeat step a	Complete nucleotide addition	1.50
(iii)	a. Sulfur in CS_2:pyridine: TEA (95:95:10; v/v/v)[c]	Oxidation	1.00
	b. CS_2	Wash	0.50
	c. CH_3OH	Wash	0.50
	d. CH_2Cl_2	Wash	0.50
(iv)	a. NMI:THF (30:70; v/v)[d] acetic anhydride:lutidine: THF (2:2:15; v/v/v)	Capping reaction	0.50
	b. CH_2Cl_2	Wash	0.50

[a]Multiple washes with the same solvent are possible.
[b]For each micromole of deoxynucleoside attached to silica, 0.48M tetrazole (0.125 mL) and 0.15M deoxynucleoside phosphorothioamidite (0.125 mL) are premixed in acetonitrile. During this coupling step, activated nucleotide and tetrazole are flushed from the lines leading into the reaction chamber. This procedure reduces contaminating phosphorothioate internucleotide linkages.
[c]5% sulfur by weight in CS_2:pyridine:TEA (95:95:10; v/v/v); TEA, triethylamine.
[d]NMI, N-methylimidazole; THF, tetrahydrofuran.

Because of these various limitations, we have developed a novel base-labile sulfur-protecting group, the β-thiobenzoylethyl, and incorporated it into the 2'-deoxynucleoside 3'-phosphorothioamidite synthons. This protecting group combines the best features of the two previously employed for dithioate DNA synthesis, since it has stability toward oxidation similar to the β-cyanoethyl group and can be removed during the ammonolysis deprotection procedure. The mechanism by which it is removed, however, involves selective nucleophilic displacement similar to removal of the 2,4-dichlorobenzyl group with

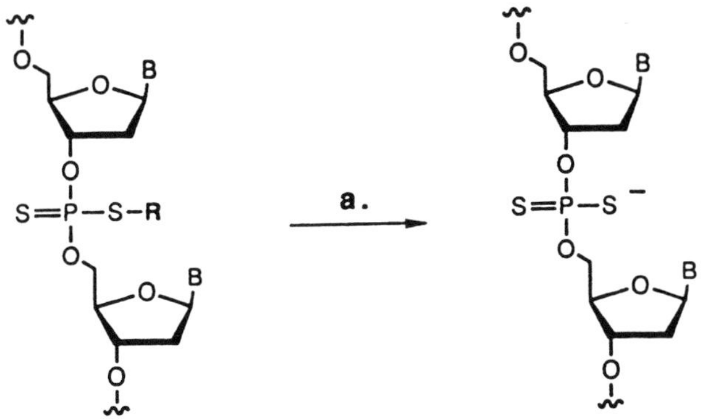

Phosphorodithioate DNA

Scheme 3. Deprotection of dithioate DNA.

R = β-cyanoethyl (**a.** = NH$_4$OH, 1 h RT, 15 h 55°C)
R = dichlorobenzyl (**a.** = thiophenol/TEA/dioxane, 5 h; then NH$_4$OH, 1 h RT, 15 h 55°C)
R = β-thiobenzoyl ethyl (**a.** = NH$_3$/EtOH/benzene, 1 h RT, 15 h 55°C)

thiophenolate. The selectivity of this reaction yields a low level of phosphorothioate contamination (2–5%) comparable to the results obtained with 2,4-dichlorobenzyl. Another feature of β-thiobenzoylethyl is that it can be considered a protected protecting group. This is because ammonolysis leads to rapid deacylation of the pendant thiol to yield a thiolate intermediate, which, via intramolecular nucleophilic displacement, then causes elimination of ethylene sulfide to give the deprotected phosphorodithioate (Fig. 1).

A series of β-thioethyl acyl phosphorothioamidites have been prepared and tested by synthesizing octathymidine deoxyoligonucleotides. Among those tested and analyzed by ^{31}P-NMR, the benzoyl group was found to yield the best results. More stable acyl groups (i.e., pivaloyl) gave higher levels of monothioate contamination (8–10%) during ammonolysis. Presumably this is because of slower deprotection relative to nonspecific hydrolysis of the phosphorodithioate triester. Levels of monothioate contamination have recently been reduced by using more hydrophobic ammonia solutions. For example, ethanolic ammonia containing 15% benzene yields >97% phosphorodithioate linkages.

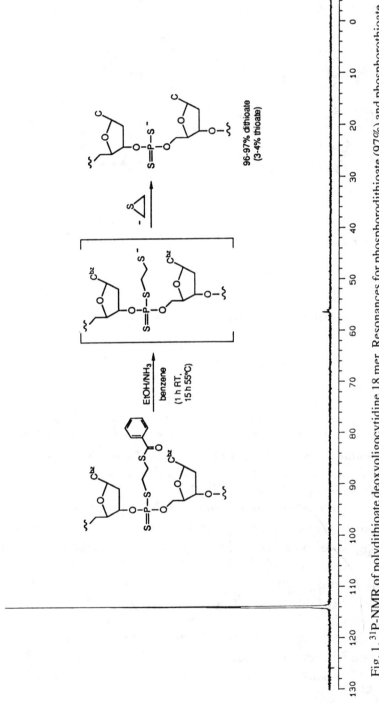

Fig. 1. ³¹P-NMR of polydithioate deoxyoligocytidine 18 mer. Resonances for phosphorodithioate (97%) and phosphorothioate (3%) DNAs are at 114 and 56 ppm, respectively. Deprotection conditions and the two-step deblocking mechanism for β-benzoylmercaptoethyl are also shown.

3.3. Deprotection of Dithioate DNA

After synthesis, the CPG column is removed from the synthesizer and dried with argon. The CPG support is removed, placed in a 1-dram vial, and ethanolic ammonia (850 µL), and benzene (150 µL) are added. The vial is sealed with a Teflon™-lined screw cap, allowed to remain at rt for 1 h, and then placed in a 55°C oven for 15 h. After removal from the oven and cooling to rt, toluene (2 mL) is first added, and the combined solution carefully removed from the CPG support by pipeting. This solution contains traces of DNA and nearly all of the benzamide. Water (1 mL) is then added to the vial containing the CPG in order to extract the dithioate DNA from the support. In the case of purine-rich sequences, water:acetonitrile (3:1) is a better solvent for the oligomers.

3.4. Purification of Dithioate DNA by High-Pressure Liquid Chromatography (HPLC)

HPLC offers a high-capacity method for purification of dithioate deoxyoligonucleotides. Our current method of choice uses a divinyl benzene/polystyrene copolymer reverse-phase column (Hamilton PRP-1) with acetonitrile in 50 mM triethylammonium acetate (pH 7) as mobile phase. Oligonucleotides are synthesized without removing the trityl-protecting group after the final coupling ("Trityl ON"). They are then injected and subjected to a linear gradient running from 15–75% acetonitrile over 66 min. Trityl ON oligomer elutes as a well-separated peak (34–40% acetonitrile) that is dependent on oligomer length. Desired fractions are collected and evaporated. Detritylation is accomplished with 80% acetic acid for 10 min on ice. After three diethylether extractions, the solutions are neutralized with an equal vol of concentrated ammonium hydroxide, and the oligomers again injected and eluted using the same gradient. The profile from the injection of a detritylated dithioate deoxyoligonucleotide solution is shown in Fig. 2A. The desired fractions are pooled, evaporated to dryness, resuspended in water, and quantitated by UV spectroscopy.

3.5. Purification of Dithioate DNA by Polyacrylamide Gel Electrophoresis (PAGE)

PAGE affords a simple method for analysis and high-resolution purification of dithioate DNA after synthesis and deprotection. This can be seen from the analytical slab gel shown in Fig. 2B. Comparison to a control (phosphodiester or phosphorothioate) can be used to estimate quali-

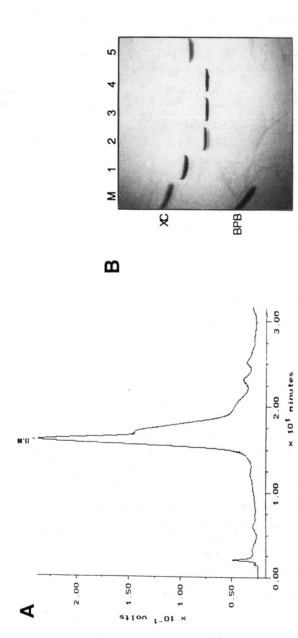

Fig. 2A. Chromatogram of an injection of phosphorodithioate DNA on a Hamilton PRP-1 reverse-phase HPLC column. After trityl ON chromatography and subsequent detritylation, 1.0 A_{260} U of 5'-TCGTCGCTGTCTCCGCTTCTT CCTGCC-3' containing exclusively phosphorodithioate linkages was injected and subjected to the gradient described in the text at a flow rate of 1.0 mL/min. Detection was at 260 nm. Phosphorodithioate oligomer elutes as a well-separated peak at 15.88 min. The chromatogram also shows a small peak at 23 min corresponding to trityl ON oligomer. B. Denaturing analytical polyacrylamide slab gel of phosphorodithioate and phosphorothioate DNA after two rounds of reverse-phase HPLC. After PAGE as described in the text, the resulting gel was visualized by UV shadowing and photographed. M: marker dyes, bromophenol blue (BPB), xylene cyanol (XC); 1: 5'-TCGTCGCTGTCTCCGCTTCTTCCTGCC-3', exclusively phosphorodithioate; 2: 5'-TCGTCGCTGTCTCCGCTTCT-3', exclusively phosphorodithioate; 3: 5'-TxCxGxTxCpGxCpTxGpTxCpTxCpCxGpTxCpCxGpCxTpGpTx CpTpCxCpGpCxTxTxCxT-3', where x = phosphorodithioate, p = phosphorothioate; 4: 5'-TxCxGxTxCpGpCxTpGpTx CCTGCC-3', exclusively phosphorothioate; 5: 5'-TCGTCGCTGTCTTCCGCTTCTT-

tatively the amount of product relative to failure sequences in the mixture. Oligonucleotides are separated on a 20% polyacrylamide, $8M$ urea slab gel ($20 \times 20 \times 0.15$ cm) after heat denaturation in 50% deionized formamide. Gels are typically run at 20–25 V/cm at constant voltage. Products can be visualized by UV shadowing, and if desired, gel material containing the oligomer cut out with a razor blade for recovery of the oligomer.

Research into the extraction of dithioate oligomers from gel slices is an ongoing process. Currently, a method involving electroelution of the oligomer onto DEAE cellulose followed by elution with ionic denaturants (sodium thiocyanate or sodium chloride/formamide) shows promise as a high-yielding procedure. The freeze/thaw method described below is effective for recovering dithioate analogs as well as conventional deoxyoligonucleotides.

A gel slice containing the desired product can be minced and immersed in about four times the volume of crushed gel. Although sterile deionized water works well for most oligomers, a 20-mM dithiothreitol solution is most effective with polydithioates. The gel slurry is frozen in dry ice until a white solid forms throughout the mixture. Incubation at 37°C with vigorous shaking for 3–4 h follows. This freeze/thaw cycle is then repeated, and the water/gel slurry is filtered through siliconized glass wool. The resulting gel material is then subjected to another cycle through the above-described procedure and the eluants pooled. For oligomers of eight or more bases, ethanol precipitation is effective. Triplicate precipitation with 0.3M sodium acetate, pH 5–7, and 4, 3.5, and 3 vol of ethanol sequentially follows. For very short oligomers, the gel eluant is evaporated in order to reduce the total volume by approximately three-fourths. Desalting on a Sephadex G-25 column (60×1.5 cm diameter) with 50 mM triethylammonium bicarbonate in 30% aqueous acetonitrile as mobile phase works well for separating dithioate oligomer from contaminants. Oligomer elution can be monitored by UV absorbance and the desired fractions evaporated. After following the described procedure, the amount of oligomer is determined spectrophotometrically.

4. Discussion

The degree of difficulty in purifying deoxyoligonucleotide analogs prepared by these methods depends largely on the extent of dithioate modification within the molecule. Oligomers bearing ≤50% dithioate linkages are easily purified using polyacrylamide gel electrophoresis to separate the product DNA from other components in the crude mixture. Standard gel extraction procedures for conventional phosphodiesters give good results for the isolation of these analogs. Oligomers containing small percentages of dithioate linkages can also be routinely purified by conventional HPLC methods (reverse phase and ion exchange) with excellent results.

Problems using these standard procedures are encountered when phosphorodithioates become the predominant linkage. Since the sulfur substituents interact strongly with certain metal cations, the quality of the deionized water used in these procedures is very important. It is particularly important to keep oligomers containing exclusively dithioate linkages in the correct salt form in order to maintain good solubility in aqueous solutions. We find that if the phosphorodithioate-linked oligomers are maintained as amine salts (ammonium or triethylammonium), they are fully water soluble. It is also possible to optimize gel extraction and chromatographic techniques, so that polyphosphorodithioate deoxyoligonucleotides can be purified to homogeneity.

In the course of learning how to purify dithioate oligomers, some insights were gained into their physical properties. They are retained to a greater extent on reverse-phase columns than their unmodified and phosphorothioate counterparts, which indicates a greater degree of hydrophobicity. This is not surprising considering the chemical modification. At the same time, all attempts at ion-exchange chromatography (under conditions where reverse-phase effects should be negligible) have been unsuccessful. This is apparently owing to very strong ionic interactions with the column-linked amines. Dithioate oligomers easily exchange only under strong denaturing conditions (urea or formamide). In denaturing PAGE gels, polydithioate oligomers migrate with mobility comparable to the unmodified oligomers of the same length and sequence. Only a slight retardation, approximately equivalent to one difference in nucleotide length, is observed. In contrast, these same polydithioate oligomers migrate much more slowly on nondenaturing gels and exhibit mobility comparable to natural DNA duplexes of the same length (even homopolymers behave in this manner). Perhaps this unusual property is indicative of a defined structure, such as a rigid rod, under nondenaturing conditions. However, no structural analysis has been performed to date on polydithioates. These analogs also cannot be generally characterized as "sticky," which is a term that has sometimes been applied to sulfur-containing oligomers. For example, even though dithioate oligomers bind with much higher affinity to polymerases than do natural DNAs *(14)*, this is not a general phenomenon with all proteins. Even with DNA-binding proteins, such as *lac* repressor, cro repressor, and catabolite activator protein, we have observed (by nitrocellulose filter and gel-binding assays) that polydithioate operators have much lower affinity for their respective repressors or activator protein than do normal DNAs. Moreover, our recent data also suggest that dithioate oligomers are very poor competitive inhibitors of several nucleases. Clearly, the biophysical and biochemical properties of this analog must be more thoroughly understood before we can determine how to use dithioate DNA most effectively in biochemistry and biology.

Acknowledgment

This research was supported by the National Institutes of Health (Grants GM25680 and GM21120), and by an American Cancer Society Postdoctoral Fellowship (PF3465) to W. T. Wiesler. This is paper 37 in a series on Nucleotide Chemistry. Paper 36 is ref. *15*.

References

1. Beaucage, S. L. and Iyer, R. P. (1992) Advances in the synthesis of oligonucleotides by the phosphoramidite approach. *Tetrahedron* **48**, 2223–2311.
2. Englisch, U. and Gauss, D. H. (1991) Chemically modified oligonucleotides as probes and inhibitors. *Angewandte Chemie Int. Edition* **30**, 613–722.
3. Uhlmann, E. and Peyman, A. (1990) Antisense oligonucleotides: a new therapeutic principle. *Chem. Rev.* **90**, 544–584.
4. Goodchild, J. (1990) Conjugates of oligonucleotides and modified oligonucleotides: A review of their synthesis and properties. *Bioconjugate Chemistry* **2**, 165–187.
5. Stein, C. A. (1989) Phosphorothioate oligodeoxynucleotide analogues, in *Oligodeoxynucleotides. Antisense Inhibitors of Gene Expression* (Cohen, J., ed.) Macmillan Press, London, pp. 97–117.
6. Ts'o, P. O. P., Miller, P. S., Aurelian, L., Murakami, A., Agris, C., Blake, K. R., Lin, S.-B., Lee, B. L., and Smith, C. C. (1987) An approach to chemotherapy based on base sequence information and nucleic acid chemistry, in *Biological Approaches to the Controlled Delivery of Drugs.* Annals of the New York Academy of Sciences. vol. 507, New York Academy of Sciences, New York, pp. 220–241.
7. Stein, C. A. and Cohen, J. S. (1988) Oligodeoxynucleotides as inhibitors of gene expression: a review. *Cancer Res.* **48**, 2659–2668.

8. Lesnikowski, Z. J., Jaworska, M., and Stec, W. J. (1990) Octa(thymidine methanephosphonates) of partially defined stereochemistry: synthesis and effect of chirality at phosphorus on binding to pentadecadeoxyriboadenylic acid. *Nucl. Acids Res.* **18,** 2109–2115.
9. Stec, W. J., Grajkowski, A., Koziolkiewicz, M., and Uznanski, B. (1991) Novel route to oligo(deoxyribonucleoside phosphorothioates) Stereocontrolled synthesis of P-chiral oligo(deoxyribonucleoside phosphorothioates). *Nucl. Acids Res.* **19,** 5883–5888.
10. Brill, W. K.-D., Tang, J.-Y., Ma, Y.-X., and Caruthers, M. H. (1989) Synthesis of oligodeoxynucleoside phosphorodithioates via thioamidites. *J. Am. Chem. Soc.* **111,** 2321,2322.
11. Nielsen, J., Brill, W. K.-D., and Caruthers, M. H. (1988) Synthesis and characterization of dinucleoside phosphorodithioates. *Tetrahedron Lett.* **29,** 2911–2914.
12. Grandas, A., Marshall, W. S., Nielsen, J., and Caruthers, M. H. (1989) Synthesis of deoxycytidine oligomers containing phosphorodithioate linkages. *Tetrahedron Lett.* **30,** 543–546.
13. Porritt, G. M. and Reese, C. B. (1989) Nucleoside phosphonodithioates as intermediates in the preparation of dinucleoside phosphorodithioates and phosphorothioates. *Tetrahedron Lett.* **30,** 4713–4716.
14. Caruthers, M. H., Beaton, G., Cummins, L., Dellinger, D., Graff, D., Ma, Y.-X., Marshall, W. S., Sasmor, H., Shankland, P., Wu, J., and Yau, E. K. (1991) Chemical and biochemical studies with dithioate DNA. *Nucleosides and Nucleotides* **10,** 47–59.
15. Caruthers, M. H., Beaton, G., Wu, J. V., and Wiesler, W. T. (1992) *Methods in Enzymology*, vol. 211 (Lilley, D. M. J. and Dahlberg, J. E., eds.) Academic Press, NY, pp. 3–20.
16. Dahl, O. (1991) Preparation of nucleoside phosphorothioates, phosphorodithioates and related compounds. *Sulfur Reports*, **11,** and references cited therein, 167–192.
17. Beaton, G., Dellinger, D., Marshall, W. S., and Caruthers, M. H. (1991) Synthesis of oligonucleotide phosphorodithioates, in *Oligonucleotides and Analogues, A Practical Approach* (Eckstein, F., ed.) IRL Press, Oxford, pp. 109–135.
18. Brill, W. K.-D., Nielsen, J., and Caruthers, M. H. (1991) Synthesis of deoxydinucleoside phosphorodithioates. *J. Am. Chem. Soc.* **113,** 3972–3980.
19. Caruthers, M. H., Barone, A. D., Beaucage, S. L., Dodds, D. R., Fisher, E. F., McBride, L. J., Matteucci, M. D., Stabinski, Z., and Tang, J.-Y. (1987) Chemical synthesis of deoxyoligonucleotides by the phosphoramidite method, in *Methods in Enzymology, vol. 154* (Wu, R., ed.) Academic Press, NY, pp. 287–313.

CHAPTER 10

Oligodeoxyribonucleotide Phosphotriesters

Maria Koziolkiewicz and Andrzej Wilk

1. Introduction

The common interest in the synthesis of oligo(nucleoside *O*-alkyl phosphates) was posed first by the generally accepted view that alkylation of DNA by several alkylating agents may also occur at internucleotide phosphate nonbridging oxygen atom(s) inducing DNA misfunction *(1)*. Perturbations in backbone conformation, partial charge neutralization of the backbone, and steric interference by the esterifying alkyl groups on the protein–nucleic acid interactions were all proposed to explain changes in the DNA biochemistry on phosphate alkylation *(2)*. Thus, the availability of synthetic "DNA-triesters" as models for studying stability of the phosphate-alkylated DNA fragments under the physiological conditions became a challenging task for oligonucleotide chemists. Furthermore, recent increasing interest in antisense oligonucleotide analogs *(3a,b)* brought the additional impact to development of reliable technology of the synthesis of oligonucleotide phosphotriesters.

In both synthetic methods, commonly used in oligonucleolide chemistry, i.e., triester approach *(4)* and phosphoramidite approach *(5)*, aryl (*o*-chlorophenyl) or alkyl (methyl, 2-cyanoethyl) groups are used for protection of the phosphate or phosphite moiety, respectively. These groups are, however, cleaved during the postsynthetic work-up. Obviously, the use of the synthons containing alkoxy group attached to the phosphorus, which will survive the deprotection of nucleic bases and cleavage from the solid support, other than *O*-methyl or *O*-(2-cyanoethyl), allows introduction of the triester moiety at a preselected position. The phosphoramidite method, introduced by Letsinger et al. *(6)* and elegantly developed by Caruthers *(7)*, offers a more versatile approach, and its common use for automated synthesis promotes it as the method of choice also for the synthesis of triester analogs. In order to introduce the *O*-alkyl group at the preselected position in the oligonucleotide chain, altered synthons are required, such as (protected nucleoside)3'-*O*-(O-alkyl)*N*,*N*-diisopropylphosphoramidites. The number of *O*-alkyl derivatives are reported: methyl *(8)*, ethyl *(9)*, trifluoroethyl *(10)*, isopropyl *(11)*, neopentyl *(12a,b)*.

Perhaps the most dramatic problem, which was only recently solved, was that of the synthesis of oligo(nucleoside) *O*-methylphosphates (Fig. 1, $R=CH_3$). Generally used protective groups for the amino functions, and the linkers between synthesized oligonucleotides and the solid support, require for their removal conditions under which *O*-methyl groups at the phosphates are also hydrolyzed via an attack of the nucleophilic reagents on the carbon atom. The method for the synthesis of dinucleotide *O*-methyl phosphates, developed by Buck *(8,13)*, was found to give rise to partial hydrolysis of the resulting *O*-methylated oligomer. Thus, the results were discredited because of inadequate characterization of the material used in the biological assays. The real breakthrough has been achieved recently by Alul et al. *(14)*. Instead of succinyl ester linkage between 3'-hydroxyl of the "starting" nucleoside and the solid support, an oxalyl-type linker (Fig. 2), which can be cleaved within a few minutes at room temperature with ammonia solution in methanol, has been introduced.

The synthesis of the oligonucleotides, bearing *O*-ethyl phosphate(s) at desired positions, also posed some difficulties, because the conditions, exemplified in routine protocol for release of oligonucleotides from the solid support, deprotection of bases, and the release of "unwanted" *O*-methyl phosphate blocking groups appeared to be too severe, if 5'-DMT(protected nucleoside)-3'-*O*(-*O*-ethyl)*N*,*N*-diisopropylphosphoramidites had been used as the synthons in the synthesis of *O*-ethyl phosphorotriesters. However, modification of the conditions of cleavage from the support and deprotection *(25% NH_3 aq., 36 h, 25°C)* allowed the researchers to obtain oligonucleotide with *O*-ethyl phosphate group at preselected position *(11)*.

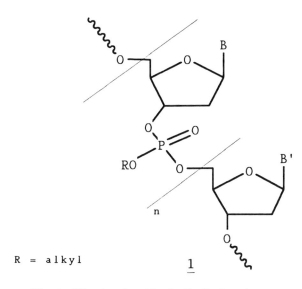

R = alkyl

Fig. 1. Oligo(nucleoside O-alkyl) phosphate.

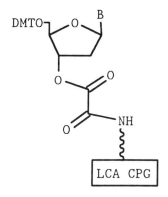

Fig. 2. Base-labile oxalyl-type linker to the solid support.

Improvement has been brought about by modification of the protecting groups blocking exoamino functions. Uznanski et al. (10) have introduced isopropoxyacetylated nucleosides (N^6-IPA-A, N^4-IPA-C, N^2-IPA-G), which, when converted into corresponding 5'-DMT (nucleoside^{IPA})-3'-O-(O-2-cyanoethyl)-N,N-diisopropylphosphor-

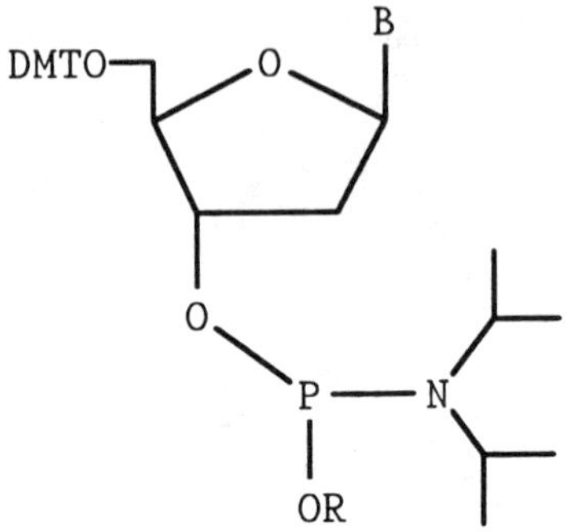

B = Ade^{Bz}, Gua^{iBu}, Cyt^{Bz}, Thy, Ade^{IPA}, Gua^{IPA}, Cyt^{IPA}

R = CH$_3$, CH$_3$CH$_2$, i-C$_3$H$_7$, CF$_3$CH$_2$, CNCH$_2$CH$_2$

Fig. 3. Nucleoside 3'-(O-alkyl)N,N-diisopropylphosphoramidites.

amidites and 5'-DMT-(nucleoside^{IPA})-3'-O-(O-ethyl)-N,N-diisopropyl-phosphoramidites, can be successfully used for the routine automated DNA synthesis.

It was found that the cleavage of oligonucleotide from the support, deprotection of the nucleobases (removal of IPA-blocking groups), and deprotection of 2-cyanoethyl phosphates were completed within 2 h of treatment of the solid-support-bound oligonucleotide with 25% aqueous ammonia; under these conditions, the O-ethyl phosphate moiety(ies) remains intact. Application of the isopropoxyacetylated synthons to the synthesis of the natural oligonucleotides, and oligonucleotides with O-ethyl (or O-trifluoroethyl) phosphates at preselected positions, has also been demonstrated (*10;* Fig. 3). The work of Abramova and Lebedev on per-ethylated oligonucleotides up to 4 mers, where detailed characteristics and stereochemistry are presented, should be noted (*15*).

Synthesis of the oligo(nucleoside O-isopropyl phosphates) required preparation of 5'-DMT-nucleoside (*N*-protected with standard conventional groups) 3'-O-(O-isopropyl)N,N-diisopropylphosphoramidites. They appeared to be easily incorporable into the growing oligonucleotide chain at preselected position(s), and conditions required for the removal of *N*-protective groups (N^6-benzoyl of adenosine, N^4-benzoyl of cytidine, and N^2-isobutyryl of guanosine), and O-methyl or O-2-

cyanoethyl groups under treatment of the oligonucleotides with PhS⁻ or 25% NH_3 aq., respectively, did not cause observable removal of desired O-isopropyl groups from the preselected phosphate(s) *(11)*. Analysis of O-isopropyl dinucleotides by means of the FAB mass spectrometry *(16)* has shown the potential of this method in the field of DNA analogs.

DNA triesters are, by virtue of asymmetry of tetracoordinated phosphorus, P-chiral analogs, and their diastereoisomers differ in the orientation of O-alkyl substituent (inwards or outwards DNA helix) *(17)*. Assignment of the orientation is equivalent to the determination of absolute configuration at a phosphorus atom of modified internucleotide bond.

In recent years, several methods have been applied to this purpose. Hamblin and Potter *(18)* used ^{31}P-NMR technique to establish the configuration at the P-chiral center of methyl ester of d[T$_p$T]. Weinfeld and Livingston *(19)* assigned the absolute configuration of dinucleotides d[N$_{p(o)(oEt)}$T] by means of circular dichroism.

The method developed in the authors' laboratory is based on the stereochemical correlation between three classes of analogs of dinucleotides, namely phosphorothioates, O-ethyl phosphorothioates, and O-ethylphosphates *(9)*. Stereospecific dealkylation of diastereoisomers of the dinucleoside O-alkyl phosphorothioate allows the correlation of their configuration with phosphorothioates. On the other hand, stereospecific oxidation leads to O-alkyl phosphates (Fig. 4).

The emphasis of this chapter is focused on the oligo(nucleoside O-alkyl phosphates), although since model studies may require the use of dinucleoside(3',5')O-alkyl phosphates, the experimental section includes examples of their preparations in the solution, involving condensation of 5'-DMT nucleoside 3'-O-alkyl-N,N'-diisopropyl phosphoramidite with 3'-blocked nucleoside in the presence of tetrazole, followed by oxidation of the internucleotide O-alkyl phosphite function with I_2/H_2O and subsequent removal of the terminal 5'-OH- and 3'-OH-protective groups under standard conditions. The synthesis of the "dimer blocks" can be easily used for isotopically labeled analogs (if water [^{18}O-H_2O or [^{17}O]-H_2O in the oxidation step is employed). Labeled dinucleoside O-alkylphosphates, after their separation into R_p and S_p diastereoisomeric species, could be incorporated via the "dimer blocks" technique (*vide supra*) into the oligonucleotides. In this way, diastereoisotopomeric oligonucleotides of known absolute configuration at P-chiral centers were obtained and used in elegant enzymatic studies *(20)*.

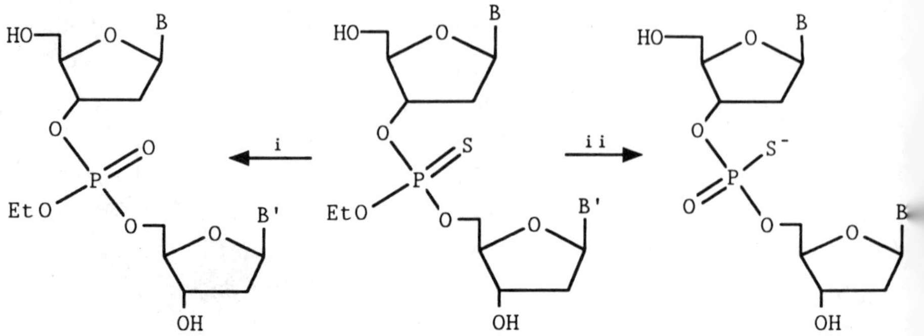

i   3% H_2O_2 in CH_3CN/H_2O (1:1), 48h.
ii $PhSH/Et_3N$/Dioxane (20:40:40) or 25% NH_3 aq. 36h.

Fig. 4. Scheme of chemical correlation of absolute configuration at phosphorus in *O*-ethyl dinucleotides.

2. Materials

1. 2'-Deoxynucleosides.
2. Controlled pore glass (CPG) support.
3. Isopropoxyacetic anhydride.
4. Phosphorus trichloride.
5. Diisopropylamine.
6. *N,N,N',N'*-tetraisopropyl-*O*-(2-cyanoethyl)phosphordiamidite.
7. Methanol.
8. Ethanol.
9. Isopropanol.
10. 3-Hydroxypropionitrile.
11. 2,2,2-Trifluoroethanol.
12. Tetrazole.
13. Acetonitrile.
14. Aqueous ammonia (25%).
15. Deionized water.
16. Acetic acid (20%).
17. Methylene chloride.
18. Triethylamine.
19. Benzene.
20. Triethylammonium bicarbonate (0.1*M*) (TEAB).
21. C_{18} RP HPLC column.

22. HP TLC plates 60 F_{254} (Merck, Germany).
23. Eppendorf tubes (1.5 mL).
24. Rubber septa (14 mm).
25. Snake venom phosphodiesterase (SVPDE).
26. Alkaline phosphatase.
27. Nuclease P1.
28. Oligonucleotide synthesis reagents.

3. Methods

Oligonucleotides are synthesized on an Applied Biosystems 380B DNA synthesizer. HPLC analyses are performed using LDC Milton Roy system. NMR spectra are recorded on Bruker MSL 300. For transfer of moisture-sensitive chemicals, gas-tight syringes (Hamilton, USA) should be used. Rubber septa (14 mm) (Aldrich, USA) are fitted for all reactions in which chemicals are added by means of the syringe. Water containing solutions of oligonucleotides is concentrated in the vacuum rotary concentrator (Speed Vac). Oligonucleotides containing O-alkyl phosphate moieties are dissolved in aqueous acetonitrile solution (usually 10%).

3.1. Preparation of the Nucleoside Components

3.1.1. General Method for the Protection of 2'-Deoxynucleosides with Isopropoxyacetic Anhydride

1. Add trimethylchlorosilane (250 mmol) to the suspension of dry deoxynucleoside (70 mmol) (the nucleosides dried by coevaporation with pyridine) in anhydrous pyridine (80 mL) with stirring at room temperature.
2. After 0.5 h, add isopropoxyacetic anhydride (75 mmol), and continue stirring for an additional 2 h.
3. Add anhydrous benzene (50 mL), filter off the sediment of pyridinium hydrochloride, wash with benzene (20 mL), and concentrate combined filtrates under reduced pressure to approx 40 mL.
4. Add acetonitrile (50 mL) and water (30 mL). TLC monitoring (acetonitrile:water 85:15, or methylene chloride:methanol 80:20) is used in order to determine a completion of removal of the trimethylsilyl groups. Usually, for dC^{IPA} and dG^{IPA}, 2 h are required, whereas for dA^{IPA}, overnight hydrolysis is necessary. Concentrate the reaction mixture and crystallize the oily residue from water. Some physicochemical characteristics are presented in Table 1.

Table 1
Physicochemical Characteristics of dA^{IPA}, dG^{IPA}, and dC^{IPA}

Compound	mp, °C	R_f b	R_f c	$t_{1/2}{}^a$, min	Yield, %
dA^{IPA}	134–135	0.22	0.52	20	76
dG^{IPA}	>200 (dec.)	0.14	0.50	50	81
dC^{IPA}	157–158	0.24	0.59	10	87

^{a}Half-time of deprotection in pyridine:29% ammonia solution (80:20).
^{b}Chloroform:methanol (90:10) (Merck silica gel 60 plates).
^{c}Acetonitrile:water (85:15) (Merck silica gel 60 plates).

3.1.2. Synthesis of Phosphitylating Agent

1. Add anhydrous ethanol (23g, 0.5 mol) dropwise to vigorously stirred phosphorus trichloride (100 g, 0.7 mol) at –30°C over 2 h.
2. Degass the solution, and remove the residues of PCl$_3$ under the reduced pressure. Distill the product through the short column (bp 30°C/20 mm Hg, yield 60 g/82%).
3. Add diisopropylamine (97 g, 1.7 mol) dropwise to agitated solution of ethyl phosphorodichloridite (60 g, 0.4 mol) in anhydrous pentane *(21)*, and maintain the temperature below 40°C. Leave overnight.
4. Filter off the sediment of diisopropylamine hydrochloride, wash with pentane (200 mL), and concentrate pooled fractions of the filtrate under the reduced pressure. The residue is distilled under high vacuum (bp 85°C/0.5 mm Hg), yield 90 g (89%).

Caution: *O*-alkyl phosphorodichloridites are extremely sensitive to moisture, and reaction with water may lead to spontaneous combustion. Some physicochemical characteristics are presented in Tables 2 and 3.

3.1.3. Phosphitylation

1. Add *N,N,N',N'*-tetraisopropyl-(*O*-alkyl-phosphordiamidite) (0.11 mol) in one batch under an atmosphere of argon to a stirred suspension of adequately protected 5'-dimethoxytrityl 2'-deoxynucleoside (0.1 mol) in methylene chloride (20 mL).
2. Add tetrazole in portions (0.1 mol) (we use glass adapter, which enables adding tetrazole in portions without opening the flask, but the 0.5*M* solution in acetonitrile can be used) at room temperature, and stirring is continued for 2 h (for dG, the reaction time is extended to 12 h).
3. Concentrate the reaction mixture to dryness. Dissolve the resultant residue in methylene chloride (containing 10% of triethylamine), and purify on the silica gel column (20 × 50 mm). Evaporate combined fractions

Table 2
Characteristics of O-Alkyl Phosphordichloridites

No.	Compound	bp	^{31}P-NMR, ppm, C_6D_6
1a	CH_3OPCl_2	48–49°C/100 mm Hg	180.3
1b	$C_2H_5OPCl_2$	30°C/20 mm Hg	178.0
1c	$NCCH_2CH_2PCl_2$	56–57°C/0.05 mm Hg	179.9
1d	$CF_3CH_2PCl_2$	55–58°C/170 mm Hg	181.4
1e	$(CH_3)_2CHPCl_2$	65–67°C/100 mm Hg	174.0

Table 3
Characteristics of (O-Alkyl)N,N-Diisopropylphosphoramidites

No.	Compound	bp	^{31}P-NMR
2a	O-methyl	64–65°C/0.5 mm Hg	122.6
2b	O-ethyl	85°C/0.5 mm Hg	124.8
2c	2-Cyanoethyl	107°C/0.05 mm Hg	122.4
2d	Trifluoroethyl	58–60°C/0.5 mm Hg	129.6

containing desired product to dryness. The product is obtained as an amorphous foam. Some physiochemical characteristics are presented in Tables 4 and 5.

3.2. Dinucleoside Phosphotriester Synthesis

3.2.1. Synthesis in the Solution

1. Dry adequately protected nucleoside O-alkyl phosphoramidite (1 mmol), 3'-acetyl nucleoside (1 mmol) under high vacuum for 1 h.
2. Add anhydrous acetonitrile (0.5 mL).
3. Add dropwise the solution of tetrazole (140 mg, 2 mmol) in anhydrous acetonitrile (3 mL) with stirring.
4. After 0.5 h, the reaction is completed (TLC monitoring using chloroform:methanol 9:1 as eluting system).
5. Add diisopropylamine (1 mmol), and filter off the sediment of diisopropylammonium tetrazolide. This operation leads to quite a pure solution of protected dinucleoside O-alkyl phosphite. The product is oxidized by means of 0.1M iodine solution in 2.6 lutidine:water:THF (10:1:40 v/v/v).
6. Add 25% aqueous ammonia (2 mL) in order to remove 3'-acetyl-protecting group (TLC monitoring as above).
7. Purify the product on the silica gel column (20 × 50 mm) using gradient of methanol in chloroform (0–10%).

Caution: This method is suitable for O-alkyl groups resistant to aqueous ammonia: ethyl, isopropyl, neopentyl.

Table 4
^{31}P NMR Characteritics of 5'-Dimethoxytrityl Nucleoside
3'-O-(N,N-Diisopropyl)-O-Alkyl-Phosphoramidites

Base	Eluting system for OCH$_3$ esters	^{31}P-NMR for OCH$_3$ esters	Eluting system for OC$_2$H$_5$ esters, ppm	^{31}P-NMR for OC$_2$H$_5$ esters CH$_2$Cl$_2$ + 10%C$_6$D$_6$
dA^{bz}	Benzene:hexane:Et$_3$N (20:10:2, v/v/v)	148.67 146.47	Benzene:hexane:Et$_3$N (20:15:2, v/v/v)	153.1, 152.8 (47:53)
dG^{ibu}	Benzene:CH$_2$Cl$_2$:Et$_3$N (5:15:2, v/v/v)	148.47 148.33	Benzene:CH$_2$Cl$_2$:Et$_3$N (5:15:2, v/v/v)	155.8, 15.55 (37:63)
dC^{bz}	Benzene:hexane:Et$_3$N (20:10:2, v/v/v)	148.74 148.33	Benzene:hexane:Et$_3$N (20:15:2, v/v/v)	153.3, 152.9 (49:51)
dT	Benzene:hexane:Et$_3$N (20:10:2, v/v/v)	148.94 148.54	Benzene:hexane:Et$_3$N (20:15:2, v/v/v)	153.1, 152.6 (56:44)

Table 5
^{31}P NMR Characteristics of 5'-Dimethoxytrityl Nucleoside
3'-O-(N, N-Diisopropyl)-O-Alkyl-Phosphoramidites
Isopropoxyacetyl Protecting Groups

Base	Eluting system for OC$_2$H$_5$, OCH$_2$CH$_2$CN, OCH$_2$CF$_3$ esters	^{31}P-NMR for OC$_2$H$_5$ esters	^{31}P-NMR for OCH$_2$CH$_2$CN esters	^{31}P-NMR for OCH$_2$CF$_3$ esters
dA^{iPA}	Benzene:Et$_3$N (8:2, v/v)	148.20 147.93	149.73 149.40	153.69 153.40
dG^{iPA}	CH$_2$Cl$_2$:Et$_3$N (10:1, v/v)	148.22 147.99	149.89 149.46	153.57 153.39
dC^{iPA}	Benzene:Et$_3$N (8:2, v/v)	148.28 147.77	149.66 149.52	153.67 153.28
dT	Benzene:Et$_3$N (8:2, v/v)	148.24 147.37	– –	153.71 153.21

3.2.2. Synthesis on the Solid Support

1. Twenty-μmol scale synthesis of dinucleoside O-alkyl phosphates is performed manually on 10-μmol DNA synthesis column (Applied Biosystems Inc., cat. no. 400393). Large-scale synthesis is possible because of high loading of some of our solid supports (50 μmol/g) and the fact

that for the manual synthesis, one can fill the column completely, which is not recommended for the automated mode, and still provide efficient mixing (and thus contact of the reagents with the solid support).

2. Treat 5'-OH-nucleoside bound to LCA CPG with a 1.0-mL acetonitrile solution of 5'-DMT-nucleoside 3'-O-(O-alkyl)N,N-diisopropylphosphoramidite (100 mg, 120 µmol) and 1H tetrazole (25.2 mg, 360 µmol) for 15 min, and wash thoroughly with acetonitrile.
3. Oxidize (0.1M iodine solution in 2.6 lutidine:water:THF [10:1:40 v/v/v], 2.5 mL, 1 min), and wash with acetonitrile.
4. Cleave the synthesized dinucleotide from the support with 25% NH_3 aq. for 2 h at 25°C.
5. Wash additionally the column with acetonitrile, and concentrate the solution to dryness.
6. Separation into individual diastereoisomers can be accomplished on HPTLC plates 60 F_{254} (10 × 20 cm, Merck-art. 15696). The plate is developed four times using $CHCl_3$:CH_3OH:$(C_2H_5)_2O$ (20:3:10) as an eluting system. Visualization by means of UV light enables cutting off from the plate. Elute the gel with $CHCl_3$:CH_3OH (10:1), and evaporate the extacts to dryness.
7. Analytical samples (0.5 A_{260} U) of both diastereoisomers are deprotected (20% acetic acid for 0.5 h), and their purity is checked by means of HPLC in Table 6.

3.3. Synthesis of O-Ethylphosphotriester Analog of Decanucleotide d[GGGAATTCCC] Containing O-Ethyl Function Between Adjacent Thymidines

1. An automated 1-µmol scale synthesis is performed on a commercially available Applied Biosystems column using standard cycle. Corresponding nucleoside O-alkyl phosphoramidite is placed in one of the additional reservoirs, and the corresponding code is inserted in the sequence. In the present example, 5'-O-dimethoxytritylthymidine-3'-O-(O-ethyl) N,N-diisopropylphosphoramidite is placed in the reservoir no. 5 (concentration is exactly the same as recommended by the manufacturer), and the sequence d[GGGAA5TCCC] entered from the sequence editor. However, when only a limited amount of the exceptional building block is available (*vide infra*), we use the manual coupling, inserting synthesis interruption code (here: Interruption Ahead—Base 5, Step 2).
2. In both cases, 5'-DMT derivative of base-protected decanucleotide 3f is obtained as the final product. The cleavage of the oligonucleotide from the support is achieved by treatment of the solid-support-bound oligonucleotide with 25% NH_3 aq. for 2 h at 25°C. However, instead of

Table 6
Absolute Configuration at P-Chiral Atom of Dinuleoside O-Ethyl Phosphates

Bases	d[N$_{p(O) (OEt)}$N'] **5**	d[N$_{p(S) (OEt)}$N'] **6**	d[N$_{p(S)}$N'] **7**
N = A, N' = A	Fast–R_p Slow–S_p	Fast–S_p* Slow–R_p*	Fast–R_p Slow–S_p
N = A, N' = G	Fast–R_p Slow–S_p	Fast–S_p* Slow–R_p*	Fast–R_p Slow–S_p
N = A, N' = T	Fast–S_p Slow–R_p	Fast–S_p Slow–R_p	Fast–R_p Slow–S_p
N = C, N' = C	Fast–R_p Slow–S_p	Fast–S_p* Slow–R_p*	Fast–R_p Slow–S_p
N = G, N' = A	Fast–R_p Slow–S_p	Fast–S_p* Slow–R_p*	Fast–R_p Slow–S_p
N = G, N' = G	Fast–R_p Slow–S_p	Fast–S_p* Slow–R_p*	Fast–R_p Slow–S_p
N = T, N' = C	Fast–R_p Slow–S_p	Fast–Sp* Slow–R_p*	Fast–R_p Slow–S_p
N = T, N' = T	Fast–S_p Slow–R_p	Fast–S_p Slow–R_p	Fast–R_p Slow–S_p

*Chromatographic mobility of 5'-DMT derivatives.

standard deprotection procedure (usually 25% NH$_3$ aq. 24 h at 55–60°C), more mild conditions are used for the removal of 2-cyanoethyl, benzoyl, and isobutyryl groups, namely 25% NH$_3$ aq. for 36 h at 25°C. This change is necessary for the preservation of ethyl groups modifying preselected internucleotide bonds. Under these conditions, 90% of the internucleotide O-ethyl phosphate triesters remains intact, whereas under standard conditions, undesired deethylation reaches 50%, as estimated by the ratio of deethylated products vs target sequence in HPLC chromatogram.

3. After 36 h, the solution of oligonucleotide is concentrated in a vacuum centrifuge, the resulting residue is dissolved in the solution of CH$_3$CN:H$_2$O (1:9), and its further purification is performed by means of HPLC on a µBondapak C$_{18}$ column (300 × 7.8 mm) with the gradient 5–30% CH$_3$CN/0.1M TEAB (triethylammonium bicarbonate), pH 7.4 (exponent 0.25), for 20 min followed by isocratic separation, at a flow rate of 3.5 mL/min. Under these conditions, the separation of 5'-DMT d[GGGAAT$_{p(o)(oEt)}$TCCC] into its diastereoisomers is observed: "Fast"-isomer is eluted at 16.50 min, and "slow"-isomer is eluted at 18.20 min.

4. Separated diastereoisomers, after removal of solvents, are detritylated (by treatment with 20% aqueous solution of CH$_3$COOH [20 min, 25°C]), and the resulting solution is concentrated in the vacuum centrifuge. The residue is dissolved in a solution of CH$_3$CN:H$_2$O (1:9) (0.5 mL), and then

Table 7
Relative Chromatographic Mobilities
and Absolute Configurations at P-Chiral Atom of Decanucleotides 3

Comp. sequence 5' -> 3'	HPLC mobility of 5'-DMT derivatives	R. T., min.	Absolute configuration	HPLC mobility of 5'-HO derivatives	R. T., min.
d[G*GGAATTCCC]	Fast	19.0[a]	R_p		11.0[b]
	Slow	21.0	S_p		11.0
d[GG*GAATTCCC]	Fast	19.5	R_p		10.5
	Slow	21.0	S_p		10.5
d[GGG*AATTCCC]	Fast	20.0	R_p	Slow	10.7
	Slow	21.0	S_p	Fast	10.5
d[GGGA*ATTCCC]	Fast	18.5	R_p	Slow	11.0
	Slow	20.0	S_p	Fast	10.5
d[GGGAA*TTCCC]	Fast	19.0	R_p	Slow	11.5
	Slow	22.0	S_p	Fast	11.0
d[GGGAAT*TCCC]	Fast	16.0	R_p	Slow	11.5
	Slow	18.0	S_p	Fast	11.3
d[GGGAATT*CCC]	Fast	16.0	R_p	Fast	11.0
	Slow	18.0	S_p	Slow	11.2
d[GGGAATTC*CC]	Fast	16.0	R_p		11.0
	Slow	18.0	S_p		11.0
d[GGGAATTCC*C]	Fast	19.0	R_p		10.7
	Slow	20.0	S_p		10.7

*Internucleotide O-ethyl triester.
[a]Retention time in min. HPLC conditions as described in text for 5'-DMT derivatives.
[b]Retention time in min. HPLC conditions as described in text for 5'-HO derivatives.

repeatedly purified on a C_{18} column with the linear gradient 5–30% CH_3CN/ 0.1M TEAB, pH 7.4, 1. 25%/min at a flowrate of 3.5 mL/min. Under these conditions, "fast"-DMT-3f gives "slow"-3f (retention time 11.50 min), and "slow"-DMT-3f gives "fast"-3f (retention time 11.25 min). Although all O-ethylphosphotriester analogs of the decamer d[GGGAATCCC] are separable into diastereoisomers as 5'-DMT derivatives, their 5'-HO derivatives differ in chromatographic properties. Some of them appeared to be inseparable under the above-described conditions (3a,b, 3 h, 3i), whereas diastereoisomers of 3c–g have shown reversed elution order compared to their 5'-DMT precursors in Table 7.

For the separation of the 3 into diastereoisomers, the HPLC of 5'-DMT derivatives (first-step chromatography) is more effective than the second-step chromatography. However, it should be noted that HPLC of O-ethyl phosphotriester analogs of the oligonucleotides is not always so effective, as in this case. Successful separation of the backbone-modified analogs of DNA into diastereoisomers mainly depends on the sequence of the nucleotides, and on the nature and position of the modified internucleotide bond. Although under the above-described conditions, all O-ethyl phosphotriester analogs of the decanucleotide d[GGGAATTCCC] appeared to be separable into the diastereoisomers, diastereoisomeric separation of the decamer d[AAGAAT$_{(oET)}$TCCC] *(4)* (sequence related to d[GGGAATTCCC]) was unsuccessful. The change of first two 5'-end dG residues for two dA residues completely alters chromatographic properties of the decanucleotide 4.

In this case, separated diastereoisomer of d[AAGAAT$_{p(o)(oEt)}$TCCC] can be obtained by the alternative approach *(vide infra)* based on the synthesis of 5'-DMT d[T$_{p(o)(oEt)}$T], its separation into diastereoisomers, their *in situ* phosphitylation, and the introduction of activated isomers of 5'-DMT d[T$_{p(o)(oEt)}$T] at the preselected position of growing oligonucleotide chain.

3.3.1. In Situ Phosphitylation and Synthesis of Oligonucleotide

1. Dissolve each of the diastereoisomers (5 µmol) independently in dry acetonitrile (30 µL), and to each of these samples, add O-(2-cyanoethyl) N,N,N',N'-tetraisopropylphosphordiamidite (carefully calculated 5% molar excess and measured by means of the 10-µL Hamilton gas-tight syringe) and tetrazole (0.35 mg, 8 µmol) in acetonitrile (30 µL).
2. After 0.5 h, this solution can be used as substrate for manual coupling, enabling insertion of the dinucleotide O-alkyl block into the growing chain of oligodeoxyribonucleotide synthesized on the solid support. If pure diastereoisomers are used, the resulting oligonucleotide bears O-alkyl modification at a predetermined position and defined configuration at the phosphorus atom.
3. Automated synthesis of oligonucleotide is interrupted after at the appropriate cycle, and the column is washed and dried (if the flow of dry argon is not sufficient, we apply 1 min drying under high vacuum). Subsequently, to the solution of the *in situ* phosphitylated dinucleotide, tetrazole (0.5M in acetonitrile, 60 µL) is added, and the resulting mixture is introduced by means of the syringe into the column with solid support. Coupling time is prolonged to 10 min because of rather small molar excess of the "dimer block" (only fivefold*)*, but it is sufficient and the coupling yield (trityl assay) is usually not lower than 98%. After thorough washing with acetonitrile, the automated synthesis is resumed and completed.

3.4. Synthesis of Dinucleoside O-Alkyl Phosphorothioates

The synthesis of these compounds is performed as described above in Section. 3.2. with the following modification: Regular oxidation by means of 0.1M iodine solution in 2,6 lutidine:water:THF (10:1:40) is replaced by treatment of the intermediate phosphite triester with a saturated solution of elemental sulfur in lutidine (1 mL, 6 h, 25°C). Deprotection and purification require no modifications. Correlation of the HPLC elution order with configuration at the phosphorus is shown in the Table 6.

4. Discussion

The presented material summarizes the authors' efforts and experience in preparation of DNA triesters, focused mostly on oligonucleotides bearing a single O-alkyl phosphate function within a construct, utilized for studies on DNA–protein interactions. Detailed protocols, including the methodology for the assignment of the configuration, are elaborated and can be applied to any sequence with a predetermined modification of the phosphate moiety, including a predetermined sense of chirality. However, still unsolved is the problem of the synthesis of oligonucleotides bearing more than one modification at neighboring (juxtaposed) positions. In the case of oligo(nucleoside phosphorothioates), such a goal has been reached recently *(21)*. Also partial success has been achieved in the synthesis of small oligo(nucleoside methanephosphonates) *(22)*. Methodology developed for the synthesis of the oligo(nucleoside methanephosphonates) can also be used for the synthesis of oligo(nucleoside O-alkyl phosphates) [Fig. 5]).

It would be necessary to synthesize and separate the diastereoisomers of 5'-DMT-O-nucleoside 3'-O-(O-alkyl-O-[4-nitrophenyl]) phosphates and use these synthons for the reaction with 5'-OH nucleotides(sides), where the 5'-OH group would be activated by, e.g., t-BuMgCl. One can assume that nucleophilic substitution at the phosphorus should occur with defined (presumably inversion) stereochemistry *(23)*. Such methodology will suffer from the same limitation as that elaborated for the synthesis of oligo(nucleoside methanephosphonates)—difficulty in automation, necessity of work in solution, low yield, and laborious purification of dimers, trimers, and so on.

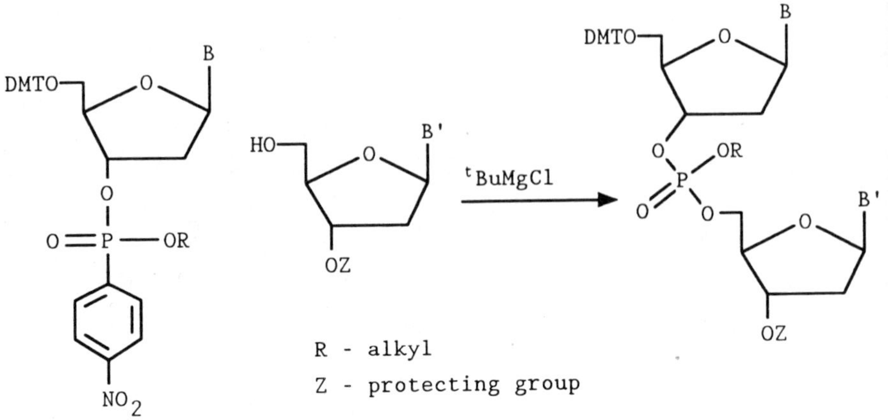

Fig. 5. Scheme of possible stereo-controlled synthesis of O-alkylated oligonucleotides.

Much more promising seems to be another, so far hypothetical, assumption that an access to stereo-defined oligo (nucleoside phosphorothioates) will allow stereospecific conversion of phosphorothioate function to O-alkyl functions, for example, by selective S-alkylation and conversion of S-alkyl phosphates into O-alkyl phosphates (24). However, one has to take into account potential problems with selective S-alkylation in the presence of other nucleophilic centers, mostly on the nucleic bases.

Some earlier efforts in this direction are reported by Stec et al. (25), who tried to convert internucleotide phosphorothioate into corresponding phosphorofluoridates, further convertible into dinucleoside O-alkyl phosphates. Unfortunately, 2,4-dinitro fluorobenzene, used in their studies, caused nonstereospecific conversion.

Undoubtedly, in spite of several difficulties, the problem can be solved, and the success depends on the amount of effort involved. Although the function and ability of oligo(nucleoside O-alkyl phosphates) in modulation of gene expression (besides problematic oligo[nucleoside O-methyl phosphates] [8,13]) are still not elucidated, their stereo-controlled synthesis is still a challenging task for the oligonucleotide chemists.

Acknowledgment

The authors are indebted to W. J. Stec for his inspiration and valuable advice in writing this chapter.

References

1. Singer, B. (1982) Mutagenesis from a chemical perspective, nucleic acid reactions, repair, translation, and transcription. *Basic Live Sci.* **20**, 103–107.
2. Lawrence, D. P., Wenqiao, C., Zon, G., Stec, W. J., Uznanski, B., and Broido, M. J. (1987) NMR studies of backbone-alkylated DNA: Duplex stability, absolute stereochemistry, and chemical shift anomalies of prototypal isopropyl phosphotriester modified octanucleotides, (R_p,R_p)—and (S_p,S_p)-{d-[GGA(iPr)ATTCC]}$_2$ and -{d-[GGAA(iPr)TTCC]}$_2$. *J. Biochem. Struct. & Dyn.* **4**, 757–783.
3a. Cohen, J. S., ed. (1989) Oligonucleotides—antisense inhibitors of gene expression, *Topics in Molecular and Structural Biology*. MacMillan Press.
3b. Uhlmann, E. and Peyman, A. (1990) Antisense oligonucleotides. *Chem. Rev.* **90**, 543–584.
4. Letsinger, R. L. and Mahadevan, V. (1965) Oligonucleotide synthesis on a polymer support. *J. Am. Chem. Soc.* **87**, 3526, 3527.
5. Matteucci, M. and Caruthers, M. H. (1981) Synthesis of deoxyoligonucleotides on a polymer support. *J. Am. Chem. Soc.* **103**, 3185–3191.
6. Letsinger, R. L. and Lunsford, W. B. (1976) Synthesis of thymidine oligonucleotides by phosphite triester intermediates. *J. Am. Chem. Soc.* **98**, 3655–3659.
7. Caruthers, M. H. (1985) Gene synthesis machines: DNA chemistry and its uses. *Science* **230**, 281–285.
8. Koole, L. N., Moody H. N., Broeders, N. L. H. L., Quaedflieg, P. J. L. M., Kuijpers, W. H. A., van Genderen, M. H. P., Coenen, A. J. J. M, van der Wal, S., and Buck, H. M. (1989) Synthesis of phosphate-methylated DNA fragments using 9-fluorenylmethoxycarbonyl as transient base protecting group. *J. Org. Chem.* **54**, 1657–1664.
9. Guga, P. J., Koziolkiewicz, M., Okruszek, A., Uznanski, B., and Stec, W. J. (1987) DNA-triesters—The synthesis and absolute configuration assignments at P-stereogenic centres. *Nucleosides & Nucleotides* **6**, 111–119.
10. Uznanski, B., Grajkowski, A., and Wilk, A. (1989) Isopropoxyacetic group for convenient base protection during solid-support synthesis of oligodeoxyribonucleotides and their triester analogs. *Nucl. Acids Res.* **17**, 4863–4871.
11. Stec, W. J., Zon, G., Gallo, K. A., Byrd, R. A., Uznanski, B., and Guga, P. J. (1985) Synthesis and absolute configuration of P-chiral O-isopropyl oligonucleotide triesters. *Tetr. Lett.* **26**, 2191–2194.
12a. Han, J. F., Asseline, U., and Thuong, N. T. (1991) Octathymidylates involving alternating neopentylophosphothionoester-phosphodiester linkages with controlled stereochemistry at the modified P-center. *Tetrahedron Lett.* **32**, 2497, 2498.
12b. Lancelot, G., Guesnet, J.-L., Asseline, U., and Thuong, N. T. (1988) NMR studies of complex formation between the modified oligonucleotide d(T*TCTGT) covalently linked to an acridine derivative and its complementary sequence d(GCACAGAA). *Biochemistry* **27**, 1265–1273.
13. Moody, H. M., van Genderen, M. H. P., Koole, L. H., Kocken, H. J. M., Meijer, E. M., and Buck, M.H. (1989) Regiospecific inhibition of DNA duplication by antisense phosphate-methylated oligodeoxynucleotides. *Nucl. Acids. Res.* **17**, 4769–4782.

14. Alul, R. H., Signman, C. H. N., Zhang, G., and Letsinger, R. L. (1991) Oxalyl-CPG: a labile support for synthesis of sensitive oligonucleotide derivatives. *Nucl. Acids Res.* **19**, 1527–1532.
15. Abramova, T. V. and Lebedev, A. V. (1983) Investigation of non-ionic diastereomeric analogs of oligonucleotides. Synthesis and separation of diastereoisomers of di- and tetrathymidylate ethyl esters. *Bioorg. Chim.* **9**, 823–830.
16. Phillips, L. R., Gallo, K. A., Zon, G., Stec, W. J., and Uznanski, B. (1985) Fast atom bombardment mass spectra of O-isopropyloligodeoxyribonucleotide triesters. *Org. Mass Spec.* **20**, 781–792.
17. Zon, G., Summers, M. F., Gallo, K. A., Shao, K. L., Koziolkiewicz, M., Uznanski, B., and Stec, W. J. (1987) Stereochemistry of oligodeoxyribonucleotide phosphotriesters, in *Biophosphates and Their Analogs—Synthesis Structure, Metabolism and Activity* (Bruzik, K. S. and Stec, W. J., eds.) Elsevier, Amsterdam, p. 165–168.
18. Hamblin, M. R. and Potter, B. V. L. (1985) *E. coli* Ada regulatory protein repairs the S_p diastereoisomer of alkylated DNA. *FEBS Lett.* **189**, 315–317.
19. Weinfeld, M. and Livingston, D. C. (1986) Synthesis and properties of oligodeoxyribonucleotides contaning an ethylated internucleotide phosphate. *Biochemistry* **25**, 5083–5091.
20. Potter, B. V. L., Eckstein, F., and Uznanski, B. (1983) A stereospecifically ^{18}O-labelled deoxydinucleoside phosphate block for incorporation into an oligonucleotide. *Nucl. Acids Res.* **11**, 7087–7090.
21. Stec, W. J., Grajkowski, A., Koziolkiewicz, M., and Uznanski, B. (1991) Novel route to oligo(deoxyribonucleoside phosphorothioates). Stereo controlled synthesis of P-chiral oligo(deoxyribonucleoside phosphorothioates). *Nucl. Acids Res.* **19**, 5883–5888.
22. Lesnikowski, Z. J., Jaworska, M., and Stec, W. J. (1990) Octa(thymidine methanephosphonates) of partially defined stereochemistry: synthesis and effect of chirality at phosphorus on binding to pentadecadeoxyriboadenylic acid. *Nucl. Acids Res.* **18**, 2109–2115.
23. Lesnikowski, Z. J., Wolkanin, P., and Stec, W. J. (1987) Determination of absolute configuration at phosphorus in diastereoisomers of thymidylyl (3',5') thymidylyl methanephosphonate, in *Biophosphates and Their Analogs—Synthesis Structure, Metabolism and Activity* (Bruzik, K. S. and Stec, W. J., eds.) Elsevier, Amsterdam, pp. 189–192.
24. Stec, W. J. (1973) Organophosphorus compounds of sulphur and selenium. 1. Stereochemistry of silver-ion promoted solvolysis of some halogenoanhydrides and thioloesters of organophosphorus acids. *Bull. Acad. Polon. Sci., Ser. Sci. Chim.* **21**, 709–803.
25. Stec, W. J., Zon, G., and Uznanski, B. (1985) Reversed-phase high-performance liquid chromatographic separation of diastereomeric phosphorothioate analogs of oligodeoxyribonucleotides and other back-bone-modified congeners of DNA. *J. Chromatogr.* **326**, 263–280.

CHAPTER 11

Oligonucleoside Boranophosphate (Borane Phosphonate)

Barbara Ramsay Shaw, Jon Madison, Anup Sood, and Bernard F. Spielvogel

1. Introduction

A number of short modified synthetic oligodeoxynucleotides have shown promise as agents for gene therapy (1–8). Among these, two classes of oligonucleotides with modified backbones (the phosphorothioates [9] and methylphosphonates [10]) have emerged as primary candidates. We have synthesized another homologous type of nucleotide called a boranophosphate, or borane phosphonate (11), which can be viewed as a "hybrid" of two types of oligonucleotide molecules, i.e., the normal phosphodiester and the methylphosphonate (see Fig. 1). In addition, the boranophosphate is negatively charged like the phosphorothioate, another well-studied antisense agent.

The boranophosphate might offer an ideal antisense agent. The borane moiety (—BH_3) in the boranophosphate is isoelectronic with oxygen in naturally occurring phosphodiesters, and isoelectronic and isosteric with the —CH_3 group in methylphosphonates. The boronated oligonucleotides carry the same negative charge as normal phosphodiesters and phosphorothioates and, like them, are soluble in aqueous solutions; however, the distribution of charge density would differ somewhat. Furthermore, since the —BH_3 group in boranophosphates is isoelectronic as well as isosteric with the —CH_3 group, it might be expected that boronated oligonucleotides also would exhibit some desirable properties of the methylphosphonate nucleotides, which form

From: Methods in Molecular Biology, Vol. 20: Protocols for Oligonucleotides and Analogs
Edited by: S. Agrawal Copyright ©1993 Humana Press Inc., Totowa, NJ

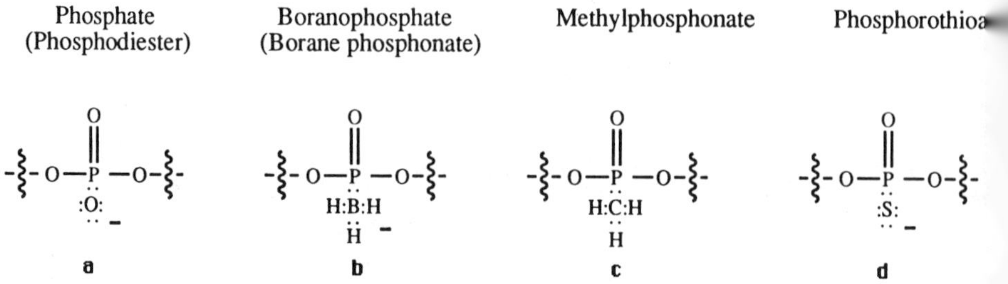

| Phosphate | Boranophosphate | Methylphosphonate | Phosphorothioate |
| (Phosphodiester) | (Borane phosphonate) | | |

$$\begin{array}{cccc} \text{O} & \text{O} & \text{O} & \text{O} \\ \| & \| & \| & \| \\ -\text{O}-\text{P}-\text{O}- & -\text{O}-\text{P}-\text{O}- & -\text{O}-\text{P}-\text{O}- & -\text{O}-\text{P}-\text{O}- \\ :\ddot{\text{O}}:^- & \text{H:B:H} & \text{H:C:H} & :\ddot{\text{S}}:^- \\ & \text{H}^- & \text{H} & \\ \text{a} & \text{b} & \text{c} & \text{d} \end{array}$$

Fig. 1. Internucleotide linkage modifications.

stable hybrids with DNA and may enter cells through passive transport mechanisms. Finally, and potentially important for drug delivery, we have shown that the boranophosphates are nuclease resistant *(12)*, like the phosphorothioates and methylphosphonates. Short boranophosphate oligomers exhibit a high degree of stability to spleen phosphodiesterase and snake venom phosphodiesterase, indicating that they could have long half-lives in the cell.

1.1. Boron Neutron Capture Therapy

An additional property of boron is the ability of one of its isotopes to absorb thermal neutrons. Boron-10 has a high cross-section for thermal neutrons and, on capturing a neutron, it fissions to generate a lithium-7 nucleus and energetic α particle, which are highly destructive with a relatively short (10–14 mm) path. If boron can be localized specifically in infected cells (as may be the case in viral diseases or overexpression), then these cells may be readily destroyed by using boron neutron capture therapy (BNCT) *(13)* without affecting normal cells nearby. For a radiotoxic reaction, BNCT requires a reasonable concentration of boron-10 (i.e., about 5 ppm) and a source of low-energy thermal neutrons. The inherent advantage is that the boron and the neutron flux may be manipulated independently to produce the desired radiation effect in the target cells with little damage to other tissues. One goal of our work is to design boronated nucleosides *(14)* and oligonucleotides *(11)*, which may be used to transport boron to tumor tissue selectively.

The synthetic methods used to prepare short boranophosphate oligodeoxynucleotides are reviewed below. Approaches to synthesizing longer boron-containing oligomers by solid-phase methods are discussed.

2. Materials and Methods

2.1. Materials

5'-O-DMT-thymidine–3'-(methyl-N,N-diisopropyl)phosphoramidite (Sigma, St. Louis, MO), 3'-acetylthymidine (Sigma), tetrazole (Aldrich, Milwaukee, WI), and dimethylsulfide-borane (Aldrich) were obtained commercially. Tetrazole was sublimed prior to use. Acetonitrile used in the reactions was dried by refluxing over CaH_2 and then distilled onto 4-Å molecular sieves. HPLC solvents were obtained from Baxter Scientific Products (IL), and were filtered and degassed prior to use.

HPLC purifications are performed on a Bio-Rad (Richmond, CA) gradient processor system 3.8 consisting of an Apple IIe computer, an HP 3392A integrator, two Bio-Rad model 1330 pumps, a Bio-Rad gradient mixer, a Bio-Rad 7125 syringe loading injector, and a Bio-Rad 1305A UV monitor operating at 260 nm. A Rainin Dynamax microsorb 5 μ, C–18 (21.4 × 250 mm) prep size column is used. HPLC-grade water and acetonitrile are used as mobile phase A and B, respectively. ^{1}H- and ^{13}C-NMR spectra are obtained on a Varian XL–300 spectrometer. ^{31}P-NMR spectra are obtained either on a Varian (Palo Alto, CA) XL–300 or a JEOL (Tokyo, Japan) FX 90Q spectrometer. ^{11}B-NMR spectra are obtained on a JEOL FX 90Q spectrometer. Acetone-d^6 is used as solvent unless indicated otherwise.

2.2. Methods

2.2.1. Dithymidyl Boranophosphate, Methyl Ester, 1

Tetrazole (0.50 g) is dissolved in freshly dried CH_3CN (15 mL) under argon. To this solution, 5'-O-DMT-thymidine phosphoramidite (1.00 g, 95% pure) dissolved in CH_3CN (10 mL) is added by a syringe. The amidite vial is rinsed with another 3–4 mL of CH_3CN. 3'-Acetylthymidine (0.40 g) is added to the reaction mixture, and the mixture is stirred at room temperature. After 15 min, dimethylsulfide•borane (470 μL, 3.3 Eq) is added, and the mixture is stirred for 3–4 min. A small portion of the reaction mixture is taken in $CDCl_3$ for ^{31}P-NMR, which shows complete disappearance of phosphite resonances and, after a large number of accumulations, the appearance of a broad peak at ~118 ppm for boranophosphate. After 4 h, solvent is removed from the reaction mixture at room temperature under reduced pressure. The residue is purified by flash chromatography on silica gel followed by HPLC on a Dynamax C-18 column using a gradient of 25–100% B in 38 min at a flow rate of 9.2 mL/min. Yield: 417 mg, 51.6% based on the amount of pure amidite. ^{1}H-NMR: *see* Fig. 2. ^{11}B-NMR: $\delta = -44.0$ ppm, br. peak; ^{13}C-NMR:* $\delta =$ 12.49 ppm, s, 5-CH$_3$'s; 20.81 ppm, s, CH$_3$ of 3'-OAc; 37.12 ppm, s, 2'-C; 39.27 ppm, s, 2'-C; 54.22 ppm, 2d, POCH$_3$; 62.22 ppm, s, 5'-C$_A$; 66.93 ppm, 2d, 5'-C$_B$; 74.75 ppm, d, 3'-C$_B$; 79.36 ppm, 2d, 3'-C$_A$; 83.21 ppm, 2d, 4'-C$_B$; 85.39 ppm, d, 4'-C$_A$; 85.46 ppm, s, 1'-C; 86.65 ppm, 2s, 1'-C; 111.08 and 111.50 ppm, 2s, 2 C-5; 136.24 ppm and 136.78 ppm, 2s, 2 C-6; 151.24 and 151.34 ppm, 2s, 2 C-2; 164.31 ppm, s, 2 C-4 and 171.03 ppm, s, carbonyl C of OAc; ^{31}P-NMR: $\delta = 118.2$ ppm, br. peak. FAB mass spectral data: m/e 601 (MH$^+$). Anal. calculated for BC$_{23}$H$_{34}$N$_4$O$_{12}$P: %C, 46.02, %H, 5.71, %N, 9.33, %P, 5.16; Found: %C, 46.01, %H, 5.94, %N, 9.02, %P, 4.97.

2.2.2. Dithymidyl Boranophosphate, Ammonium Salt, 2 (TpBT) (i.e., the Dinucleoside Borane Phosphonate, Ammonium Salt)

Dimer **1** (40.7 mg, 0.068 mmol) is taken in conc. ammonium hydroxide (10 mL) in a sealed tube. The mixture is shaken overnight at room temperature. The tube is cooled in ice and opened to atmosphere. After allowing the ammonia to escape, the solution is lyophilized to give a white solid. Yield of crude product: 42.0 mg. ^{1}H-NMR of crude product shows product, ammonium acetate, and a small amount of unidentified impurity. A fraction of crude product is purified on reverse-phase HPLC by a stepwise gradient of 0–10% B in 5 min and 10–35% B in 25 min at a flow rate of 9.2 mL/min. Yield: 38.6%. ^{1}H-NMR (D$_2$O): *see* Fig. 3; ^{11}B-NMR: $\delta = -40.5$ ppm, m, ^{1}J$_{B,H}$ and ^{1}J$_{B,P}$ are not measured. ^{31}P-NMR: $\delta = 93.8$ ppm, br. q. peak. FAB mass spectral data: m/e 545 (M + 2H)$^+$ where M is dithymidyl boranophosphate anion.

---

*In many cases, NMR peaks for similar carbons from two-nucleoside units have identical chemical shifts, and therefore, only a single peak is observed. Peaks resulting from two diastereomers are resolved only for certain carbons. Peak assignments are tentative, and only where absolutely certain are the peaks from two-nucleoside units distinguished. Subscript A indicates carbon from 5'-nucleoside, and B indicates carbon from 3'-nucleoside.

Oligonucleotides with a Boronated Backbone 229

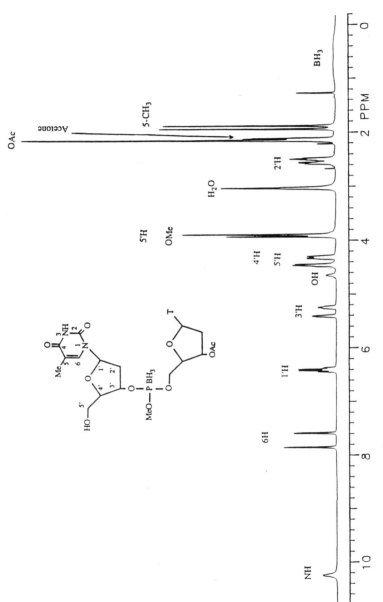

Fig. 2. ¹H-NMR of dinucleoside boranophosphate methyl ester 1.

230 Shaw et al.

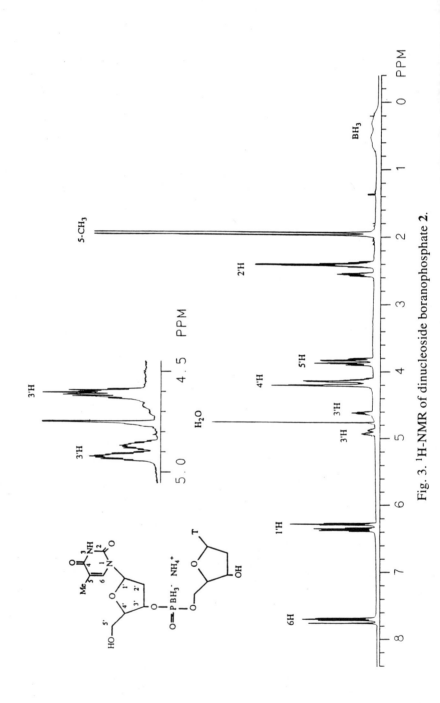

Fig. 3. ¹H-NMR of dinucleoside boranophosphate **2**.

2.2.3. Trithymidyl Bis-(Boranophosphate), 3

Trimer **3** is prepared by the same procedure as described for dimer **1** using dimer **1** in place of 3'-acetylthymidine and allowing the coupling to proceed for 40 min. The product is purified by flash chromatography followed by reverse-phase HPLC using a stepwise gradient of 25–52.5% B in 20 min and 52.5–100% B in 20 min at a flow rate of 9.2 mL/min. ^{1}H-NMR: see Fig. 4; ^{11}B-NMR: $\delta = -44.0$ ppm, br. peak; ^{13}C-NMR:[†] see * and [†], $\delta = 12.50$ and 12.55 ppm, 5 CH$_3$'s; 20.85 ppm, s, CH$_3$ of OAc; 37.05, 38.56, and 39.35 ppm, 2'-C; 54.37 ppm, POCH$_3$'s; 62.48, 66.35, and 67.19 ppm, 5'-C; 74.72, 77.92, and 79.53 ppm, 3'-C; 83.28, 83.93, and 85.55 ppm, 4'-C; 85.81, 86.09, and 86.70 ppm, 1'-C; 111.12 and 111.47 ppm, C-5; 136.34–136.70 ppm, C-6; 151.25 and 151.44 ppm, C-2; 164.19 ppm, C-4 and 170.98 ppm, CO of OAc. ^{31}P-NMR: $\delta = 118.5$ ppm, br. peak. FAB mass spectral data: m/e 917 (MH$^+$).

3. Synthesis of Oligonucleotides with a Boronated Backbone

The method employed for the synthesis of boronated oligonucleotides is similar to the phosphoramidite method used for the synthesis of normal oligomers, except that the oxidation step has been replaced by reaction with dimethylsulfide-borane, Scheme 1. Yields have not been optimized.

3.1. Dimers

The boranophosphate, **2**, and the boranophosphate methyl ester, **1**, were the two types of boronated TpT dimers prepared. The methyl ester was prepared by reacting 5'-DMT-thymidine phosphoramidite with 3'-acetylthymidine in the presence of tetrazole, which results in the formation of an intermediate phosphite *(15)*. The phosphite is then converted to the dithymidyl boranophosphate methyl ester, **1**, by reaction with 3.3 Eq of dimethyl sulfide-borane (Scheme 1). Both reactions can be easily followed with ^{31}P-NMR. In the first step, the amidite peaks at 148.7 and 148.4 ppm disappear within the time required for recording the spectrum, and are replaced with sharp phosphite peaks at 140.4 and 139.8 ppm. In the second reaction, the phosphite peaks dis-

---

[†]Because of the presence of several diasteromers, in many cases, several peaks are observed for each set of carbons. Chemical shifts given are for each set of peaks rather than a single peak.

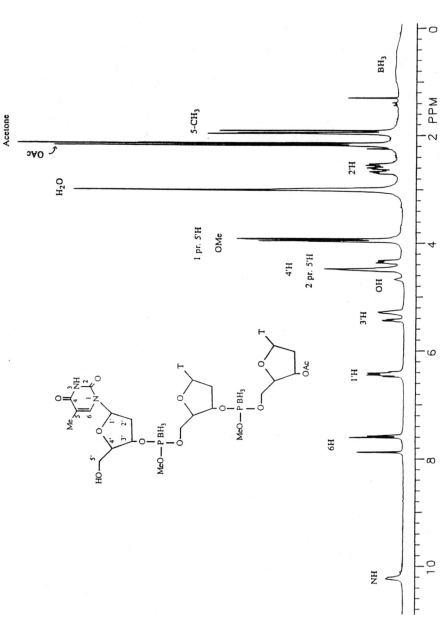

Fig. 4. ^{1}H-NMR of trinucleoside boranophosphate methyl ester **3**.

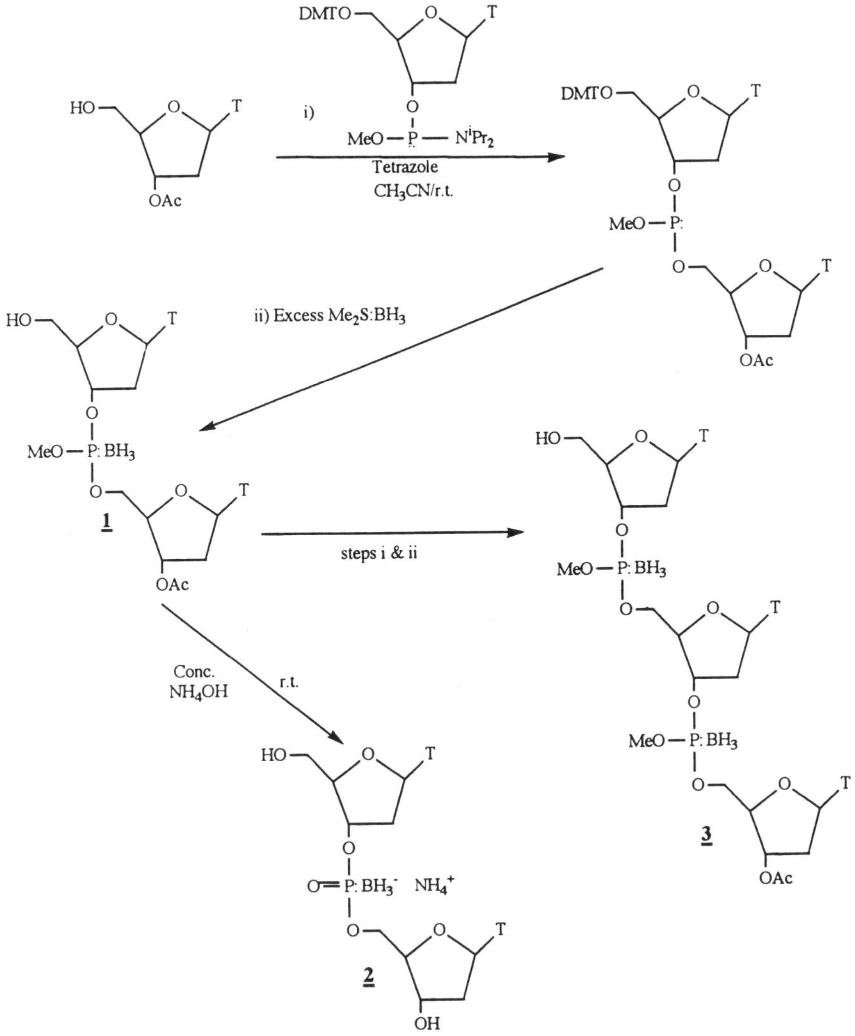

Oligonucleotides with a Boronated Backbone

Scheme 1. Synthesis of boronated di- and trinucleotides with borane bonded to phosphorous *(11)*.

appear within 5–10 min and, after a large number of accumulations, a broad quartet-shaped peak for boranophosphate phosphorus appears at 118 ppm. *(See* Fig. 5.) The quartet shape is the result of the coupling of phosphorus-31 to boron-11, which has a spin of 3/2. The broadness of

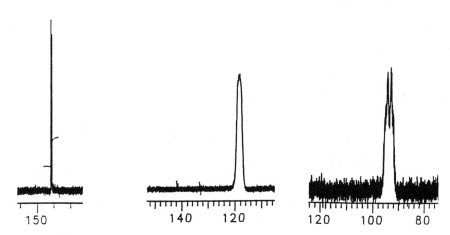

Fig. 5. ^{31}P-NMR of the phosphite (left), boranophosphate methyl ester (middle), and boranophosphate (right).

the peak is a result of the high quadrupole moment of boron, and the overlap of peaks is owing to two diastereomers. The boronation of phosphite was completed in <5 min (the time required for observing disappearance of phosphite peaks by ^{31}P-NMR).

The oxidation step with dimethylsulfide-borane offers two advantages. In addition to forming the phosphite-borane adduct, the excess Me$_2$S•BH$_3$ in the presence of tetrazole also removes the 5'-DMT-protecting group, thereby eliminating the extra 5'-deprotection step required for further chain elongation. If the method were to be used for solid-phase synthesis, it might be possible to combine the boronation step with detritylation by adding additional tetrazole during the boronation step. In this case, however, the capping step would have to precede the oxidation step.

Dimer **1**, the boranophosphate methyl ester, was purified by flash chromatography followed by reverse-phase HPLC to give an overall yield of 52%. The compound was characterized by ^{1}H-NMR (including COSY, HOHAHA [Homonuclear Hartman Hann] techniques), ^{11}B-NMR, ^{13}C-NMR, ^{31}P-NMR, and FAB mass spectroscopy as well as elemental analysis. Dimer **1** was either hydrolyzed to yield the dimer boranophosphate, **2**, or it served as the precursor for further condensation to give **3**.

The boranophosphate dimer, **2** (TpBT), was prepared by treating the 3'-acetyl dinucleoside boranophosphate methyl ester, **1**, with concentrated NH$_4$OH at room temperature for 18 h. This results in the hydrolysis of the MeO-group and 3'-acetate to give the ammonium salt of the dinucleoside boranophosphate, **2** (Scheme 1). The reaction proceeds without any detectable degradation of the internucleotide or glycosidic bonds, as determined by ^{1}H-NMR of the crude product. The deprotected dimer is very soluble in water compared to its protected counterpart, which is sparingly soluble in aqueous medium. The crude product was purified by HPLC, and characterized by ^{1}H-NMR (including COSY-correlated spectroscopy), ^{11}B-NMR, ^{31}P-NMR, and FAB mass spectroscopy. In the ^{1}H-NMR of **2**, two sets of peaks (for two diastereomers) are observed for most protons (Fig. 3), whereas only a single set of peaks (for both diastereoisomers) is observed in the case of **1**. The diastereomers have been separated by further HPLC and are being characterized.

3.2. Trimer

Reaction of dimer **1** with 5'-DMT-thymidine phosphoramidite, followed by Me$_2$S•BH$_3$, results in the formation of trimer **3** (trinucleoside boranophosphate methylester, **3**) *(see* Scheme 1) in which both internucleotide linkages are boranophosphate methyl ester groups. This reaction was carried out with equal molar amounts of **1** and DMT-thymidinephosphoramidite, and the yield of **3** was only 22.4% prior to and 5.2% after HPLC. Yet a large proportion (ca. 66%) of dimer **1** was recovered unreacted. Clearly, the procedure must be optimized to obtain a 99%+ coupling yield before it could be used for practical synthesis of oligonucleotides. In a second experiment performed in an NMR tube, 5 Eq of amidite resulted in >95% conversion.

3.3. Progress Toward Solid-Phase Synthesis

Studies have been carried out to determine the compatibility of the boranophosphates with DNA synthesis reagents. The results are shown in Table 1. The boranophosphate linkage is stable under solid-phase DNA synthesis conditions, and the boronating agent does not interfere with the solid support. It is also apparent that a mixed backbone, where some of the phosphorus atoms are unmodified and others are converted to boranophosphates, could also be synthesized. Yet, in order to incorporate boranophosphates into DNA by solid-phase synthesis, two problems remain to be solved. A boronating agent that will not reduce bases other than T must be found, and the 5'-hydroxyl-deprotection

Table 1
Studies Toward the Development of a Solid-Phase Method
for the Synthesis of Oligonucleotide Boranophosphates

Experiment	Results
Stability of solid support linkage in the presence of boronating agent	Stable
Stability of boranophosphate linkage in a completely unprotected thymidine boranophosphate dimer in the presence of	
Cl_3CCOOH/CH_2Cl_2	Stable
I_2/H_2O/Lutidine/THF	Stable
Ac_2O/Lutidine/THF	Stable
N-methylimidazole	Stable
Tetrazole	Stable
Conc. NH_4OH/ 55°C	Stable

conditions must be modified because the trityl cation reacts with phosphite boranes. Considerable progress has been made toward solving the latter problems.

4. Physical, Chemical, and Nuclear Properties of Boranophosphates

4.1. Stability of the P–BH₃ Bond

4.1.1. Acid-Base Hydrolysis

The internucleotide boranophosphate group is very stable toward basic or acidic hydrolysis. Thus, heating the phosphite borane **1** or boranophosphate **2** at 55°C overnight in concentrated NH_4OH (conditions used for deprotection of bases in normal oligonucleotide synthesis) does not result in any change (other than formation of the stable boranophosphate **2** when **1** is the reactant, as in Scheme 1). ^{31}P-NMR shows no effect on the P–B bond. The ^{11}B-NMR also shows an intact P–BH₃ group. Thus, the P–BH₃ group should be stable under conditions normally used for the deprotection of bases in the solid-phase synthesis of oligodeoxynucleotides.

The P–BH₃ group is also remarkably stable under acidic conditions. When **1** is shaken at room temperature overnight in a mixture of 1N HCl and MeOH (1:1 v/v), <10% of the phosphite-borane group is hydrolyzed to phosphate (by ^{11}B- and ^{31}P-NMR). The 3'-acetate group is hydrolyzed, although the POMe group appears to remain intact (by ^{1}H-NMR). Since the conditions used here are very harsh, the P–BH$_3$ group should be stable under pH conditions prevailing inside the body or required for deprotection of the 5'-hydroxyl group during solid-phase synthesis.

The boranophosphate backbone is also very stable under physiologically relevant conditions. In incubations without exonuclease at 37°C, no degradation of the TpBT dimer **2** was observed (in 10 m*M* Tris-HCl, pH 7.5, and 10 m*M* MgCl$_2$) even after 5 d. In contrast, under the same conditions, the normal TpT phosphodiester compound underwent 50% decomposition after 237 h *(12)*. Thus, the boranophosphate linkage **2** (Fig. 1b) under physiological conditions shows greater stability than the normal phosphodiester linkage (Fig. 1a).

4.1.2. Nuclease Stability

The boranophosphate internucleotide linkage in dimer **2** is quite stable toward cleavage by exonucleases. Enzymatic hydrolysis studies were performed on TpT in vitro (normal dinucleotide), and a mixture of R and S boronated diastereomers of TpBT **2** to ascertain the in vitro exonuclease resistance of the internucleotide linkage *(12)*. Under conditions where normal dithymidylyl phosphate is >97% cleaved, dimer **2** is >92% stable to both calf spleen and snake venom phosphodiesterase. The boronation confers considerable resistance to these two particular exonucleases. The extent of protection shows that both the R and S chiral forms exhibit nuclease resistance. These experiments are now being repeated for the R and S stereoisomers of the dimer, which we have separated by HPLC. In summary, the P—BH$_3$ group possesses sufficient hydrolytic stability to survive use in biological systems.

4.2. Solubility and Permeability

The charge on the boranophosphate (Fig. 1b), like the charge on the normal phosphate backbone (Fig. 1a), permits water solubility. However, charge is diffused over the –BH$_3$ group, and the borane group (like the –CH$_3$ group) would be expected to have some lipophilic character. Moreover, the –BH$_3$ group (which has replaced an isoelectronic

$$:\underset{..}{O}: \quad VS \quad H:\underset{H}{\underset{..}{B}}:H$$

Fig. 6. Valence bond structure of oxygen and borane (–BH$_3$).

oxygen in the phosphate) lacks the three available electron pairs of the oxygen that otherwise would form hydrogen bonds with water (Fig. 6). Because the charge is diffused over the borane group and many boron derivatives (e.g., boron analogs of amino acids, peptides, and so on) show a propensity toward amphiphilic character, we anticipate that boronated oligomers will be more membrane permeable than the normal phosphate oligomers.

4.3. Toxicity

No acute toxicity data are available. However, compounds were tested by Iris H. Hall (Univ. of North Carolina) for hypolipidemic activity in mice at 8 mg/kg/d for 16 d. At this dose, no observable toxicity was present.

4.4. "Hybrids" of Normal Phosphates and Methylphosphonates

Boranophosphate oligomers are hybrids of normal phosphates and methylphosphonates, two of the widely used antisense molecules (Fig. 1). The borane (–BH$_3$) group in boranophosphates is isoelectronic to oxygen and has a negative charge like normal phosphates, yet it is sterically very different from oxygen (Fig. 2). It is also isosteric and isoelectronic to –CH$_3$ in methylphosphonates, but the methylphosphonates are neutral species. The boranophosphates are also negatively charged like the phosphorothioates and phosphorodithioates *(16)*.

4.4.1. Environment Around Phosphorus Atom

The electronic environment around the phosphorus in the boronated oligonucleotide triesters **1** and **3** appears to be similar to that in phosphorodithioate oligonucleotides as demonstrated by similar ^{31}P chemical shifts (118 ppm for the boronated compound vs 113.0 ppm for the $^-$S–P=S phosphorodithioate *[16]*). By contrast, the ^{31}P shift for normal $^-$O–P=O is approx 0 ppm.

Table 2
Summary of Hypochromicity Values and Molar Absorptivities

Compounds	$\varepsilon_{max} \times 10^{-3}$	λ_{max}	% Hypochromicity	Stacking interaction	Reference
pT[c]	9.2	265	–	–	This study[a]
TpT[c]	8.5	267	8	Good	This study[a]
TpBT, 2	9.1	267	1	Fair	This study[a]
TBp(OMe)TOAc, 1	9.9	267	–7	Perturbed to some degree	This study[a]
pT	9.6	265	–	–	Ref. 17[b]
TpT	9.2	267	4	Good	Ref. 17[b]
Tp(OMe)T	9.1	267	5	Good	Ref. 17[b]
Tp(OEth)T	9.0	267	6	Good	Ref. 17[b]

[a]In 10 mM Tris-HCl, 10 mM MgCl$_2$, pH 7.5.
[b]In water at 28°C, pH 7.
[c]pT monomer and TpT dimer were purchased from Sigma.

4.4.2. Conformational Studies

Thymidyl (3'-5') thymidyl boranophosphate, TpBT, **2**, is a dinucleotide monophosphate with a borane group on the phosphodiester linkage replacing a nonbridging oxygen. The boranophosphates are charged similarly to the phosphorothioates and normal phosphates, but are structurally similar to the methylphosphonates. Because of the modification of the internucleotide linkage with the borane group, these nucleotide analogs should have a number of interesting biophysical properties. In preliminary studies, the electronic and conformational effects of the borane group have been examined by UV spectroscopy. Molar absorptivities and λ_{max} for the dithymidyl boranophosphate methyl ester **1** and the boranophosphate **2** analogs are compared in Table 2 with previous data by Miller et al. *(17)* on normal phosphates and phosphotriester analogs. For all the dimer analogs, including the boranophosphate, the λ_{max} shifts from 265 nm (for the pT monomer) to 267 nm. For all the dimers, the hypochromicity changes by under 8%. On closer inspection, the reported values for hypochromicities of the normal TpT in water and of the two phosphotriester dimers are somewhat greater than the TpBT boranophosphate **2**, which exhibits almost no hypochromicity, and the boranophosphate methyl ester **1**, which has a negative hypochromicity.

An exact comparison with the triesters cannot be made since the borane compounds were studied in a higher ionic strength medium than the other nucleotides and compared to a different TpT sample. Nevertheless, it is clear that among the two boronated nucleotides, the boranophosphate methyl ester **1** shows a greater change in UV absorption and is in the opposite direction. The greater perturbation for TBp(OMe)TOAc **1** may be indicative of less stacking, or more disruption for that structure, than for the boranophosphate dimer **2**, and is consistent with the idea that greater disruption of stacking would occur owing to the phosphotriester/borane combination on the backbone when compared to the borane substitution alone as in TpBT. To summarize, the boranophosphate TpBT dimer exhibits the same λ_{max} as the normal TpT and phosphotriester *(17)* dimers, and behaves like other modified nucleotides by following a similar trend in hypochromicity. Further conformational studies are in progress.

4.5. Nucleotides as Carriers of Boron-10

4.5.1. Destruction of Tumors Using Boron Neutron Capture Therapy (BNCT)— A Double-Pronged Approach for Therapy

Boronated oligonucleotides may offer a unique advantage over other modified oligonucleotides, because they can be used for boron neutron capture therapy. Boron-10 itself is nonradioactive, but in the presence of slow neutrons, the boron will fission into a lithium-7 nucleus and α particle. The energy generated is sufficient to destroy the cell that captures the neutron. There is little damage to surrounding cells because: (1) the Li and α particles generated are large by radiation standards and have a relatively short track in tissue (only 10–14 mm), and (2) thermal neutrons have such low energy that minimal ionization of tissue components would occur *per se*. Thus, damage can be localized to those tissues that incorporate the most boron and are targeted by the neutron beam. With boronated oligonucleotides acting as antisense agents, it may be possible to accumulate the boronated oligomers selectively in abnormal cells, such as those infected by a virus, and then to destroy the targeted cells by using boron neutron capture therapy.

The design and synthesis of boron compounds that will specifically localize in the nuclei of tumor cells and bind to DNA would be very desirable and may be quite feasible. A number of types of modified nucleic acids are known to be incorporated in the cell nucleus. A boron-containing thymidine analog, 5-dihydroxy-boryl-2'-deoxyuridine, not only incorporates into the cell nucleus, but also significantly radio-sensitizes the cell *(18)*. Thus it is reasonable to expect that a similar effect may be observed with boron-containing oligonucleotides. Another important consideration for BNCT efficacy is the amount of boron present in the cell. Various calculations estimate that for BNCT to be effective, the amount of boron in the cell should be approx 5–20 µg/g of tissue if it were distributed uniformly over the cell. If, however, the boron were targeted *(19,20)* to the RNA or DNA of the cell (e.g., if boronated oligonucleotides were able to form duplexes or triplexes by an antisense or antigene effect), then the required effective concentration would be considerably reduced. It is anticipated that borane modification of the phosphate backbone of oligonucleotides could provide the requisite amount for selective destruction, since it has been calculated that antisense oligomers can be taken up by cells to give concentrations calculated at 4 µM *(21)*. Additional tumor preference or specificity could be gained by attaching various tumor-seeking groups, such as prophyrins *(22)* or other hydrophobic groups *(21,23)* to the 5'- or 3'-ends.

5. Prospective

We have shown the feasibility of preparing novel boronated (and stable) oligomers as precursors to "antisense" molecules. These oligonucleotides can be envisioned as "hybrids" of the normal O-oligo (charged DNA backbone) and the methylphosphonate backbone. They should be quite useful in elucidating the properties and mechanism of action of "antisense" molecules. The —BH_3 group in boranophosphates imparts resistance to nucleases *(11,12)*, making them much more stable to nuclease degradation than normal phosphodiesters. These boronated oligonucleotides should provide valuable information permitting rational design of potent new antisense and antigene therapeutics. Boronated oligonucleotides themselves constitute an entirely new class of potent therapeutics that may permit a combination therapy of antisense therapy and boron neutron capture therapy.

Acknowledgment

This work was supported by grants NP-741 and a Faculty Research Development Award from the American Cancer Society to B. R. S. and by NIH 1 R43 GM41510 to B. F. S.

References

1. Miller, P. S. and Ts'o, P. O. P. (1987) A new approach to chemotherapy based on molecular and nucleic acid chemistry: Matagen (masking tape for gene expression). *Anticancer Drug Design* **2**, 117–128.
2. Sarin, P. S., Agrawal, S., Civeira, M. P., Goodchild, J., Ikeuchi, T., and Zamecnik, P. C. (1988) Inhibition of acquired immunodeficiency syndrome virus by oligodeoxynucleoside methylphosphonates. *Proc. Natl. Acad. Sci. USA* **85**, 7448–7451.
3. Cohen, J. S. (ed.) (1989) *Topics in Structural and Molecular Biology: Oligo-Deoxynucleotides: Antisense Inhibitors of Gene Expression.* CRC Press, Boca Raton, FL.
4. Agrawal, S., Goodchild, J., Civeira, M. P., Thornton, A. H., Sarin, P. S., and Zamecnik, P. C. (1988) Oligodeoxynucleoside phosphoramidates and phosphorothioates as inhibitors of immunodeficiency virus. *Proc. Natl. Acad. Sci. USA* **85**, 7079–7083.
5. Cazenave, C. and Helene, C. (1991) Antisense oligonucleotides, in *Antisense Nucleic Acids and Proteins; Fundamentals and Applications* (Mol, J. N. M. and van der Krol, A. R., eds.) Marcel Dekker, New York, pp. 47–93.
6. Wickstrom, E. (1991) *Prospects for Antisense Nucleic Acids Therapy of Cancer and AIDS.* Wiley-Liss, New York.
7. Matsukura, M., Zon, G., Shinozuka, K., Robert-Guroff, M., Shimada, T., Stein, C. A., Mitsuya, H., Wong-Staal, F., Cohen, J. S., and Broder, S. (1989) Regulation of viral expression of human immunodeficiency virus *in vitro* by an antisense phosphorothioate oligodeoxynucleotide against *rev (art/trs)* in chronically infected cells. *Proc. Nat. Acad. Sci. USA* **86**, 4244–4248.
8. Zon, G. (1990) Innovations in the use of antisense oligonucleotides. *Ann. NY Acad. Sci.* **616**, 161–172.
9. Burgess, P. M. J. and Eckstein, F. (1979) Diastereomers of 5'-O-adenosyl-3'-O-uridyl phosphorothioate: Chemical synthesis and enzymatic properties. *Biochemistry* **18**, 592–596.
10. Miller, P. S., Yano, J., Yano, E., Carroll, C., Jayasaman, K., and Ts'o, P. O. P. (1979) Nonionic nucleic acid analogues. Synthesis and characterization of dideoxyribonucleoside methylphosphonate. *Biochemistry* **18**, 5134–5143.
11. Sood, A., Shaw, B. R., and Spielvogel, B. F. (1990) Boron-containing nucleic acids: 2. Synthesis of oligodeoxynucleoside boranophosphates. *J. Am. Chem. Soc.* **112**, 9000,9001.
12. Madison, J., Sood, A., Spielvogel, B. F., and Shaw, B. R. (unpublished data).
13. Barth, R. F., Soloway, A. H., and Fairchild, R. G. (1990) Boron neutron capture therapy for cancer. *Scientific American* **263**, 100–107.
14. Sood, A., Spielvogel, B. F., and Shaw, B. R. (1989) Boron-containing nucleic acids: Synthesis of cyanoborane adducts of 2'-deoxynucleosides. *J. Am. Chem. Soc.* **111**, 9234,9235.
15. Beaucage, S. L. and Caruthers, M. H. (1981) Deoxynucleoside phosphoramidites—A new class of key intermediates for deoxypolynucleotide synthesis. *Tetrahedron Lett.* **22**, 1859–1862.

16. Brill, K.-D. W., Nielsen, J. and Caruthers, M. H. (1991) Synthesis of deoxydinucleoside phosphorodithioates. *J. Am. Chem. Soc.* **113,** 3979,3980.
17. Miller, P. S., Fang, K. N., Konda, N. S., and Ts'o, P. O. P. (1971) Syntheses and properties of adenine and thymine nucleoside alkyl phosphotriesters, the neutral analogs of dinucleoside monophosphates. *J. Am. Chem. Soc.* **93,** 6657–6665.
18. Laster, B. H., Schinazi, R. F., Fairchild, R. G., Popenoe, E. A., and Sylvester, B. (1985) *Neutron Capture Therapy*, Proc. 2nd Int. Sym. (pub. 1986), 46–54.
19. Kobayashi, T. and Konda, K. (1984) Boron-10 dosage in cell nucleus for neutron capture therapy—Boron selective dose ratio. *Proceeding of the First International Symposium on Neutron Capture Therapy*, October 12–14, 1983, BNL Report No. 51730, pp. 120–127.
20. Gabel, D., Larsson, B., and Rowe, W. R. (1984) Biological effect of the B-10 [n,α]-Li^7 reaction simulation by Monte Carlo calculations. *Proceedings of the First International Symposium on Neutron Capture Therapy*, October 12–14, 1983, BNL Report No. 51730, pp. 128–133.
21. Saison-Behmoaras, T., Tocque, B., Rey, I., Chassignol, M., Nguyen, T. T., and Helene, C. (1991) Short modified antisense oligonucleotides directed against Ha-*ras* point mutation induce selective cleavage of the mRNA and inhibit T24 cells proliferation. *EMBO J.* **10,** 1111–1118.
22. Doan, T. L., Perraouault, L., and Helene, C. (1986) Targeted cleavage of polynucleotides by complementary oligonucleotides covalently linked to ironporphyrins. *Biochemistry* **25,** 6736–6739.
23. Leonetti, J.-P., Mechti, N., Degols, G., and Lebleu, B. (1991) Nuclear accumulation of microinjected antisense oligonucleotides. *Nucleosides and Nucleotides* **10,** 537–539.

CHAPTER 12

Oligonucleotide Phosphorofluoridates and Fluoridites

Wojciech Dabkowski, Jan Michalski, and Friedrich Cramer

1. Introduction

Fluoro derivatives of phosphorus are of great importance in the chemistry and biochemistry of both elements. The incorporation of fluorine into biomolecules has frequently resulted in a remarkable change of biological properties. Simple phosphofluoridates and their structural analogs are the classical inhibitors of serine proteases *(1a)*. It can be envisaged that a combination of nucleoside fragments with the phosphorofluoridate moiety could result in new properties with respect to selectivity of interaction with the active site of the serine hydroxy group. On the other hand, nucleotides containing a P—F linkage when incorporated into oligonucleotides could be used in the studies of biological functions of nucleic acids, including possibilities connected with the chirality of the phosphorofluoridate moiety. New phosphorofluoridate analogs of nucleotides could also be useful in studying the metabolic processes *(1b)* by noninvasive NMR imaging techniques taking advantage of 100% abundance of ^{19}F nucleus and its high inherent NMR sensitivity.

2. Nucleoside Monoesters of Monofluorophosphoric Acid

In 1963 Wittmann prepared the nucleoside phosphorofluoridates **1** by reaction of the corresponding nucleoside phosphates with 2,4dinitrofluorobenzene *(2a)* as depicted in Scheme 1. A number of nucleosid-3'-yl

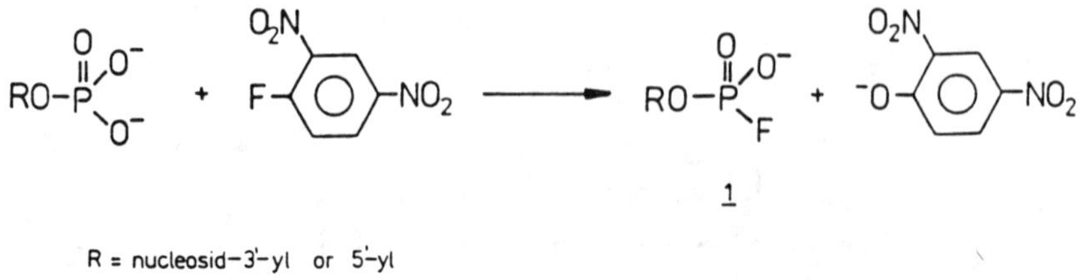

Scheme 1.

R = nucleosid-3'-yl or 5'-yl

or 5'-yl phosphorofluoridates were obtained in a ribo and deoxyribo series *(3a–h)*. Examination of the reaction in detail revealed that the initial product was the 2,4-dinitrophenyl ester of nucleotide. This reacts with the fluoride anion to produce the phosphorofluoridate **1** *(3e,4a–c)*. Consequently, the concerted mechanism originally suggested by Wittmann seems to be improbable *(2a,b)*. In the presence of pyridine, the reaction takes a different course, and the formation of sym-pyrophosphate structure is observed. Wittmann's procedure has been successfully employed by other authors and provides the best access to biologically interesting compounds of the type **1** *(4b,c)*. Alternative routes to **1** are also available.

Condensation of nucleotide with fluoride anion *(3e,5a,b)* and condensation of nucleoside with monofluorophosphoric acid *(6)* with the aid of dicyclohexylcarbodiimide give the compound **1** in low yield.

O,O,S-phosphates, in which a nucleosyl residue is attached either to the oxygen or to the sulfur atom, react with large excess of n-tetrahydrofuranepyridine-water system. The compounds of type **1** are formed as major products (Scheme 2). Yields estimated by ^{31}P-NMR spectroscopy are 79–80%, e.g. *(7)*. This reaction likely proceeds via the formation of difluoridate intermediate.

Partial hydrolysis of nucleoside phosphorodifluoridates (Scheme 3) proceeds in a very high yield *(8)*. The preparation of nucleosid-3'-yl or 5'-yl phosphorodifluoridates is described elsewhere in this chapter.

Wittmann's procedure is not suitable for synthesis of P^3-fluoro-P^1-5'-adenosine **3** and guanosine triphosphates and diphosphates *(9)*. Syntheses of this type of compound involve formation of the mixed anhydride **2** and its condensation with the corresponding diphosphates (Scheme 4) *(9,10a–c)*.

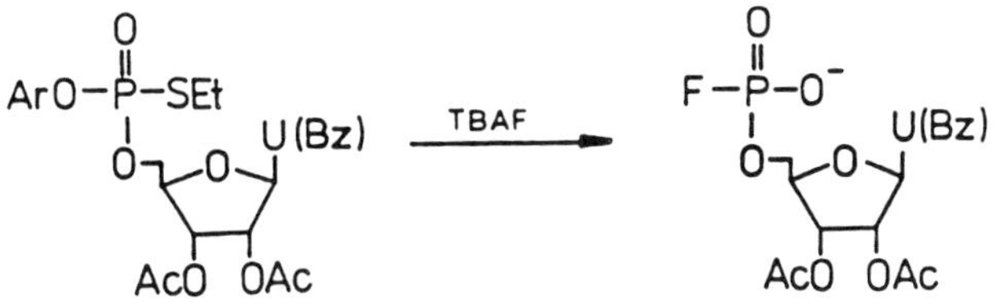

Scheme 2.

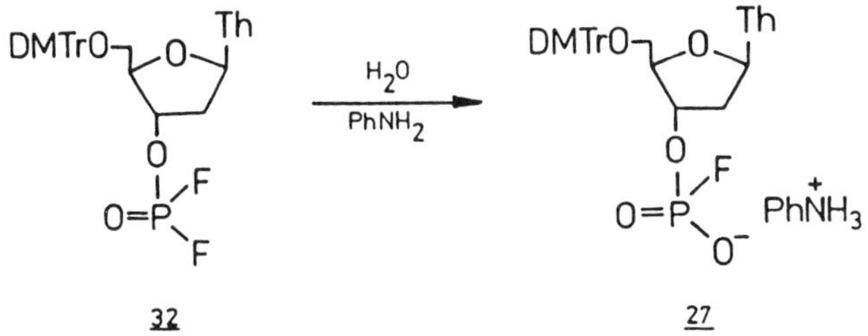

Scheme 3.

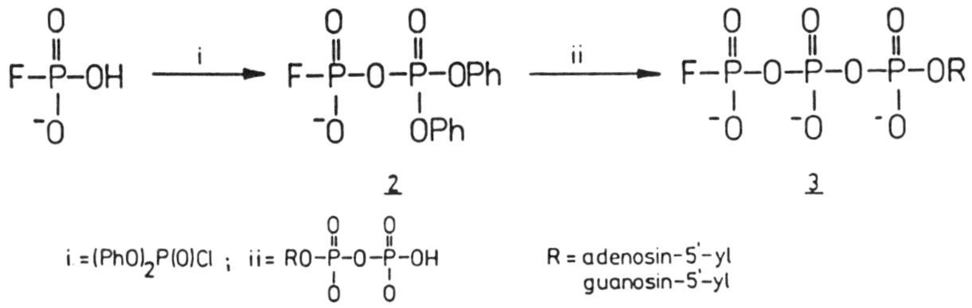

Scheme 4.

i = $(PhO)_2P(O)Cl$; ii = $RO-\overset{O}{\underset{O^-}{P}}-O-\overset{O}{\underset{O^-}{P}}-OH$

R = adenosin-5'-yl
guanosin-5'-yl

Nucleoside phosphorofluoridates salts **1** are stable compounds resistant to hydrolysis, but unstable in the presence of a vicinal hydroxy group in ribonucleosides *(7)*. Compound **1** has been used in nucleotide synthesis under somewhat drastic conditions. Suitably protected nucleotides and nucleosides react in the presence of tertiary butoxide as the base and the fluoride as the leaving group of the phosphate *(3c,g)*. Biological properties of **1** and **3** were described by various authors *(2,3b,d,h,5b,6,8,10a–c)*.

3. Nucleoside Diesters of Monofluorophosphoric Acid

Numerous pathways to phosphorofluoridates **5** and their structural analogs have previously been described *(11a–e)*. The majority of them are not suitable for the application to sensitive molecules, like nucleotides. Recently, a number of methods have been devised that fulfill requirements of high yield and selectivity.

3.1. Reaction of Nucleoside Trimethylsilyl Phosphites with Sulfuryl Chloride Fluoride

Trimethylsilyl phosphites **4** react with sulfuryl chloride fluoride in a fully chemoselective manner *(11d)* (Scheme 5). Phosphorofluoridates **5** of high purity are formed in almost quantitative yield. The side products are volatile and readily separable. Trimethylsilyl esters of the dinucleoside phosphites can be prepared from hydrogen phosphonates *(12a,b)* (Scheme 6) or more conveniently by phosphitylation procedures *(13)* (Schemes 7 and 8).

Phosphitylating reagents **6** *(14)* and **7** *(15)* are available allowing synthesis of phosphites **4** containing one or two nucleoside residues.

$$(CF_3CH_2O)_2P\text{-}OSiMe_3 \qquad (Pr^i_2N)_2P\text{-}OSiMe_3 \qquad (1)$$
$$\textbf{6}\textbf{7}$$

Phosphitylation of suitably protected nucleosides by the trimethylsilyl ester **6** proceeds in the presence of pyridine. Synthesis of the intermediate trimethylsilyl esters **8** *(16a,c)* and **9** *(16a,c)* and their conversion by the sulfuryl chloride fluoride into fluoridates **10** *(16a,c)* and **11** *(16a,c)* are presented in Scheme 7. Trimethylsilyl phosphites **8** and **9** prepared *in situ* are treated in a dry pyridine solution with the excess of sulfuryl chloride fluoride at –50°C, using a simple vacuum-line technique. The reaction can also be performed in other nonprotic solvents in the presence of pyridine. The methodology described in Scheme 7 is limited to synthesis of mononucleoside phosphorofluoridates **10** and **11**, because the phosphites **8** and **9** are poor phosphitylating reagents *(14,16c)*. This limitation can be avoided by employing the reagent **7**, which allows the selective formation of trimethylsilyl esters **12** *(13,16a–c)* and **13** *(13,16a–c)* as well as the esters **14** *(13,16a–c)*, which contain two

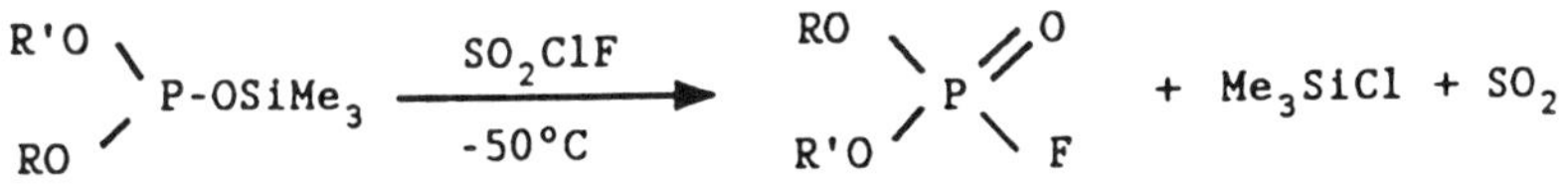

Scheme 5.

$$\underset{\underset{RO}{R'O}}{\diagdown}P-OSiMe_3 \xrightarrow[-50°C]{SO_2ClF} \underset{\underset{R'O}{RO}}{\diagdown}P\underset{F}{\overset{O}{\diagup\!\!\!\diagdown}} + Me_3SiCl + SO_2$$

$$\quad\;\; 4 \qquad\qquad\qquad\qquad\qquad 5$$

RO, R'O: alkyl, aryl, alkoxy, aryloxy

Scheme 5.

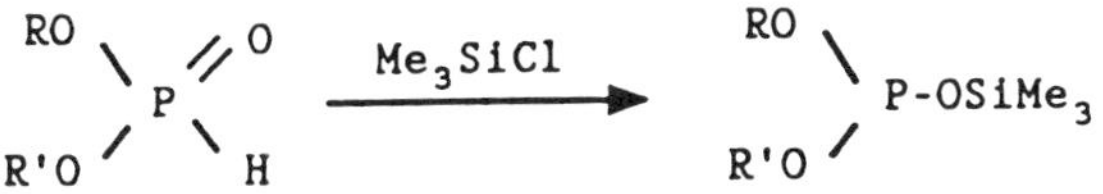

Scheme 6.

$$\underset{\underset{R'O}{RO}}{\diagdown}P\underset{H}{\overset{O}{\diagup\!\!\!\diagdown}} \xrightarrow{Me_3SiCl} \underset{\underset{R'O}{RO}}{\diagdown}P-OSiMe_3$$

$$\qquad\qquad\qquad\qquad\qquad\qquad 4$$

R: nucleosid-3'-yl, R': nucleosid-5'-yl

Scheme 6.

nucleoside residues. Phosphitylations presented in Scheme 8 involve activation by tetrazole. Strictly anhydrous conditions and sublimed tetrazole are necessary to secure high yields of the phosphites. The reaction with sulfuryl chloride fluoride leading to fluoridates **15** *(16a,c)*, **16** *(16a,c)*, and **17** *(16a,c)* proceeds under the same conditions that are used for the reaction mentioned in Scheme 7.

In every P—F nucleotide **10, 11, 15, 16,** and **17,** a characteristic doublet of doublets has been observed in the ^{31}P-NMR spectroscopy, and 1:1 ratios of corresponding diastereoisomers were noted *(16a,c)*. This means that there is no kinetic selection or thermodynamic preference favoring a particular diastereoisomer.

3.2. Reaction of P-Azolides Derived from Nucleosides with Benzoyl Fluoride

Tetracoordinate organophosphorus compounds containing azolide ligands *(17a–f)* **18** and **19** react in a fully chemoselective manner with benzoyl fluoride *(18)*. The corresponding P—F compounds **20** are formed in quantitative yield. Nucleosid-3'-yl or 5'-yl phosphorofluoridates **20** and difluoridates **24** and **26** are available via imidazolidates

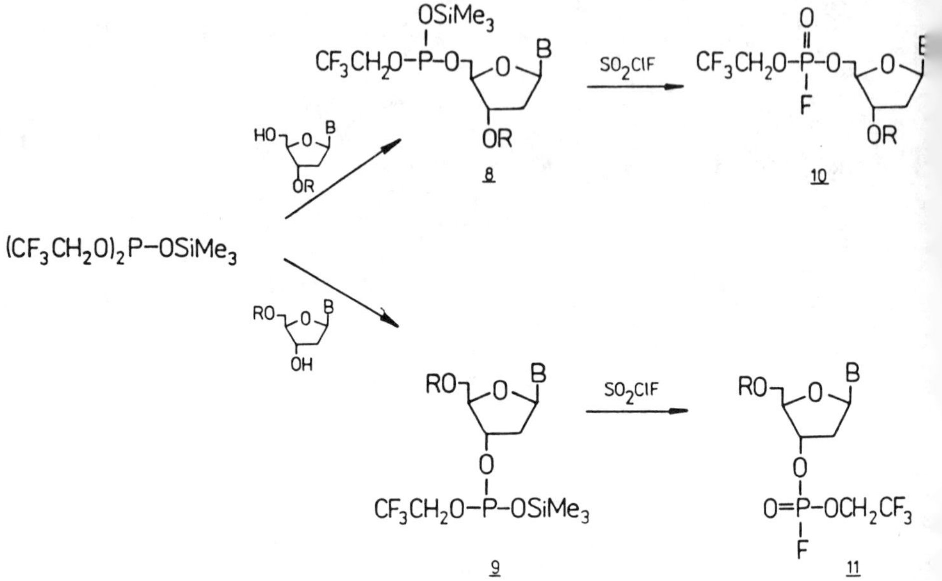

Scheme 7.

	R		BH
10 a:	SiMe₂Buᵗ		N⁶-benzoyladenine
b:	Ac		thymine
c:	4,4'-dimethoxytrityl		N⁶-benzoyladenine
d:	9-phenylxanthen-9-yl		thymine

	R		BH
11 a:	4,4'-dimethoxytrityl		N⁶-benzoyladenine
b:	Ac		thymine

19 and diimidazolidates **23** and **25**. The transformation of commercially available phosphorodichloridates into diimidazolides **18** is almost quantitative *(17a–f)*. The same high efficiency applies to the final step **19** → **20**. However, the intermediate step **18** → **19** procedes in moderate yield, which determines the final yield of **20** (Scheme 9).

A similar situation is observed in the sequence of reactions presented in Scheme 10. In both cases, the necessary amount of fluorinating reagent PhCOF must be estimated from ^{31}P-NMR spectra.

The reaction of the 3'-protected thymidine with trisimidazoylphosphine oxide **21** *(17a,19a)* and sulfide **22** *(19a)* affords ca. 40% of diimidazolidates **23** and **25** contaminated with unchanged materials. Prior to the reaction with benzoyl fluoride, unchanged nucleoside is neutralized by silylation *(8)*. The amount of benzoyl fluoride was calculated from ^{31}P-NMR spectra taking into account compounds **23** and **25** as

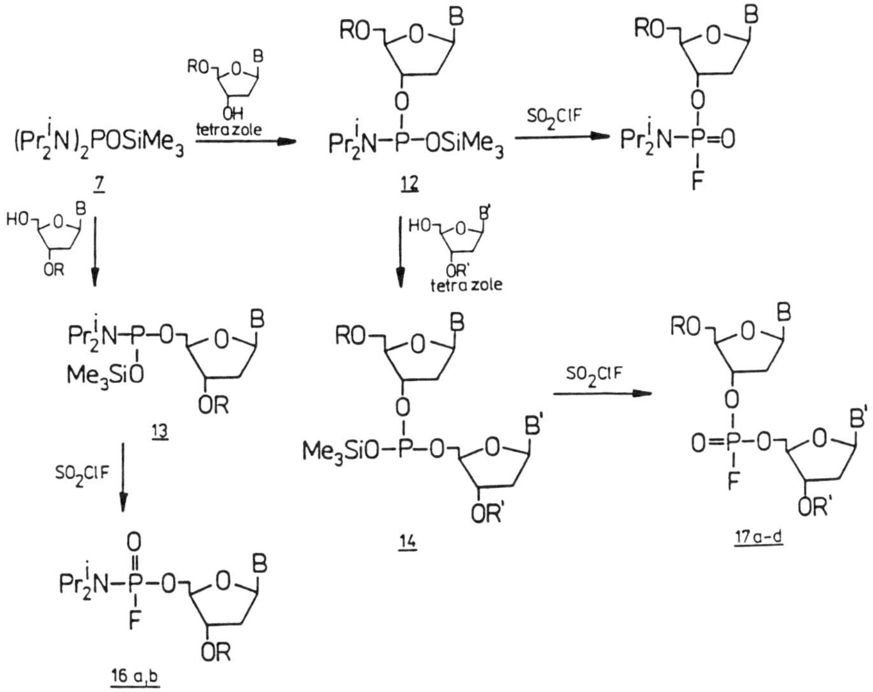

Phosphorofluoridates and Fluoridites   251

Scheme 8.

well as unchanged **21** and **22**. The volatile oxide and sulfide of trifluorophosphine were removed by evaporation, and the difluoridates **24** and **26** were finally purified by silica gel chromatography. The total yield of pure **24** and **26** is below 40%.

The difluorides **24** and **26** react with nucleosides in the presence of tertiary amine. One fluorine atom can be exchanged to form **17** in almost quantitative yield. This reaction requires catalysis by cesium fluoride *(8)* (Schemes 11,12).

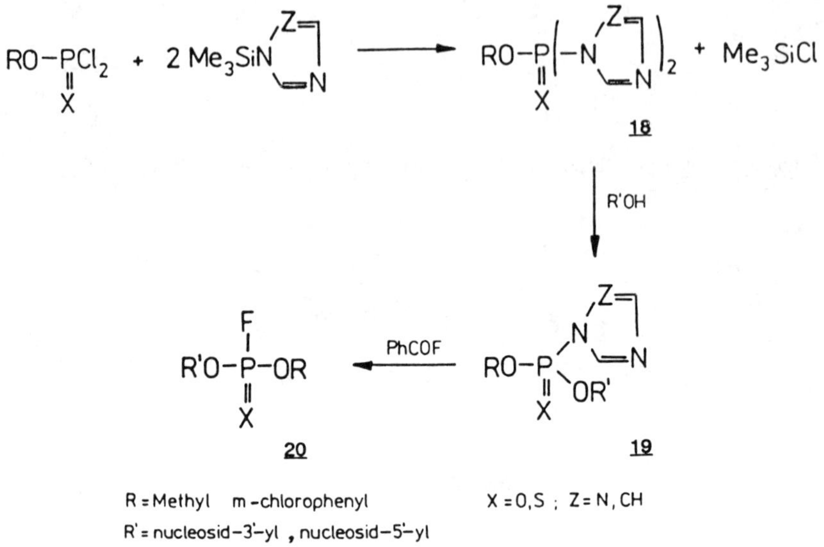

Scheme 9.

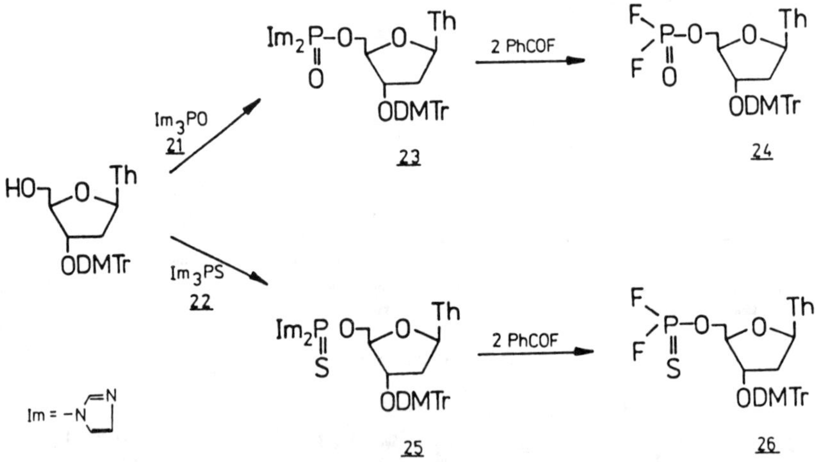

Scheme 10.

The compound **26** undergoes similar transformation into a thioanalog of **17**. The dinucleoside phosphorofluoridothionate **28** is formed in high yield.

Another strategy leading to dinucleoside phosphorofluoridates is shown in Scheme 13. In this example, trimethylsilylphosphite **14a** reacts at room temperature with oxalyl diimidazolide to give phosphoroami-

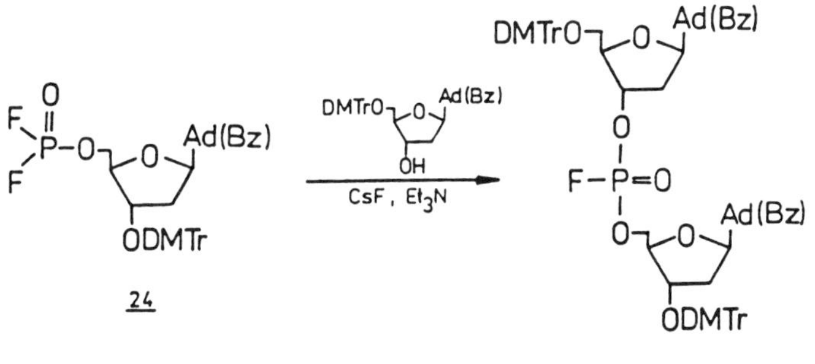

Phosphorofluoridates and Fluoridites

Scheme 11.

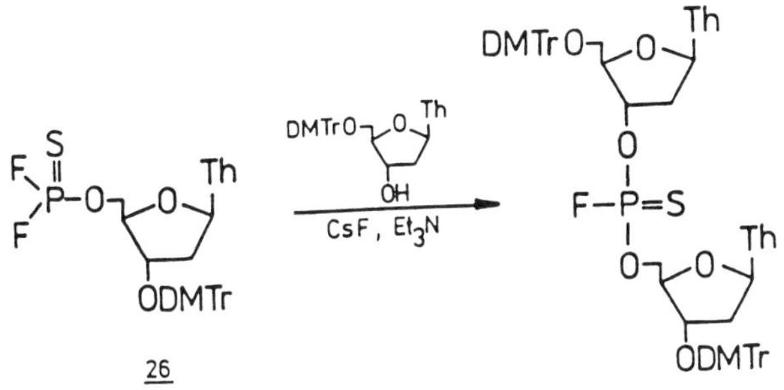

Scheme 12.

dazolidate **29** in very good yield *(16c)*. Final fluorination with benzoyl fluoride affords the derived fluoridate **17a** *(8)*. Both reactions can be combined in one flask procedure. (Scheme 13).

3.3. Synthesis of Nucleoside Phosphorofluoridates via Tricoordinate P—F Intermediates

An alternative route to nucleoside phosphorofluoridates is based on tricoordinate P—F intermediates. Trisimidazoyl phosphine *(20a–d)* was introduced as a reagent to prepare nucleoside phosphorodiimidazolidites *(20b–d)*. The diimidazolidite **30**, when allowed to react with two equivalents of benzoyl fluoride, is transformed *in situ* into phosphorodifluoridite **31** *(8)* The difluoridite **31** undergoes oxidation by air or

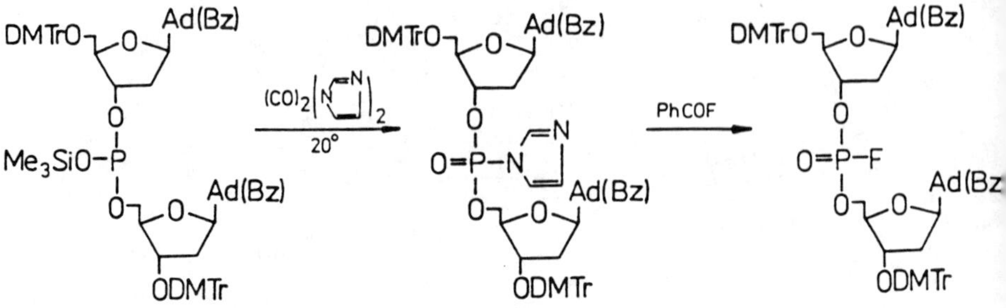

Scheme 13.

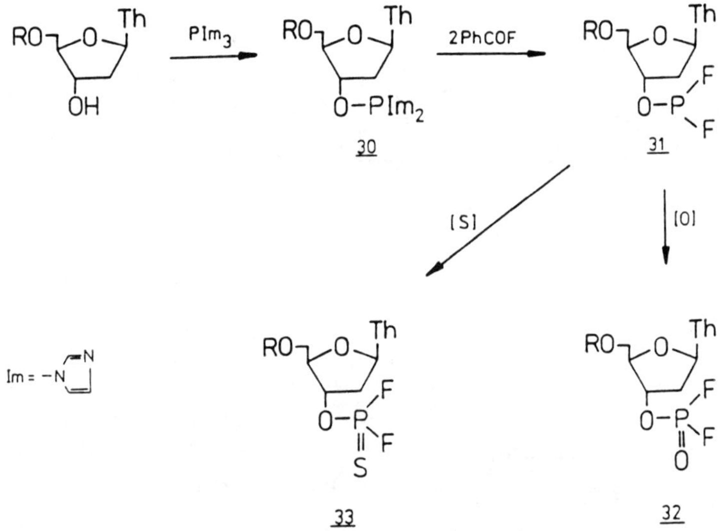

Scheme 14.

elemental sulfur to form **32** and **33**, respectively. All reactions involving tricoordinate P—F intermediates must be carried out in oxygen and moisture-free surroundings (Scheme 14).

Phosphorodifluoridite **31** itself proved to be capable of selective condensation with one equivalent of nucleoside to give dinucleoside phosphorofluoridite **34** in excellent yield *(8)*. This is shown in Scheme 15. The phosphorofluoridite **34** undergoes oxidation by air or elemental

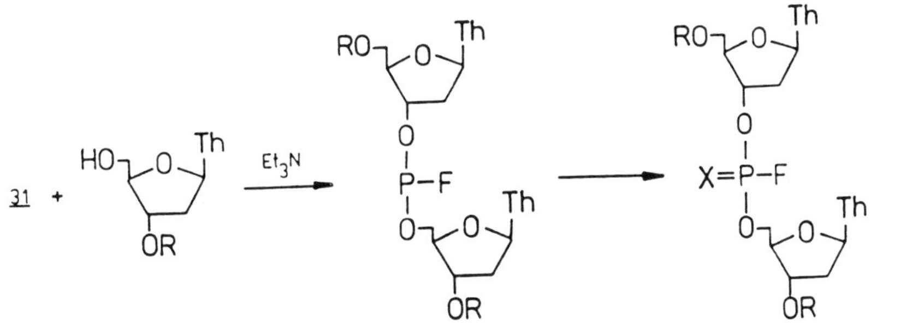

Phosphorofluoridates and Fluoridites

Scheme 15.

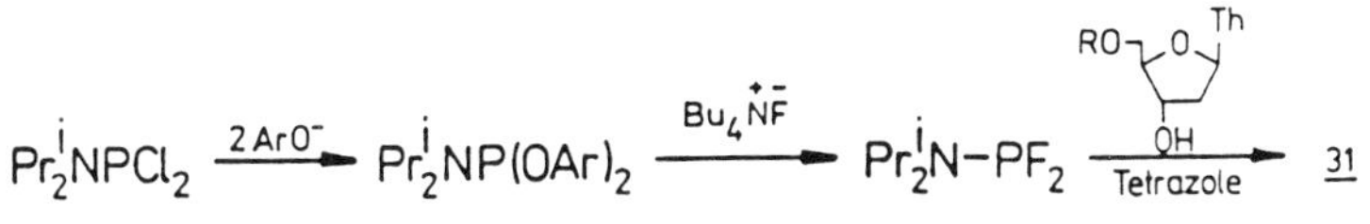

Scheme 16.

sulfur to give **17** or its sulfur analog, in excellent yield *(8)*. The difluoridite **31** can also be conveniently prepared by transformations *(8)* that are shown in Scheme 16.

3.4. Synthesis of Diastereoisomeric Dinucleoside Phosphorofluorides

Neutral dinucleoside phosphorofluorides prepared by methods described in the previous part of this chapter are formed as 1:1 mixture of diastereoisomers. It is known that thioesters **35** undergo highly selective transformation into the phosphorofluoridates when allowed to react with sulfuryl chloride fluoride *(11b)* (Schemes 17,18).

Pure diastereoisomers of O,O-dinucleoside-O-methyl thiophosphate **36a,b** R_p and S_p are readily available *(21a,b)*, and their reaction with sulfuryl chloride fluoride is either fully stereoselective (**36a**, R_p) or highly selective (**36b**, S_p) *(22)*.

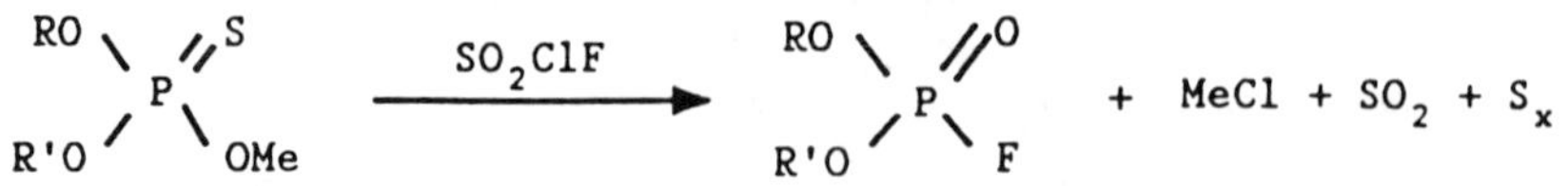

Scheme 17.

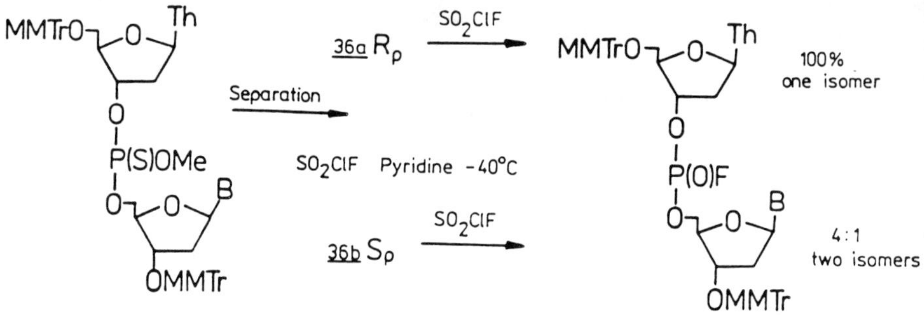

36

B	^{31}P	J_{P-F} Hz
Th	-8.9	975.8
Ad(bz)	-8.5	976.0

B	^{31}P	J_{P-F} Hz	
Th	-8.9;-9.4	975.7; 970.1	1:4
Ad(bz)	-8.4;-8.9	976.0; 980.1	1:3

Scheme 18.

4. Final Remarks

The relative low ability of P—F structures to undergo nucleophilic displacement at the phosphorus center has its origin in unique bonding properties of the fluorine atom. The combination of very high electronegativity, short bond length, and very high bond energy (ca. 117 kcal/mol) contributes toward this behavior.

For example, nucleophilic displacements at the tetracoordinate phosphorus center on models containing two potential departing groups, chlorine and fluorine, show that the former is replaced by hydroxide anion, alkoxy anions, and amines without breaking of the P—F bond *(23)*. These properties are also expressed in a relatively high chemical stability of the tetracoordinate structures—PF(O)O⁻ >P(O)F and even the tricoordinate >P—F, and are likely to be of importance in synthetic applications and biological interactions.

A preliminary illustration of special properties of >P(O)F compounds in nucleotide chemistry is the ability of 3'- and 5'-substituted dinucleoside phosphorofluoridates to undergo deprotection under stan-

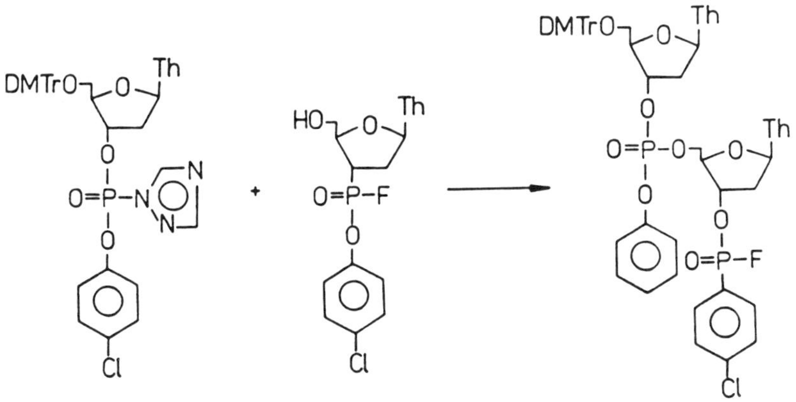

Scheme 19.

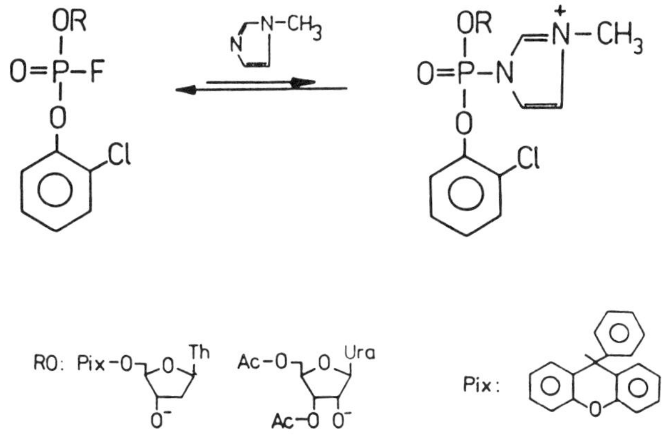

Scheme 20.

dard protocol to give the corresponding dinucleoside phosphorofluoridates without affecting the phosphorus fluorine bond *(18)*. Another demonstration of a stability of P—F bond is shown in Scheme 19. The internucleoside linkage is formed in the reaction of triazolidate **37** with nucleoside fluoridate **38** without affecting the P—F bond. However, the phosphorylating properties of nucleoside phosphorofluoridate can be enhanced in the presence of a large amount of *N*-methyl imidazole or similar highly nucleophilic amines (Scheme 20).

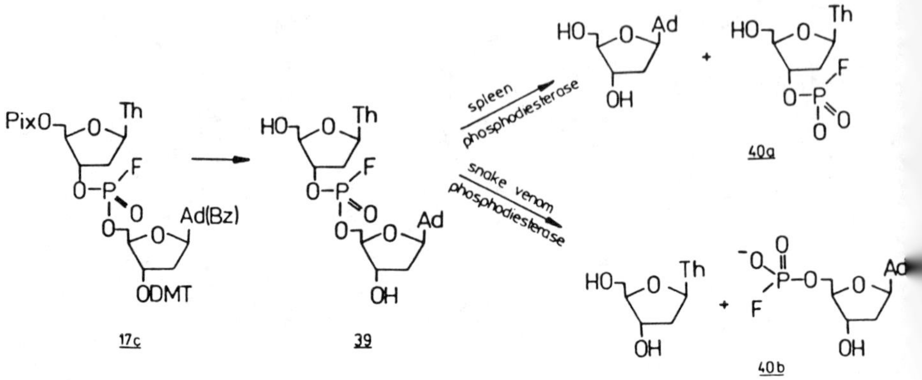

Scheme 21.

The dinucleoside phosphorofluoridates **39** undergo chemoselective hydrolysis in the presence of spleen phosphodiesterase and snake venom phosphodiesterase. Compound **17c** was readily deprotected under standard conditions *(24a–c)* to give the corresponding phosphorofluoridate **39** without affecting the phosphorus–fluorine bond. Enzymatic hydrolysis of compound **39** provides additional proof of the structure of compound **17** and shows the relatively high stability of the latter compound in aq. medium. The monofuoridates **40a,b** (Scheme 21) are identical with compounds prepared recently by Sund et al. *(7)*.

The mono- and dinucleoside phosphorofluorides described in this chapter can be considered as building units for oligonucleotides of biological interest. Our preliminary studies show that the phosphorus–fluorine bond is able to survive both deprotection and coupling procedures.

References

1a. Walsh, C. (1979) *Enzymatic Reaction Mechanisms.* Freeman and Company, San Francisco.
1b. Filler, R. and Kobayashi, Y. (1983) *Biomedical Aspects of Fluorine Chemistry.* Elsevier Biomedicinal, Tokyo.
2. Wittmann, R. (1963) *Chem. Ber.* **96**, 771.
2b. Clark, V. M. and Hutchinson, D. W. (1968) *Progress in Organic Chemistry.*
3a. Borden, R. K. and Smith, M. (1966) *J. Org. Chem.* **31**, 3241.
3b. Sporn, M. B., Berkowitz, D. M., Glinski, R. P., Ash, A. B., and Stevens, C. L. (1969) *Science* **164**, 1408.
3c. von Tigerstrom, R. and Smith, M. (1970) *Science* **167**, 1266.
3d. Kucerova, Z. and Skoda, J. (1971) *Biochem. Biophys. Acta* **247**, 194.

3e. Johnson, P. W., von Tigerstrom, R., and Smith, M. (1975) *Nucl. Acids Res.* **2**, 1745.
3f. von Tigerstrom, R., Jahnke, P., and Smith, M. (1975) *Nucl. Acids Res.* **2**, 1727.
3g. Von Tigerstrom, R., Jahnke, P., Wylie, V., and Smith, M. (1975) *Nucl. Acids Res.* **2**, 1737.
3h. Withers, S. G. and Madsen, N. B. (1980) *Biochem. Biophys. Res. Comm.* (1980) **97**, 513.
4a. Johnson, P. W. and Smith, M. (1971) *Chem. Comm.* 379.
4b. Wilson, J. W. and Chung, V. (1989) *Arch. Biochem. Biophys.* **269**, 517.
4c. Percival, M. D. and Withers, S. G. (1992) *J. Org. Chem.* **57**, 811.
5a. Pitzele, B. S. (1970) Doctoral Dissertation. Department of Chemistry, Washington Univ., St. Louis, Mo.
5b. Kun, E., Zimber, P. H., Chang, A. Y., Puschendorf, B., and Grunicke, H. (1975) *Proc. Natl. Acad. Sci. USA* **72**, 1436.
6. Nichol, A. W., Nomura, A., and Hampton, A. (1967) *Biochemistry* **6**, 1008.
7. Sund, Ch. and Chattopadhyaya, J. (1989) *Tetrahedron* **45**, 7523.
8. Dabkowski, W. (unpublished results).
9. Eckstein, F., Bruns, W., and Parmeggiani, A. (1975) *Biochemistry* **14**, 5225.
10a. Haley, B. and Yount, R. G. (1972) *Biochemistry* **11**, 2863.
10b. Vogel, H. J. and Bridger, W. A. (1982) *Biochemistry* **21**, 394.
10c. Monasterio, O. and Timasheff, S. N. (1987) *Biochemistry* **26**, 6091.
11a. For synthesis of phosphorofluoridates, see: *Organischen Phosphorverbindungen*, Parts I and II in *Methoden der Organischen Chemie (Houben Weyl)* (Regita, M., ed.) (1982) Georg Thieme Verlag, Stuttgart.
11b. Lopusinski, A. and Michalski, J. (1982) *J. Am. Chem. Soc.* **104**, 290.
11c. Lopusinski, A. and Michalski, J. (1982) *Angew. Chem.* **24**, 302.
11d. Stec, W. J., Zon, G., and Uznanski, B. (1985) Phosphorofluoridate of the type 5 is presumably formed in the reaction of a dinucleoside phosphorothioate with 2,4-dinitrofluorobenzene. *J. Chromatogr.* **32b**, 263.
11e. Dabkowski, W. and Michalski, J. (1987) *J. Chem. Soc. Chem. Commun.* 755.
12a. Kume, A., Fujii, M., Sekine, M., and Hata, T. (1984) *J. Org. Chem.* **49**, 2139.
12b. de Vroom, E., Spierenburg, M. L., Dreef, C.E., van der Marel, G.A., and van Boom, J. H. (1987) *Recl. Trav. Chim. Pays-Bas.* **106**, 65.
13. Dabkowski, W., Michalski, J., and Wang, Q. (1991) *Nucleosides & Nucleotides* **10**, 601.
14. Imai, K., Ito, T., Kondo, S., and Tataku, T. (1985) *Nucleosides & Nucleotides* **4**, 669.
15. Dabkowski, W., Cramer, F., and Michalski, J. (1987) *Tetrahedron Lett.* **28**, 3559.
16a. Dabkowski, W., Cramer, F., and Michalski, J. (1988) *Tetrahedron Lett.* **29**, 3301.
16b. Dabkowski, W., Michalski, J., and Wang, Q. (1990) *Angew. Chem. Int. Ed. Engl.* **29**, 522.
16c. Dabkowski, W., Cramer, F., and Michalski, J. (1990) *J. Chem. Soc. Perkin Trans. I* **141**, 7.
17a. Cramer, F., Schaller, H., and Staab, H. A. (1961) *Chem. Ber.* **94**, 1621.
17b. Katagiri, N., Itakura, K., and Narang, S. A. (1975) *J. Am. Chem. Soc.* **97**, 7332.
17c. Matteucci, M. D. and Caruthers, M. H. (1981) *J. Am. Chem. Soc.* **103**, 3185.

17d. Dabkowski, W., Skrzypczynski, Z., Michalski, J., Piel, N., McLaughlin, L., and Cramer, F. (1984) *Nucl. Acids Res.* **12,** 9123.
17e. Dabkowski, W., Michalski, J., and Skrzypczynski, Z. (1986) *Phosphorus and Sulfur* **26,** 321.
17f. Sonveaux, E. (1986) *Bioorg. Chem.* **6,** 159, and references cited therein.
18. Dabkowski, W., Cramer, F., and Michalski, J. (1987) *Tetrahedron Lett.* **28,** 3561.
19a. Kraszewski, A. and Stawinski, J. (1980) *Tetrahedron Lett.* **21,** 2935.
19b. Eckstein, F. (1966) *J. Am. Chem. Soc.* **88,** 4292.
20a. Birkofer, F. and Ritter (1961) *Angew. Chem.* **73,** 134.
20b. Shimidzu, T., Yamana, K., Murakami, A., and Nakamichi, K. (1980) *Tetrahedron Lett.* **21,** 2717.
20c. Froehler, B. C., Ng, P. G., and Matteucci, M. D. (1986) *Nucl. Acids Res.* **14,** 5399.
20d. Garegg, P.J., Regberg, T., Stawinski, J. (1986) *Chem. Scr.* **26,** 59.
21a. Koziolkiewicz, M., Uznanski, B., and Stec, W. J. (1986) *Chemica Scrpt.* **26,** 251.
21b. Eckstein, F. and Gish, G. (1989) *Trends in Biochemical Sciences* **14,** 97 and references cited therein.
22. Michalski, J., Dabkowski, W., Lopusinski, A., and Cramer, F. (1991) *Nucleosides & Nucleotides* **10,** 283.
23. Stolzer, C. and Simon, A. (1960) *Chem. Ber.* **93,** 1323.
24a. Michelson, A. M. and Todd, A. R. (1953) *J. Chem. Soc.* 951.
24b. Chattopadhayaya, J. B. and Reese, C. B. (1978) *J. Chem. Soc. Chem. Commun.* 639.
24c. Jones, R. A. (1984) *Olgonucleotide Synthesis, A Practical Approach* (Gait, M. J., ed.) IRL Press, Oxford.

CHAPTER 13

α-Oligodeoxynucleotides

François Morvan, Bernard Rayner, and Jean-Louis Imbach

1. Introduction

Monomeric units of nucleic acids DNA or RNA are characterized by a β configuration at the anomeric center of the sugar moiety (Fig. 1). Natural constitutive nucleosides of this configuration are readily commercially available, and from such starting building blocks, numerous phosphate-backbone-modified oligonucleotides have been proposed, the most well known being the methylphosphonate and the phosphorothioate series *(1)*.

Starting from the nucleosides of α configuration, the corresponding α-DNA, α-RNA, or phosphate-backbone-modified α-oligonucleotides can be reached using the same chemistry procedures. Since as the α-nucleosidic synthons are not commercially available (except α-dT), however, we have to synthesize them. Therefore, we will describe here the procedure for the chemical synthesis of α-DNA (the most studied series of α-oligonucleotides) from the α-deoxynucleosides obtainment to any corresponding sequence of α-oligodeoxynucleotides containing the four natural bases.

2. Materials

1. Melting points are determined on a Buchi 510 apparatus and are uncorrected.
2. Ultraviolet spectra (UV) are recorded on a Cary 118 spectrophotometer; optical rotations are measured in a 1-cm cell on a Perkin-Elmer model 241 spectropolarimeter in the indicated solvents.

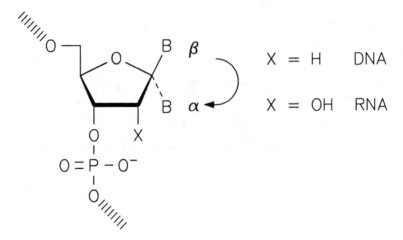

Fig. 1. Monomeric structure of oligonucleotides.

3. Proton nuclear magnetic resonance spectra are recorded on a Bruker WB UM 360 or a Bruker SP200 spectrometer with tetramethylsilane as internal standard; the signals are described as: s, singulet; d, doublet; t, triplet; q, quartet, m, multiplet; br, broad; Ar-H, aromatique.
4. ^{31}P-NMR spectra are recorded on a Bruker SP200 spectrometer with ^{1}H broad-band decoupling. Chemical shifts are expressed downfield from external 85% H_3PO_4 in D_2O.
5. Mass spectra are determined with a Jeol JMS-DX 300 instrument. Polyethylene glycol 400 or thioglycerol is used as the FAB matrix.
6. Elemental analysis is determined by the Service Central d'Analyse du CNRS, Division de Vernaison, France.
7. Thin-layer chromatography (TLC) is performed on precoated silica gel sheets 60 F_{254} (E. Merck, Darmstadt, Germany), and flash column chromatography is performed on silica gel 60H (Merck). Unless otherwise specified, elution is carried out with an increasing proportion of methanol (0–10%) in methylene chloride.
8. High-performance liquid chromatography (HPLC) analysis is carried out on an XL ODS C_{18}, 3-μm (70 × 4.6 mm) column (Beckman Instruments Inc., Altex Division, San Ramon, CA). HPLC purification is carried out on a Nucleosil C_{18}, 5-μm (250 × 10 mm) column (SFCC/Shandon, Eragny, France) with an increasing proportion of acetonitrile (5–15%) in 0.05M triethylammonium acetate, pH 7, as eluant. A Waters U6K injector, Model 510 pumps, an M 720 gradient controller, and a Waters 990 photodiode array detector are employed.

9. Acetonitrile, pyridine, and diisopropylethylamine are dried by heating, under reflux, with powdered CaH_2 for 16 h; these solvents are then distilled at atmospheric pressure and stored over 4-Å molecular sieves *(4)*.
10. Triethylamine (Fluka Chemie AG, Buchs, Switzerland, ref. 90340) cyclohexane (Merck ref. 9666), and methylene chloride (Merck ref. 6050), are used during the purification of phosphoramidite derivatives, which are lyophilized from benzene (Merck ref. 1783).
11. α-Phosphoramidites are stable at –20°C and under argon over 1 yr.
12. α-Oligodeoxynucleotides are synthesized on an Applied Biosystems model 381A DNA synthesizer.

3. Methods

3.1. Synthesis of Base-Protected α-Deoxynucleosides

3.1.1. 6-N-Benzoyl-α-Deoxyadenosine

α-Deoxyadenosine is synthesized by glycosylation reaction between 3,5-di-*O*-*p*-nitrobenzoyl-1-chloro-2-deoxyribofuranose **1a** and 6-*N*-adenine in the presence of molecular sieve in dry methylene chloride *(2)* (Fig. 2). The fully protected α-anomer **2a** is isolated by fractionated crystallization, and then sugar-protecting groups are selectively removed to afford **3a** (34.8%).

3',5'-Di-*O*-*p*-nitrobenzoyl–6-*N*-benzoyl-α-deoxyadenosine (**2a**)

To a vigorously stirred suspension of powdered 6-*N*-benzoyladenine (31.7 g, 132.5 mmol) and 62 g of molecular sieve (type 4) in dry methylene chloride, **1a** (31.7 g, 70.3 mmol) is added. The suspension is stirred at room temperature for 3 d. The mixture is then filtered through celite, and the solids are thoroughly washed with methylene chloride. The filtrate and washings are combined and evaporated to dryness. The residue is dissolved in a hot solution of ethyl acetate:absolute ethanol (1:1 v/v) and left standing for slow cooling to yield the greater part of α-anomer. The filtrate is evaporated, and the remaining α-anomer **2a** is purified by flash silica gel column chromatography using increasing proportions of acetone (0–10%) in methylene chloride—19.9 g (43%), mp 115–117°C. ^{1}H-NMR (CDCl$_3$) δ 3.14 (m,2H,H$_{2'}$, and H$_{2''}$); 4.69 (m,2H,H$_{5'}$, and H$_{5''}$); 5.11 (m,1H,H$_{4'}$); 5.76 (m,1H,H$_{3'}$); 6.74 (m,1H,H$_{1'}$); 7.47–8.82 (m,15H,H$_2$,H$_8$, and Ar-H); 8.90 (br s, 1H,NH).

6-*N*-benzoyl-α-deoxyadenosine (**3a**)

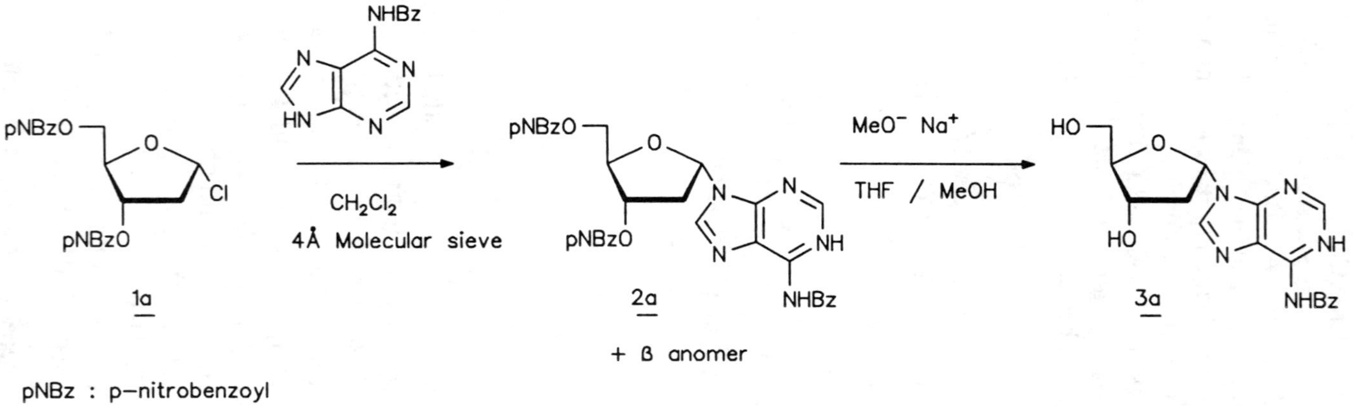

Fig. 2. Synthesis of 6-*N*-benzoyl-α-2'-deoxyadenosine.

pNBz : p–nitrobenzoyl

The fully protected α-nucleoside 2a (19.9 g, 30.4 mmol) is dissolved in tetrahydrofuran (300 mL) and dry methanol (150 mL), and treated with 2N sodium methoxide in methanol (150 mL) at room temperature for 30 min. The solution is neutralized with Dowex 50W X2 (150 mL, pyridinium form, suspended in methanol) and filtered. The filtrate is evaporated to dryness. The residue is treated with a solution of methanol:water 1:1 v/v (150 mL) from which methyl p-nitrobenzoate crystallizes. The mixture is filtered, and the filtrate is evaporated to dryness. Remaining water is eliminated by evaporation with absolute ethanol, and 3a is crystallized from acetone—8.75 g (81%), mp 163–164°C. Anal. Calc. for $C_{17}H_{17}N_5O_4$: C, 57.46; H, 4.82; N, 19.71; Found: C, 57.30; H, 4.86; N, 19.75. UV λ_{max} (in H_2O) (nm[ε]): 280 (21,500); λ_{max} (pH2.1) (nm[ε]): 253 (10,900), 284 (24,000). $[\alpha]_D^{20}$ +59° (c = 1, DMF). ^{1}H-NMR (DMSO-d_6) δ 2.44 (m,1H,$H_{2''}$); 2.8 (m,1H,$H_{2'}$); 3.49 (m,2H,$H_{5'}$, and $H_{5''}$); 4.21 (m,1H,$H_{4'}$); 4.37 (m,1H,$H_{3'}$); 4.86 (t,1H,$OH_{5'}$); 5.54 (d, 1H,$OH_{3'}$); 6.50 (dd, 1H,$H_{1'}$,J = 2.58 and 7.57); 7.52–8.06 (m,5H,Ar-H); 8.69 and 8.74 (2s,2H,H_2, and H_8); 11.14 (br s, 1H,NH).

3.1.2. 4-N-Benzoyl-α-Deoxycytidine

α-Deoxycytidine is synthesized by self-anomerization reaction of fully protected β-deoxycytidine 1b in the presence of silylating reagent and Lewis acid *(3)* (Fig. 3). Fully protected α-anomer 2b is isolated by crystallization, and then sugar-protecting groups are selectively removed to afford 3b (44%) after crystallization.

3',5'-Di-*O*-acetyl-4-*N*-benzoyl-α-deoxycytidine (2b)

Compound 1b* *(3,4)* (25 g, 60.2 mmol) is suspended in dry acetonitrile (375 mL) and mixed with *bis*-trimethylsilylacetamide (BSA) (15 mL, 61.3 mmol). The mixture is stirred under heating at 70°C for 10 min, and then TMS-triflate (13 mL, 71.7 mmol) is added to the resulting clear solution. After 3 h at 70°C, the content of α-anomer reaches a maximum, and the solvent is removed under a reduced pressure. The residue is redissolved in methylene chloride (750 mL), and treated with cold saturated aqueous sodium bicarbonate (750 mL). The resulting suspension is filtered through a large filter paper, and the solids are thoroughly washed with methylene chloride (3 × 150 mL).

---

*1b is synthesized from deoxycytidine following a selective N-benzoylation with benzoyl anhydride in dry DMF (4) and then O-acetylated with acetic anhydride in dry pyridine (3).

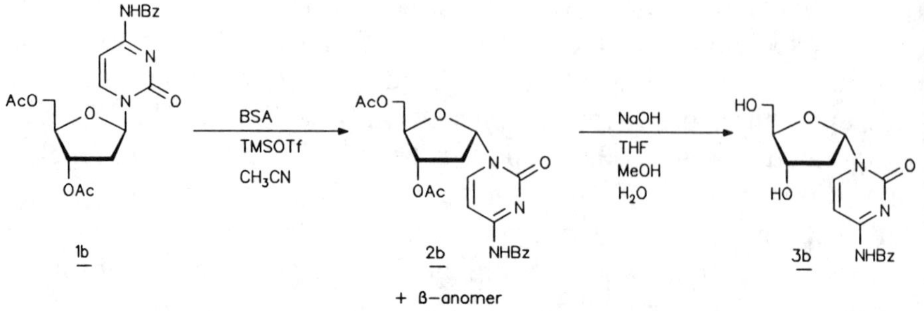

Fig. 3. Synthesis of 4-N-benzoyl-α-2'-deoxycytidine.

The aqueous layer is separated and extracted with methylene chloride (4 × 120 mL). The combined organic layers are dried over anhydrous sodium sulfate and evaporated to dryness. The α-anomer 2b is crystallized from toluene—12.4g (50%) mp 178–179. Anal. Calc. for $C_{20}H_{21}N_3O_7$: C, 57.82; H, 5.11; N, 10.12; Found: C, 58.01; H, 5.10; N, 9.95. UV λ_{max} (in EtOH) (nm): 271, 305. Rf_α = 0.31 and Rf_β = 0.38 (methylene chloride:methanol, 19:1, v/v), ^{1}H-NMR (CDCl$_3$) δ 1.97 (s,3H,CH$_3$); 2.13 (s,3H,CH$_3$); 2.16–3.02 (m,2H,H$_{2'}$, and H$_{2''}$); 4.13 (m,2H,H$_{5'}$, and H$_{5''}$); 4.68 (t,1H,H$_{4'}$); 5.15 (d,1H,H$_{3'}$) 6.23 (dd,1H,H$_{1'}$, J = 2.0 and 7.0); 7.32–8.40 (m,7H,H$_5$,H$_6$,Ar-H) 8.70 (s,1H,NH).

4-N-benzoyl-α-deoxycytidine (3b)

To a cooled (0°C) solution of 2b (12.4 g, 29.9 mmol) in a mixture of tetrahydrofuran (205 mL), methanol (175 mL), and water (41 mL) are added cooled 2N of aqueous sodium hydroxide solution (45 mL). The clear solution is stirred for 5 min at 0°C, then neutralized (pH 6) with Dowex 50W X2 (pyridinium form,135 mL), and filtered. The filter cake is washed with methanol (3 × 100 mL); the combined filtrate and washing are evaporated, and 3b is crystallized from water—8.72 g (88%) mp 131–132°C. UV λ_{max} (in EtOH) (nm[ε]): 257 (21,800), 302 (10,200); λ_{min} (in EtOH) (nm) 229, 282; λ_{max} (0.01N HCl) (nm[ε]): 256 (19,600), 304 (13,100); λ_{min} (0.01N HCl) (nm) 230, 280. $[\alpha]_D^{21}$ −69.3°C (c = 1.01, EtOH). Anal. Calc. for $C_{16}H_{17}N_3O_5$: C, 58.0; H, 5.17; N, 12.68; Found: C, 57.88; H, 5.02; N, 12.58. ^{1}H-NMR (DMSO-d$_6$) δ 2.08 (m,1H,H$_{2'}$); 2.73 (m,1H,H$_{2'}$); 3.52 (m,2H,H$_{5'}$, and H$_{5''}$); 4.23–4.50 (m,2H,H$_{3'}$, and H$_{4'}$); 6.12 (pseudo-d,1H,H$_{1'}$); 7.40 (d,1H,H$_5$); 7.63–8.15 (m,5H,Ar-H); 8.31 (d,1H,H$_6$).

3.1.3. 2-N-Palmitoyl- and 2-N-Isobutyryl-α-Deoxyguanosines

The method described in the literature for the synthesis of α-deoxyguanosine is unsatisfactory because of low yield (16%) and multiple steps *(5)*. Therefore, we have developed two procedures for its synthesis. The first, which we have used since 1986, consists of a transglycosylation reaction between 2-*N*-palmitoyl-guanine *(6)* and 3',5'-di-*O*-acetyl-4-*N*-benzoyl-2'-deoxycytidine as sugar donor in the presence of silylating reagent and Lewis acid (Fig. 4). The reaction gives a mixture of α- and β-7-N and α- and β-9-N isomers *(7)*. The 7-N isomers are in a greater part at the beginning of the reaction and then are converted slowly into 9-N isomers. After 6 h, the ratio 9-N:7-N seems to reach a maximum. The 9-N isomers are separated from the residual 7-N isomers by silica gel column chromatography. Then, α-9-N anomer is obtained by fractionated crystallization. The fully protected α-anomer 2c is then selectively deprotected to afford 3c (29%).

The second procedure consists of a glycosylation reaction between silylated guanine and 1-*O*-acetyl-3,5-di-*O*-*p*-nitrobenzoyl-2-deoxy-D-ribofuranose (1d) under phase transfer conditions using potassium iodine and dibenzo-18-crown-6 *(8)* (Fig. 5). Then the exocyclic amino group of guanine is acylated with isobutyryl chloride. The α-anomer 2d is obtained by crystallization, and its sugar-protecting groups are selectively removed to afford 3d (26.6%).

3',5'-di-*O*-acetyl–2-*N*-palmitoyl-α-deoxyguanosine (2c)

Compounds 1b (15 g, 36 mmol) and 2-*N*-palmitoylguanine *(6)* (37.4 g, 96 mmol) are suspended in dry acetonitrile (220 mL), and mixed with BSA (74 mL, 30 mmol). The mixture is stirred under heating at 70°C for 1 h, and then TMS-triflate (8 mL, 47 mmol) is added. After 6 h, the solvent is evaporated, and the residue dissolved in methylene chloride (750 mL) is poured in cold saturated aqueous sodium bicarbonate (750 mL). The resulting suspension is filtered through a large filter paper, and solids are thoroughly washed with methylene chloride (3 × 150 mL). The aqueous layer is separated and extracted with methylene chloride (4 × 120 mL). The combined organic layers are dried over anhydrous sulfate and evaporated to dryness. The residue is dissolved in chloroform and applied to a column of silica gel (250 g).

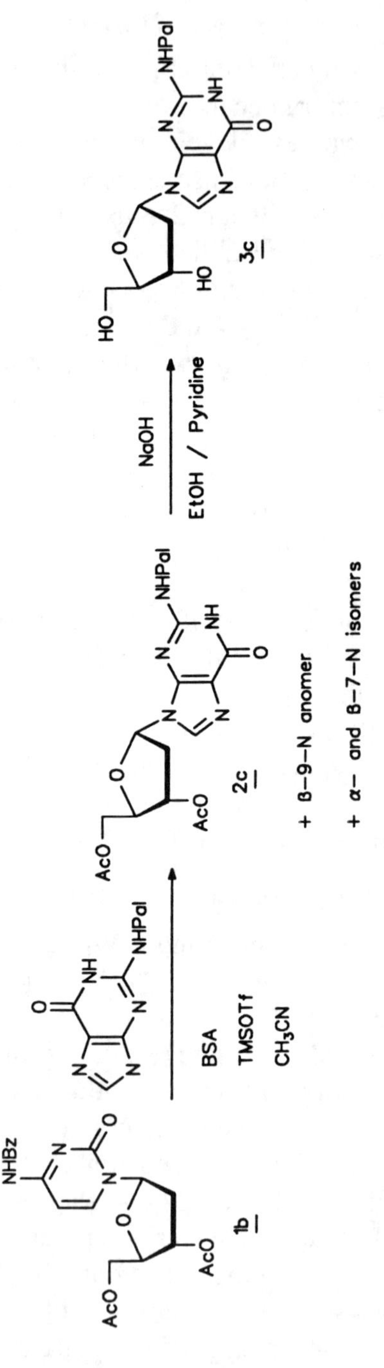

Pal : palmitoyl ($CH_3-(CH_2)_{14}-CO-$)

Fig. 4. Synthesis of 2-*N*-palmitoyl-α-2'-deoxyguanosine.

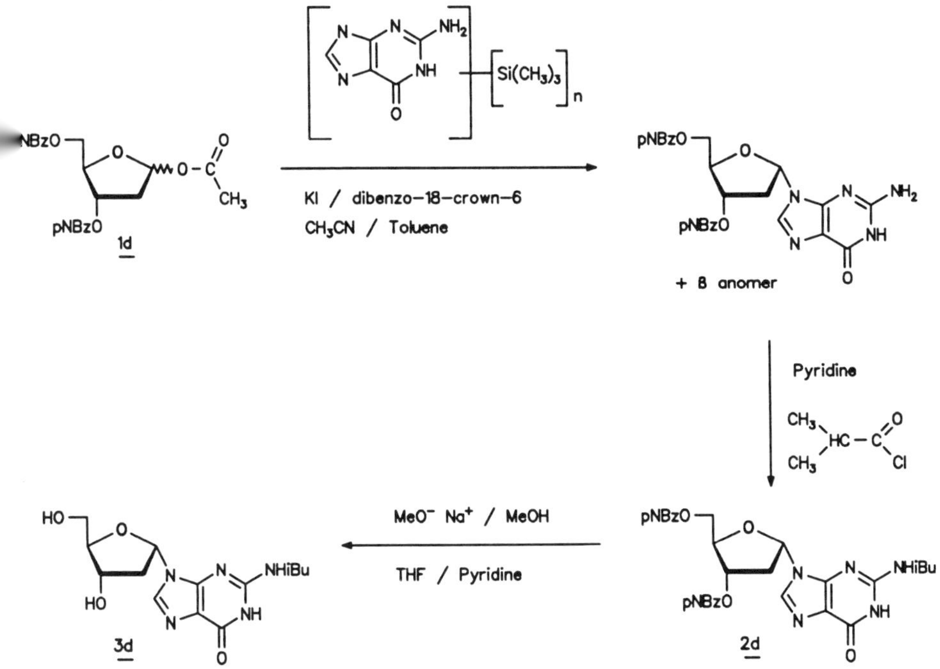

α-Oligodeoxynucleotides

Fig. 5. Synthesis of 6-*N*-isobutyryl-α-2'-deoxyguanosine.

The 9-N isomers are separated from the residual 7-N isomers by elution with chloroform† containing an increasing proportion of methanol (0–5%). The α-9-N isomer 2c is crystallized from ethanol—6.37 g (30%), TLC Rf_{9-N} = 0.40, Rf_{7-N} = 0.56 methylene chloride:methanol 92.5:7.5 v/v, mp 129–130°C. Anal. Calc. for $C_{30}H_{47}N_5O_7$: C, 61.10; H, 8.03; N, 11.88; Found C, 60.95; H, 8.19; N, 12.06. UV λ_{max} (in EtOH) (nm[ε]): 258 (18,300); 279 (13,900). $[\alpha]_D^{20}$ + 19°C (c = 1, DMF). ^{1}H-NMR (CDCl$_3$) δ 0.86 (t,3H,[CH$_2$]$_{14}$—C$\underline{H}_3$); 1–20–1.40 (m,24H—[C$\underline{H}_2$]$_{12}$—CH$_3$); 1.71 (m,2H,—C[=O]—CH$_2$C$\underline{H}_2$—); 2.03 and 2.10 (2s,6H,2CH$_3$—C[=O]—); 2.48 (t,2H,—C[=O]—C$\underline{H}_2$—[CH$_2$]$_{13}$–); 2.56 (m,1H,H$_{2''}$); 2.84 (m,1H,H$_{2'}$); 4.18–4.28 (m,2H,H$_{5'}$, and H$_{5''}$); 4.52 (m,1H,H$_{4'}$); 5.28 (m,1H,H$_{3'}$); 6.23 (dd,1H,H$_{1'}$, J = 1.89 and 7.47); 7.92 (s,1H,H$_8$); 8.77 and 12.00 (2br s,2H,2 NH).

2-*N*-palmitoyl-α-deoxyguanosine (3c)

---

†The separation of 7-N and 9-N isomers by chromatography or silica is better with chloroform than methylene chloride as eluent.

To a cooled solution (0°C) of 2c (6.37 g, 10.8 mmol) in pyridine (130 mL) and ethanol (54 mL) is added a cooled 2N aqueous sodium hydroxide solution (54 mL, 108 meq). The mixture is vigorously stirred for 10 min. Quickly a gum will occur. After neutralization with acetic acid (6.18 mL, 108 meq), the gum is dissolved in methylene chloride:ethanol 2:1 v/v (750 mL). The organic layer is washed with water (3 × 200 mL) and evaporated to dryness. After coevaporation with toluene, 3c is crystallized from ethanol—5.30 g (97%) mp 210°C (dec.). Anal. Calc. for $C_{30}H_{47}N_5O_7$: C, 61.10; H, 8.03; N, 11.88. Found: C, 60.95; H, 8.19; N, 12.06. $[\alpha]_D^{20}$ + 37°C (c = 1,CHCl$_3$). UV λ_{max} (in EtOH) (nm[ε]): 254 (16,060); 259 (16,060); 280 (12,000): λ_{min} (nm): 256, 270. ^{1}H-NMR (20% CD$_3$OD in CDCl$_3$) δ 0.88 (t,3H,C$\underline{H}_3$—[CH$_2$]$_{14}$); 1.26–1.39 (m,24H,C$\underline{H}_3$—[CH$_2$]$_{12}$—); 1.67–1.73 (m,2H,C[=O]—CH$_2$—C$\underline{H}_2$); 2.46 (t,2H,C[=O]—C$\underline{H}_2$—[CH$_2$]$_{13}$); 2.59 (m,1H,H$_{2''}$); 2.82 (m,1H,H$_{2'}$); 3.63–3.74 (m,2H,H$_{5'}$, and H$_{5''}$); 4.30 (pseudo q,1H,H$_{4'}$); 4.47 (m,1H,H$_{3'}$); 6.23 (dd,1H,H$_{1'}$, J = 7.8 and 3.6); 8.12 (s, 1H,H$_8$).

2-N-isobutyryl-3',5'-di-O-p-nitrobenzoyl-α-deoxyguanosine (2d)

To a mixture of KI (4.23, 25.5 mmol) and dibenzo-18-crown-6 (1.83 g, 5.09 mmol; previously dried by azeotropic distillation with dry toluene) in dry toluene (84 mL) is added, under argon atmosphere, a solution of silylated guanine *(9)* (7.7 g, 50.9 mmol) in dry acetonitrile (60 mL) and a solution of 1d†† *(10,11)* (12.0 g, 25.5 mmol) in dry acetonitrile (60 mL). After 20 h of stirring at 80°C, the mixture is cooled (–10°C), and dry pyridine (20 mL) and isobutyryl chloride (9.0 mL, 85.9 mmol) are added. The mixture is then stirred at room temperature for 4 h, and then a solution of methanol:pyridine (1:4, v/v, 50 mL) is added. The heterogenous mixture is filtered through celite, and solids are thoroughly washed with methylene chloride (3 × 50 mL). The organic layer is evaporated to dryness. The residue dissolved in methylene chloride (300 mL) is poured on saturated aqueous sodium bicarbonate and is extracted with methylene chloride (2 × 300 mL). The combined organic layers are dried over sodium sulfate and evaporated to dryness. The two anomers are obtained by flash column chromatography on a silica

---

††1d is synthesized from 2-deoxy-D-ribose by following the reported procedure described in refs. *10* and *11*.

gel using increasing proportions of methanol (0–5%) in methylene chloride as eluent. Both α- and β-anomers are crystallized together from methylene chloride, and the α-anomer 2d is extracted from the mixture by washing with hot methylene chloride. 2d is then recrystallized from methylene chloride 5.2 g (32%). ^{1}H-NMR (CDCl$_3$) δ 1.23 and 1.24 (2s,6H,CH$_3$); 2.84 (m,1H,CH); 3.13 (m,2H,H$_{2'}$ and H$_{2''}$); 4.66 (m,2H,H$_{5'}$ and H$_{5''}$); 5.10 (m,1H,H$_{4'}$); 5.73 (m,1H,H$_{3'}$); 6.59 (pseudo d,1H,H$_{1'}$, J = 3.78); 7.91–8.30 (m,8H,Ar-H); 8.70 (s,1H,H$_8$); 10.05 and 12.35 (2br s,2H,NH).

2-N-isobutyryl-α-deoxyguanosine (3d)

To a solution of 2d (5.2 g 8.18 mmol) in dry tetrahydrofuran:pyridine:methanol (2:3:4, v/v/v) is added 1.0M sodium methoxide in methanol (10 mL). After 5 min under stirring, the solution is neutralized with Dowex 50W X2 (60 mL, pyridinium form, suspended in methanol). The mixture is filtered, and resine is washed with methanol (2 × 30 mL). The filtrate is evaporated, and the methyl p-nitrobenzoate is removed by trituration with diethyl ether (3 × 50 mL). Then 3d is crystallized from ethanol:acetone (110 mL, 1:10, v/v), 3d crystallizes with 1/3 mol of acetone—2.4 g (83%), mp 132–134°C. UV λ$_{max}$ (in H$_2$O) (nm[ε]) 259 (15,900), 279 (11,300); λ$_{min}$ (nm) 225, 270. FAB (>0) Thioglycerol: m/z: 338 [M + H]$^+$; m/z: 222 [G^{ibu} + H]$^+$. FAB (<0) m/z: 336 [M–H]$^-$; m/z: 220 [G^{ibu} – H]$^-$. ^{1}H-NMR (DMSO-d$_6$) δ 1.05 (2s,6H,2CH$_3$) 2.30–2.75 (m,3H,CH,H$_{2'}$ and H$_{2''}$); 3.4 (m,2H,H$_{5'}$ and H$_{5''}$); 4.13 (m,1H,H$_4$); 4.33 (m,1H,H$_{3'}$); 4.83 (t,1H,OH$_{5'}$); 5.41 (d,1H,OH$_{3'}$); 6.18 (dd, 1H,H$_{1'}$, J = 2.24 and 7.49); 8.22 (s, 1H,H$_8$); 11.7 and 12.05 (2s,2H,NH).

3.1.4. α-Thymidine

α-Thymidine is synthesized by glycosylation reaction between the 1-chloro-3,5-di-O-toluoyl-2-deoxy-D-ribofuranose 1e *(9)*, and silylated thymine *(9)* (Fig. 6). The α-nucleoside 2e is isolated by fractionated crystallization, and then protecting groups are removed to afford α-thymidine 3e (47%).

3',5'-Di-O-p-toluoyl-α-thymidine (2e)

To a solution of 1e (23.6 g, 60.7 mmol), which had been allowed to stand in dry acetonitrile (700 mL) for 30 min, a solution of 2,4-bis-trimethylsilyloxythymine (16.4 g, 60.7 mmol) in 130 mL of dry

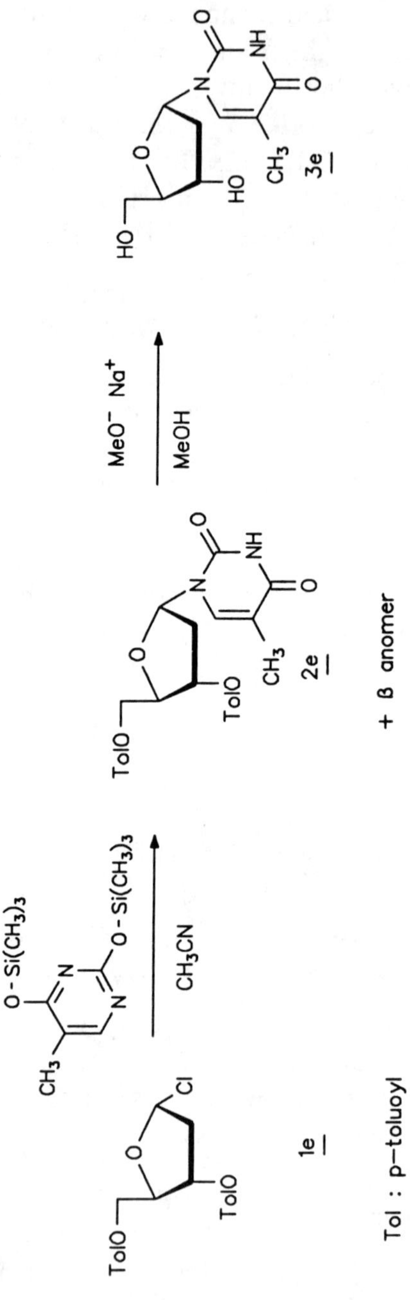

Fig. 6. Synthesis of α-thymidine.

acetonitrile is added dropwise during 30 min under inert atmosphere and magnetic stirring. The mixture is left to stand for 16 h at room temperature. The reaction mixture is then evaporated under a reduced pressure to give an oily residue. First, the greater part of β-anomer is eliminated by crystallization from methanol. From the evaporated mother liquor, α-anomer 2e is crystallized from ethanol. Thus, by fractionated crystallization, we can isolated most of α-anomer. The remaining α-anomer mixed with the β can be isolated by flash column chromatography on silica gel using increasing proportions of methanol (1–10%) in methylene chloride as eluent—14.7 g (51%), mp 147–148°C $Rf_\alpha = 0.24$ and $Rf_\beta = 0.32$ (methylene chloride:ethyl acetate, 4:1, v/v). Anal. Calc. for $C_{26}H_{26}N_2O_7$: C,65.26; H,5.48; N,5.85; Found C,65.01; H,5.52; N,6.05. ^{1}H-NMR (CDCl$_3$) δ 1.88 (s,3H,5-CH$_3$); 2.13–3.05 (m,8H,H$_{2'}$,H$_{2''}$, and toluoyl-CH$_3$); 4.50 (d,2H,H$_{5'}$, and H$_{5''}$); 4.89 (t,1H,H$_{4'}$); 5.59 (d,1H,H$_3$); 6.36 (dd,1H,H$_{1'}$,J = 2.0 and 6.8); 7.19–7.98 (m,9H,H$_6$ Ar-H); 8.47 (s,1H,NH).

α-Thymidine (3e)

Compound 2e (13.7 g, 28.8 mmol) is dissolved in dry methanol (450 mL), and 50 mL of 1.0M sodium methoxide in methanol are added. The solution is stirred 45 min at room temperature (Rf$_{3e}$ = 0.12, Rf$_{2e}$ = 0.61, methylene chloride:methanol, 9:1, v/v). The reaction mixture is treated with distilled water (15 mL), and the solution is neutralized (pH 6) with 50 mL of 1.0M aqueous hydrochloric acid. The mixture is evaporated under a reduced pressure. The residue is dissolved in distilled water (200 mL) and evaporated once more. The residue is dissolved in distilled water (400 mL), and methyl p-toluoylate is extracted with methylene chloride (3 × 300 mL). The aqueous layer is evaporated to dryness, and α-thymidine 3e is dissolved in dry methanol. The sodium chloride provided from neutralization is insoluble and is eliminated by filtration. The filtrate solution is then evaporated, and α-thymidine 3e is crystallized from ethanol— 6.49 g (93%), mp 185–187°C. Anal. Calc. for $C_{10}H_{14}N_2O_5$: C, 49.58; H, 5.83; N, 11.56; Found C, 49.69; H, 5.77; N, 11.44. UV λ$_{max}$ (in H$_2$O) (nm[ε]) 268 (9880) $[\alpha]_D^{18}$ = +4.4 (c = 1, H$_2$O). 1H-NMR (D$_2$O) δ 1.93 (s,3H,5-CH$_3$); 2.15–2.83 (m,2H,H$_{2'}$, and H$_{2''}$); 3.63–3.79 (m,2H,H$_{5'}$, and H$_{5''}$); 6.18–6.24 (dd, 1H,H$_{1'}$,J = 2.9 and 7.3) 7.79 (s, 1H,H$_6$).

3.2. Synthesis of Base-Protected 5'-O-(4,4'-Dimethoxytrityl)- α-Deoxynucleosides (4a–e)

The four base-protected α-nucleosides 3a–e are converted into the respective 5'-O-dimethoxytrityl derivatives 4a–e (Fig. 7). 4,4'-Dimethoxytrityl chloride (895 mg, 2.64 mmol) is added to a solution of N-acyl α-2'-deoxyribonucleoside 3a–e (2.2 mmol) in dry pyridine (11 mL). In the case of N-2-palmitoyl-α-deoxyguanosine 3c, this compound is suspended in pyridine:methylene chloride (1:1, v/v, 43 mL), and 4,4'-dimethoxytrityl chloride is added in three portions over 30 min. The reaction mixture is stirred at room temperature for 1–4 h, and methanol (2 mL) is added. After an additional 10 min of stirring, the reaction mixture is poured on 5% aqueous hydrogencarbonate solution (60 mL), and the products are extracted with methylene chloride (4 × 40 mL). The combined organic layers are evaporated to dryness under reduced pressure, and the resulting residue is fractionated by flash column chromatography (60 g) using increasing proportions of methanol (0–6%) or acetone (0–50%) in methylene chloride as eluent. Fractions containing pure 5'-O-dimethoxytrityl 4a–e derivative are combined and evaporated to dryness, and the residue is precipitated from petroleum ether (<40°C) to afford an amorphous powder.

5'-O-dimethoxytrityl-6-N-benzoyl-α-deoxyadenosine (4a)

Yield 96% as an amorphous powder. Anal. Calc. for $C_{38}H_{35}N_5O_6$, 0.5 H_2O: C, 68.46; H, 5.44; N, 10.50. Found: C, 68.36; H, 5.52; N, 10.44. ^{1}H-NMR (CDCl$_3$) 2.58 (m,1H,$H_{2'}$); 3.11–3.78 (m,3H,$H_{2''}$,$H_{5'}$, and $H_{5''}$); 3.80 (s,6H,O—CH$_3$); 4.49 (m,2H,$H_{3'}$, and $H_{4'}$); 6.39 (m,2H,$H_{1'}$, and $OH_{3'}$); 6.82–8.05 (m,18H,Ar-H); 8.21 and 8.81 (2s,2H,H_2, and H_8); 9.09 (br s,1H,NH).

5'-O-dimethoxytrityl-4-N-benzoyl-α-deoxycytidine (4b)

Yield 90% as colorless crystals from methylene chloride, mp 123–125°C. Anal. Calc. for $C_{37}H_{35}N_3O_7$: C, 70.13; H, 5.57; N, 6.63. Found: C, 69.80; H, 5.65; N, 6.40. ^{1}H-NMR (CDCl$_3$) δ 2.67 (m,1H,$H_{2'}$); 2.84 (m,1H,$H_{2''}$); 3.14–3.33 (m,2H,$H_{5'}$, and $H_{5''}$); 3.80 (s,6H,O—CH$_3$); 4.12 (m,1H,$OH_{3'}$); 4.50–4.58 (m,2H,$H_{3'}$, and $H_{4'}$); 6.18 (pseudo d,1H,$H_{1'}$, J = 6.7); 6.80–7.64 (m,18H,Ar-H); 7.89 and 8.05 (2d,2H,H_5, and H_6); 8.83 (br s, 1H,NH).

5'-O-dimethoxytrityl-2-N-palmitoyl-α-deoxyguanosine (4c)

α-Oligodeoxynucleotides

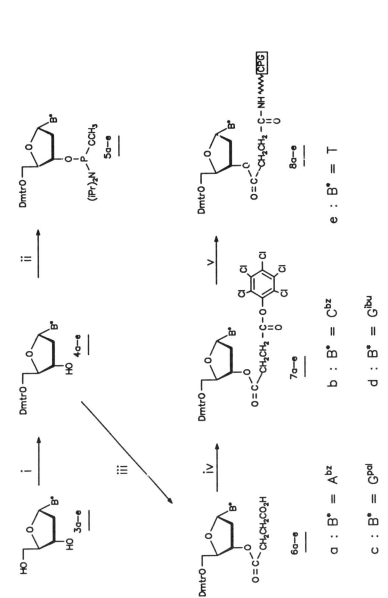

Fig. 7. Synthesis of α-2'-deoxynucleoside phosphoramidites and α-2'-deoxynucleoside derivatized solid supports. Reaction conditions (i) 4,4'-dimethoxytrityl chloride, pyridine; (ii) chlorodiisopropylaminomethoxyphosphine, ethyldiisopropylamine, triethylamine, methylene chloride; (iii) succinic anhydride, dimethylaminopyridine, triethylamine, methylene chloride; (iv) pentachlorophenol, dicyclohexylcarbodiimide, 1,2-dimethoxyethane; (v) long chain alkylamine controlled pore glass, pyridine.

Yield 89% as an amorphous powder. Anal. Calc. for $C_{47}H_{65}N_5O_7$: C, 69.95; H, 7.52; N, 8.68. ^{1}H-NMR (CDCl$_3$) δ 0.88 (t,3H,C$\underline{H}_3$—[CH$_2$]$_{14}$); 1.25 (br s,24H,CH$_3$—[C$\underline{H}_2$]$_{12}$—); 1.68 (m,2H,C[=O]—CH$_2$—C$\underline{H}_2$); 2.46 (t,2H,C[=O]—C$\underline{H}_2$—[CH$_2$]$_{13}$); 2.73–2.96 (m,2H,H$_{2'}$ and H$_{2''}$); 3.10–3.22 (m,2H,H$_{5'}$ and H$_{5''}$); 3.78 (s,6H,O—,CH$_3$); 4.49 (m,2H,H$_{3'}$, and H$_{4'}$); 5.38 (br d, 1H,OH$_{3'}$); 6.19 (pseudo d, 1H,H$_{1'}$, J = 6.7); 6.80–7.43 (m,13H,Ar-H); 7.98 (s, 1H,H$_8$); 9.19 and 12.00 (2br s,2H,NH).

5'-O-dimethoxytrityl-2-N-isobutyryl-α-deoxyguanosine (<u>4d</u>)

Yield 80% as an amorphous powder. Anal. Calc. for $C_{35}H_{37}N_5O_7 \cdot H_2O$: C, 63.92; H, 5.98; N, 10.65; Found: C, 64.12; H, 5.80; N, 10.43. ^{1}H-NMR (CDCl$_3$) δ 1.24 and 1.27 (2s,6H,CH$_3$); 2.68 (m,2H,H$_{2'}$ and H$_{2''}$); 2.92 (m,1H,CH); 3.22 (m,2H,H$_{5'}$ and H$_{5''}$); 3.78 (s,6H,CH$_3$O); 4.48 (m,2H,H$_{3'}$, and H$_{4'}$); 5.42 (br s, 1H,OH$_{3'}$); 6.21 (dd,1H,H$_{1'}$,J = 1.88 and 8.01); 6.80–7.43 (m,13H,Ar-H); 7.97 (s,1H,H$_8$); 9.18 and 12.04 (2 br s,2H,NH).

5'-O-dimethoxytrityl-α-thymidine (<u>4e</u>)

Yield 95% as an amorphous powder. Anal. Calc. for $C_{31}H_{32}N_2O_7$, H$_2$O: C, 67.26; H, 6.01; N, 5.06. Found: C, 67.33; H, 6.02; N, 4.90. ^{1}H-NMR (CDCl$_3$) δ 1.87 (s,3H,5-CH$_3$); 2.30 (m,1H,H$_{2''}$); 2.65–2.8 (m,1H,H$_{2'}$); 3.1–3.25 (m,2H,H$_{5'}$ and H$_{5''}$); 3.31 (d,1H,OH$_{3'}$); 3.78 (s,6H,O—CH$_3$); 4.40-4.47 (m,2H,H$_{3'}$, and H$_{4'}$); 6.1 (pseudo d, 1H,H$_{1'}$, J = 5.7); 6.79–7.52 (m,13H,Ar-H); 9.4 (br s, 1H,NH).

3.3. Synthesis of α-Deoxynucleoside Phosphoramidites (<u>5a–e</u>)

The protected α-nucleosides <u>4a–e</u> are converted into their corresponding phosphoramidites <u>5a–e</u> following a procedure that was originally developed by McBride and Caruthers for natural β-nucleosides *(12)* (Fig. 7). 5'-O-Dimethoxytrityl-N-acyl-α-deoxynucleoside <u>4a–e</u> (1 mmol) is dissolved in anhydrous methylene chloride (3 mL) and diisopropylethylamine (0.7 mL, 4 mmol) in a round-bottom flask preflushed with argon. Chlorodiisopropylaminomethoxyphosphine (0.26 mL, 1.3 mmol) is added dropwise by syringe to the solution under argon at room temperature. After 10 min stirring, the solution is diluted with ethyl acetate (35 mL), and extracted with 5% aqueous hydrogencarbonate (60 mL) and brine (4 × 80 mL). The organic phase is dried over anhydrous sodium sulfate and evaporated to dryness. The resulting foam is fractionated by flash column chromatography using triethylamine:methylene chloride:cyclohexane (1:19:80–1:99:0, v/v/v) as eluent. The appropriate fractions are combined and evaporated to dryness. The residue is precipitated from cold hexane (−78°C) or lyophilyzed from benzene affording a colorless powder. Their purity is checked by ^{31}P-NMR and is >98%.

5a: (90%), ^{31}P-NMR (CD$_3$CN): δ 150.21, 150.38, m/z: 817 [M + H]$^+$.
5b: (92%), ^{31}P-NMR (CD$_3$CN) δ: 149.77, 150.62, m/z: 793 [M + H]$^+$.
5c: (86%) ^{31}P-NMR (CD$_3$CN) δ: 150.15, 150.46, m/z: 967 [M + H]$^+$.
5d: (77%) ^{31}P-NMR (CD$_3$CN) δ: 150.45, 150.11, m/z: 801 [M + H]$^+$.
5e: (91%) ^{31}P-NMR (CD$_3$CN) δ (ppm) in CD$_3$CN: 150.24, 150.40, m/z: 704 [M + H]$^+$.

3.4. Synthesis of α-Deoxynucleoside-Derivatized Solid Supports

The preparation of the α-deoxynucleoside-derivatized supports is achieved via formation a succinyl linkage between the 3'-hydroxyl group of an α-deoxynucleoside and the amino group of the solid support (Fig. 7). Base protected 5'-O-dimethoxytrityl deoxynucleosides 4a–e are derivatized to the corresponding 3'-hemisuccinates 6a–e with succinic anhydride in the presence of *N,N*-dimethylaminopyridine *(13)* and further reacted with pentachlorophenol in the presence of dicyclohexylcarbodiimide *(14)*. The resulting 3'-pentachlorophenyl succinates 7a–e are obtained in 83–93% overall yield after purification by flash column chromatography and reacted with the amino group of the long chain alkylamine on controlled pore glass beads *(15)*. Unreacted amino groups are capped by treatment with a mixture of acetic anhydride, 2,6-lutidine, and *N,N*-dimethylaminopyridine in tetrahydrofuran. The amount of loaded nucleoside to the solid support 8a–e is around 27 µmol/g from the calculation of released dimethoxytrityl cation by 0.1*M* 4-toluenesulfonic acid in acetonitrile.

3.5. Solid-Phase Synthesis of α-Oligodeoxynucleotides

3.5.1. Elongation Cycle

The α-oligonucleotides are synthesized on an Applied Biosystems Synthesizer model 381A employing the cycle described in Table 1. The synthesis is carried out using a column with 1 µmol of immobilized α-deoxynucleoside 8a–e. The reactions consisted of 5'-O-

Table 1
Steps Involved in One Elongation Cycle
for the Synthesis of α-Oligodeoxynucleotides

Step	Solvent or reagent	Time, s
Wash and flush	CH_3CN, argon	52
Detritylation	3% TCA in CH_2Cl_2	100
Wash and flush	CH_3CN, argon	225
Coupling	For A^{bz}, C^{bz}, T, and G^{ibu}: 0.1M amidite (20 Eq) in CH_3CN + 0.5M tetrazole (100 Eq) in CH_3CN	45
	For G^{pal}: 0.05M amidite (10 Eq) in CH_3CN^a + 0.5M tetrazole (100 Eq) in CH_3CN	75 (×2)
Flush	Argon	10
Capping	6.5% MeIm in THF + Ac_2O/ lutidine/THF (1/1/8)	143
Flush	Argon	20
Oxidation	0.1M I_2 in THF/Pyr/H_2O (40/10/1)	55
Wash	CH_3CN	100

^{a}The solution was kept at 37°C. MeIm: N-methyl-imidazole.

detritylation, condensation, capping, and oxidation (Fig. 8). Owing to the low solubility of α-deoxyguanosine phosphoramidite 5c in acetonitrile, it is used as a 0.05M solution gently warmed (37°C) instead of 0.1M at room temperature. Furthermore, the coupling reaction is duplicated during the course of 5c incorporation in order to keep the coupling yield as high as possible. Another possibility to avoid this difficulty is to dissolve 5c in a solution of dry acetonitrile:tetrahydrofuran (9:1, v/v). In this case, the coupling step is not duplicated. Since we developed the synthesis of α-deoxyguanosine phosphoramidite protected with isobutyryl group 5d, its good solubility in acetonitrile allows us to use the same synthesis cycles as for β-oligonucleotide synthesis.

3.5.2. Deprotection

After synthesis, the oligonucleotide anchored on the solid support is then treated with thiophenol:triethylamine:dioxane (1:2:2, v/v/v, 0.5 mL) for 30 min at room temperature. After washing with metha-

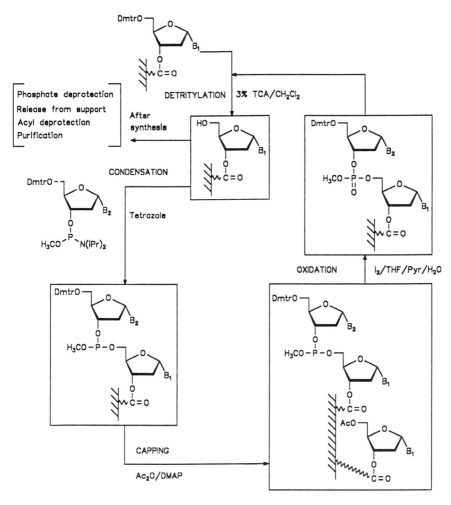

Fig. 8. Elongation cycle of α-oligodeoxynucleotides.

nol (10 mL), the oligonucleotide is cleaved from the support by treatment with aqueous concentrated ammonia for 1.5 h at room temperature, and the supernatant (2 mL) is kept overnight in sealed tubes at 55°C.

3.5.3. Purification

After filtration, the solution is evaporated and the crude α-oligonucleotide is purified by C_{18} reverse-phase HPLC using increasing proportions of acetonitrile (7–15%) in 0.05M triethylammonium acetate, pH 7, as eluent. The fractions containing the pure α-oligonucleotide are combined and evaporated. The residue is dissolved in water (500 µL), lyophilized, and stored frozen at –20°C.

4. Discussion

In order to synthesize α-oligodeoxynucleotides, we first needed the α-deoxynucleosides corresponding to the four natural bases in sufficient amount (multigram scale) since they are not—except α-dT—commercially available. Thus, we scaled up procedures already described or developed new ones, and selected those that gave the best results with respect to efficiency.

Although they always produced anomeric mixtures, glycosylation reactions between 2-deoxy-D-ribofuranose derivatives and pyrimidines or purines appeared to be the best compromise. Base- and sugar-protecting groups are choosen among those that allowed easy separation of the desired α-anomer by fractionated crystallization from the reaction mixture. In each case, when the exocyclic amino function of the base is required to be protected during the glycosylation reaction, the protecting group is selected in such a way that it could be kept in place during deprotection of the hydroxyl functions of the isolated α-anomers and will be suitable for the synthesis of α-oligodeoxynucleotides.

Reactions between chlorosugar <u>1a</u> and 6-N-benzoyladenine afforded, as already described *(2)*, the crystallized α-anomer <u>2a</u>, the 3'- and 5'-hydroxyls of which are selectively deprotected with sodium methoxide in a mixture of methanol and tetrahydrofuran. Thus, 6-N-benzoyl-α-deoxyadenosine (<u>3a</u>) is obtained in 34.8% (8.75 g) overall yield.

Self-anomerization reaction *(3)* with fully protected deoxycytidine <u>1b</u> is readily scaled up, and the corresponding α-anomer <u>2b</u> is crystallized from toluene and then treated with 0.19M sodium hydroxide at 0°C for 5 min. After neutralization with Dowex 50W (pyridinium form), 4-N-benzoyl-α-deoxycytidine (<u>3b</u>) is obtained as colorless crystals in 44% overall yield.

2-N-protected α-deoxyguanosine is obtained originally by transglycosylation reaction *(7)* between fully protected deoxycytidine <u>1b</u> and 2-N-palmitoylguanine as sugar donor and sugar acceptor, respectively. This reaction produced unavoidably 7-N-(3',5'-di-O-acetyl-2'-deoxy-α [and β]-D-ribofuranosyl)-2-N-palmitoylguanines as the kinetic isomers *(7)* along with the expected 9-N-isomers. Even after an extended reaction time (6 h), these undesired isomers are present in a significant amount and have to be removed from the crude reaction mixture of 3',5'-di-O-acetyl-2-N-palmitoyl-α (and β)-deoxyguanosines, from which the desired α-anomer 2c is isolated by fractionated crystallization in 30% yield. Alkaline treatment of 2c afforded 2-N-palmitoyl-α-deoxyguanosine (3c) as colorless crystals in 29% overall yield. Because of the poor solubility of the corresponding phosphoramidite derivative 5c in acetonitrile, the solvent of choice for coupling reactions in phosphoramidite method, isobutyryl, is introduced instead of palmitoyl for the protection of exocyclic amino function in α-deoxyguanosine through a new synthetic procedure involving a phase-transfer reaction *(8)*. Thus, 1-O-acetyl-3,5-di-O-nitrobenzoyl- 2-deoxy-D-ribofuranose (1d) *(10,11)* is reacted with trimethylsilylated guanine *(9)* in the presence of potassium iodide and dibenzo-18-crown-6 in a mixture of toluene and acetonitrile at 80°C. Then pyridine and isobutyryl chloride are added to the cooled reaction mixture. After a work-up, 2-N-isobutyryl-3',5'-di-O-p-nitrobenzoyl-α-deoxyguanosine 2d is obtained together with the corresponding β-anomer and is isolated by fractionated crystallization. Subsequent alkaline treatment affords 2-N-isobutyryl-α-deoxyguanosine (3d) as colorless crystals. Although the overall yield of this latter procedure is slightly lower (26.5%) than that of the previous one (29%), this is largely overcome by an easier working up and the obtainment of a more soluble phosphoramidite derivative 5d.

α-Thymidine could be either purchased (Sigma, T 3763) or produced by glycosylation reaction between 1-chloro-3,5-di-O-p-toluoyl-2-deoxy-D-ribofuranose and trimethylsilylated thymine *(9)*. α-Thymidine derivative 2e is obtained as the major product together with the corresponding β-anomer and is isolated by fractionated crystallization. Subsequent treatment with sodium methoxide in methanol yielded the desired α-thymidine in 47% overall yield. This procedure is easily scaled up and can yield up to 24 g of α-thymidine in a simple batch.

Although obtainment of anomerically pure α-deoxynucleoside derivatives is recommended at the earliest step of the synthesis, we noticed that traces of nucleoside derivatives could be completely removed during silica gel purification of the 5'-dimethoxytrityl derivatives 4α–e, since the corresponding β-deoxynucleoside derivatives are in each case markedly more polar and are eluted later. This late purification step avoids the need for repeated recrystallization in the previous steps with unavoidable loss of material and ensures a high isomeric purity. Preparation of the α-deoxynucleoside 3'-O-methylphosphoramidites 5a–e and the α-deoxynucleoside-derivatized solid supports 8a–e is straightforward and very similar to that previously reported for the β-deoxynucleosides.

Last, the synthesis of α-oligodeoxynucleotides requires the same reagents and the same elongation cycles as for β-oligodeoxynucleotide synthesis. Likewise, α-oligodeoxynucleotides are deprotected and purified by following the standard procedures.

In conclusion, the synthesis of α-oligodeoxynucleotides containing the four natural bases is as easy as that of β-oligodeoxynucleotides. Furthermore, α-oligodeoxynucleoside phosphorothioates can be obtained with ease by substitution of the oxidation step with the sulfurization step either with elemental sulfur *(16)* or ^{3}H-1,2-benzodithiole-3-one 1,1-dioxyde *(17)* by following the standard conditions.

Abbreviations

Ac, acetyl; BSA, *N,O-bis*-trimethylsilyl-acetamide; Bz, benzoyl; DMF, dimethylformamide; Ibu, isobutyryl; THF, tetrahydrofuran; TMS-triflate, trimethylsilyl trifluoromethanesulfonate; Tol: toluoyl; Pal, palmitoyl.

References

1. Uhlmann, E. and Peyman, A. (1990) Antisense oligonucleotides: A new therapeutic principle. *Chem. Rev.* **90,** 543–584.
2. Ness, R. K. (1968) 2'-Deoxyadenosine and its α-D-anomer, in *Synthetic Procedure in Nucleic Acid Chemistry* (Zorbach, W. W. and Tipson, R. S., eds.) Interscience Publishers, New York, pp. 183–187.
3. Yamaguchi, T. and Saneyoshi, M. (1984) Synthetic nucleosides and nucleotides. XXI. On the synthesis and biological evaluations of 2'-deoxy-α-D-ribofuranosyl nucleosides and nucleotides. *Chem. Pharm. Bull.* **32,** 1441–1450.
4. Bhat, V., Ugarkar, B. G., Sayeed, V. A., Grimm, K., Kosora, N., Domenico, P. A., and Stocker, E. (1989) A simple and convenient method for the selective *N*-acylations of cytosine nucleosides. *Nucleosides & Nucleotides* **8,** 179–183.
5. Robins, M. J. and Robins, R. K. (1969) Purine nucleosides. XXIV. A new method for the synthesis of guanine nucleosides. The preparation of 2'-deoxy-α- and β-guanosines and the corresponding N^2-methyl derivatives. *J. Org. Chem.* **34,** 2160–2163.
6. Furukawa, Y. and Honjo, M. (1968) A novel method for the synthesis of purine nucleosides using Friedels-Crafts catalysts. *Chem. Pharm. Bull.* **16,** 1076–1080.

7. Watanabe, K. A., Hollenberg, D. H., and Fox, J. J. (1974) Nucleosides LXXXV. On mechanisms of nucleoside synthesis by condensation reaction. *J. Carbohydrate Nucleosides Nucleotides,* **1,** 1–37.
8. Azymah, M., Chavis, C., Lucas, M., and Imbach, J.-L. (1989) Efficient synthesis of 1,2-*seco* and 1,2-*seco* 2-*nor* pyrimidine and purine nucleosides. *Tetrahedron Lett.* **30,** 6165–6168.
9. Hubbard, A. J., Jones, A. S., and Walker, R. T. (1984) An investigation by 1H NMR spectroscopy into the factors determining the β:α ratio of the product in 2'-deoxynucleoside synthesis. *Nucl. Acids Res.* **12,** 6827–6837.
10. Bhat, C. C. (1968) 2-Deoxy-3,5-di-*O*-*p*-toluoyl-D-erythro-pentosyl chloride, in *Synthetic Procedures in Nucleic Acid Chemistry* (Zorbach, W. W. and Tipson, R. S., eds.) Interscience Publishers, New York, pp. 521,522.
11. Smejkal, J. and Sorm, F. (1964) Nucleic acids components and their analogues. LIII. Preparation of 1-(2'-deoxy-β-L-ribofuranosyl)-thymine, "L-thymidine." *Collect. Czech. Chem. Commun.* **29,** 2809–2813.
12. McBride, L. J. and Caruthers, M. H. (1983) An investigation of several deoxynucleoside phosphoramidites useful for synthesizing deoxyoligonucleotides. *Tetrahedron Lett.* **24,** 245–248.
13. Chow, F., Kempe, T., and Palm, G. (1981) Synthesis of oligodeoxyribonucleotides on silica gel support. *Nucl. Acids Res.* **9,** 2807–2817.
14. Miyoshi, K., Miyake, T., Hozumi, T., and Itakura, K. (1980) Solid-phase synthesis of polynucleotides. II. Synthesis of polythymidylic acids by the block coupling phosphotriester method. *Nucl. Acids Res.* **8,** 5473–5489.
15. Gough, G. R., Brunden, M. J., and Gilham, P. T. (1981) Recovery and recycling of synthetic units in the construction of oligodeoxyribonucleotides on solid supports. *Tetrahedron Lett.* **22,** 4177–4180.
16. Stec, W. J., Zon, G., Egan, W., and Stec, B. (1984) Automated solid-phase synthesis, separation, and stereochemistry of phosphorothioate analogues of oligodeoxyribonucleosides. *J. Am. Chem. Soc.* **106,** 6077–6079.
17. Iyer, R. P., Egan, W., Regan, J. B., and Beaucage, S. L. (1990) 3H-1,2-benzodithiole-3-one 1,1-dioxide as an improved sulfurizing reagent in the solid-phase synthesis of oligodeoxyribonucleoside phosphorothioates. *J. Am. Chem. Soc.* **112,** 1253,1254.

CHAPTER 14

Stereospecific Synthesis of P-Chiral Analogs of Oligonucleotides

Wojciech J. Stec and Zbigniew J. Lesnikowski

1. Introduction

As has been emphasized in corresponding chapters of this book, isotopic or elemental replacement of any of two nonbridging oxygens attached to internucleotide phosphorus atom(s) by substituent X (Fig. 1) creates, by virtue of asymmetry, new center(s) of chirality and results in the formation of m diastereoisomers of the oligonucleotide's congener. The number m is determined by the number n of modified phosphates, according to formula $m = 2^n$. With the exception of diastereoisotopomers (X= ^{17}O or ^{18}O), which are nonseparable by any so-far-developed separation technique, separation of diastereoisomeric oligonucleotides containing modified phosphates (Fig. 1, X = S; Se; C_2H_5O-; CH_3-; ArNH–; R_2N-) has been achieved by means of chromatographic techniques. Effectiveness of the separation depends, however, on the number of modifications within the oligonucleotide chain and is practically limited to n=1,2. Achievements in this field have been exhaustively summarized by Zon [1].

The choice of synthetic method leading to stereo-defined oligonucleotide analogs must depend on the requirements of the number n of modified groups and their relative position within the oligonucleotide chain. The synthesis of oligonucleotide bearing single modification ($n = 1$), performed by the nonstereoselective phosphor-amidite method, leading to the mixture of two diastereoisomers, may be followed by attempts

From: *Methods in Molecular Biology, Vol. 20: Protocols for Oligonucleotides and Analogs*
Edited by: S. Agrawal Copyright ©1993 Humana Press Inc., Totowa, NJ

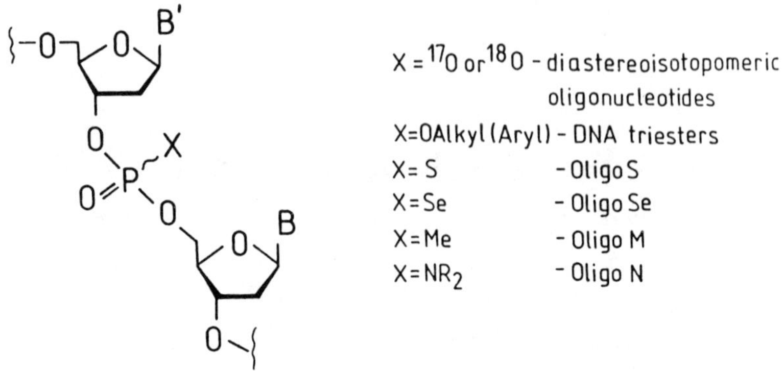

Fig. 1. Modifications to the DNA phosphate groups.

$X = {}^{17}O$ or ${}^{18}O$ – diastereoisotopomeric oligonucleotides
X = OAlkyl (Aryl) – DNA triesters
X = S – Oligo S
X = Se – Oligo Se
X = Me – Oligo M
X = NR_2 – Oligo N

at separation by reversed-phase (RP) HPLC. Two-step separation is recommended, first at the level of the oligonucleotide bearing the 5'-O-DMT group (elsewhere necessary for isolation of desired product from "failed" sequences). If not successful, separation of the diastereoisomers should be tried after removal of the 5'-O-DMT group (2). Several oligonucleotide analogs bearing one or two modified internucleotide phosphates have been successfully separated into diastereoisomers (2–10).

However, severe difficulties encountered during separation of diastereoisomeric species, especially if $n > 2$, called for the choice of another strategy and introduction of new reagents to the protocol of synthesis of oligonucleotides via the solid-support phosphoramidite method. These new reagents, consisting of "dimeric" synthons (1) (Fig. 2), are available via synthesis and separation of precursors (1a–e, Z = H), followed by their phosphitylation with O-β-cyanoethyl-N,N,N',N'-tetraisopropylphosphordiamidite (11–13). Postsynthetic workup of desired modified oligonucleotide (after last cycle of synthesis) depends on the nature of X and Y ligands at phosphorus in 1. Introduced internucleotide methanephosphonate function (1c, Y = O, X = CH_3) requires mild conditions for deblocking of exoamino functions (at nucleobases Cyt, Gua, and Ade) and cleavage from the support. The ethylenediamine/ethanol mixture is recommended since 25% NH_4OH may cause the cleavage of the oligonucleotide chain at site of modification, especially in the case of longer modified oligomers (14). Introduction of dimeric synthons with internucleotide O-alkyl phosphorothioate func-

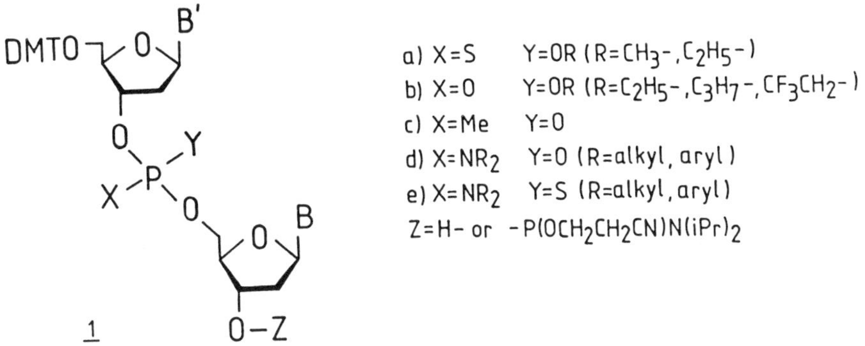

a) X=S   Y=OR (R=CH$_3$-, C$_2$H$_5$-)
b) X=O   Y=OR (R=C$_2$H$_5$-, C$_3$H$_7$-, CF$_3$CH$_2$-)
c) X=Me   Y=O
d) X=NR$_2$ Y=O (R=alkyl, aryl)
e) X=NR$_2$ Y=S (R=alkyl, aryl)
Z=H- or -P(OCH$_2$CH$_2$CN)N(iPr)$_2$

Fig. 2. Modified dinucleotide "synthon."

tions (**1a**) (as an alternative precursor of oligonucleotide monophosphorothioate) to the synthetic protocol results in preparation of oligonucleotide bearing O-alkyl phosphorothioate group. In the case of **1a** (R = C$_2$H$_5$), cleavage from support, phosphates (except P(S)OEt), and backbone/base deprotection is achievable with concentrated NH$_4$OH (48 h, 25°C). Diastereoisomerically pure oligonucleotide bearing at preselected position internucleotide function P(S)OEt must be heated at 70°C for 20 h in conc. NH$_4$OH, which effectively deprotects phosphorothioate function without epimerization at phosphorus atom *(6)*. Preparation of DNA triester analogs **1b** (R = C$_2$H$_5$, CF$_3$CH$_2$) via dimer-block technique is perhaps the most laborious because it requires the modification of the overall synthetic procedure. A recommended strategy has been presented in Chapter 10 by Koziolkiewicz and Wilk in this book. The dimer-block approach is especially useful for the synthesis of oligonucleotides bearing a larger number *n* of P-chiral modifications under the condition that each of these modified groups is separated by at least one "regular" internucleotide phosphate function. A number of such oligonucleotides with multiple modifications have been obtained: methanephosphonate *(15)*, phosphorothioate *(11)*, and phosphomorpholidate *(13)*. P-chiral groups of known sense of chirality have been introduced in alternating positions *(11,13,16–20)*.

Perhaps the most difficult problem is posed by the requirement of the introduction of stereo-defined modification at neighboring positions. Although the examples of partially diastereoselective syntheses of dinucleoside methanephosphonates *(21)*, phosphorothioates *(22–24)*, and *O*-aryl phosphates *(25)* are numerous, the finding of a general solution to the stereocontrolled synthesis of oligonucleotide analogs, bearing at juxtaposed positions desired modifying phosphorothioate or methanephosphonate functions, requires further efforts and is a challenging problem to organic chemists. Recent results from authors' laboratory kindled some hopes that stereocontrolled synthesis of methanephosphonate- and phosphorothioate-oligonucleotide analogs (oligoM and oligoS, respectively) is possible. [$R_PR_PR_P$]- and [$S_PS_PS_P$]-tetra(thymidine methanephosphonates) (**5**) have been obtained *(26)*. The process of chain extension is based on a stereospecific reaction between diastereoisomerically pure [R_P]- or [S_P]-5'-*O*-MMT-thymidine-3'-*O*-[*O*-(4-nitrophenyl)methane-phosphonate] (**2**) and the 5'-OH group of 3'-protected thymidine (first step) *(27)* or 5'-OH group of corresponding dimer and trimer. This reaction is most effectively catalyzed by the *t*-butyl magnesium halide, originally introduced to oligonucleotide chemistry by Hayakawa et al. *(28)*. Although the stereospecificity of single coupling is high (above 96%), and the yields of coupling (65%) as well as reaction times (around 4 h/coupling) are moderate *(29)*, still this procedure leading to stereoregular oligo(nucleoside methanephosphonate)s is laborious because of the necessity of purification of intermediates, such as **3**, **4**, or **5** (Fig. 3).

Condensation of appropriately 5'- and 3'-protected tetramers **5**, gave octamers **6**, with six among seven internucleotide methanephosphonate functions of precisely defined stereochemistry and high diastereoisomeric purity. Those, when used in melting studies, have shown a dramatic difference in T_m parameters *(30)*. To our best knowledge, there are no other published methods of preparation of P-stereoregular oligo(nucleoside methanephosphonate)s, if $n > 2$. Similar strategy leading to [R_PR_P]- and [S_PS_P]-tri(thymidine phosphorothioate) (**7**) has been used with 5'-*O*-monomethoxytritylthymidine-3'-*O*-[*S*-(2-nitrobenzyl)-*O*-(4-nitrophenyl)phosphorothioates] as substrates *(31)*. The extension of this methodology to stereospecific synthesis of phosphorothioate analogs of oligoribonucleotides is illustrated by the only recently published synthesis of [S_PS_P]- and [R_PR_P]-isomers of tri(uridine phosphorothioate) *(32)*.

These otherwise promising results have been overwhelmed by a new strategy recently developed for the synthesis of oligo(nucleoside phosphorothioate)s *(33)*. It has been proven that reaction of 5'-*O*-

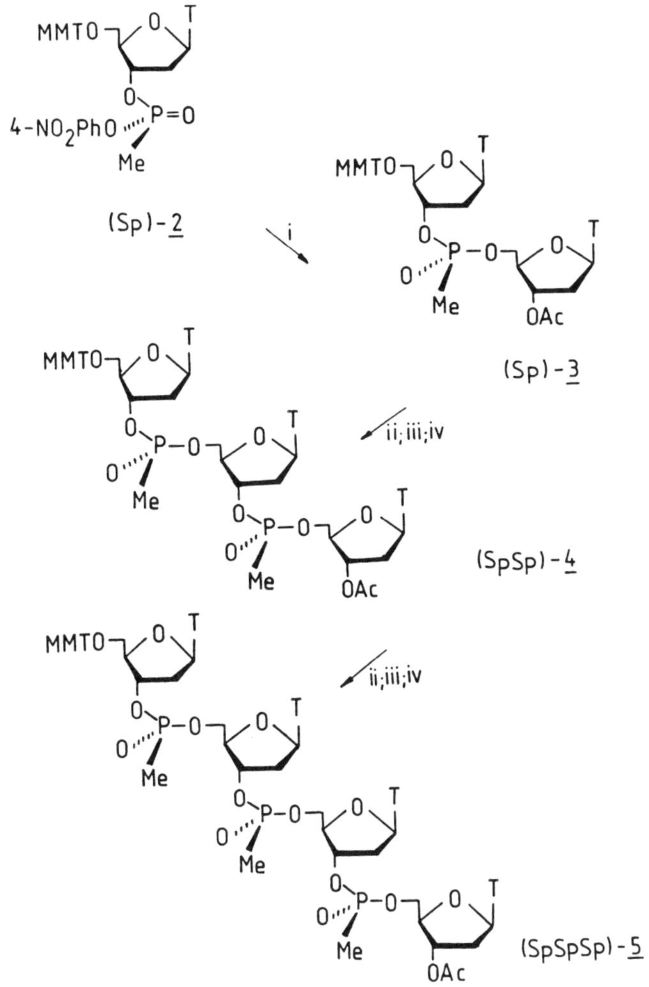

Fig. 3. Synthesis of individual homochiral di-, tri-, and tetra(nucleoside methane-phosphonates). i = 3'-O-acetylthymidine/t-BuMgCl; ii = 80% acetic acid; iii = t-BuMgCl; iv = [S$_P$]-2.

dimethoxytritylnucleoside 3'-O-(2-thiono-1.3.2-oxathiaphospholanes) (**8**) (*N*-protected in case of Ade, Gua, and Cyt) with 5'-hydroxyl function of support-bound nucleoside(tide), performed in the presence of 1,8-diazabicyclo(5.4.0) undec-7-cne (DBU) as a catalyst, leads to the formation of internucleotide phosphorothioate function in >95%

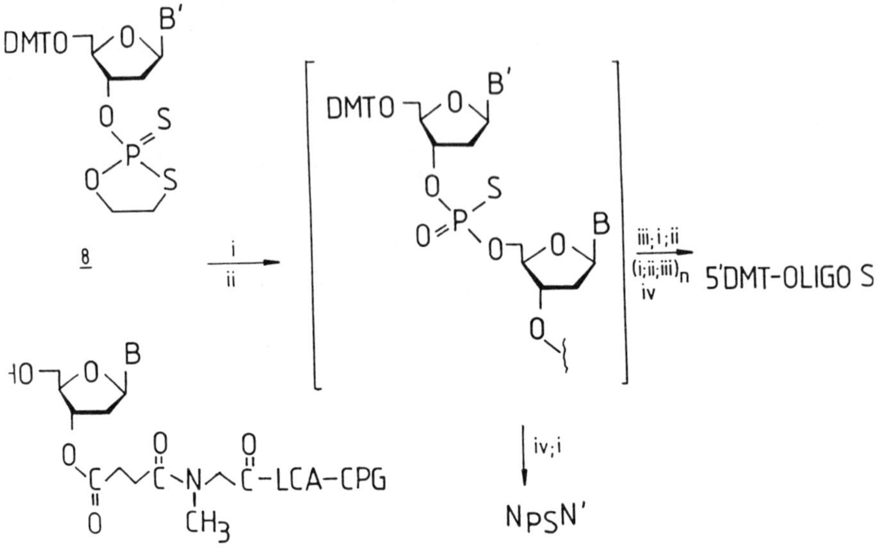

Fig. 4. Synthesis of individual homochiral oligo(nucleoside phosphorothioates). i = DBU/acetonitrile; ii = Ac$_2$O/2,6-lutidine, N-methyl imidazole; iii = dichloroacetic acid/methylene chloride; iv = 25% ammonium hydroxide.

yield. Such a chemistry has been applied for the automated synthesis of oligo(nucleoside phosphorothioate)s. Because the commercially available "succinyl" linker, conventionally used for an attachment of the oligonucleotide to the support matrix, is unstable to DBU, the oxathiaphospholane method required modification. Fortunately, a suitable resin-oligonucleotide linker **9** that is resistant to DBU, but is hydrolyzed in <1 h at room temperature by conc. NH$_4$OH, has been designed by T. Brown et al. *(34)*.

It was found that by using "succinic-sarcosinyl"-type linker **9**, the solid-phase automated synthesis of dodeca(deoxyadenosine phosphorothioate) (d[(A$_{PS}$)$_{11}$A]) can be performed via the oxathiaphospholane method with an average of 95% effectiveness of a single coupling step. Purity of this material, as indicated by polyacrylamide gel electrophoresis and HPLC, is comparable with that of sample prepared by phosphoramidite-stepwise sulfurization or H-phosphonate-tandem sulfurization methods *(35)*. ^{31}P-NMR spectrum did not contain any signal within the phosphate region. The synthesis is outlined within Fig. 4, and synthetic protocol is given in Table 1.

Table 1
Chemical Steps for One Synthesis Cycle

Reagent or solvent		Purpose	Time, min
Dichloroacetic acid in CH_2Cl_2 (2:98, v/v)	2 mL	Detritylation	1.5
Acetonitrile	5 mL	Wash	2
Activated nucleotide in acetonitrile[a]		Coupling	10
Acetonitrile	5 mL	Wash	2
Acetic anhydride/lutidine in THF (10:10:80, v/v)	1 mL	Capping	1
N-methylimidazole in THF (16:84, v/v)	1 mL		
Acetonitrile	5 mL	Wash	1

[a]For 1 µmol synthesis scale, 2M DBU in pyridine (150 µL) and 0.1M 5'-O-DMT-deoxynucleoside-3'-O-(2-thio-1,3,2-oxathiaphospholane) in acetonitrile (50 µL) is used.

As anticipated, diastereoisomeric oxathiaphospholanes **8** have appeared separable into pure diastereoisomers, and their use in the synthesis has also proven, as expected on the basis of earlier mechanistic studies *(36)*, that the ring-opening process is fully stereospecific. In careful model studies, the dimers **10** (d[T_{PS}T], d[A_{PS}A)], d[C_{PS}C], and d[G_{PS}G]) were obtained with diastereoisomeric purity >99%. Enzymatic analysis of [all-R_P]- and [all-S_P]-d([C_{PS}]$_4$C) (**14**), prepared with the use of separated diastereoisomers of **8** (B = C^{iBu}), confirmed that chain extension above dimer level is also stereospecific.

This introduction, and the examples of synthetic procedures given below, are limited to presentation of methods developed in the authors' laboratory; space limitation does not allow the presentation of results of all research groups contributing to the synthesis of oligonucleotide analogs of a defined sense of chirality at phosphorus. These contributions are acknowledged by pertinent citations.

2. Materials

The commercial solvents and reagents of highest purity are recommended. Since the content of water is a critical factor for several procedures given below, dry solvents or solvents dried by conventional methods were stored under nitrogen in tightly closed containers over CaH_2 (triethylamine, pyridine, acetonitrile). Tetrahydrofurane, also kept over CaH_2, was distilled before use over $LiAlH_4$. Argon and nitrogen were dried by passing through a column containing dessicant silica gel, followed by another horizontal column half-filled with P_2O_5/glass wool. 4-Nitrophenol was recrystallized from CCl_4. 5'- and 3'-protected nucleosides were prepared according to standard literature methods *(37)*. Nucleoside and nucleotide components were usually dried under high vacuum conditions, if not specified otherwise.

2.1. Chemicals and Supplemental Materials

Item
- Nucleosides
  1. 2'-*O*-Deoxyadenosine.
  2. 2'-*O*-Deoxycytidine.
  3. 2'-*O*-Deoxyguanosine.
  4. Thymidine.
- Miscellaneous reagents
  5. Acetic acid.
  6. Acetyl chloride.
  7. *t*-Butyldimethylsilyl chloride.
  8. *t*-Butylmagnesium chloride.
  9. Calcium hydride.
  10. 1,3-Diazabicyclo (5.4.0)undec-7-ene (DBU).
  11. Dicyclohexylcarbodiimide (DCC).
  12. Diisopropylamine.
  13. 4,4'-Dimethoxytriphenylmethyl chloride (DMTCl).
  14. *N*-Fmoc-sarcosine.
  15. Lithium aluminum hydride.
  16. Magnesium sulfate, anhydrous.
  17. 2-Mercaptoethanol.
  18. Methylphosphonic dichloride.
  19. 4-Methoxytriphenylmethylchloride.
  20. 4-Nitrophenol.
  21. Phosphorus trichloride.
  22. Phosphorus pentoxide.
  23. Sulfur.
  24. 1H-Tetrazole.
  25. 1.2.4-Triazole.
  26. Triethylamine.
- Enzymes
  27. Alkaline phosphatase (EC 3.1.3)
  28. Nuclease P1 (EC 3.1.30.1).
  29. Snake-venom phosphodiesterase (EC 3.1.4.1).

Others
30. Acetic anhydride/lutidine/THF.
31. LCA-controlled pore glass.
32. HPTLC plates.
33. *N*-methyl imidazole/THF.
34. Silica gel for column chromatography.
35. TLC plates.

2.2. Solvents

Any reputable supplier can be used. However, wherever possible, use analytical-grade solvents. Those required are acetonitrile, benzene, carbon tetrachloride, chloroform, dichloromethane, dimethylformamide, ethyl acetate, *n*-hexane, methanol, *n*-pentane, pyridine, and tetrahydrofurane.

3. Methods

3.1. Stereocontrolled Synthesis of Oligo(Nucleoside Methanephosphonates) (OligoM)

3.1.1. Synthesis of [O-(4-Nitrophenyl)] Methanephosphonochloridate (11)

Freshly distilled (*see* Note below) methanephosphonic dichloride (8.00 g, 60.0 mmol), and 4-nitrophenol (6.95 g, 50.0 mmol) are placed in a 50-mL, two-neck, round-bottomed flask, heated with oil bath, and equipped with thermometer and reflux condenser with drying tube. A magnetically stirred reaction mixture is warmed from room temperature to 180°C in the course of 7 h and maintained at this temperature for the next 18 h. Unreacted methanephosphonic dichloride is removed by distillation under reduced pressure of a water respirator and crude product is fractionated by distillation using an oil vacuum pump; bp 120°C/0.01 mm Hg (lit: bp 138–141°C/0.2 mm Hg) *(38)*. Yield 30%; ^{31}P-NMR: δ 34.2 ppm (benzene).

Note: Methanephosphonic dichloride was distilled under reduced pressure, bp 54°C/10 mm Hg, ^{31}P-NMR: d42.7 (benzene). This product is highly toxic, corrosive, and moisture-sensitive, and it may crystallize in the condenser during distillation. All operations have to be performed under an effective hood.

Warning: The distillation of the final product **11** should be performed under sufficiently low pressure. Since the product may crystallize during distillation, an air condenser is recommended. Wear appropriate eye protection.

3.1.2. Synthesis of Monomers, 5'-O-Monomethoxytrityl- (2'-O-Deoxyribonucleoside) 3'-O-[O-(4-Nitrophenyl) Methanephosphonates]

3.1.2.1. SYNTHESIS OF 5'-O-MONOMETHOXYTRITYLTHYMIDINE 3'-O-[O-(4-NITROPHENYL)METHANEPHOSPHONATE] (2) (OPTION I)

To a threefold molar excess of methanephosphonic dichloride (3.0 g, 22.5 mmol) in pyridine (22 mL), a solution of 5'-O-monomethoxytritylthymidine (3.8 g, 7.5 mmol) in the same solvent (18 mL) is added dropwise over a period of 1 h at room temperature. The reaction mixture is stirred next for an additional 1.0 h, and then a ninefold molar excess of 4-nitrophenol (9.5 g, 68.0 mmol) in pyridine (34 mL) is added. After 1 h (TLC control; $CHCl_3:CH_3OH= 96:4$), the reaction is quenched with 50% aqueous pyridine (8.0 mL) followed by H_2O (200 mL).

The resulting emulsion is extracted with $CHCl_3$ (3 × 120 mL), and organic fraction is dried over $MgSO_4$. The solvent is evaporated under reduced pressure yielding dry foamy residue. Crude product is then purified and diastereoisomers are separated by silica gel column chromatography (*see* Section 3.1.2.3).

Note: Slow, dropwise addition of 5'-protected nucleoside to $MeP(O)Cl_2$ solution is essential in order to avoid the formation of symmetrical *bis*-[3'-O-(5'-O-monomethoxytritylthymidylyl)] methanephosphonate.

3.1.2.2. SYNTHESIS OF 5'-O-MONOMETHOXYTRITYL- (2'-O-DEOXYRIBONUCLEOSIDE)-3'-O-[O-(4-NITROPHENYL) METHANEPHOSPHONATE] (2, B = THY,CYT,ADE,GUA) (OPTION II)

To [O-(4-nitrophenyl)]methanephosphonochloridate (11, 0.35 g, 1.5 mmol) and triazole (0.42 g, 6.0 mmol), acetonitrile (5.0 mL) followed by triethylamine (0.31 g, 3.0 mmol) are mixed together under nitrogen, with magnetic stirring. After 15 min at room temperature, the resultant clear solution of [O-(4-nitrophenyl)]methanephosphonotriazolidate is added under nitrogen to a solution of dry 5'-O-monomethoxytrityl-2'-O-deoxyribonucleoside (1 mmol) in acetonitrile (5.0 mL). The reaction progress is monitored by TLC ($CHCl_3:CH_3OH= 9:1$). After 3 h, the postreaction mixture is concentrated under reduced pressure. The oily residue is redissolved in $CHCl_3$ (15 mL). The chloroform solution of crude product is washed with H_2O (2 × 10 mL). Organic fraction is separated, dried over $MgSO_4$, and then concentrated under reduced pressure.

P-Chiral Analogs

The oily residue is dried under high vacuum yielding crude **2** (foamy solid), which is purified and separated into individual isomers as described below. The yields, and chromatographic and ^{31}P-NMR characteristics, are presented in Table 2.

3.1.2.3. SEPARATION OF DIASTEREOISOMERS OF 5'-O-MONOMETHOXYTRITYL-(2'-O-DEOXYRIBONUCLEOSIDE)-3'-O-[O-(4-NITROPHENYL) METHANEPHOSPHONATE] (**2**, B = THY,CYT,ADE,GUA)

Standard conditions for separation of diastereoisomers **2** are as follows:

Glass column (30 × 3 cm) with porous-glass bed support.
Kieselgel 60 (230–400 mesh, Merck, Darmstadt, Germany, Art. 9385).
Sample/silica gel ratio ca. 1:30 (w/w).
Eluting solvent system: monomer obtained according to option I, B = Thy: from CHCl$_3$ to CHCl$_3$:CH$_3$OH=98:2; monomers obtained according to Option II, B = Thy: CHCl$_3$:CH$_3$OH=98:2; B = Ade: CHCl$_3$:CH$_3$OH=94:6; B = Cyt,Gua: CHCl$_3$:CH$_3$OH:CH$_3$COOH= 100:3:3.

Note: For thymidine monomer **2** (B = Thy) prepared according to option I, eluting of the column first with CHCl$_3$ is necessary in order to wash out an excess of 4-nitrophenol fouling the crude product.

3.1.3. Synthesis of Protected Dimers, 5'-O-Monomethoxytritylnucleosidylyl (3',5') (3'-O-Protected Nucleosidylyl) Methanephosphonate

3.1.3.1. SYNTHESIS OF [S$_P$]- AND [R$_P$]-ISOMERS OF 5'-O-MONOMETHOXYTRITYL-NUCLEOSIDYLYL (3',5') (3'-O-PROTECTED NUCLEOSIDYLYL) METHANEPHOSPHONATE (**3**, B = THY,CYT,ADE)

A 2.15*M* solution of freshly prepared *t*-butylmagnesium chloride in dry THF (0.462 mL, 1 mmol) is added under nitrogen to a solution of 3'-O-acetylthymidine or 3'-O-(*t*-butyldimethylsilyl)-2'-O-deoxyribonucleoside (1 mmol) in anhydrous pyridine (1.0 mL). To the resulting suspension of 5'-hydroxyl-activated nucleoside (25% molar excess), a solution of individual [R$_P$]- or [S$_P$]-isomer of suitable monomer **2** (0.75 mmol) in anhydrous pyridine (2.0 mL) is added after 15 min. The reaction mixture is stirred for 4 h at room temperature, and then left overnight without access to moisture. The reaction is quenched with water (20.0 mL), and the product is extracted with CHCl$_3$ (2 × 20 mL).

Table 2
Characteristics of Mono- (2) and Di(2'-O-Deoxyribonucleoside Methanephosphonate) (3)

B	Monomers (2)				Dimers (3)				
	Configuration	TLC, R_f	^{31}P-NMR[e]	Yield[a] %	Configuration	HPTLC, R_f	^{31}P-NMR,[e] ppm	HPLC,[f] Rt	Yield, %
Thy	Sp	0.48[b]	28.65	53	Sp	0.12[b]	32.72	12.59	60
	Rp	0.38[b]	29.81		Rp	0.16[b]	32.14	11.84	66
Ade	Sp	0.39[b]	28.99	45	Sp	0.35[c]	31.87	12.07	50
	Rp	0.34[b]	28.97		Rp	0.37[c]	31.83	10.94	50
Cyt	Sp	0.24[b]	29.15	43	Sp	0.07[c]	32.34	7.44	60
	Rp	0.17[b]	29.34		Rp	0.16[c]	32.41	7.44	60
Gua	Sp	0.56[c]	29.11	74	Sp	0.53	31.94	7.30	20
	Rp	0.46[c]	29.03		Rp	0.61[d]	33.96	7.30	17

[a] Yield of [Rp] + [Sp] isomers.
[b] Developing system: 6% MeOH in CHCl$_3$; TLC was performed on silica gel 60 F$_{254}$ plates (Merck), and HPTLC on silica gel 60 F$_{254}$ plates (Merck).
[c] Developing system: CHCl$_3$:CH$_3$OH = 9:1.
[d] Developing system: CH$_3$CN:H$_2$O = 9:1.
[e] In CDCl$_3$ (except G$_{PMe}$G- in C$_5$D$_5$N).
[f] Rt parameters for 3'- and 5'-deprotected compounds 5 (5'), ODS Hypersil 5 μm, 30 cm × 4.6 mm column, gradient 5–20% CH$_3$CN in 0.1M TEAB; 1.0% CH$_3$CN/min.

The combined organic fractions are dried over MgSO$_4$. Solvents are evaporated and corresponding product [S$_P$]- or [R$_P$]-**3** purified by column silica gel chromatography (solvent system: from CHCl$_3$ to CHCl$_3$:CH$_3$OH=97:3). The yields, and chromatographic and ^{31}P-NMR characteristics, are presented in Table 2.

Note: After 4 h the degree of nucleoside phosphonylation is usually 80–90%. Commercially available *t*-BuMgCl (e.g., from Aldrich, Milwaukee, WI) sometimes contains crystalline magnesium salts, which are difficult to redissolve. In such a case, the *t*-BuMgCl concentration should be redetermined.

3.1.3.2. SYNTHESIS OF [S$_P$]- AND [R$_P$]-ISOMERS
OF 5'-*O*-MONOMETHOXYTRITYL-2'-*O*-DEOXYRIBOGUANOSINYL
(3',5') (3'-*O*-*T*-BUTYLDIMETHYLSILYL-2'-*O*-DEOXYRIBOGUANOSINYL)
METHANEPHOSPHONATE (**3**, B = GUA)

A 2.15*M* solution of freshly prepared *t*-butylmagnesium chloride in dry THF (0.173 mL, 0.375 mmol) is added under nitrogen to a solution of 3'-*O*-(*t*-butyldimethylsilyl)-2'-*O*-deoxyriboguanosine (0.048 g, 0.125 mmol) in dry THF (0.2 mL). To the resulting suspension of 5'-hydroxyl-activated nucleoside a solution of individual [R$_P$]- or [S$_P$]-isomer of monomer **2** (B = Gua, 0.074 g, 0.1 mmol) in dry THF (0.4 mL) is added after 0.5 h. The reaction mixture is left overnight with stirring at room temperature (TLC control: CH$_3$CN:H$_2$O= 9:1) and the solvent is evaporated under reduced pressure. The residue is redissolved in CHCl$_3$ (5 mL) and washed with water (3 × 5 mL). Organic fraction is dried over MgSO$_4$ and the solvent is evaporated. The crude product is purified chromato-graphically using CH$_3$CN:H$_2$O= 9:1 as a solvent system. The yields, and chromatographic and ^{31}P-NMR characteristics, are presented in Table 2.

*3.1.4. Synthesis of Protected Trimers: [S$_P$S$_P$]-
and [R$_P$R$_P$]-Isomers of 5'-O-Monomethoxytritylthymidylyl
(3',5')Thymidylyl (3',5') (3'-O-Acetylthymidylyl)
Di(methanephosphonate) (4, B = Thy)*

A 2.15*M* solution of freshly prepared *t*-butylmagnesium chloride in dry THF (0.157 mL, 0.341 mmol) is added under dry nitrogen to a solution of 5'-deprotected [S$_P$]- or [R$_P$]-isomer of **3** (B = Thy, 0.200 g, 0.341 mmol) in dry pyridine (2.0 mL). To the resulting suspension of 5'-hydroxyl-activated dinucleotide is added a solution of indi-

Table 3
Characteristics of Homochiral Diastereoisomers of Di-(**3**), Tri-(**4**), and Tetra (Thymidine Methanephosphonate) (**5**, B = Thy)

Compound	Yield,[a] %	TLC,[b] R_f	^{31}P-NMR[d]
[Sp]-**3**	60	0.42	32.72
[Rp]-**3**	66	0.57	32.14
[SpSp]-**4**	55	0.29	33.10
			32.79
[RpRp]-**4**	55	0.33	32.52
			32.39
[SpSpSp]-**5**	40	0.11	33.36
			33.19
			32.66
[RpRpRp]-**5**	38	0.20	33.19
			32.85
			32.74
[SpSpSp]-**5'**	73	0.46[c]	35.03[e]
			34.89
			34.58
[RpRpRp)-**5'**	64	0.40[c]	34.71[e]
			34.45

[a]Yield of the products after purification.
[b]Developing solvent system: $CHCl_3:CH_3OH=9:1$.
[c]Developing solvent system: $CH_3CN:H_2O=8:2$.
[d]In $CDCl_3$.
[e]In CD_3OD.

vidual [R$_P$]- or [S$_P$]-isomer of **2** (B = Thy, 0.243 g, 0.341 mM) in dry pyridine (1.0 mL). Further procedures and product isolation are as described for the synthesis of **3**. For the chromatographic purification, the solvent system from $CHCl_3$ to $CHCl_3:CH_3OH=93:7$, is used. The yields and ^{31}P-NMR characteristics of products **4** are presented in Table 3.

3.1.5. Synthesis of Protected Tetramers: [S$_P$S$_P$S$_P$]- and [R$_P$R$_P$R$_P$]-Isomers of 5'-O-Monomethoxytritylthymidylyl (3',5')Thymidylyl (3',5')Thymidylyl (3',5')-(3'-O-Acetylthymidylyl) Tri(methanephosphonate) (5, B = Thy)

A 2.15M solution of freshly prepared *t*-butylmagnesium chloride in dry THF (0.078 mL, 0.169 mmol) is added under dry nitrogen to a solution of 5'-deprotected [S$_P$S$_P$]- or [R$_P$R$_P$]-isomer of **4** (0.150 g, 0.169 mmol) in dry pyridine (0.5 mL). To the resulting suspension of 5'-hydroxyl-activated trinucleotide is added a solution of the [R$_P$]- or [S$_P$]-isomer of **2** (0.120 g, 0.169 mmol) in dry pyridine (0.75 mL). A further procedure is as described for **4**. For the chromatographic purification, the solvent system from $CHCl_3$ to $CHCl_3:CH_3OH=9:1$ is used. The yields, and TLC and ^{31}P-NMR characteristics of products **5**, are presented in Table 3.

3.1.6. Deprotection of OligoM 3, 4, and 5

3.1.6.1. REMOVAL OF 5'-O-MONOMETHOXYTRITYL GROUP

Individual isomers of **3**, **4**, or **5** (0.4 mmol) are treated with 80% aqueous acetic acid (10 mL) at room temperature. The reaction progress is monitored by TLC ($CHCl_3:CH_3OH=9:1$). After 1–3 h, AcOH is removed by evaporation under reduced pressure (oil pump). The oily residue is disolved in pyridine and dropped into hexane. The precipitate is filtered off, washed with *n*-pentane, and dried under reduced pressure. The yields of 5'-deprotected oligonucleotides are 95–100%. They are used for the following coupling reactions without further purification.

Note: (a) If the deprotection proceeds too slowly (TLC control), the reaction mixture can be heated at +60°C. (b) Usually twice-repeated precipitation yields tritanol-free product. If necessary, 5'-deprotected oligonucleotides can be purified by flash chromatography (Kieselgel 60, 230–400 mesh, Merck, Art. 9385) using $CHCl_3:CH_3OH= 9:1$, as a solvent system.

3.1.6.2. REMOVAL OF 3'-PROTECTING GROUP

Individual isomers of 5'-deprotected **3, 4,** or **5** , prepared as above (ca. 0.05 g) are dissolved in CH_3OH (0.75 mL) [or CH_3CN (0.75 mL) for B = Gua] and treated with conc. NH_4OH (0.5 mL, 25%, w/w) in Eppendorf tubes at room temperature. The reaction progress is monitored by TLC ($CHCl_3:CH_3OH=8:2$), and when completed (ca. 0.5 h), the reaction mixture is concentrated. The solid residue is redissolved in 85% $CH_3CN:H_2O$ (v/v) solution and purified by HPLC under the following condition: ODS Hypersil 5 µm, 4.6 mm × 30 cm, gradient from 8–40% CH_3CN in $0.1M$ triethylammonium bicarbonate (TEAB, pH 7.0), 1.5% CH_3CN/min. The retention times of deprotected diastereoisomers of di-, tri-, and tetra(thymidine methanephosphonate) **3'**, **4'**, and **5'** originating from **3**, **4**, and **5,** respectively, are given in Table 4.

The crude, 3',5'-deprotected oligonucleotides may be also purified by silica gel column chromatography using Kieselgel 60, 230–400 mesh and $CHCl_3:CH_3OH=8:2$ as eluting solvent. The 3'-*O-t*-butyldimethylsilyl group of 5'-deprotected dinucleotides, B = Ade,Cyt,Gua, is removed by treatment with tetra-*n*-butylammonium fluoride under standard conditions *(39)* and then purified by HPLC.

Table 4
HPLC Characteristics of Deprotected Diastereoisomers
of Di-, Tri-, and Tetra(Thymidine Methanephosphonate) (3',4', and 5')

Oligomer	Structure	R_t
[Sp]-3'	d ($T_{PMe}T$)	7.23
[Rp]-3'	d ($T_{PMe}T$)	6.93
[SpSp]-4'	d ($T_{PMe}T_{PMe}T$)	9.43
[RpRp]-4'	d ($T_{PMe}T_{PMe}T$)	9.02
[SpSpSp]-5'	d ($T_{PMe}T_{PMe}T_{PMe}T$)	10.50
[RpRpRp]-5'	d ($T_{PMe}T_{PMe}T_{PMe}T$)	10.26

^{a}ODS Hypersil 5 μm, 30 cm × 4.6 mm column, gradient from 8 to 40% CH_3CN in 0.1M TEAB (pH 7), 1.5% CH_3CN/min, flow rate 1.5 mL/min.

3.2. Stereocontrolled Synthesis of Oligo(Nucleoside Phosphorothioates) (OligoS)

3.2.1. Synthesis of 2-Chloro-1.3.2-Oxathiaphospholane (12)

Into the mixture of pyridine (79.1 g, 1.0 mol), and benzene (400 mL) 2-mercaptoethanol (39.1 g, 0.5 mol) and phosphorus trichloride (68.7 g, 0.5 mol) are added with stirring at room temperature. Stirring is continued for 0.5 h, pyridinium chloride is filtered off, and filtrate is condensed under reduced pressure. Raw product **12** is purified via distillation under reduced pressure, and the fraction boiling at 70–72°C/20 mm Hg is collected. ^{31}P-NMR: δ 205.0 ppm (benzene). Yield 72% [Lit. *(40)*, bp 84–85°C/12 mm Hg].

3.2.2. Synthesis of N,N-Diisopropylamino-1.3.2-Oxathiaphospholane (13)

Into the solution of **12** (28.5 g, 0.2 mol) in *n*-pentane (300 mL) diisopropylamine (40.5 g, 0.4 mol) is added dropwise with stirring at room temperature. After 0.5 h, diisopropylamine hydrochloride is removed by filtration, solvent is evaporated under reduced pressure, and product **13** is distilled. Fraction collected at 70°C/0.1 mm Hg is shown by means of ^{31}P-NMR to be a homogenous product. Yield: 70%. ^{31}P-NMR: δ 147.8 ppm (benzene). M.S.: m/z 207 (M+, E.I., 15 eV).

3.2.3. Synthesis of Monomers 5'-O-Dimethoxytrityl-(2'-O-Deoxyribonucleoside)-3'-O-(2-Thiono-1.3.2-Oxathiaphospholanes) (8)

The mixture of corresponding 5'-*O*-dimethoxytrityl-2'-*O*-deoxyribonucleoside (5'-*O*-DMT-Ade[Bz], -Gua[iBu], -Cyt[Bz] or -Thy)] (10 mmol), and 1H-tetrazole (0.77 g, 11 mmol) is dried under high-vacuum for

5 h, and then dissolved in dichloromethane (25 mL). Into this solution the compound **13** (2.28 g, 11 mmol) is added dropwise during 10 min and the resulting mixture is maintained at room temperature, with stirring, for 2 h. Elemental sulfur, previously dried on vacuum line for several hours (0.48 g, 15 mmol), is added in one portion, and stirred reaction mixture is left overnight. Unreacted sulfur is filtered off, and solvent is evaporated on rotatory evaporator. Residue is dissolved in chloroform (3 mL) and applied on silica gel 230–400 mesh column (30 × 6 cm, 170 g of silica). Elution is performed first with $CHCl_3$ (200 mL) and then with $CHCl_3:CH_3OH=97:3$, v/v. Isolation is controlled by HPTLC. Collected fractions containing corresponding product **8** are pooled and concentrated giving white foamy solid. According to ^{31}P-NMR assay compounds **8a-d** consist of the mixture of two diastereoisomers in the ratio approx 55:45.

Isolation of pure diastereoisomers of **8** is illustrated by an example of **8c**: product **8c** (1 g) is dissolved in ethyl acetate (4 mL) and introduced on the column (30 × 6 cm) with silica gel 60H (200 g, Merck, Art. No.7736). Diastereoisomers are eluted with ethyl acetate and fractions of 15 mL are collected. Elution of products is controlled by HPTLC (threefold development in ethyl acetate; detection: HCl spray). Fractions containing separated diastereoisomers (FAST-**8c**, fractions 61–73 and SLOW-**8c**, fractions 87–98), are concentrated to dryness under reduced pressure, and residue is characterized by means of ^{31}P-NMR and HPLC using Lichrospher Si100, 5 µm (30 cm × 7.8 mm), with ethyl acetate (flow rate 3 mL/min) as an eluent. Fast-**8c**: 250 mg (yield 25%), δ_{31P} 104.31 ppm (benzene), 100% diastereoisomeric purity (d.p.) according to ^{31}P-NMR, and HPTLC; Slow-**8c**: 180 mg (yield 18%) δ_{31P} 104.26 ppm (benzene), 100% d.p. Interfractions 74–86 are recycled for repeated isolations of diastereoisomeric forms of **8c**. Yields and physicochemical characteristics of separated diastereoisomers **8** are presented in Table 5.

3.2.4. Control of Stereospecificity of Reaction Between Diastereoisomerically Pure 8 and 5'-OH Nucleosides Under Conditions of Solid-Phase Automated Synthesis

An Applied Biosystems (Foster City, CA) model 380B automated DNA synthesizer was employed using the manufacturer's columns (1 µmol scale). Manufacturer's program used routinely for the syn-

Table 5
Physicochemical Characteristics of Separated Diastereoisomers of **8**

B		^{31}P-NMR^a	Rf^b	Yield, %^{c}
8a A^{Bz}	Fast	103.23	0.34	90
	Slow	103.18	0.31	
8b G^{iBu}	Fast	104.52	0.22	85
	Slow	104.17	0.20	
8c C^{Bz}	Fast	104.31	0.27	89
	Slow	104.16	0.22	
8d T	Fast	104.27	0.59	92
	Slow	104.23	0.57	

^{a}In C$_6$D$_6$, H$_3$PO$_4$ an ext. standard.
^{b}Ethyl acetate as a developing solvent system, HPTLC silicagel 60 F$_{254}$ plates (Merck).
^{c}Preparative yield of diastereoisomeric mixtures.

thesis of oligonucleotides via 2-cyanoethylphosphoramidite method has been modified according to protocol presented in Table 1. Respective reservoirs (Nos. 1–4) are filled up with solutions of pure diastereoisomers of **8a–d** in acetonitrile (0.1M), whereas 1H-tetrazole reservoir (No. 9) does contain 2M solution of DBU in pyridine. When synthesis is completed, acidic detritylation is performed followed by ammoniacal cleavage from solid support and base deprotection. Raw products **10** are analyzed and purified by reverse-phase HPLC using an ODS Hypersil (5 µm) column (30 cm × 4.6 mm) that is eluted with the linear gradient of acetonitrile: 5–20% CH$_3$CN/0.1M TEAB (triethylammonium bicarbonate); 0.75%/min, flow rate 1.5 mL/min. HPLC profiles of resulting dimers **10** are shown in Fig. 5, whereas stereospecificities, diastereoisomeric purities, and retention times, together with the yield assigned from measurement of UV absorptions, are given in Table 6. As is evident from Table 6, slow-eluting diastereoisomers of **8** always give diastereoisomers of **10** of [R$_P$]-configuration, whereas from fast-eluting precursors, corresponding dimers of [S$_P$]-configurations are formed.

P-Chiral Analogs

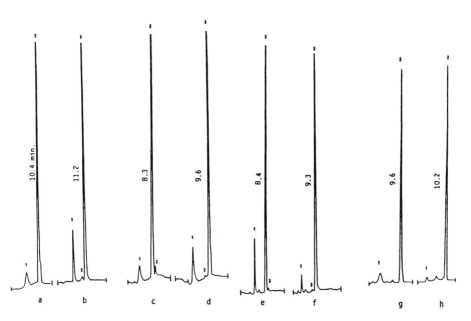

Fig. 5. HPLC profiles of dinucleoside (3',5')phosphorothioates (crude products): (a) d($A_{PS}A$): 1,—dA; 2. [R_P]-d($A_{PS}A$); (b) d($A_{PS}A$): 1,—dA; 2.-[R_P]-d($A_{PS}A$); 3. 3. [S_P]-d($A_{PS}A$); (c) d($G_{PS}G$): 1,—dG; 2. [R_P]-d($G_{PS}G$); 3. [S_P]-d($G_{PS}G$); (d) d($G_{PS}G$): 1,—dG; 2. [R_P]-d($G_{PS}G$); 3. [S_P]-d($G_{PS}G$); (e) d($C_{PS}C$): 1,—dC; 2. [R_P]-d($C_{PS}C$); 3. [S_P]-d($C_{PS}C$); (f) d($C_{PS}C$): 1,—dC; 2. [R_P]-d($C_{PS}C$); 3. [S_P]-d($C_{PS}C$); (g) d($T_{PS}T$): 1,—dT; 2. [R_P]-d($T_{PS}T$); (h) d($T_{PS}T$): 1,—dT; 2. [S_P]-d($T_{PS}T$); HPLC analysis was performed on ODS-Hypersil (5 µM) column with the linear gradient of acetonitrile: 5–20% CH_3CN/0.1M TEAB (triethylammonium-bicarbonate); 0.75%/min, 1.5 mL/min.

3.2.5. Synthesis of 5'-O-Dimethoxytritylnucleosides Bound to Solid Matrix via "Succinic-Sarcosinyl" Linker (9)

1. Long chain alkylamine controlled pore glass (LCA-CPG, Sigma, St. Louis, MO, cat. no. L-8638, 500A, 80–130 mesh, 2 g) and N-Fmoc-sarcosine (Bachem Bioscience Inc., Philadelphia, PA, prod. no. B-1720, 0.5 g, 1.6 mmol) are mixed together and dried under high vacuum for 3 h. Dry dimethylformamide (5 mL), pyridine (0.5 mL), and dicyclohexylcarbodiimide (DCC, 0.5 g, 2.4 mmol) are added, and the whole mixture, placed in a tightly closed vial (7.4 mL), is gently shaken for 12 h. Supension of the solid support is transferred into sintered glass

Table 6
Stereospecificities of the Formation
of Dinucleoside (3'-5') Phosphorothioates (**10**)
in Reaction Between Corresponding **8** and 5'-OH Nucleosides
Bound to Solid Support, with DBU as the Catalyst

Substrates			Product	
8		d.p.[a]	**10**[b]	d.p.,[a]
A^{Bz}	Fast	100%	(Sp)-d ($A_{PS}A$)	99.4%
	Slow	100%	(Rp)-d ($A_{PS}A$)	99.5%
C^{Bz}	Fast	100%	(Sp)-d ($C_{PS}C$)	99.3%
	Fast	95%[c]	(Sp)-d ($C_{PS}C$)	95.0%
	Slow	100%	(Rp)-d ($C_{PS}C$)	99.5%
	Slow	95%[c]	(Rp)-d ($C_{PS}C$)	95.0%
G^{iBu}	Fast	100%	(Sp)-d ($G_{PS}G$)	99.3%
	Slow	100%	(Rp)-d ($G_{PS}G$)	98.0%
T	Fast	100%	(Sp)-d ($T_{PS}T$)	99.0%
	Slow	100%	(Rp)-d ($T_{PS}T$)	100.0%

[a]d.p.—Diastereoisomeric purity determined by HPLC.

[b]All diastereoisomeric **10** were identified via HPLC by coinjections with genuine samples prepared according to ref. 2.

[c]Prepared as appropriate mixtures of formerly separated diastereo-isomers.

funnel, solvent removed by suction, and the support is washed three times with methanol:acetonitrile:pyridine (1:1:1, v/v, 3 × 20 mL). Residual solvents are removed under high vacuum, and N-Fmoc-sarcosinylated LCA-CPG is suspended in 10% solution of piperidine in pyridine (v/v, 10 mL). Removal of Fmoc N-protecting group is maintained for 0.5 h, and N-sarcosinylated LCA-CPG is filtered off and washed with methanol:acetonitrile:pyridine (1:1:1, v/v, 3 × 20 mL), and subsequently dried under high vacuum for 5 h.

2. The product obtained according to Step 1 (0.5 g), together with corresponding 3'-O-succinylated 5'-O-DMT-dA^{Bz}, 5'-O-DMT-dG^{iBu}, 5'-O-DMT-dC^{Bz}, or 5'-O-DMT-T, is dried under high vacuum for 2 h. Then DMF (2 mL), pyridine (0.2 mL), and DCC (50 mg) are added, and the resulting mixture is shaken at room temperature in a tightly closed vial for 12 h. Suspension of the solid support is transferred into a sintered glass funnel, and washed three times with methanol: acetonitrile:pyridine (1:1:1, v/v, 3 × 20 mL) and finally with acetonitrile (3 × 10 mL). After

drying with the flow of dry nitrogen, the support is treated with acylating reagent (*N*-methyl imidazole:THF, 1 mL, Applied Biosystems reagent cat. no. 400785, and acetic anhydride/lutidine:THF, 1 mL, Applied Biosystem reagent cat. no. 400607) for 15 min. After a thorough wash with methanol:acetonitrile:pyridine (1:1:1, v/v, 3 × 10 mL), and acetonitrile (3 × 10 mL), obtained solid support is dried under high vacuum. Loading of the support with the corresponding nucleoside unit (as determined by trityl assay), is as follows: **9**, B = dA^{Bz} —42.7 µmol/g; **9**, B = dG^{iBu}—46.7 µmol/g; **9**, B = dC^{Bz} —31.6 µmol/g; **9**, B = dT—35.0 µmol/g.

3.2.6. Stereocontrolled Synthesis of [R$_P$R$_P$R$_P$R$_P$]-Penta (2'-O-Deoxyribocytidine Phosphorothioate) (14)

The aforementioned synthesizer ABI 380B was employed using the cycle for coupling as used for synthesis of **10** with the modification that the homemade column containing **9** (B = C^{Bz}) was used. Solution of SLOW-**8c** (100 mg) in acetonitrile (1.2 mL) is used for synthesis of **14**. Compound [all-R$_P$]-**14** is isolated by two-step reversed-phase HPLC of tritylated and detritylated products. Analysis of both 5'-DMT-protected and 5'-OH-deprotected compounds by reversed-phase HPLC (ODS Hypersil (5 µm) column, 30 cm × 4.6 mm, flow rate 1.5 mL/min) gives rise to single peaks. For dimethoxytritylated compound: rt=20.40 min (5–30% CH$_3$CN/0.1M TEAB, t = 20 min, exponent 0.25). For detritylated [all-R$_P$]-**14**: rt=11.80 min, 5–20% CH$_3$CN/0.1M TEAB, 0.75%/min. Preparative yield: 14%.

3.2.7. Stereocontrolled Synthesis of [S$_P$S$_P$S$_P$S$_P$]-Penta (2'-O-Deoxyribocytidine Phosphorothioate) (14)

Diastereoisomer [S$_P$S$_P$S$_P$S$_P$]-**14** is prepared in an identical manner using the solution of FAST-**8c**: rt for 5-*O*-DMT-derivative 20.7 min; detritylated [all-S$_P$]-**14**, rt=12.00 min (conditions as above). Yield: 15% (preparative).

3.2.8. Diastereoisomeric Purity Control of [R$_P$R$_P$R$_P$R$_P$]-and [S$_P$S$_P$S$_P$S$_P$]-Penta (2'-O-Deoxyribocytidine Phosphorothioate) (14)

Diastereoisomeric purity of the title oligonucleotides has been checked by means of two nucleolytic enzymes, snake venom phosphodiesterase (SVPDE) and nuclease P1 of known diastereoselectivity toward dinucleoside phosphorothioates. Although SVPDE hydrolyzes only

R_P isomers of dinucleoside phosphorothioates *(41,42)*, nuclease P1 is known to catalyze the hydrolysis of their S_P isomers *(43)*. Compounds [all-R_P]-**14** and [all-S_P]-**14** are in separated experiments incubated with SVPDE and, independently, with nuclease P1.

1. 0.5 A_{260} U of [all-R_P]-**14** is dissolved in 250 µL of the buffer containing 0.1*M* Tris-Cl (pH 8.5) and 15 m*M* $MgCl_2$, and incubated with svPDE (Sigma, no. P6877, 100 µg) at 37°C for 24 h. A large amount of the phosphodiesterase is necessary because of low enzyme activity toward oligo(deoxyribonucleoside phosphorothioate)s. Then alkaline phosphatase (Sigma, no. P9761, EC.3.1.3., 1 µg) is added to the sample and digestion is continued for 1 h at 37°C. After heat denaturation (1 min, 100°C), digestion mixture is centrifuged (1 min, 13,000 rpm), and supernatant is analyzed by HPLC *(see* Fig. 6a).

Note: Please notice that despite large excess of the enzyme HPLC analysis of digestion mixture of [all-R_P]-**14** indicates the presence of [R_P]-d(C_{PS}C).

2. Oligonucleotide [all-S_P]-**14** incubated with SVPDE under analogous conditions appears to be completely resistant toward this enzyme (Fig. 6b).

3. The digestion of [all-R_P]-**14** and [all-S_P]-**14** by nuclease P1 is performed as follows: 0.5 A_{260} U of each oligonucleotide is dissolved in 250 µL of the buffer containing 0.1*M* Tris-Cl (pH 7.2), and 1 m*M* $ZnCl_2$, and incubated with nuclease P1 (Sigma, no. 8630, 10 µg) for 24 h, and then with alkaline phosphatase (1 µg) for 1 h at 37°C. After heat denaturation and centrifugation (conditions as above) supernatant is analyzed by HPLC. Oligonucleotide [all-R_P]-**14** appears to be resistant to hydrolytic action of nuclease P1 (Fig. 6c), whereas [all-S_P]-**14**, under above conditions, is completely digested to dC (Fig. 6d).

4. Discussion

The discovery of antiviral activity of oligo(nucleoside methanephosphonate)s *(44)* and oligo(nucleotide phosphorothioate)s *(45–47)*, so far chemically prepared by the nonstereocontrolled methods, attracted the attention of several research establishments to the search for stereospecific synthesis of those classes of oligonucleotide analogs. Since their routine synthesis via phosphoramidite or any other approach leads to the mixture of *m* diastereoisomers, the question may be asked whether desired antiviral activity is owing to all *m* components of diastereoisomeric mixture or to the fraction possessing the proper sense of chirality at each modified phosphate. Moreover, since the seminal works of

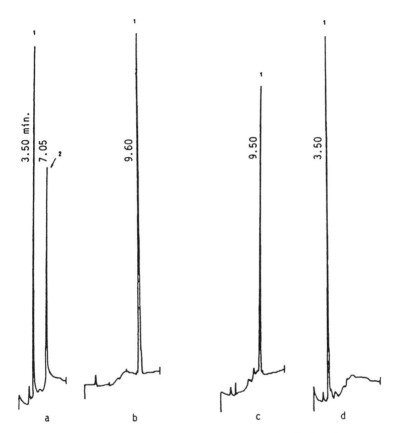

Fig. 6. HPLC analysis of the digestion mixtures obtained after the treatment of [R$_P$]-6 and [S$_P$]-6 with snake venom phosphodiesterase (SVPDE), nuclease P1 (nP1), and alkaline phosphatase (AP) (a) [R$_P$]-d([C$_{PS}$]$_4$C) treated with SVPDE/AP: 1.—dC; 2.-[R$_P$]-d(C$_{PS}$C); (b) [S$_P$]-d([C$_{PS}$]$_4$C) treated with SVPDE/AP: 1.—undigested substrate; (c) [R$_P$]-d([C$_{PS}$]$_4$C) treated with nP1: 1.—undigested substrate; (d) [S$_P$]-d([C$_{PS}$]$_4$C) treated with nP1/AP: 1— dC.

Eckstein *(41)* and Benkovic *(42)*, it is known that nucleolytic activity of several nucleases is diastereoselective toward internucleotide phosphorothioates, and the rate of degradation of oligo(nucleoside phosphorothioate) may depend on the sense of chirality at phosphorus within phosphorothioate moieties. Thus, stereocontrolled synthesis of oligoS with a predetermined sense of chirality at phosphorus asymmetric centers can be considered as a vital strategy to the delivery of the corresponding oligonucleotide-congener resistant to nuclease-assisted degradation to the biological target. Moreover, legislative procedure, requiring "as exact as possible" definition of material submitted for clinical studies, together with hopes that a given stereodefined oligonucleotide analog may possess better therapeutic effects than the mixture of diastereoisomers, are key factors accelerating the search for new stereocontrolled methods of synthesis of P-chiral oligonucleotide congeners. In addition, the stereocontrolled method of any new molecule containing several chiral centers constitutes a challenging problem to organic chemists.

As presented above, the first stereocontrolled methods of syntheses of oligoM and oligoS kindled some hope that progress could be achieved if more efforts would be given to solving this general problem. Although further studies on development of this methodology are in progress, one has to realize that as high as 99% stereospecificity of a single step of synthesis does not yield a diastereoisomerically pure oligonucleotide. The chemical synthesis of 27-mer oligo(nucleoside phosphorothioates) leads to the material of at best 75% d.p. The oligo(nucleoside phosphorothioate) prepared via a polymerase-based process utilizing dNTPαS substrates is believed to be a 100% [all-R_P] diastereoisomerically pure *(48,49)*. The pivotal limitation of enzymatic synthesis is caused, however, by the fact that oligo(nucleoside phosphorothioate)s of only [all-R_P] configurations are achievable. Thus the polymerase of reverse stereoselectivity has to be found or, perhaps (?), genetically engineered. Preliminary results on enzymatic synthesis of [all-S_P] oligo(nucleoside methanephosphonate)s have been announced. d(NTPαCH_3) was used as a substrate of terminal deoxynucleotidyltransferase in the presence of appropriate oligonucleotide as an initiator *(50)*.

In addition to the importance of oligonucleotide analogs containing modified all internucleotide functions within a given oligonucleotide chain, in the introduction to this chapter the synthesis of oligonucleotides bearing modification of a single phosphate at a preselected position has been briefly mentioned. Such oligonucleotides, as demonstrated in a number of works, constitute indispensable tools for elucidation of the phenomenon of interactions between nucleic acids themselves or interactions between nucleic acids and other groups of biomolecules, especially proteins *(3,5,12,18,51–53)*.

P-Chiral Analogs 309

Because of space limitations, insufficient data concerning the problem of assignment of absolute configuration at modified chiral phosphorus centers are given here, but pertinent references will help the dedicated researcher to find out established methodologies, or develop his or her own new approaches.

References

1. Zon, G. (1990) Purification of synthetic oligodeoxyribonucleotides, in *High-Performance Liquid Chromatography in Biotechnology* (Hancock, W. S., ed.) John Wiley, New York, pp. 301–349.
2. Stec, W. J., Zon, G., Egan, W., and Stec, B. (1984) Automated solid-phase synthesis, separation, and stereochemistry of phosphorothioate analogues of oligodeoxyribonucleotides. *J. Am. Chem. Soc.* **106,** 6077–6079.
3. Connolly, B. A., Potter, B. V. L., Eckstein, F., Pingoud, A., and Grotjahn, L. (1984) Synthesis and characterization of an octanucleotide containing the EcoRI recognition sequence with a phosphorothioate group at the cleavage site. *Biochemistry* **23,** 3443–3453.
4. Stec, W. J. and Zon, G. (1984) Synthesis, separation, and stereochemistry of diastereomeric oligodeoxyribonucleotide having a 5'-terminal internucleotide phosphorothioate linkage. *Tetrahedron Lett.* **25,** 5275–5278.
5. Slim, G. and Gait, J. M. (1991) Configurationally defined phosphorothioate containing oligoribonucleotide in the study of the mechanism of cleavage of hammerhead ribozymes. *Nucl. Acids Res.* **19,** 1183–1188.
6. LaPlanche, A. L., James, T. L., Powell, C., Wilson, W. D., Uznanski, B., Stec, W. J., Summers, M. F., and Zon, G. (1986) Phosphorothioate-modified oligodeoxyribonucleotides. III. NMR and UV spectroscopic studies of the R_PR_P-S_PS_P and R_P-S_P duplexes [d(GG$_S$AATTCC)]$_2$, derived from diastereomeric O-ethyl phosphorothioate. *Nucl. Acids Res.* **14,** 9081–9093.
7. Lawrence, D. P., Wenqiao, Ch., Zon, G., Stec, W. J., Uznanski, B., and Broido, M. S. (1987) NMR studies of backbone-alkylated DNA: duplex stability, absolute stereochemistry, and chemical shift anomalies of prototypal isopropyl phosphotriester modified octanucleotides, [R_PR_P]- and [S_PS_P]-{d[GGA(iPr)ATTCC]}$_2$ and -d-[GGAA (iPr)TTCC]}$_2$. *J. Biomol. Structure and Dynamics* **4,** 757–783.
8. Stec, W. J., Zon, G. and Uznanski,B. (1985) Reversed-phase high-performance liquid chromatographic separation of diastereomeric phosphorothioate analogues of oligo-deoxyribonucleotide and other backbone-modified congeners of DNA. *J. Chromatogr.* **326,** 263–280.
9. Stec, W. J., Zon, G., Egan, W., Byrd, R. A., Phillips L. R., and Gallo, K. A. (1985) Solid-phase synthesis, separation, and stereochemical aspects of P-chiral methane- and 4,4'-dimethoxytriphenylmethanephosphonate analogues of oligodeoxyribonucleotides. *J. Org. Chem.* **50,** 3908–3913.

10. Shibanova, E. V., Filippov, S. A., Esipov, D. S., Korobko, V. G., and Dobrynin, V. N. (1991) Enzymatic ligation of synthetic DNAs carrying point amidophosphate internucleotide linkage modification. *Bioorg. Khim.* **17,** 99–106.
11. Cosstick, R. and Eckstein, F. (1985) Synthesis of d(GC), and d(CG) octamers containing alternating phosphorothioate linkages: effect of the phosphorothioate group on the B-Z transition. *Biochemistry* **24,** 3630–3638.
12. Connolly, B. A., Eckstein, F., and Pingoud, A. (1984) The stereochemical course of the restriction endonuclease EcoRI-catalyzed reaction. *J. Biol. Chem.* **259,** 10,760–10,763.
13. Ozaki, H., Kitamura, M., Yamana, K., Murakami A., and Shimidzu, T. (1990) Syntheses and properties of oligothymidylate analogues containing stereoregulated phosphoromorpholidate and phosphodiester linkages in an alternating manner. *Bull. Chem. Soc. Jpn.* **63,** 1929–1936.
14. Miller, P. S., Agris, C. H., Murakami, A., Reddy, P. M., Spitz, S. A., and Ts'o, P. O. P. (1983) Preparation of oligodeoxyribonucleoside methylphosphonates on a polystyrene support. *Nucl. Acids Res.* **11,** 6225–6241.
15. Amirkhanov, N. V. and Zarytova, V. F. (1989) Reactive oligonucleotide derivatives bearing methylphosphonate groups. II. Synthesis of stereoregular octathymidylates containing an alkylating 4-(N-2-chloroethyl-N-methylamino) benzylamine group and alternating methylphosphonate residues. *Bioorg. Khim.* **15,** 259–276.
16. Miller, P. S., Dreon, N., Pulford, M. S., and McParland, B. K. (1980) Oligothymidylate analogues having stereoregular, alternating methylphosphonate/phosphodiester backbones. *J. Biol. Chem.* **225,** 9659–9665.
17. Miller, P. S., Annan, N. D., McParland, K. B., and Pulford, S. M. (1982) Oligothymidylate analogues having stereoregular, alternating methylphosphonate/phosphodiester backbone as primers for DNA polymerase. *Biochemistry* **21,** 2507–2512.
18. Noble, S. A., Fisher, E. F., and Caruthers, M. H. (1984) Methylphosphonates as probe of protein-nucleic acid interactions. *Nucl. Acids Res.* **12,** 3387–3404.
19. Abramova, T. V., Vorob'ev, Yu. N., and Lebedev, A. V. (1986) Investigations of diastereomers of nonionic oligonucleotide analogues. IV. Influence of configuration at asymmetric phosphorus atoms in ethyl esters of deoxyoligothymidylate on the stability of their complexes with poly(dA). *Bioorg. Khim.* **12,** 1335–1347.
20. Hau, J. F., Asseline, U., and Thuong, N. T. (1991) Octathymidylates involving alternating neopentylphosphothionotriester-phosphodiester linkages with controlled stereochemistry at the modified P-center. *Tetrahedron Lett.* **32,** 2497–2498.
21. Loschner, T. and Engels, J. (1989) One pot R_P-diastereoselective synthesis of dinucleoside methylphosphonates using methyldichlorophosphine. *Tetrahedron Lett.* **30,** 5587–5590.
22. Fuji, M., Ozaki, K., Kume, A., Sekine, M., and Hata, T. (1986) Acylphosphonates. 5. A new method for stereospecific generation of phosphorothioate via aroylphosphonate intermediate. *Tetrahedron Lett.* **26,** 935–938.

23. Fuji, M., Ozaki, K., Sekine, M., and Hata, T. (1987) Acylphosphonates. 7. A new method for stereospecific and stereoselective generation of dideoxyribonucleoside phosphorothioates via the acylphosphonate intermediate. *Tetrahedron* **43**, 3395–3407.
24. Cosstick, R. and Williams, D. M. (1987) An approach to the stereoselective synthesis of Sp-dinucleoside phosphorothioates using phosphotriester chemistry. *Nucl. Acids Res.* **15**, 9921–9943.
25. Ohtsuka, E., Tozuka, Z., and Ikehara, M. (1981) New condensing reagent for stereospecific synthesis of dinucleoside monophosphate aryl esters. *Tetrahedron Lett.* **22**, 4483–4486.
26. Lesnikowski, Z. J., Jaworska, M., and Stec, W. J. (1988) Stereoselective synthesis of P-homochiral oligo(thymidine methanephosphonates). *Nucl. Acids Res.* **16**, 11,675–11,689.
27. Lesnikowski, Z. J., Wolkanin, P. J., and Stec, W. J. (1987) Determination of absolute configuration at phosphorus in diastereoisomers of thymidylyl (3',5')thymidylyl methanephosphonate, in *Biophosphates and Their Analogues Synthesis, Structure, Metabolism and Activity* (Bruzik, K. S. and Stec, W. J., eds.) Elsevier, Amsterdam, pp. 189–194.
28. Hayakawa, Y., Uchiyama, N., and Noyori, R. (1984) A convenient method for the formation of internucleotide linkage. *Tetrahedron Lett.* **25**, 4003–4006.
29. Lesnikowski, Z. J., Jaworska-Maslanka, M. M., and Stec, W. J. (1991) Stereospecific synthesis of P-chiral di(2'-O-deoxyribonucleoside) methanephosphonates. *Nucleosides and Nucleotides* **10**, 733–736.
30. Lesnikowski, Z. J., Jaworska, M., and Stec, W. J. (1990) Octa(thymidine methanephosphonates) of partially defined stereochemistry: synthesis and effect of chirality at phosphorus on binding to pentadecadeoxyriboadenylic acid. *Nucl. Acids Res.* **18**, 2109–2115.
31. Lesnikowski, Z. J. and Jaworska, M. M. (1989) Studies of stereospecific formation of P-chiral internucleotide linkage. Synthesis of [R_PR_P]- and [S_PS_P]-thymidylyl (3',5')thymidylyl (3',5')thymidine di(O,O-phosphorothioate) using 2-nitrobenzyl group as a new S-protection. *Tetrahedron Lett.* **30**, 3821–3824.
32. Lesnikowski, Z. J. (1992) The Stereocontrolled synthesis of thiooligoribonucleotide: [R_PR_P]- and [S_PS_P]-$U_{PS}U_{PS}U$. *Nucleosides and Nucleotides*, **11**, 1621-1638.
33. Stec, W. J., Grajkowski, A., Koziolkiewicz, M., and Uznanski, B. (1990) Novel route to oligo(deoxyribonucleoside phosphorothioates). Stereocontrolled synthesis of P-chiral oligo(deoxyribonucleoside phosphorothioate). *Nucl. Acids Res.* **19**, 5883–5888.
34. Brown, T., Pritchard, C. E., Turner, G., and Salisbury, S. A. (1989) A new base-stable linker for solid-phase oligonucleotide synthesis. *J. Chem. Soc. Chem. Commun.* **89**, 891–893.
35. Zon, G. and Stec, W. J. (1991) Phosphorothioate oligonucleotides, in *Oligonucleotides and Their Analogues: A Practical Approach* (Eckstein, F., ed.) IRL Press, Oxford, pp. 87–108.

36. Okruszek, A., Guga, P., and Stec, W. J. (1991) Stereochemistry of the reaction of ribonucleoside cyclic 3',5'-phosphorothioate with oxiranes. *Heteroatom Chem.* **2,** 561–568.
37. Gait, M. J. (ed.) (1984) *Oligonucleotide Synthesis: A Practical Approach*, IRL Press, Oxford.
38. De Ross, A. M. (1959) The preparation of some alkyl *p*-nitrophenyl methylphosphonates. *Rec. Trav. Chim. Pay-Bas.* **78,** 145–149.
39. Ogilvie, K. K., Beaucage, S. L., Entwistle, D. W., Thompson, E. A., Quilliam, M. A., and Westmore, J. B. (1976) Alkylsilyl groups in nucleoside and nucleotide chemistry. *Nucleosides and Nucleotides* **3,** 197–227.
40. Martynov, N. W., Kruglyak, Yu. L., Leybovskaya, G. A., Khromova, Z. J., and Strukov, O. G. (1969) Phosphorylation of oximes. IV. Reaction of 1.3.2-dioxa- and 1.3.2-oxathiaphospholanes with α-haloidonitrosoalkanes. *Zh. Obshch. Khim.* **39,** 996–998.
41. Burgers, P. M. J., Eckstein, F., and Hunneman, D. H. (1979) Stereochemistry of hydrolysis by snake venom phosphodiesterase. *J. Biol. Chem.* **254,** 7476–7478.
42. Bryant, F. R. and Benkovic, S. J. (1979) Stereochemical course of the reaction catalyzed by nucleotide phosphodiesterase from snake venom. *Biochemistry* **18,** 2825–2828.
43. Potter, B. V. L., Connolly, B. A., and Eckstein, F. (1983) Synthesis and configurational analysis of a dinucleoside phosphate isotopically chiral at phosphorus. Stereochemical course of *penicillium citrum* nuclease P1 reaction. *Biochemistry* **22,** 1369–1377.
44. Miller, P. S. (1991) Oligonucleoside methylphosphonates as antisense reagents. *Biotechnology* **9,** 358–362, and references therein.
45. DeClercq, E., Eckstein, F., and Merigan, T. C. (1969) Interferon induction increased through chemical modification of a synthetic polyribonucleotide. *Science* **165,** 1137–1139.
46. Matsukura, M., Zon, G., Shinozuka, K., Stein, C. A., Mitsuya, M., Cohen, J. S., and Broder, S. (1988) Synthesis of phosphorothioate analogues of oligodeoxyribonucleotides and their antiviral activity against human immunodeficiency virus (HIV). *Gene* **72,** 343–347.
47. Stein, C. A. and Cohen, J. S. (1989) Phosphorothioate oligonucleotide analogues, in *Oligonucleotides: Antisense Inhibitors of Gene Expression* (Cohen, J. S., ed.) Macmillan, vol. 12 of *Topics in Molecular and Structural Biology*, pp. 98–117.
48. Eckstein, F. (1985) Nucleoside phosphorothioates. *Ann. Rev. Biochem.* **54,** 367–402.
49. Latimer, L. J. P., Hampel, K., and Lee, J. S. (1989) Synthetic repeating sequence DNAs containing phosphorothioates: nuclease sensitivity and triplex formation. *Nucl. Acids Res.* **17,** 1549–1561.
50. Higuchi, H., Endo, T., and Kaji, A. (1990) Enzymic synthesis of oligonucleotides containing methylphosphonate internucleotide linkages. *Biochemistry* **29,** 8747–8753.

51. Koziolkiewicz, M. and Stec, W. J. (1992) Application of phosphate-backbone-modified oligonucleotides in the studies on Eco RI endonuclease mechanism of action. *Biochemistry* **31,** 9460-9466.
52. von Tol, H., Buzayan, J. M., Feldstein, P. A., Eckstein, F., and Bruening, G. (1990) Two autolytic processing reactions of a satellite RNA proceed with inversion of configuration. *Nucl. Acids Res.* **18,** 1971–1975.
53. Lesser, D.R., Grajkowski, A., Kurpiewski, M.R., Koziolkiewicz, M., Stec, W. J. and Jen-Jacobson, L. (1992) Stereoselective interaction with chiral phosphorothioates at the central DNA kink of the Eco RI endonuclease-GAATTC complex. *J. Biol. Chem.* in press.

CHAPTER 15

Oligonucleotide Analogs with Dimethylenesulfide, -sulfoxide, and -sulfone Groups Replacing Phosphodiester Linkages

Zhen Huang, K. Christian Schneider, and Steven A. Benner

1. Introduction

As demonstrated by this volume, analogs of oligonucleotides retaining the molecular recognition properties of natural oligonucleotides, but having altered physical, chemical, and biological properties are synthetic targets of some interest. Much of the interest comes from the possibility that such analogs might allow the sequence-specific control of the expression of encoded genetic information in vivo using what has come to be known as the "antisense" strategy *(1–5)*.

Using encoded information as the target for biologically active molecules has many advantages, at least in principle, over using expressed information (i.e., proteins) as the target. Encoded information is present in fewer copies than expressed information, implying that fewer molecules are necessary to achieve a desired biological effect if they interact with encoded information than if they interact with expressed information. Further, molecules that target encoded information can be designed from readily available sequence data; in contrast, three-dimensional structural information is invariably needed before molecules targeting expressed information can be designed.

*This contribution is dedicated to Frank H. Westheimer, on the occasion of his 80th birthday.

From: *Methods in Molecular Biology, Vol. 20: Protocols for Oligonucleotides and Analogs*
Edited by: S. Agrawal Copyright ©1993 Humana Press Inc., Totowa, NJ

However, the most attractive aspect of targeting encoded information via an antisense strategy is that very little sophistication is needed to design a molecule that can bind to an oligonucleotide; the Watson-Crick rules of base pairing are all that one must know, and these are learned at a very early stage in modern biochemical education. This simple design principle contrasts sharply with the virtual absence of design principles to guide the biological chemist as he or she seeks molecules that bind to a protein.

The idea of using complementary oligonucleotides to block the biological action of natural oligonucleotides is, at one level, obvious. It was, we believe, first proposed by Alexander Rich in 1962 in a remarkable paper that also outlines a possible role for RNA-based catalysis in the first forms of life *(6)*. What is not obvious is how such an idea can be implemented in a way that makes it useful. Despite well over a decade of work in many laboratories, starting with pioneering studies on analogs of nucleic acids in the 1970s by Miller, T'so, and their colleagues *(7–11)*, continuing with work of Zamecnik and his coworkers with natural nucleic acids *(5)*, and more recently with work in many other laboratories on many other structures, the goal of a biologically useful antisense oligonucleotide analog targeting encoded information remains disappointingly and frustratingly elusive. As a result, the excitement of using antisense oligonucleotides or their analogs as human pharmaceuticals has faded somewhat in light of experimental difficulties that have been encountered, and as certain shortcomings of the antisense strategy have become more widely appreciated.

As usually envisioned, the target mRNA for an antisense molecule is inside a living cell and is therefore shielded from extracellular biological fluids by a membrane. It lies in uncertain physical surroundings and interacts with unknown cellular proteins. The antisense drug must act in this environment. Further, the antisense drug, to be specific for a particular target mRNA sequence, must be long, approximately 15 bases in humans. These facts suggest that antisense oligonucleotides will have difficulties reaching their target mRNA molecules. Indeed, "bioavailability" problems, which render useless many much simpler pharmaceutical molecules, seem especially severe for antisense oligonucleotides.

Second, essentially no precedents exist to predict the extent or nature of side effects that antisense drugs might create. Unlike with expressed information, small changes in the structure of encoding molecules can correspond to radical changes in physiological function. Thus, one might expect side effects of antisense drugs to be capricious. The drug targeted against the AIDS virus might, for example, also cause one to forget one's pet cat, if the sequence of the (hypothetical) nucleic acids that serves as a feline memory molecule differs by a single base from that of the target sequence in HIV.

Finally, considerations of cost and dose, including the possibility of degradation and secretion of the antisense drug, make antisense compounds less than appealing as human pharmaceuticals. Simple calculations suggest that antisense drugs, if they act solely by binding to mRNA, are readily secreted, and applied to virtually any major health problem, will bankrupt the health care budgets of most industrial nations. Although attaching groups to the antisense oligonucleotide analog that catalyze the hydrolysis of the bound sense RNA (an "easy" reaction from a chemical point of view) *(12)* to yield a "catalytic antisense" molecule offers the prospect of making the effective dose lower, this too requires new chemistry.

On the brighter side, one must not overlook the evidence, now rather compelling *(13)*, that nucleic acids play roles outside of cells as regulators of cellular differentiation, accessible to blood-borne oligonucleotides or their analogs. Tumor angiogenesis factor and other extracellular angiogenic factors containing RNA components are potential extracellular targets for complementary (the title "antisense" not being fully appropriate here) oligonucleotide analogs *(13)*. Thus, oligonucleotide analogs targeted against natural oligonucleotides may find therapeutic value even should they not be able to penetrate cellular membranes.

Further, although the most visible potential uses of antisense oligonucleotide analogs is in the treatment of human diseases, it is helpful to point out that antisense compounds should have their greatest impact not as human pharmaceuticals, but rather as research tools for understanding the physiological roles of specific genes in higher organisms. Total genome sequencing is generating vast numbers of open reading frames encoding proteins with unknown function. Recently developed sophisticated computer programs permit the organization of these sequence data *(14)*. However, it remains difficult to determine the function of the proteins encoded by these open reading frames. In higher organisms, antisense oligonucleotide analogs offer a systematic, if plodding, research tool for sorting out the function of genes whose sequences (and little else) are known, a tool that can be applied without having to solve any pharmacokinetic problems in living humans. Thus, we expect that the value of suitable antisense compounds as tools in the hands of developmental biologists should far outweigh, by any measure of scientific significance, their value as human pharmaceuticals.

In selecting molecules as synthetic goals, all chemical research involves compromise. On one hand, the molecular designer can attempt to make a molecule that, from first principles, has precisely the properties desired for a particular application. On the other hand, the molecule can be chosen because it is (or appears to be) simple to synthesize. Since chemical theory is imperfect, one rarely knows what paper structure will have precisely the properties desired for a particular application. Therefore, the chemist always has a strong incentive to follow the second route and direct his or her effort toward the molecules that appear simpler to prepare. In the antisense field, synthetic considerations are extremely relevant, since expensive monomers must be transformed into still more expensive oligomers before a molecule with interesting biological activity is in hand.

Simplicity in synthesis appears to have been the motivation for the selection of many of the oligonucleotide analogs that have been described in the literature. Oligomethylphosphonates and phosphate triesters *(15)*, to name just two of the many molecules discussed in this volume (*see* Chapters 7 and 14), are attractive if for no other reason than that they can be prepared from commercially available natural deoxynucleosides as building blocks. Although deoxynucleosides are by no means inexpensive, they appear to be less expensive than deoxynucleoside analogs that must be synthesized from scratch, at least on first inspection.

A decade ago, we decided against this strategy, setting out instead to follow the first strategy: to decide exactly what physical, chemical, and biological properties we wanted in an antisense molecule, exactly what molecules would deliver them, and to pursue these molecules

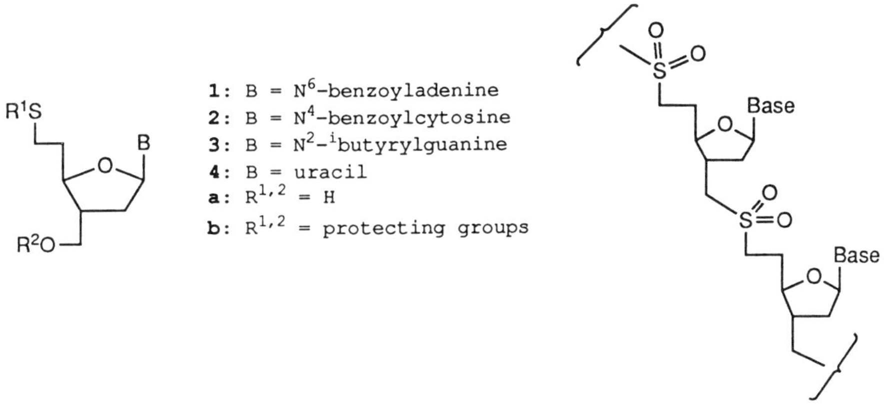

Oligonucleotide Analogs

Fig. 1.

1: B = N^6-benzoyladenine
2: B = N^4-benzoylcytosine
3: B = N^2-^{i}butyrylguanine
4: B = uracil
a: $R^{1,2}$ = H
b: $R^{1,2}$ = protecting groups

regardless of the synthetic difficulties or expense. Three issues were central to the design of the "ideal" antisense oligonucleotide analog. First, bioavailability was a key obstacle to be overcome. We had shown that sulfones (such as dimethylsulfone and sulfolane) assisted the penetration of natural oligonucleotides through cell membranes (16). The mechanism for this effect remains uncertain. However, chemical considerations relevant to the "dipolar aprotic" nature of the sulfone moiety, especially when compared with other sulfur-containing analogs (e.g., sulfonamides, sulfates, and so on) suggested that incorporating sulfone groups directly into the oligonucleotide analogs themselves was likely to be a viable strategy for obtaining membrane permeability in an oligonucleotide analog. This suggested a specific target structure for the "ideal" antisense oligonucleotide analog (Fig. 1).

Doubly attractive was the structure similarity of phosphates and sulfones. As isoelectronic species, replacing phosphates by sulfones should have minimal impact on the conformation of the nucleoside bases, implying a good "match" between an oligonucleotide analog incorporating sulfone-linking groups and a natural oligonucleotide held together by phosphate diester groups.

Further, diastereomeric purity was a prime concern. Many conceivable linking groups (e.g., phosphate triesters) introduce a new stereomeric center into an oligonucleotide analog, a center whose configuration is not easily (or cheaply) controlled. Introducing one uncontrolled stereomeric center per building block implies that oligomers with a length likely to have the desired specificity in a biological system would be complex mixtures of diastereomers, hopelessly obscuring the chemical foundation for whatever physical and biological effects might be observed. Sulfones solve this problem, since the sulfur of a sulfone group is not stereogenic.

Finally, we were concerned about the stability of our analogs to both enzymatic and, more significantly, nonenzymatic hydrolysis. Several years spent in the laboratory of F. H. Westheimer had delivered the message that neutral derivatives of phosphate esters were many orders of magnitude more sensitive to nucleophilic attack than the phosphate diester monoanions found in natural nucleic acids. Here again, dimethylene sulfones showed a distinct advantage. Their stability in virtually any aqueous environment (and in many nonaqueous environments) implied that a wide range of conditions for coupling and deprotection could be used in the synthesis of sulfone-linked oligonucleotide analogs without jeopardizing the linkage itself.

Work in our laboratories since 1981 has provided us with much opportunity to regret our decision to pursue, regardless of the synthetic obstacles, dimethylene sulfone analogs of DNA (Fig. 1) as the ideal antisense compounds. The central disadvantage of sulfone analogs is that the building blocks cannot be purchased, but rather must be assembled by total synthesis. Further, the synthesis must yield diastereomerically and enantiomerically pure compounds in substantial quantities. Thus, it was not until the late 1980s that the first building blocks for the synthesis of oligonucleotide analogs having dimethylsulfone-linking units became available in our laboratories. Shortly thereafter, the first short oligomers of these building blocks were prepared. These oligomers appear to bind to complementary oligodeoxyribonucleotides more tightly than natural oligonucleotides (Z. Huang, unpublished).

This chapter is intended to provide recipes for preparing building blocks for the synthesis of oligonucleotide analogs having dimethylsulfone-linking U, in particular, for analogs **1–4** (Fig. 1), analogs bearing functionalization appropriate for application as building blocks in the synthesis of oligonucleotide analogs having the phosphodiester groups replaced by sulfide, sulfoxide, or sulfone U. The experimental section describes two routes to these building blocks, one beginning with a Diels-Alder reaction between butadiene and an acetoxyacrylate derivative, and the other with diacetone glucose. The first route uses an enzymatic reaction to generate an optically active intermediate; the second starts with an optically active educt. The first route yields a deoxyribo analog as a building block, but as a mixture of anomeric forms, which must be separated. The second yields only the β-anomer, but of a ribo analog; the extra oxygen must be removed to create the deoxyribo analog.

The building blocks are obtained in their protected forms (**1b–4b**), since these derivatives are more stable than the corresponding unprotected ones. Conversion of the protected to unprotected forms immediately prior to coupling is achieved by standard procedures *(17)*. We illustrate the coupling procedure using a recipe for the synthesis of a dinucleotide analog; octamer U analog has been prepared by repeating this procedure (Z. Huang and T. Arslan, unpublished). Much of this work has been published previously, although not in precisely the same format and detail presented here *(18,19)*. The syntheses are outlined in Schemes 1–6.

2. Materials

^{13}C-NMR spectra were recorded on a Varian XL 300 spectrometer using attached proton test (APT) and distortionless enhancement by polarization transfer (DEPT) techniques for determination of carbon substitution; all δ values are in ppm relative to tetramethylsilane. Preparative HPLC was performed on a Knauer HPLC-column (30 × 250 mm, silica gel-Nucleosil 7 μ, 10 mL/min flow, 50–55 atm pressure). Reactions were monitored by TLC on Merck 60 F254 precoated plates, and spots were visualized with UV light or by staining with a Ce-Mo-staining reagent; for column chromatography, Fluka silica gel 60, mesh size 0.040–0.063, was used. All solvents were Fluka p.a. and were used without purification, unless mentioned otherwise. THF and diethyl ether were distilled over sodium-benzophenone; acetonitrile, dichloroethane, and pyridine were distilled over CaH_2. Reactions with air- or moisture-sensitive compounds were performed under argon atmosphere. The phrase "dried and evaporated" indicates drying with magnesium sulfate, followed by evaporation of the solvents with a Büchi rotary evaporator under house vacuum.

3. Methods
3.1. Route 1
3.1.1. (±) Trans-6-Hydroxy-3-Cyclohexene-1-Carboxylic Acid Ethyl Ester (6)

To a solution of crude 2-acetoxy-carboethoxycyclohex-4-ene (20) (5) (87.7 g, corresponding to 0.413 mol of pure material) in ethanol (1L) was added NaH (5.9 g, 55% in oil) in small portions at room temperature. The clear brown reaction mixture was stirred for 30 min at room temperature and was then neutralized with acetic acid. Most of the solvent was evaporated under reduced pressure, and the residue partitioned between water (200 mL) and CH_2Cl_2 (200 mL). The aqueous layer was extracted with CH_2Cl_2 (2 × 100 mL), and the organic phase dried and evaporated. Distillation at 95–100°C (0.5 Torr) yielded 56.0 g of an oil containing **6** (86%, determined by gas chromatography/mass spectroscopy [GC/MS] analysis) (0.283 mol, 68% yield). The material could be used for the next step without further purification. The yield of the same reaction carried out on a 1-g scale with pure starting material was 90%. ^{1}H NMR (CDCl$_3$): δ 1.28 (t, J = 7 Hz, 3 H, CH_3CH_2), 2.14–2.46 (m, 4 H, 5-H, 2-H), 2.57 (m, 1 H, 1-H), 3.03 (s, br, 1 H, OH), 4.08 (m, 1 H, 6-H), 4.19 (q, J = 7 Hz, 2 H, CH_3CH_2), 5.61 (m, 2 H, C=C–H); IR (CCl$_4$): 2582, 2933, 1732, 1182, 1078 cm^{-1}; MS, m/e 170, 152, 125, 95, 88, 79, 67. *(See* Note 1.)

3.1.2. cis-6-Benzoyloxy-3-Cyclohexene-1-Carboxylic Acid Ethyl Ester ([±]7) (21)

To a solution of PPh$_3$ (153 g, 0.566 mol), benzoic acid (72.4 g, 0.593 mol), and **6** (48.1 g, 86% pure, corresponding to 0.283 mol of pure material) in THF (2.5 L) was added DIAD (110 mL, 0.54 mol) dropwise at 0–3°C over a period of 1 h. The reaction mixture was stirred for 1 h at 0°C, and then poured into sat. Na$_2$CO$_3$ (500 mL). After separation of the aqueous phase, the mixture was dried and evaporated to a volume of ca. 700 mL. Ether (1 L) was added, and the POPh$_3$ crystallized at 0°C overnight and then was removed by filtration. The solvents were evaporated and the residue resolved chromatographically (silica gel, hexane:ethyl acetate 8:2, R$_f$ 0.30). The yield of (±)**7** as a colorless oil was 50.3 g (0.183 mol, 65% corresponding to pure **6**). ^{1}H NMR (CDCl$_3$) δ 1.17 (t, J = 7Hz, 3H, CH_3CH_2), 2.20–2.70 (m, 5H, 2-H, 5-H, 1-H), 4.12 (m, 2 H, CH_3CH_2),

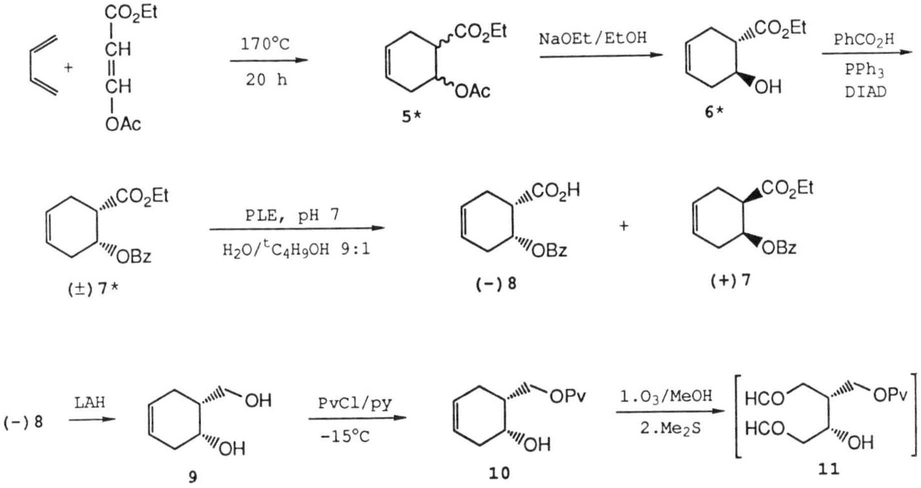

Oligonucleotide Analogs

Scheme 1.

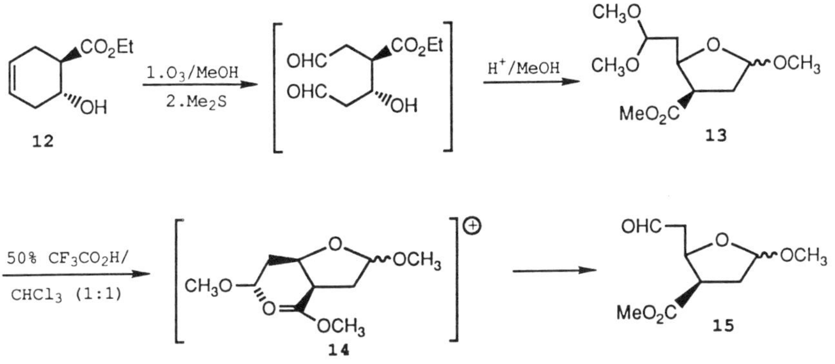

* = Only one enantiomer shown

Scheme 2.

Only one enantiomer shown

5.59–5.80 (m, 2 H, C-CH), 5.77 (m, 1 H, 6-H), 7.80 (m, 3 H, m-,p-ar-H), 8.17 (m, 2 H, o-ar-H); IR (CCl$_4$): 3039, 2982, 2881, 1730, 1252 cm^{-1}. Anal. Calc. for C$_{16}$H$_{18}$O$_4$ (274.32): C, 70.06; H, 6.61. Found: C, 69.93; H, 6.76.

3.1.3. (1S,6R)-6-Benzoyloxy-3-Cyclohexene-1-Carboxylic Acid ([−]8)

A suspension of (±)7 (65 g, 0.237 mol) in water:^{t}butanol 9:1 (4.8 L) at room temperature was hydrolyzed with pig liver esterase (Sigma, St. Louis, MO 9420 U) at pH 7. The progress of the reaction was monitored by the amount of addition of 1N NaOH with an autotitrator. After 45% of the starting material had been hydrolyzed (ca. 16 h), the reaction was stopped by addition of CH_2Cl_2 (40 mL). The reaction was extracted with ether (3 × 300 mL), and the layers were separated by centrifugation. The ether extracts were dried over $MgSO_4$, and the solvent evaporated to yield 35.5 g of an approx 10:1 mixture (ratio determined with the chiral shift reagent Eu[hfc]$_3$) (*vide infra*) of (+)-(1R,6S)-and (−)-(1S,6R)-**7**. The aqueous phase was brought to pH 2 and extracted with ether (3 × 200 mL). The ether extracts were dried and evaporated to yield crude (−)-(1S,6R)-**8** (27 g), which was used for the next reaction without further purification. A small sample was converted to its l-isoborneol ester (3 Eq DMAP, 3 Eq dicyclohexylcarbodiimide, 3 Eq triethylamine, CH_2Cl_2) for gas chromatographic (GC) analysis of enantiomeric purity. GC-MS: Temp: 200°C for 2 min and then 5°C/min gradient, retention times: isoborneol ester of (1S,6R)-**8** 11.57 min (MS, m/e 382, 229, 137, 105, 77), isoborneol ester of (1R,6S)-**8** not detectable. If the hydrolysis reaction is carried out in pure water, one obtains (1S,6R)-**8** in 63.5% enantiomeric excess, as determined by GC of the isoborneol ester: isoborneol ester of (1S,6R)-**8** 11.57 min, integrated to 81.75%, isoborneol ester of (1R,6S)-**8** 11.66 min integrated to 18.25%; MS, m/e 382, 229, 137, 105, 77. For additional NMR analysis of the enantiomeric purity, an analytical sample was esterified with diazomethane in ether to yield the methyl ester of (−)**8**: $[\alpha]_D$: −89.5° (c 6.35, acetone); ^{1}H NMR (CDCl$_3$) δ 2.44, 2.68, 2.87 (3 m, 5 H, 1-H, 2-H, 5-H), 3.67 (s, 3 H, C$\underline{H}_3$), 5.63, 5.82 (2 m, 2 H, C=C−H), 5.75 (m, 1 H, 6-H), 7.55 (m, 3 H, *m,p*-ar-H). 7.98 (dd, 2 H, *o*-ar-H); addition of the chiral shift reagent Eu(hfc)3 (2.5 mol%) in CCl$_4$:d$_6$-benzene 4:1 shifts the signal for 6-H from δ 5.75 to 7.86. The 1-H signal of the enantiomer, expected at δ 8.19, is not visible; MS, m/e (relative intensities) 229 (2.5), 138 (34), 105 (100), 79 (98), 77 (99), 51(21). (*See* Note 2.)

3.1.4. (1R,6S)-cis-6-Benzoyloxy-3-Cyclohexene-1-Carboxylic Acid Ethyl Ester ([+]7)

A mixture (ca. 10:1) of (+)-(1R,6S)- and (−)-(1S,6R)-**7** (35.5 g, 129.6 mmol) was dissolved in water:^{t}butanol 9:1 (1.35 L), and hydrolyzed with pig liver esterase (Sigma, 4000 U). The reaction was stopped, after 13 mL of 1N NaOH had been added by an autotitrator (19 h), by addition of CH_2Cl_2 (30 mL). The solution was saturated with NaCl and filtered. The clear solution was extracted with ether (3 × 200 mL), and the organic layers were dried and evaporated to yield (+)-(1R,6S)-**7** (32.3 g, 117.9 mmol, 91%). The aqueous phase was brought to pH 2 with 1M HCl and extracted with ether (3 × 200 mL). The combined organic layers were dried and evaporated to yield (−)-(1S,6R)-**8** (2.93 g, 11.9 mmol, 9%). The enantiomeric excess of (+)-(1R,6S)-**7** thus obtained was determined by ^{1}H NMR in the presence of the chiral shift reagent Eu[hfc]$_3$ (2.5 mol%) in CCl_4:d$_6$-benzene 4:1 and was >97% (no signal at 7.86). For the racemate, the signals for C-1-H were shifted from δ 5.65 to 7.86 for (−)-(1S,6R)-**7** and to 8.19 for (+)-(1R,6S)-**7**: [α]$_D$: +105.6 (c 3.1, acetone); ^{1}H NMR (CDCl$_3$) δ 1.02 (t, 3 H, C$\underline{H}_2$–C$\underline{H}_3$), 2.26–2.60 (2m, 5 H, 1-H, 2-H, 5-H), 3.95 (m, 2 H, C$\underline{H}_2$–C$\underline{H}_3$), 5.44 (m, 1 H, olefin-H), 5.65 (m, 2 H, 1 olefin-H, C$_6$-H), 7.24 (m, 3 H, m- and p-ar-H), 7.88 (m, 2 H, o-ar-H); IR (CCl$_4$) 3040, 2985, 2880, 1715, 1250, 1085 cm^{-1}; MS, m/e 51, 77, 79, 105, 123, 152, 229. Anal. Calc. for $C_{16}H_{18}O_4$ (274.32): C, 70.06; H, 6.61. Found: C, 70.19; H, 6.85.

3.1.5. (1R,6R)-6-Hydroxymethyl-3-Cyclohexene-1-ol (9)

To a suspension of LiAlH$_4$ (1.6 g, 41 mmol) in THF (140 mL) was added a solution of (−)**8** (3.32 g, 19.5 mmol) in THF (10 mL) at room temperature. The reaction mixture was stirred for 15 min at room temperature, and the reaction was then quenched with ethyl acetate (1 mL) followed by conc. HCl (9 mL). The pH was brought to 6 by adding 15% NaOH. After removal of half of the solvent under reduced pressure, and sedimentation of the inorganic material, the supernatant was decanted. The residue was continuously extracted with 100 mL ether for 20 h. The combined organic phases were dried and evaporated, and the residue chromatographed on silica gel (ethyl acetate, R$_f$ 0.28) to yield **9** (2.155 g, 16.83 mmol, 86%), which crystallized spontaneously. Recrystallization from ethyl acetate/hexane gave analytically pure, colorless crystals, which melted at 65–66°C, $[\alpha]_D$: + 5.6° (c 3.7, acetone). ^{1}H NMR (CDCl$_3$) δ 1.90–2.45 (m, 5 H, 2-H, 5-H, 6-H), 2.74 (s, 2 H, D$_2$O-exchangable, OH), 3.80 (m, 2 H, C$\underline{H}_2$OH), 4.23 (m, 1 H, 1-H), 5.58–5.73 (m, 2 H, C=C–H); IR (CCl$_4$) 3380, 3032, 2910, 1061 cm^{-1}; MS, m/e (relative intensities) 110 (40), 95 (19), 92 (47), 79 (100), 74 (40), 56 (40), 41(40). Anal. Calc. for C$_7$H$_{12}$O$_2$ (128.17): C, 65.60; H, 9.44. Found: C, 65.54; H, 9.36.

3.1.6. (1R,6R)-6-Pivaloyloxymethyl-3-Cyclohexene-1-ol (10)

To a solution of **9** (2.155 g, 16.83 mmol) in pyridine (30 mL) was added pivaloyl chloride (2.385 mL, 19.36 mmol) dropwise over a period of 1 h at –18–15°C. The reaction mixture was stirred at –18–10°C for 1 h, and was then quenched with methanol (4 mL). The solvent was evaporated and the residue purified by chromatography (hexane:ethyl acetate 7:3, R$_f$ 0.43) to yield **10** (3.40 g, 16.0 mmol, 96%) as a colorless oil. ^{1}H NMR (CDCl$_3$) δ 1.21 (s, 9 H, ^{t}bu-H), 1.68 (s br, 1 H, D$_2$O-exchangable, OH), 2.00–2.42 (m, 5 H, 2-H, 5-H, 6-H), 4.02 (m, 2 H, C$\underline{H}_2$O), 4.25 (1 H, 1-H), 5.62–5.73 (m, 2 H, C=C–H); IR (CCl$_4$) 3530, 2980, 2910, 1730, 1160 cm^{-1}; MS, m/e 158, 110, 92, 57. Anal. Calc. for C$_{12}$H$_{20}$O$_3$ (212.29): C, 67.89; H, 9.50. Found: C, 67.54; H, 9.31.

3.1.7. (2RS,4R,5R)-2-Methoxy-4-(Pivaloyloxymethyl)-5-(2-Hydroxyethyl)Tetrahydrofuran (18), and (2RS,4R,5R)-2-Methoxy-4-(Pivaloyloxymethyl)-5-(2,2-Dimethoxyethyl)Tetrahydrofuran (17)

A solution of **10** (6.327 g, 29.8 mmol) in methanol (500 mL) was cooled to –78°C and treated with a stream of ozone until it maintained a blue color (ca. 2 h). After the excess ozone was removed with a stream of dry nitrogen (45 min), dimethylsulfide (10 mL, 136 mmol) was added, and the reaction mixture slowly warmed to room temperature and stirred for 14 d in the dark. At this point, GC analysis showed that a mixture of the 2'-aldehyde **16** (R$_f$ 0.45 hexane:ethyl acetate 1:1), and the corresponding 2'-dimethylacetal **17** had formed in a ratio of ca. 3:1. The reaction mixture was cooled to 0°C, and NaBH$_4$ (2 g, 52.8 mmol) was added in small portions. After 30 min the reaction mixture was brought to pH 6 with 1M HCl. About half of the solvent was evaporated under reduced pressure, water (100 mL) was added, and the mixture extracted with CH$_2$Cl$_2$ (3 × 100 mL). The

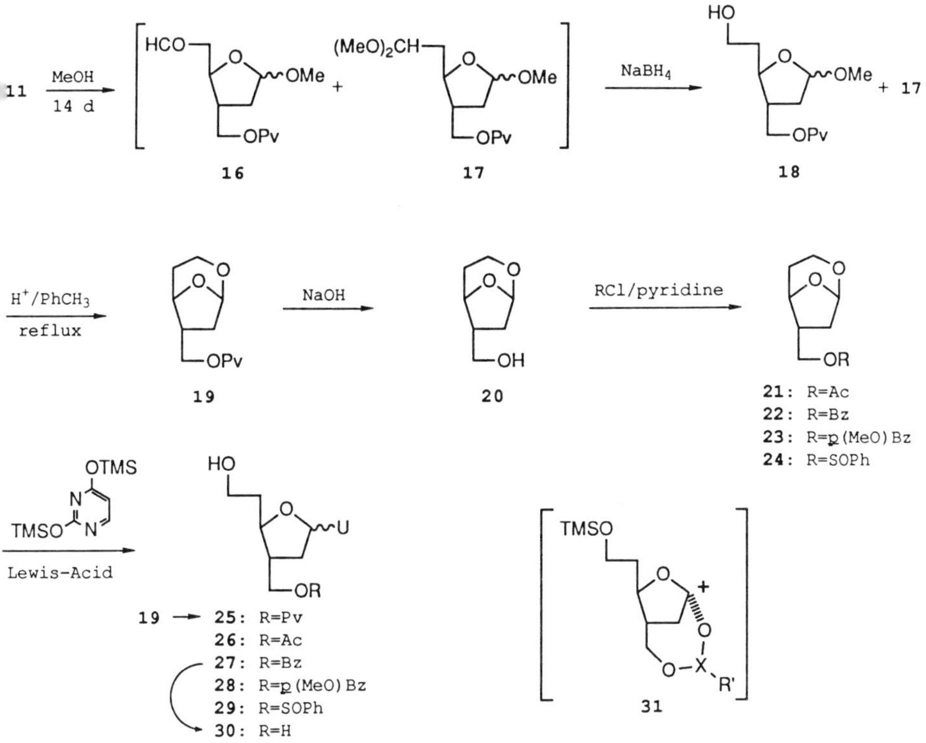

Scheme 3.

combined organic layers were dried and evaporated, and the residue chromatographed on silica gel (hexane:ethyl acetate 1:1) to yield **18** (4.851 g, 62.5%, R_f 0.22), and **17** (1.845 g, 20.3%, R_f 0.36, 0.40, 2 anomers, hexane:ethyl acetate 6:4) as colorless oils. The anomers of the latter compound could be separated by silica gel chromatography (hexane:ethyl acetate 7:3). **18** (1:1 mixture of anomers): ^{1}H NMR (CDCl$_3$) δ 1.20 (1s, 9 H, ^{t}bu-H), 1.59 (s, 1 H, D$_2$O-exchangable, OH), 1.65–2.6 (m, 5 H, 3-H, 4-H, HOCH$_2$CH$_2$), 3.32, 3.35 (2s, 3 H, α + β OCH$_3$), 3.82 (m, 2 H, HOCH$_2$), 3.95–4.19 (2m, 3 H, 5-H, CH$_2$ OPv), 4.98 (dd, J = 5 Hz, 0.5 H, 2-H of one anomer), 5.02 (dd, J = 1.5, 4 Hz, 0.5 H, 2-H of one anomer); IR (CCl$_4$) 3560, 2960, 1732, 1155 cm^{-1}; MS, m/e (relative intensities) 229 (10), 202 (8), 127 (23), 113 (67), 84 (52), 57 (100). Anal. Calc. for C$_{13}$H$_{24}$O$_5$ (260.33): C, 59.98; H, 9.29. Found: C, 59.68; H, 9.55. **17**: 1. Fraction (R_f 0.40): ^{1}H NMR (CDCl$_3$)

δ 1.20 (s, 9 H, ^{t}bu-H), 1.65–2.29 (3m, 5 H, 3-H, 4-H, HOCH$_2$CH$_2$), 3.31, 3.34, 3.36 (3 s, 9 H, OCH$_3$), 3.92 (m, 1 H, 5-H), 4.12 (m, 2 H, CH$_2$OPv), 4.61 (dd, J = 8, 4 Hz, 1 H, CH[OCH$_3$]$_2$), 5.01 (dd, J = 1.5, 5 Hz, 2-H). ^{13}C NMR δ 27.19 (^{t}bu-CH$_3$), 35.84 (C-3), 38.33 (HOCH$_2$CH$_2$), 38.78 (^{t}bu-CCH$_3$), 42.47 (C-4), 52.96, 53.39, 54.47 (OCH$_3$), 65.97 (H$_2$OPv), 77.03 (C[OMe]2), 102.39, 104.36 (C-2, HOCH$_2$), 178.36 (Me$_3$CCOO). IR (CCl$_4$) 2960, 1732, 1285, 1155 cm^{-1}; MS, m/e 241, 214, 182, 170, 138, 113, 75, 57. **17**: 2. Fraction (R$_f$ 0.38): ^{1}H NMR (CDCl$_3$) δ 1.20 (s, 9 H, ^{t}bu-H), 1.75–2.13 (m, 4 H, 3-H, HOCH$_2$CH$_2$), 2.49 (m, 1 H, 4-H), 3.33, 3.34, 3.36 (3 s, 9 H, OCH$_3$), 3.95 (m, 1 H, 5-H), 4.07 (m, 2 H, CH$_2$OPv), 4.63 (dd, J = 8, 3.5 Hz, 1 H, CH[OCH$_3$]$_2$), 4.96 (dd, J = 1.5, 5 Hz, 2-H); ^{13}C-NMR (CDCl$_3$) δ 27.18 (^{t}bu-CH$_3$), 36.71 (C-3), 38.80 (^{t}bu-CH$_3$), 40.40 (HOCH$_2$CH$_2$), 42.11 (C-4), 52.39, 53.50, 54.48 (OCH$_3$), 65.37 (C-3'), 76.66 ([OMe]$_2$), 102.45, 104.81 (C-2, HOCH$_2$), 178.32; IR (CCl$_4$) 2960, 1732, 1285, 1265, 1155 cm^{-1}; MS, m/e 241, 214, 182, 170, 138, 113, 75, 57. (*See* Note 3.)

3.1.8. (1R,5R,6R)-6-Pivaloyloxymethyl-2,8-Dioxa-(3.2.1)-Bicyclooctane (19)

A solution of **18** (1.7 g, 6.538 mmol) in toluene (60 mL) was refluxed in the presence of dry acidic cation exchange resin (Dowex 50 W8, 400 mg) for 5 h under protection from moisture. The resin was then removed by filtration, the solvent evaporated, and the residue resolved by chromatography (hexane:ethyl acetate 7:3, R$_f$ 0.28) on silica gel to yield **19** (1.25 g, 5.48 mmol, 84%) as a colorless liquid that solidified when stored at 4°C. ^{1}H NMR (CDCl$_3$) δ 1.20 (s, 9H, ^{t}bu-H), 1.57–2.38 (m, 4 H, 4-H, 7-H), 2.51 (m, 1 H, 6-H), 3.81–4.08 (m, 4 H, 3-H, CH$_2$-OPv), 4.32 (m, 1 H, 5-H), 5.44 (δ, J = 5 Hz, 1 H, 1-H); IR (CCl$_4$): 2978, 2874, 1731, 1480, 1285, 1152 cm^{-1}. (*See* Note 4.)

3.1.9. (1R,5R,6R)-6-Hydroxymethyl-2,8-Dioxa-[3.2.1.]-Bicyclooctane (20)

A solution of **19** (517 mg, 2.267 mmol) in 10M NaOH (2.26 mL, 22.6 mmol) was stirred in a mixture of THF, methanol, and H$_2$O (5:4:1, 5 mL) at room temperature for 2 h. The reaction mixture was brought to pH 6 by adding, first pyridinium-Dowex (300 mg) and then 1M HCl. After filtration of the solution and evaporation of the solvent, the residue was resolved by chromatography (ethyl acetate containing 1% triethylamine, R$_f$ 0.25) to yield **20** (293 mg, 2.034 mmol, 90%) as a viscous, colorless oil, which was used immediately for the next step.

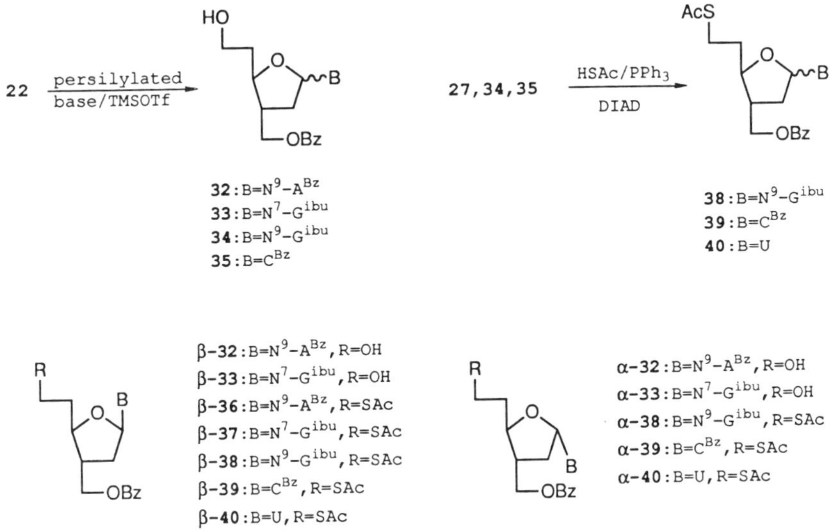

Scheme 4

22 →(persilylated base/TMSOTf)→ [sugar-B with OBz, HO-CH2]

32: B=N⁹-A^Bz
33: B=N⁷-G^ibu
34: B=N⁹-G^ibu
35: B=C^Bz

27, 34, 35 →(HSAc/PPh₃, DIAD)→ [sugar-B with OBz, AcS-CH2]

38: B=N⁹-G^ibu
39: B=C^Bz
40: B=U

β-32: B=N⁹-A^Bz, R=OH
β-33: B=N⁷-G^ibu, R=OH
β-36: B=N⁹-A^Bz, R=SAc
β-37: B=N⁷-G^ibu, R=SAc
β-38: B=N⁹-G^ibu, R=SAc
β-39: B=C^Bz, R=SAc
β-40: B=U, R=SAc

α-32: B=N⁹-A^Bz, R=OH
α-33: B=N⁷-G^ibu, R=OH
α-38: B=N⁹-G^ibu, R=SAc
α-39: B=C^Bz, R=SAc
α-40: B=U, R=SAc

Scheme 4.

3.1.10. (1R,5R,6R)-6-(Benzyloxymethyl)-2,8-Dioxa-[3.2.1]-Bicyclooctane (22)

From 20: To a solution of 20 (75 mg, 0.521 mmol) in pyridine (1 mL) was added benzoyl chloride (91 mL, 0.782 mmol) dropwise at 0°C. The reaction mixture was warmed to room temperature and stirred for 30 min. After dilution with ethyl acetate (20 mL), the crude reaction mixture was extracted with sat. CuSO₄ (5 mL) followed by sat. NaCl (15 mL). The organic phase was dried and evaporated, and the residue chromatographed on silica gel (hexane:ethyl acetate 7:3) to yield 22 (119 mg, 0.484 mmol, 93%) as a colorless oil, which crystallized on standing at room temperature. The compound was recrystallized from ethyl acetate/hexane to yield colorless crystals melting at 65–66°C, [α]_D: +5.6 (c 1.5, acetone).

From 52: A solution of azoisobutyronitrile (12 mg, 72 mol), Bu₃SnH (0.550 mmol), and 52 (140 mg, 0.3646 mmol) in toluene (4 mL) was degassed with a stream of argon for 30 min The mixture was then heated to 75°C for 1 h. The solvent was removed under reduced pressure, and the residue chromatographed on silica gel (hexane:ethyl acetate 7:3, R_f 0.27). The oil obtained was crystallized from ethyl acetate/hexane to yield 60 mg (66%) **22**, mp.: 65–66°C, $[\alpha]_D$: + 4.5 (c 1.34, acetone). **22**: ^{1}H NMR (CDCl$_3$) δ 1.32 (2mc, J = 14 Hz, 1 H, 7-H$_\alpha$), 1.78 (ddd, J = 14, 5.5, 4 Hz, 1 H, 7-H$_\beta$), 2.35 (m, 2H, 4-H), 2.66 (m, 1 H, 6-H), 3.91 (m, 2 H, 3-H), 4.23 (m, 2 H, C$\underline{H}_2$OBz), 4.44 (m, 1 H, 5-H), 5.49 (dd, J = 5.5, 1.5 Hz, 1 H, 1-H), 7.26–8.05 (3m, 5 H, ar-H); IR (CHCl$_3$): 3100–2900, 2980, 1720, 1600, 1450, 1280, 980, 810 cm^{-1}; MS, m/e 248, 230, 219, 204, 189, 176, 148, 143, 126, 123, 105. Anal. Calc. for C$_{14}$H$_{16}$O$_4$ (248.28): C, 67.73; H, 6.50. Found: C 67.83, H 6.63.

3.1.11. 1-([2RS,4R,5R]-4-[Pivaloyloxymethyl] 5-[2-Hydroxyethyl]Tetrahydrofuran-2-yl)Uracil (α + β-25)

To a solution of *bis*-TMS-uracil (62 mg, 1.1 eq), and **19** (50 mg, 0.22 mmol) in dichloroethane (1.0 mL) was added SnCl$_4$ (28 μL, 1.1 Eq dropwise). The mixture was stirred at room temperature (1 h), cooled to 0°C, and sat. NaHCO$_3$ was added. The resulting mixture was filtered through Celite, and the Celite washed with ethyl acetate. The mixture was then extracted with CH$_2$Cl$_2$ (3x), and the combined organic layers were dried and evaporated to give an oil, which was chromatographed (silica gel, chloroform:methanol 9:1, R$_f$ 0.31) to yield 36 mg (56%) α + β-**25**. ^{1}H NMR (CDCl$_3$) δ 1.19 and 1.21 (2s, 9 H, ^{t}bu-H α + β), 1.70–2.52 (m, 5 H, HOCH$_2$C$\underline{H}_2$, 3'-H, 4'-H, OH), 2.74 (m, 0.5 H, 3'-Hα), 3.84 (m, 2 H, HOCH$_2$), 3.96–4.28 (m, 3 H, C$\underline{H}_2$OPv, 5'-H), 5.77 (2 d, J = 8 Hz, 5-H), 6.06 (m, ^{1}H, 2'-H), 7.43, 7.47 (2 d, J = 8 Hz, 6-H). Anal. Calc. for C$_{16}$H$_{24}$N$_2$O$_6$ (340.38): C, 56.46; H, 7.11; N, 8.23. Found: C, 55.97; H, 6.98; N, 8.00. (*See* Note 5.)

3.1.12. General Procedure for the Conversion of an Alcohol to a Thioacetate (22)

To a solution of PPh$_3$ (2.2 Eq) in THF was added at 0°C diisopropylazodicarboxylate (DIAD) (2.2 Eq). The mixture was stirred for 15 min, by which time a thick precipitate had formed. To this suspension was added dropwise a mixture of the alcohol (1 Eq), and thioacetic acid (2.2 Eq) in THF. Stirring was continued for 1 h at 0°C, methanol added, and the solvents were removed under reduced pressure. The residue was then chromatographed. (*See* Note 6.)

3.1.13. N^6-Benzoyl-9-([2S,4R,5R]-4-[Benzoyloxymethyl] 5-[2-Hydroxyethyl]Tetrahydrofuran-2-yl) Adenine and (2R)-Isomer (α-32) and (β-32)

To a suspension of N^6-benzoyladenosine (530 mg, 1.1 Eq) in acetonitrile (7 mL) was added MSTFA (913 µL, 2.2 Eq). The mixture was stirred for 10 min. at room temperature, TMSOTf (182 µL, 0.5 Eq) was added, and the mixture stirred another 10 min. To the resulting clear solution was added dropwise **22** (500 mg, 2.06 mmol) in acetonitrile (3 mL). The mixture was stirred at 40°C for 30 min, cooled to 0°C, and sat. NaHCO₃ was added. To the resulting two-phase mixture was added potassium fluoride and a crystal of 18 crown-6, and the reaction mixture stirred overnight. The reaction mixture was then extracted three times with ethyl acetate (5 mL), and the combined organic layers were dried and evaporated to give a clear oil, which was chromatographed (silica gel, CH₂Cl₂:methanol 9:1, R_f 0.23). After evaporation of the solvents under reduced pressure, **32** (346 mg, 64%) was obtained as a tan foam. The anomers of **32** were separated by HPLC (CH₂Cl₂:THF 65:35, water saturated). The fraction eluting after 69.4 min contained β–**32**, and the fraction eluting after 76.0 min contained α-**32**, which could be crystallized from CH₂Cl₂:ether:pentane to give white needles melting at 136–137°C. α-**32**: UV (MeOH) λ max 218 (ε 26,200); 279 (20,800); ¹H NMR (CDCl₃) δ 1.91 (mc, 1 H, HOCH₂CH₂), 2.09 (mc, 1 H, HOCH₂CH₂), 2.31 (t, J = 5.5 Hz, 1 H, D₂O-exchangable, OH), 2.69 (mc, 1 H, 4'-H), 2.89 (m, 2 H, 3'-H), 3.85 (dd, J = 11, 5.5 Hz, 2 H, HOCH₂), 4.55 (m, 2 H, CH₂OBz), 4.64 (dt, J = 9 Hz, 1 H, 5'-H), 6.38 (t, J = 6.5 Hz, 1 H, 2'-H), 7.23–7.64 (6 H, m,p-ar-H), 8.00–8.05 (m, 4 H, o-ar-H), 8.14 (s, 1 H, 8-H), 8.73 (s, 1 H, 2-H), 9.12 (s, 1H, D₂O-exchangable, NH); IR (KBr) 3420, 3260, 2930, 1715, 1675, 1575, 1505, 1490, 1270, 1255, 715 cm⁻¹; MS, m/e relative intensities) 487 (<1), 383 (<1), 366 (◂1), 353 (<1), 278 (<1), 249 (<1), 239 (6), 162 (9), 108 (21), 105 (100), 77 (67). Anal. Calc. for C₂₆H₂₅N₅O₅ (487.52): C, 64.06; H, 5.17; N, 14.37. Found: C, 63.72; H, 5.15; N, 14.26. β-**34**: ¹H NMR (CDCl₃) δ 1.96–2.20 (m, 3 H, HOCH₂CH₂, OH), 2.58 (mc, 1 H, 4'-H), 2.84–3.03 (m, 2 :H, 3'-H), 3.88 (t, 2 H, HOCH₂), 4.26 (dt, J = 3.5, 8.5 Hz, 1 H, 5'-H), 4.46 (m, 2 H, CH₂OBz), 6.37 (dd, J = 2.5, 7 Hz, 1 H, 2'-H), 7.44–7.64 (m, 6 H, m,p-ar-H), 8.03–8.07 (m, 4 H, o-ar-H), 8.20 (s, 1 H, 8-H), 8.80 (s, 1 H, 2-H), 9.09 (s, 1 H, NH); IR (CHCl$_3$) 3540, 3405, 3000, 2960, 1720, 1610, 1455, 1275 cm^{-1}; MS, m/e (relative intensities) 383 (<1), 366 (<1), 353 (<1), 310 (<1), 278 (<1), 262 (<1), 248 (1.2), 220 (19), 205 (86), 145 (16), 105 (64), 43 (100). (See Notes 7 and 8.)

3.1.14. N^6-Benzoyl-9-([2R,4R,5R)-4-[Benzoyloxymethyl] 5-[2-Acetylthioethyl]Tetrahydrofuran-2-yl)adnine (β-36)

PPh$_3$ (960 mg, 2 Eq), and DIAD (740 µL, 2 Eq) in THF (15 mL) were reacted with β-**32** (861 mg, 1.768 mmol), and thioacetic acid (260 µL, 2 Eq) in THF (5 mL). After the reaction was complete (ca. 30 min), chromatographic purification (silica gel, CHCl$_3$/THF 4:1) yielded β-**36** (660 mg, 69%) as a white foam. UV (MeOH) λ$_{max}$ 218 (e 32,200), 279 (22, 700); ^{1}H NMR (CDCl$_3$) δ 1.98–2.21 (m, 2 H, 2 H, SCH$_2$CH$_2$), 2.32 (s, 3 H, COCH$_3$), 2.57 (ddd, J = 7 , 9, 13.5 Hz, 1 H, 3'-H$_\beta$), 2.78–2.99 (m,3 H, 4'-H,SCH$_2$), 3.18 (ddd, J = 5, 5, 13.5 Hz, 1 H, 3'-H$_\alpha$), 4.16 (td, J = 3.5, 8.5 Hz, 1 H, 5'-H), 4.47 (m, 2 H, CH$_2$OBZ), 6.37 (dd, J = 3, 7 Hz, 1 H, 2'-H), 7.45–7.65 (m, 6 H, m,p-ar-H), 8.04 (m, 4 H, o-ar-H), 8.29 (s, 1 H, 8-H), 8.81 (s, 1 H, 2-H), 9.01 (s, 1 H, NH); ^{13}C NMR (CDCl$_3$) δ 25.9 (HOCH$_2$H$_2$), 30.6 (CH$_3$), 35.3, 35.9 (C-3', HOCH2), 42.2 (C-4'), 64.4 (CH$_2$OBz), 82.7 (C-5'), 85.2 (C-2'), 123.9 (C-5), 127.9, 128.6, 128.8, 129.6, 132.7, 133.4, 133.7 (ar-C), 141.5 (C-8), 149.6, 151.2 (C-4, C-6), 152.5 (C-2), 164.8, 166.3, 195.5 (CO); IR (KBr) 3405 (br), 2920, 1720, 1690, 1610, 1580, 1510, 1485, 1450, 1275, 1115 cm^{-1}; MS, m/e 545, 517, 502, 470, 440, 398, 366, 352, 307, 239, 239, 141, 105. Anal. Calc. for C$_{28}$H$_{27}$N$_5$O$_5$S: C, 61.64; H, 4.99; N, 12.84. Found: C, 61.61; H, 4.80; N, 12.25.

3.1.15. N^2-Isobutyryl -9-([2RS,4R,5R]-4-[Benzoyloxymethyl] 5'[2-Hydroxyethyl]tetrahydrofuran-2-yl)Guanine (34), and N^2-Isobutyryl-7-([2RS,4R,5R]-4-[Benzoyloxymethyl] 5-[2-Hydroxyethyl]Tetrahydrofuran-2-yl)Guanine (33)

To a suspension of N^2-isobutyrylguanosine•H$_2$O (1.590 g, 1.1 eq) in acetonitrile (20 mL) was added TMSCl (1.70 mL, 2.2 eq), and HMDS (2.80 mL, 2.2 eq). The mixture was stirred for 10 min at room temperature, TMSOTf (3.00 mL, 0.5 eq) was added, and the mixture stirred another 10 min. To the resulting clear solution was added **22** (1.50 g, 6.048 mmol) dropwise in acetonitrile (5 mL). The mixture was stirred at 40°C for 15 min, cooled to 0°C, and saturated NaHCO$_3$

was added. To the resulting two-phase mixture was added potassium fluoride and a crystal of 18-crown-6, and the mixture stirred overnight. The mixture was then extracted four times with ethyl acetate (5 mL). The combined organic layers were dried and evaporated to give an oil, which was chromatographed (silica gel, CH_2Cl_2:methanol 9:1) to give two fractions that contained nucleoside analogs. Evaporation of the solvents under reduced pressure yielded **33** (843 mg, 30%, R_f 0.33) from the first fraction and **34** (1.602 g, 56%, R_f 0.26) from the second fraction, both as white foams that were slightly impure (NMR). From the fraction containing the N^7-isomers (**33**), the β-isomer could be selectively (as a ca. 9:1 mixture of anomers) crystallized from CH_2Cl_2/pentane. **33** (9:1 mixture of anomers, major anomer): ^{1}H NMR (CDCl$_3$) δ 1.24, 1.26 (2 d, 6 H, J = 2.5, 3.5 Hz, ^{i}bu-C$\underline{H}_3$), 1.98–2.13 (m, 2 H, HOCH$_2$C$\underline{H}_2$), 2.62 (m, 3 H, 3'-H, 4'-H), 2.80 (s br, 1 H, OH), 2.91 (m, 1 H, ^{i}bu-CH), 3.93 (m, 2 H, HOCH$_2$), 4.25 (td, 1 H, J = 9.8, 3 Hz, 5'-H), 4.45 (m, 2 H, C$\underline{H}_2$OBz), 6.55 (dd, 1 H, J = 5, 4 Hz, 2'-H), 7.46, 7.58, 8.03 (3 m, 5 H, ar-H), 8.22 (s, 1 H, 8-H), 9.92 (s, 1 H, N^2-H), 12.32 (s, 1 H, 1-NH); IR (CHCl$_3$) 3405, 3200, 3000, 2960, 1690, 1610, 1275 cm^{-1}; MS, m/e 425, 407, 357, 329, 281, 248, 189, 151. Anal. Calc. for $C_{23}H_{27}N_5O_6$•H_2O (487.52): C, 56.67; H, 6.00; N, 14.37. Found: C, 57.55; H, 5.92; N, 13.81. **34** (1:1 mixture of anomers): ^{1}H NMR (d$_6$–DMSO) δ 1.13, 1.15 (2 s, 6 H, ^{i}bu-C$\underline{H}_3$), 1.68–2.00 (m, 2 H, HOCH$_2$C$\underline{H}_2$), 2.42–2.86, 3.54 (2m, 6 H, 3'-H, ^{i}bu-CH, 4'-H, HOCH$_2$, OH, all unresolved), 4.08, 4.34 (2 td, 1 H, 5'-H), 4.50 (m, 2 H, C$\underline{H}_2$OBz), 6.07 (dd, 0.5 H, J = 7, 3.5 Hz, 2'-H), 6.07 (t, 0.5 H, J = 6.5 Hz, 2'-H), 7.57, 7.70, 8.01 (3 m, 5 H, ar-H), 8.22, 8.25 (2s, 1 H, 8-H), 11.66, 12.07 (2s br, 1 H, 1-NH); IR (CHCl$_3$) 3405, 3200, 3000, 2980, 1690, 1610, 1275 cm^{-1}; MS, m/e 374, 248, 221, 204, 189, 151.

3.1.16. N^2-Isobutyryl-9-([2S,4R,5R]-4-[Benzoyloxymethyl] 5-[2-Acetylthioethyl]Tetrahydrofuran-2-yl) Guanine and (2R)-Isomer (α-**38** and β-**38**).

PPh$_3$ (2.170 g, 2.2 Eq), and DIAD (1.670 mL, 2.2 Eq) in THF (40 mL) were reacted with **34** (1.70 g, 3.625 mmol) obtained from the previous reaction, and thioacetic acid (590 μL, 2.2 Eq) in THF (10 mL). Chromatography (silica gel, CHCl_3:methanol 95:5) yielded a mixture of α-**38** and β-**38** (1.160 g) as a slightly yellow foam. The anomers were separated by HPLC (CHCl$_3$/2.5% ethanal, water saturated). The first fraction contained the α-anomer of **38** (retention time 60.5 min), and the second fraction contained the β-anomer of **38** (retention time 64.7 min). Both compounds were recovered as a white foam following evaporation of the solvents. α-**38**: UV (MeOH) λ_{max} 230 (ε 21,300), 257 (18,700), 279 (sh); ^{1}H NMR (CDCl$_3$) δ 1.31, 1.33 (2d, J = 7 Hz, 6 H, ^{i}bu-H), 1.86–1.95 (m, 2 H, SCH$_2$C$\underline{H}_2$), 2.17 (s, 3 H, C$\underline{H}_3$), 2.53 (ddd, J = 13.5, 7, 5 Hz, 1 H, 3'-H$_\beta$), 2.63–2.96 (m, 4 H, 4'-H, 3'-H$_\alpha$, C$\underline{H}_2$S), 3.03 (dquar. = sept., J = 7 Hz, 1 H, sec.^{i}bu-H), 4.44 (ddd = td, J = 7.5, 4 Hz, 1 H, 5'-H), 4.62 (dd, J = 11.5, 7 Hz, 2 H, C$\underline{H}_2$OBz), 6.11 (dd, j = 8, 4.5 Hz, 1H, 2'-H), 7.48 (mc, 3 H, m,p-ar-H), 7.63 (mc, 2 H, o-ar-H), 7.73 (s, 1 H, 8-H), 9.57 (s, 1 H, N^2-H), 12.12 (s, 1 H, 1-NH), irradiation at δ 6.11 gives NOE enhancement at δ 7.73 (8-H) and 2.80 (3'-H$_\alpha$); ^{13}C NMR (CDCl$_3$) δ 19.0, 19.1 (^{i}bu-C$\underline{H}_3$), 25.7 (SCH$_2$C$\underline{H}_2$), 30.5 (C$\underline{H}_3$CO), 34.7 (C-3'), 35.3 (C-4'), 36.2 (SCH$_2$), 43.3 (^{i}bu-CH), 64.8 (C$\underline{H}_2$OBz), 82.2 (C-5'), 85.4 (C-2'), 122.3 (C-5), 129.3 (ar.-C), 128.6, 129.6, 133.7 (ar-CH), 138.5 (C-8), 147.5, 147.7 (C-2, C-6), 155.7 (C-4), 167.0, 179.5, 195.5 (CO); IR (KBr) 3420, (br), 3170 (br), 2970, 2930, 1735, 1685, 1605, 1555, 1275, 1115, 710 cm^{-1}; MS, m/e (relative intensities) 221(83), 151(18), 141 (9), 108 (47), 77 (20), 43 (100). β-**38**: UV (MeOH) λ_{max} 229 (ε 20,400), 257 (17,900), 279 (sh); ^{1}H NMR (CDCl$_3$) δ 1.27, 1.28 (2d, J = 8.5 Hz, 6 H, ^{i}bu-H), 1.97, 2.12 (2m, 2 H, SCH$_2$C$\underline{H}_2$), 2.29 (s, 3 H, C$\underline{H}_3$), 2.40 (ddd, J = 13.5, 8.5, 7.5 Hz, 1 H, 3'-H$_\alpha$), 2.72 (m, 2 H, SC$\underline{H}_2$). 2.95 (m, 2 H, 4'-H, ^{i}bu/CH), 3.11 (ddd, J = 5, 5, 13.5, 3'-H$_\beta$), 4.05 (ddd = td, J = 8, 3.5 Hz, 1 H, 5'-H), 4.39 (mc, 2 H, C$\underline{H}_2$OBz), 6.05 (dd, 7.5, 3.5 Hz, 1H, 2'-H), 7.49 (mc, 3 H, m,p-ar-H), 7.60 (mc, 2 H, o-ar-H), 7.80 (s, 1 H, 8-H), 8.87 (s, 1 H, N^2-H), 12.08 (s, 1 H, 1-NH), irradiation at δ 6.05 gives NOE enhancement at δ 7.80 (8-H), 4.05 (4'-H), 2.40 (3'-H$_\alpha$); ^{13}C-NMR (CDCl$_3$) δ 19.0, 19.1 (^{i}bu-C$\underline{H}_3$), 25.6 (SCH$_2$C$\underline{H}_2$), 30.6 (C$\underline{H}_3$CO) 34.7 (C-3'), 35.6 (C-4'), 36.2 (SCH$_2$), 41.8 (^{i}bu-CH), 64.9 (H$_2$OBz), 82.5 ≠ C-5'), 84.7 (C-2'), 121.8 (C-5), 128.5, 129.5, 133.3 (ar.-CH), 137.6 (C-8), 147.7, 147.8 (C-2, C-6), 155.8 (C-4), 166.3, 179.5, 196.0 (CO); IR (KBr) see α-anomer; MS, m/e (relative intensities). 221 (1.5), 151 (4.5), 141 (9), 108 (47), 77 (20), 43 (100). Anal. Calc. for C$_{25}$H$_{29}$N$_5$O$_8$S (527.60): C, 56.91; H, 5.54; N, 13.27; S, 6.08. Found: C, 56.65; H, 5.47; N, 13.13; S, 6.21.

3.1.17. N^4–Benzoyl-1-([2RS,4R,SR]-4-[Benzoyloxymethyl] 5-[2-Hydroxyethyl]Tetrahydrofuran-2-yl)Cytosine (35)

To a suspension of N^4-benzoylcytosine (286 mg, 1.1 Eq) in acetonitrile (3 mL) was added MSTFA (550 µL, 2.2 Eq). The mixture was stirred for 10 min at room temperature, TMSOTf (400 µL, 1.8 eq) was added, and the mixture stirred another 10 min. The resulting clear solution was heated to 40°C, and **22** (300 mg, 1.21 mmol) in acetonitrile (2 mL) was added dropwise. The mixture was stirred at 40°C for 1 h, cooled to 0°C and sat. NaHCO₃ was added. The resulting two-phase mixture, which contained a white precipitate, was extracted with ethyl acetate (3 × 3 mL). The combined organic layers were dried and evaporated to give a clear oil, which was chromatographed (silica gel, CH₂Cl₂:methanol 9:1, R_f 0.32). After evaporation of the solvents under reduced pressure, α-**35** (346 mg, 64,) was obtained slightly impure (NMR) as a tan foam. A small sample of the mixture of anomers could be separated by fractional crystallization. The first fraction contained α-**35** (90% anomerically pure); the second fraction contained a mixture of α- and β-**35** in a ratio of 1:3. α-**35**: ¹H NMR (CDCl₃) δ 1.88–2.14 (m, 4 H, 3'-H$_α$, HOCH₂C$\underline{H}$₂, OH), 2.63 (mc, 1 H, 4'-H), 3.10 (ddd, J = 6.5, 8, 14 Hz, 1 H, 3'-H$_β$), 3.91 (br. s, 2 H, HOCH₂), 4.32–4.44 (m, 3 H, 5'-H, C$\underline{H}$₂OBz), 6.11 (t, J = 6 Hz, 1 H, 2'-H), 7.40–7.64 (m, 7 H, *m,p*-ar.-H, 5-H), 7.88–8.03 (m, 5 H, *o*-ar.-H, 6-H), irradiation at δ 6.11 gives NOE enhancement at δ 8.01 (6-H), 3.10 (3'-H$_β$), 2.63 (4'-H). β-**35**: ¹H NMR (CDCl₃) δ 1.90–2.14 (m, ~4H, 3'-H, HOCH₂C$\underline{H}$₂, OH), 2.41 (m, ~1 H, 3'-H), 2.57 (m, ~1 H, 4'-H), 3.94 (br. dd, ~2 H, HOCH₂), 4.20 (td, J = 3.5, 9 Hz, ~1 H, 5'-H), 4.40 (m, ~2H, C$\underline{H}$₂OBz), 6.13 (dd, J = 6.5, 2 Hz, ~1 H, 2'-H), 7.40–7.64 (m, 7 H, *m,p*-ar-H, 5-H), 7.88–8.08 (m, 5 H, *o*-ar-H, 6-H).

3.1.18. N^4–Benzoyl-1-([2S,4R,5R]-4-[Benzoyloxymethyl] 5-[2-Acetylthioethyl]Tetrahydrofuran-2-yl)Cytosine and (2R)-Isomer (α-39 and β-39)

PPh₃ (885 mg, 2 Eq), and DIAD (665 µL, 2 Eq) in THF (20 mL) were reacted with a 1:1 mixture of anomers of **35** (732 mg, 3.625 mmol), and thioacetic acid (238 1, 2 Eq) in THF (7 mL). Chromatography (silica gel, ethyl acetate) yielded **39** (717 mg, 87%) as a slightly tan oil. The anomers were separated by HPLC (ethyl acetate:hexane:water 7:3:0.07). The first fraction contained the β-anomer (retention time

48.3 min), and the second fraction contained the α-anomer (retention time 53.3 min). The α–anomer was crystallized as white needles (mp 138–139°C) from CH_2Cl_2:ether:pentane at room temperature. The α-anomer could not be crystallized, forming a white foam on evaporation of the solvents. α-**39**: ^{1}H NMR (CDCl$_3$) δ 1.93–2.14 (m, 2 H, SCH$_2$C$\underline{H}_2$), 2.35 (s, 3 H, COCH$_3$), 2.59 (m, 1 H, 4'-H), 3.11 (m, 2 H, 3'-H), 3.64 (ddd, J = 13.5, 10, 7 Hz, 1 H, SCH$_2$), 3.88 (ddd, J = 13.5, 10, 7 Hz, 1 H, SCH$_2$), 4.24 (td, J = 3.5, 8.5 Hz, 1 H, 5'-H), 4.35 (m, 2 H, C$\underline{H}_2$OBz), 6.08 (J = 6 Hz, 1 H, 2'-H), 7.42–7.72, m, 7 H, m,p-ar-H, 5-H), 7.93 (m, 2 H, o-benzamide-H), 7.99 (m, 2 H, o-benzoate-H), 8.03 (d, J = 7.5 Hz, 1 H, 6-H); IR (CHCl$_3$) 3400, 3000, 1740, 1690, 1660, 1480, 1270, 1080 cm^{-1}; MS, m/e (relative intensities) 357 (0.4), 242 (1.1), 215 (5), 186 (7), 153 (5), 141 (30), 108 (79), 105 (100), 95 (45), 77 (58), 43 (86);-**39**: UV (MeOH) λ$_{max}$ 228 (ε 21,400), 258 (18,700), 303 (8500); ^{1}H NMR (CDCl$_3$) δ 1.97 2.24 (2m, 2 H, SCH$_2$C$\underline{H}_2$), 2.30–2.50 (s, m, 5 H, C$\underline{H}_3$, 4'-H, 3'-H$_β$), 2.59 (ddd, J = 13.5, 8, 6.5 Hz, 1 H, 3'-H$_α$), 2.94 (ddd, J = 13.5, 8.5, 7.5, 1 H, SCH$_2$), 3.29 (ddd, J = 13.5, 9.5, 7 Hz, 1 H, SCH$_2$), 4.10 (td, J = 10, 7, 3Hz, 1H, 5'-H), 4.40 (m, 2 H, C$\underline{H}_2$OBz), 6.11 (dd, J = 6.5, 3 Hz, 1 H, 2'-H), 7.44 (m, 7 H, m,p-ar-H, 5-H), 7.91 (d, J = 8 Hz, 2 H, o-benzamide-H), 8.02 (m, 2 H, o-benzoate-H), 8.20 (d, J = 7.5 Hz, 1 H, 6-H), irradiation at δ 6.11 gives NOE enhancement at δ 8.20 (6-H), 4.10 (5'-H), 2.59 (3'-H$_α$); IR (KBr) 3380, 2950, 1740, 1690, 1660, 1485, 1270, 1095, 710 cm^{-1}; MS, m/e (relative intensities) 280 (0.8), 279 (9), 278 (49), 277 (100), 205 (54), 152 (11), 105 (18), 77 (40), 45 (95). Anal. Calc. for $C_{27}H_{27}N_3O_6S$: C, 62.17; H, 5.22; N, 8.06; S, 6.15;. Found: C, 62.48; H, 4.99; N, 7.69; S, 5.98.

3.1.19. 1-([2RS,4R,5R]-4-[Benzoyloxymethyl] 5-[2-Hydroxyethyl]Tetrahydrofuran-2-yl)Uracil (27)

To a solution of bis-TMS-uracil (600 mg, 1.1 Eq), and TMSOTf (125 µL, 0.3 Eq) in acetonitrile (8 mL) was added dropwise a solution of **22** (500 mg, 2.016 mmol) in acetonitrile (2 mL). The mixture was stirred at room temperature for 15 min, cooled to 0°C, and saturated NaHCO$_3$ was added. The resulting two-phase mixture was extracted three times with ethyl acetate. The combined organic layers were dried and evaporated to give an oil, which was chromatographed (silica gel, CH$_2$Cl$_2$:methanol 9:1). Evaporation of the solvents yielded slightly impure (NMR) **27** (522 mg, 75%) as a tan foam. ^{1}H NMR (CDCl$_3$) δ 1.80–2.20 and 2.20–2.66 (2m, 5.5 H, 3'-H, 4'-H, HOCH$_2$C$\underline{H}_2$, OH), 2.82 (m, 0.5 H, 3'-H), 3.86 (m, 2 H, HOC$\underline{H}_2$), 4.11 (td, J = 3, 9 Hz, 0.5 H, 5'-H), 4.35 (td, J = 3, 9 Hz, 0.5 H, 5'-H), 4.40 (m, 2 H, C$\underline{H}_2$OBz), 5.87, 5.88 (2d, J = 8 Hz, 1 H, 5-H), 6.09 (t, J = 6.5 Hz, 0.5 H, 2'-H), 6.12 (dd, j = 4, 6.5 Hz, 0.5 H, 2'-H), 7.47 (m, 3 H, *m,p*-ar-H), 7.60 (m, 2H, o-ar-H), 8.02 (2d = t, J = 8 Hz, 1 H, 6-H), 9.60 (s br., 1 H, NH); IR (CHCl$_3$) 3380, 3000, 1720, 1690, 1460, 1450, 1270, 1110 cm^{-1}.

3.1.20. 1-([2S,4R,5R]-4-[Benzoyloxymethyl] 5-[2-Acetylthioethyl]Tetrahydrofuran-2-yl)Uracil and (2R,4S,5S)-Isomer (α-**40** and β-**40**)

PPh$_3$ (2.122 g, 2.2 Eq), and DIAD in THF (40 mL) were reacted with thioacetic acid (570, 2.2 Eq) in THF (10 mL), and **27** (1.347 g, 3.625 mmol). Chromatography (silica gel, ethyl acetate) yielded a mixture of the anomers of **40** (1.273 g, 78%) as a slightly yellow foam. The anomers were separated by HPLC (ethyl acetate:hexane:water 7:3:0.07). The first fraction contained the α-anomer (retention time 51.5 min), and the second fraction contained the β-anomer (retention time 56.9 min). The α-anomer was cystallized as white needles (mp. 103–104°C) from CH$_2$Cl$_2$:ether:pentane at room temperature. The α-anomer crystallized as a white microcrystalline material (mp: 105–106°C) from ethyl acetate/hexane at room temperature.

α-**40**: ^{1}H NMR (CDCl$_3$) δ 1.88–2.10 (m, 3 H, 3'-H, SCH$_2$C$\underline{H}_2$), 2.32 (s, 3 H, SOCH$_3$), 2.55 (mc, 1 H, 4'-H), 2.87 (ddd, J = 13.5, 7, 6.5 Hz, 1 H, 3'-H$_\alpha$), 2.94–3.14 (m, 2 H, SCH$_2$), 4.19 (ddd = td, J = 8.5, 8.5, 1.5 Hz, 1 H, 5'-H), 5.77 (d, J = 8 Hz, 1 H, 5-H), 6.06 (dd = t, J = 6.5, 6.5 Hz, 1 H, 2'-H), 7.46 (m, 3 H, p-ar-H) 7.58 (d, J = 8 Hz, 1 H, 6-H), 7.99 (m, 2 H, o-ar-H), 9.13 (s, 1 H, NH), irradiation δ at 6.06 gives NOE enhancement at δ 7.46 (*m*-ar-H), 5.77 (5-H), 2.87 (3'-H$_\alpha$), 2.55 (4'-H); IR (KBr) 3410 (br), 3040, 1740, 1700, 1675, 1460, 1260, 1100 cm^{-1}; MS, m/e (relative intensities) 307 (12), 185 (12), 143 (17), 108 (15), 105 (69), 77 (34), 43 (100). Anal. Calc. for C$_{20}$H$_{22}$N$_2$O$_6$S (418.47): C, 57.40; H, 5.13; N, 6.69; S, 7.66. Found: C, 57.27; H, 5.17; N, 6.54; S, 7.41. β-**40**: UV (MeOH) λ$_{max}$ 229 (ε 21,400); ^{1}H NMR (CDCl$_3$) δ 1.91–2.17 (2m, 2 H, SCH$_2$C$\underline{H}_2$), 2.28 (m, 1 H, 3'-H), 2.33 (s, 3 H, C$\underline{H}_3$), 2.42 (m, 2 H, 3'-H$_\alpha$, 4'-H), 2.89, 3.18 (2mc, 2 H, SCH$_2$), 3.99 (ddd = td, J = 4.5, 4.5, 1.5 Hz, 1 H, 5'-H), 4.38 (mc, 2 H, C$\underline{H}_2$OBz), 5.81 (d, J = 8 Hz, 5-H),

6.09 (dd, J = 6.5, 4 Hz, 1 H, 2'-H), 7.26 (m, 2 H, m-ar-H), 7.39 (m, 1 H, p-ar-H), 7.56 (d, J = 8 Hz, 6-H), 8.02 (m, 2 H, o-ar-H), 9.04 (s, 1 H, NH), irradiation at δ 6.09 gives NOE enhancement at δ 8.02 (o-ar-H), 7.56 (6-H), 3.99 (5'-H), 2.42 (3'-H$_\alpha$); IR (CHCl$_3$) 3390, 3010, 1690, 1715, 1450, 1270, 1115 cm^{-1}; MS, m/e (relative intensities) 307 (16), 185 (22), 143 (23), 108 (55), 95 (32), 77 (53), 43 (100). Anal. Calc. for C$_{20}$H$_{22}$N$_2$O$_6$S (418.47): C, 57.40; H, 5.13; N, 6.69; S, 7.66. Found: C, 57.27; H, 5.25; N, 6.64; S, 7.75.

3.1.21. N^2-Isobutyryl-7-([2R,4R,SR]-4-[Benzoyloxymethyl] 5-[2-Acetylthioethyl]Tetrahydrofuran-2-yl)Guanine (β-37)

PPh$_3$ (116 mg, 2 Eq), and DIAD (84)11, 2.2 Eq) in THF (1.0 mL) was reacted with β-**33** (100 mg, 0.2130 mmol), and thioacetic acid (31 μL, 2 Eq) in THF (1.2 mL). Chromatography (silica gel, CH$_2$Cl$_2$:methanol 95:5) yielded **37** (96 mg, 89%). UV (MeOH) λ$_{max}$ 223 (ε 34,,500), 264 (15600); ^{1}H NMR (CDCl$_3$) δ 1.22 (2d = t, J = 7 Hz, 6 H, ^{i}bu-C$\underline{H}_3$), 1.99–2.17 (m, 2 H, SCH$_2$C$\underline{H}_2$), 2.32 (s, 3 H, C$\underline{H}_3$CO), 2.50–2.64 (m, 3 H, 3'-H, ^{i}bu-CH, 4'-H), 3.00 (m, 2 H, SCH$_2$), 3.21 (ddd = sept., J = 5, 8.5, 14 Hz, 1 H, 3'-H$_\alpha$), 4.12 (dt, J = 3, 9 Hz, 1 H, 5'-H), 4.37, 4.45 (2dd, J = 5, 11 Hz, 2H, C$\underline{H}_2$OBz), 6.53 (t, J = 5 Hz, 1 H, 2'-H), 7.46 (m, 3 H, m,p-ar-H), 7.59 (m, 2 H, o-ar-H), 8.18 (s, 1 H, 8-H), 10.52 (s, 1 H, N^2-H), 12.39 (s, 1 H, 1-NH); ^{13}C-NMR (CDCl$_3$) d 19.1, 19.2 (^{i}bu-C$\underline{H}_3$), 26.1 (SCH$_2$C$\underline{H}_2$), 30.6 (C$\underline{H}_3$CO), 35.2 (C-3'), 35.9 (C-4'), 38.3 (SCH$_2$), 41.6 (^{i}buCH), 64.2 (H$_2$OBz), 83.0 (C-5'), 84.7 (C-2'), 128.6, 129.6, 133.4 (ar.-CH), 140.7 (C-8), 148.1 (C-2), 153.1 (C-6), 158.0 (C-4), 166.3 (NHCO), 180.1 (PhCO), 195.5 (C$\underline{H}_3$CO).

3.2. Route 2

3.2.1. 3-Deoxy-3-Benzoyloxymethyl-1:2-O-Isopropylidene-α-D-Allofuranose (45)

To a solution of **44** (24.1 g, 87.9 mmol), prepared from diacetone glucose by the method of Mazur et al. *(23)*, and DMAP (2 mg) in pyridine (50 mL) was added benzoyl chloride (15.0 mL, 1.5 eq) slowly at 0°C. The reaction mixture was allowed to warm to room temperature and was stirred for 3 h. Methanol (20 mL) was then added, and the solvents removed under reduced pressure. The residue was suspended in ethyl acetate and extracted twice with 10% HCl. The organic layer was washed with saturated NaHCO$_3$ and water, dried, and evap-

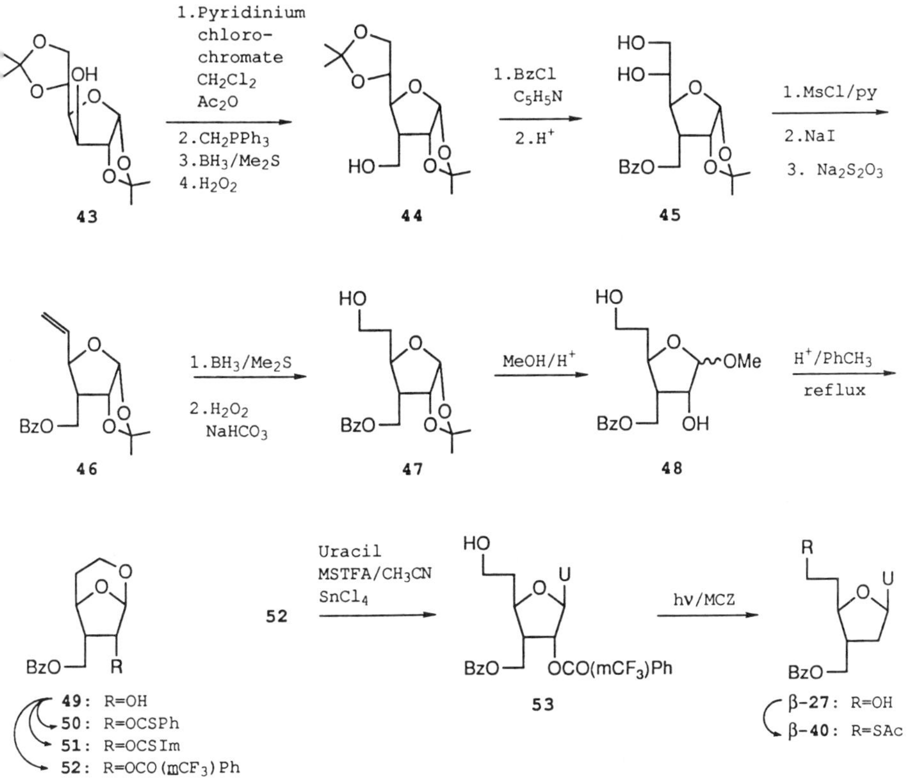

Oligonucleotide Analogs

Scheme 5.

orated. The benzoate (3-deoxy-3-benzoyloxymethyl-1:2,5:6-di-O-isopropylidene-α-D-allofuranose) obtained was sufficiently pure to be used for the next step. An analytical sample was prepared by TLC (1 mm plate, EtOAc:hexane 2:8, R_f 0.35); ^{1}H NMR (CDCl$_3$) δ 1.32, 1.33, 1.42, 1.53 (4s, 12 H, C$\underline{H}_3$), 2.43 (dddd = sept., J = 4, S, 10, 10 Hz, 1 H, 3-H), 3.88(dd, J = 10, 7 Hz, 1 H, 6-H), 3.96–4.25 (m, 3 H, 4-H, SH, 6-H), 4.50 (dd, J = 11, 10 Hz, 1 H, 3'-H), 4.75 (dd, 11, 5 Hz, 1 H, 3'-H), 4.82 (t, J = 4 Hz, 1 H, 2-H), 5.82 (d, J = 3.5 Hz, 1 H, 1-H), 7.45 (m, 2 H, m-ar-H), 7.58 (m, 1 H, p-ar-H), 8.07 (m, 2 H, o-ar-H); IR (CCl$_4$) 2990, 1755, 1385, 1375, 1275 cm^{-1}; MS, m/e 363, 305, 277, 219, 181, 155, 105, 77. Anal. Calc. for C$_{20}$H$_{26}$O$_7$ (378.42): C, 63.48; H, 6.93. Found: C, 63.50; H, 6.93.

The crude benzoate was dissolved in a mixture of MeOH (630 mL), CHCl$_3$ (210 mL), and 1.6% H$_2$SO$_4$ (100 mL), and stirred for 48 h. The reaction mixture was neutralized with saturated NaHCO$_3$, the solvent removed *in vacuo*, and the residue partitioned between ethyl acetate and water. The water layer was extracted with ethyl acetate, and the combined organic layers dried and evaporated. The residue was chromatographed on silica gel (ethyl acetate:hexane 7:3, R$_f$ 0.38) to yield **45** (24.3 g, 82% for two steps) as a clear oil that solidifies while standing at 4°C. The waxy material softens at 40°C and melts at 50–53°C. ^{1}H NMR (CDCl$_3$) δ 1.33, 1.52 (2s, 6 H, C$\underline{H}_3$), 1.88 (br.s, 1 H, OH), 2.48 (m, 1 H, 3-H), 3.16 (br.s, 1 H, OH), 3.80 (m, 3H, 6-H, 5-H), 4.00 (dd, J = 10, 5.5 Hz, 1 H, 4-H), 4.55 (dd, J = 11, 9 Hz, 1 H, 3'-H), 4.70 (dd, J = 11, 5 Hz, 1 H, 3-H), 4.80 (dd, J = 4, 3.5 Hz, 1 H, 2-H), 5.83 (d, J = 3.5 Hz, 1 H, 1-H), 7.45 (m, 2H, *m*-ar-H), 7.58 (m, 1 H, *p*-ar-H), 8.07 (m, 2 H, *o*-ar-H); IR (CCl$_4$) 3480 (br.), 2990, 1725, 1385, 1375, 1275 cm^{-1}; MS, m/e, 323, 219, 155, 105, 77. Anal. Calc. for C$_{17}$H$_{22}$O$_7$ (338.36): C, 60.35; H, 6.55. Found: C, 60.27; H, 6.53.

3.2.2. 3,5,6-Trideoxy-5,6-Didehydro-3-Benzoyloxymethyl-1:2-O-Isopropylidene-α-D-Allofuranose (**46**)

To a solution of **45** (24.0 g, 71.0 mmol) in pyridine (100 mL) was added slowly at –10°C methylsulfonyl chloride (17.0 mL, 3 Eq). The reaction mixture was allowed to warm to room temperature and was stirred for 2 h.

Methanol (20 mL), was added, and most of the solvent removed under reduced pressure. The residue was partitioned between ethyl acetate and 10% HCl. The aqueous layer was extracted three times with ethyl acetate; the combined organic layers were washed with sat. NaHCO$_3$ and brine, and dried and evaporated to yield the dimesylate (3-deoxy-3-benzoyloxymethyl-5,6-di-*O*-methylsulfonyl-1:2-*O*-isopropylidene-α-D-allofuranose) (29.0 g) as a viscous oil. An analytical sample was prepared by silica gel chromatography (ethyl acetate:hexane 1:1, R$_f$ 0.40). ^{1}H NMR (CDCl$_3$) δ 1.33, 1.52 (2s, 6 H, CCH$_3$), 2.59 (m, 1 H, 3-H), 3.04, 3.14 (2s, 6 H, SO$_2$C$\underline{H}_3$), 4.29 (dd, J = 10, 5 Hz, 1 H, 6-H), 4.46 (dd, J = 12, 6 Hz, 1 H, 3'-H), 4.60 (m, 3H, 4-H, 3'-H, 6-H), 4.82 (dd, J = 4, 3.5 Hz, 1 H, 2-H), 5.02 (m, 1 H, 5-H), 5.85 (d, J = 3.5 Hz, 1 H, 1-H), 7.45 (m, 2 H, *m*-ar-H), 7.58 (m, 1 H, *p*-ar-H), 8.07 (m, 2 H, *o*-ar-H); IR (CCl$_4$) 2990, 1745, 1375, 1270, 1240,

1180 cm^{-1}; MS, n/e 479, 219, 155, 123, 105, 77. Anal. Calc. for C$_{19}$H$_{26}$O$_{11}$S$_{2}$ (494.54): C, 46.15; H, 5.30. Found: C, 46.27; H, 5.22.

To a solution of the crude dimesylate in ethylmethylketone (250 mL) was added NaI (30 g), and the mixture was heated at reflux for 15 h (24). The solvent was then removed *in vacuo*, and the residue partitioned between ethyl acetate and sat. aqueous Na$_2$S$_2$O$_3$. The aqueous phase was extracted two times with ethyl acetate and the combined organic layers were washed with brine. The solvent was dried and evaporated and the residue chromatographed on silica gel (hexane:ethyl acetate 8:2, R$_f$ 0.36) to yield **46** (20.3 g) as a clear oil which was ca. 95% pure by NMR (corresponding to an 89% yield after two steps). An analytical sample was prepared by a second chromatographic purification on silica gel. ^{1}H NMR (CDCl$_3$) δ 1.36, 1.56 (2s, 6 H, CH$_3$), 2.30 (m, 1 H, 3-H), 4.38 (dd, J = 10, 7 Hz, 1 H, 4-H), 4.43 (dd, J = 11, 6 Hz, 1 H, 3'-H), 4.56 (dd, J = 11, 8 Hz, 1 H, 3'-H), 4.81 (dd, J = 4, 3.5 Hz, 1H, 2-H), 5.27 (dm, J = 11 Hz, 1 H, 6-H$_E$), 5.41 (dm, J = 18 Hz, 1 H, 6-Hz), 5.83 (dd, J = 10, 7 Hz, 1 H, 5-H), 5.91 (d, 1 H, 1-H), 7.45 (m, 2 H, *m*-ar-H), 7.58 (m, 1 H, *p*-ar-H), 8.07 (m, 2 H, *o*-ar-H); IR (CCl$_4$) 2990, 2960, 2940, 1730, 1450, 1375, 1365, 1270 cm^{-1}; MS, m/e 290, 190, 125, 105, 77. Anal. Calc. for C$_{17}$H$_{20}$O$_5$ (304.34): C, 67.09; H, 6.62. Found: C, 66.94; H, 6.59.

3.2.3. 3,5-Dideoxy-3-Benzoyloxymethyl-1:2-O-Isopropylidene-α-D-Allofuranose (47)

The olefin obtained above (20.0 g, 62.5 mmol, calculated on the basis of a content of 95% **46**) was dissolved in THF (200 mL), the solution cooled to 0°C, and BH$_3$•Me$_2$S (15 mL) added. The reaction mixture was stirred for 24 h at 0–4°C, and cooled to –10°C, and methanol (20 mL) was added carefully. After 1 h at 0°C, water (100 mL), NaHCO$_3$ (35 g) and then 30% H$_2$O$_2$ (60 mL, dropwise) were added while the temperature was maintained at 0°C. The mixture was allowed to warm to room temperature and was stirred for 2 h. After half of the solvents were removed under reduced pressure, the mixture was extracted three times with ethyl acetate. The organic layer was washed with saturated Na$_2$S$_2$O$_3$ and brine, dried and evaporated. The residue was chromatographed (silica gel, ethyl acetate:hexane 6:4, R$_f$ 0.40) to yield 13.8 g (68%) of **47** as an oil. ^{1}H NMR (CDCl$_3$) δ 1.35, 1.54, (2s, 6 H, CH$_3$), 1.74–1.86 (m, 1 H, 5-H), 2.04–2.15 (m, 1 H, H-5), 2.06 (s, 1 H, OH), 2.22–2.34 (m, 1 H, 3-H),

3.85 (m, 2 H, 6-H), 4.20 (td, J = 11, 3 Hz, 1 H, 4-H), 4.43 (dd, J = 11, 7 Hz, 1 H, 3'-H), 4.60 (dd, J = 11, 7.5 Hz, 1 H, 3'-H), 4.77 (dd, J = 4, 3.5 Hz, 1H, 2-H), 5.87 (d, J = 3.5 Hz, 1 H, 1-H), 7.45 (m, 2H, m-ar-H), 7.58 (m, 1 H, p-ar-H), 8.07 (m, 2 H, ar-H); IR (CCl_4) 3540, 2990, 2960, 2940, 1730, 1450, 1375, 1365, 1275 cm^{-1}; MS, m/e 307, 247, 219, 185, 125, 105, 77.

3.2.4. (1R,5R,6R,7R)-6-Benzoyloxymethyl-7-Hydroxy-2,8-Dioxa-(3.2.1.)-Bicyclooctane (49)

To a solution of **47** (2.246 g, 6.976 mmol) in methanol was added dry acidic cation exchange resin (Dowex 50 W8, 2.0 g), and the mixture heated at reflux for 5.5 h. The resin was removed by filtration, and the methanol evaporated to leave **48** (2.029 g) as an oil. The oil was dissolved in toluene (25 mL), and added to a refluxing suspension of dry cationic exchange resin (Dowex 50 W8, 13 g) in toluene (250 mL), and the mixture refluxed for 15 min. After being filtered, the toluene solution was washed with saturated $NaHCO_3$ and brine, dried, and evaporated. The residue was chromatographed (silica gel, ethyl acetate:hexane 6:4) to leave **49** (1.250 g, 69%) as a clear oil, which was crystallized from ethyl acetate/hexane (white needles, mp: 96–97°C). **48** (ca. 9:1 mixture of anomers, major anomer): ^1H NMR ($CDCl_3$) δ 1.78–2.04 (m, 2 H, $HOCH_2C\underline{H}_2$), 2.5 (br., OH), 2.56 (m, 1 H, 3-H), 3.38 (s, 3 H, OCH_3), 3.89 (t, J = 7 Hz, 2 H, $HOCH_2$), 4.10–4.28 (m, 3 H, 4-H, 3'-H), 4.80 (dd, J = 11, 9.5 Hz, 1 H, 2-H), 4.88 (s, 1 H, 1-H), 7.45 (m, 2 H, m-ar-H), 7.58 (m, 1 H, p-ar-H), 8.07 (m, 2 H, o-ar-H). **49**: ^1H NMR ($CDCl_3$) δ 1.30 (dm, J = 13 Hz, 1 H, 4-H), 2.29 (d, J = 6.5 Hz, 1 H, OH), 2.38 (m, 1 H, 6-H), 2.73 (m, 1 H, 4-H), 3.82 (td, J = 12, 4 Hz, 1 H, 3-H), 3.94 (dd, J = 12, 6.5 Hz, 1 H, 3-H), 4.49 (dd, J = 11, 4 Hz, 1H, ($C\underline{H}_2OBz$)), 4.57 (s, 1 H, 5-H), 4.72 (dd, J = 11, 7 Hz, 1 H, $C\underline{H}_2OBz$), 4.65 (t, J = 6.5 Hz, 1 H, 7-H), 5.23 (s, 1 H, 1-H), 7.44 (m, 2 H, m-arH), 7.60 (m, 1 H, p-ar-H), 8.05 (m, 2 H, o-ar-H); IR (KBr) 3470 (br.), 2960, 1710, 1290, 1270, 1255, 1140, 1085, 710 cm^{-1}; MS, m/e 135, 123, 105, 77. Anal. Calc. for $C_{14}H_{16}O_5$ (264.28): C, 63.63; H, 6.10. Found: C, 63.78; H, 6.19.

3.2.5. (1R,5R,6R,7R)-6-Benzoyloxymethyl-7-([4-Trifluoromethyl]Benzoyloxy)-2,8-Dioxa-(3.2.1.)-Bicyclooctane (52)

To a solution of **49** (272 mg, 1.030 mmol), and DMAP (5 mg) in pyridine (20 mL) was added at 0°C m-trifluoromethylbenzoyl chloride (240 1, 1.5 eq). The mixture was warmed to room temperature and stirred for 3 h. Methanol (2 mL) was added, the solvents removed *in vacuo*, and the residue chromatographed (silica gel, hexane:ethyl acetate 7:3) to yield **52** (396 mg, 88%) as a clear oil. ^{1}H NMR (CDCl$_3$) δ 1.43 (dm, J = 13 Hz, 1 H, 4-H), 2.47 (m, 1 H, 6-H), 3.04 (quar., J = 7.5 Hz, 1 H, 4-H), 4.02 (m, 2 H, 3-H), 4.51 (m, 2 H, CH$_2$OBz), 4.64 (m, 1 H, 5-H), 5.41 (s, 1 H, 1-H), 6.04 (d, J = 8 Hz, 1 H, 7-H), 7.34, 7,53, 7.77, 7.86, 8.17, 8.20 (6m, 9 H, ar-H); IR (CCl$_4$) 2960, 2870, 1740, 1450, 1335, 1270, 1245 cm^{-1}; MS, m/e 436, 364, 275, 263, 173, 145, 105. Anal. Calc. for C$_{22}$H$_{19}$O$_6$F$_3$ (436.38): C, 60.55; H, 4.39. Found: C, 60.59; H, 4.53.

3.2.6. 1-([2R,3R,4R,5R]-3-[4-Trifluoromethyl Benzoyloxy]-4-[Benzoyloxymethyl]-5-[Hydroxyethyl] Tetrahydrofuran-2-yl)Uracil (53)

A mixture of **52** (98 mg, 0.211 mmol), uracil (26 mg, 1.1 eq), and MSTFA in acetonitrile (2.0 mL) was stirred at 80°C until a clear solution had formed (ca 30 min). To this solution was added SnCl$_4$ (37 µL) at room temperature, and the solution stirred for 40 h. The reaction mixture was cooled to 0°C and sat. NaHCO$_3$ (1 mL) was added. The mixture was then extracted three times with ethyl acetate. The combined organic layers were dried and evaporated, and the residue chromatographed (silica gel, ethyl acetate) to yield **53** (110 mg, 95%) as a colorless oil. ^{1}H NMR (CDCl$_3$) δ 1.98–2.26 (3 H, HOCH$_2$CH$_2$, OH), 3.07 (m, 1 H, 4'-H), 3.93 (m, 2 H, HOCH$_2$), 4.45 (td, J = 3.5, 9 Hz, 1 H, 5'-H), 4.56 (m, 2 H, CH$_2$OBz), 5.79 (d, J = 8 Hz, 1 H, 5-H), 5.81 (d, J = 2 Hz, 1 H, 2'-H), 5.83 (dd, J = 6.5, 2 Hz, 3'-H), 7.38, 7.55, 7.83, 7.94, 8.08, 8.24 (6m, 10 H, 6-H, ar-H), 9.03 (s, 1 H, NH); IR (KBr) 3430, 3120, 3060, 2950, 1720, 1690, 1618, 1600, 1582, 1450, 1250, 1130, 815, 755, 714, 693 cm^{-1}; MS, m/e 190, 173, 145, 125, 111, 105, 95, 77.

3.2.7. 1-([2R,4R,SR]-4-[Benzoyloxymethyl] 5-[Hydroxyethyl]Tetrahydrofuran-2-yl)Uracil (β-27)

Benzoate **53** (77 mg, 0.14 mmol) and *N*-methylcarbazole (13 mg, 1.1 Eq) were dissolved in isopropanol:water (9:1, degassed by saturating with argon for 30 min, 140 mL), and the mixture irradiated at ca. 10°C for 90 min with a 400-W high-pressure mercury lamp through a Pyrex filter *(25)*. The solvents were removed under reduced pressure, and the residue chromatographed (silica gel, CH$_2$Cl$_2$:methanol

9:1). After evaporation of the solvent, the residue was crystallized from ethyl acetate to give β-**27** (43 mg, 85%) as colorless prisms, mp. 157–158°C. ^{1}H NMR (CDCl$_3$) δ 1.80–2.58 (5m, 6 H, 3'-H, 4'-H, HOCH$_2$CH$_2$, OH), 3.89 (m, 2 H, HOCH$_2$), 4.11 (td, J = 9, 3 Hz, 1 H, 5'-H), 4.21 (m, 2 H, CH$_2$OBz), 5.77 (dd, J = 8, 2 Hz, 1 H, 5-H), 6.12 (dd, J = 7, 3.5 Hz, 1 H, 2'-H), 7.48 (m, 3 H, m-ar-H, 6-H), 7.61 (m, 1 H, p-ar-H), 8.02 (m, 3 H, NH, o-ar-H); IR (KBr) 3510, 3460, 3110, 3060, 2990, 1727, 1680, 1450, 1220, 1123, 772, 710 cm^{-1}; MS, m/e 249, 127, 112, 109, 105, 9S, 77. Anal. Calc. for C$_{18}$H$_{20}$N$_2$O$_6$ (360.37): C, S9.99; H, 5.59; N, 7.77. Found: C, 60.03; H, 5.62; N, 7.68.

3.2.8. 1-([2R,4R,5R]-4-[Benzoyloxymethyl]5-[Acetylthioethyl]Tetrahydrofuran-2-yl)Uracil (β-**40**)

PPh$_3$ (39 mg, 2 Eq), and DIAD (30 μL, 2 Eq) in THF (1.5 mL) were reacted with β-**27** (27 mg, 75 μmol), and thioacetic acid (11.5 μL, 2 eq) in THF (0.75 mL). TLC (1-mm silica gel plate, CH$_2$Cl$_2$:methanol 96:4) to yield β-**40** (21 mg, 67%) as a clear oil (analytical data *vide supra*).

3.2.9. N^6-Benzoyl-9-([2R,4R,5R]-4-[Hydroxymethyl]5-[2-Mercaptoethyl]Tetrahydrofuran-2-yl)Adenine (**54**)

To a solution of **36** (300 mg, 0.550 mmol) in THF:methanol (1:1:2.0 mL) was added 1M NaOH (1.7 mL), and the mixture was stirred at 0°C for 1.5 h. Pyridinium-Dowex® was added, the mixture filtered and the solvent evaporated. The resulting oil was chromatographed (silica gel, CH$_2$Cl$_2$:methanol 9:1) to yield **54** (197 mg, 90%) as a white foam. ^{1}H NMR (CDCl$_3$) δ 1.98–2.10 and 2.45–2.82 (2 m, 8 H, OH, 3'-H, 4'-H, HS-CH$_2$-CH$_2$), 3.82 (m, 2 H, HO-CH$_2$), 4.19 (m, 1 H, 5'-H), 6.33 (dd, J = 3, 6.5 Hz, 1'-H), 7.50–7.65 (m, 3 H, m,p-ar-H), 8.04 (m, 2 H, o-ar-H), 8.18 (s, 1 H, 8-H), 8.81 (s, 1 H, 2-H), 9.07 (s br., 1 H, NH); IR (KBr) 3620, 3400, 3000, 2940, 1710, 1610, 1560, 1455 cm^{-1}.

3.2.10. N^6-Benzoyl-9-([2R,4R,5R]-4-[Hydroxymethyl]5-[2-Methylthioethyl]Tetrahydrofuran-2-yl)Adenine (**55**)

To a solution of **54** (30 mg, 75.1 mol) in DMF (0.2 mL) was added Cs$_2$CO$_3$ (25.0 mg, 1.1 eq.), and the mixture was stirred at 0°C for 2 h. Acetic acid (10 μL) was added, and the solvent removed *in vacuo*. The resulting oil was chromatographed (1 mm silica gel TLC plate, CH$_2$Cl$_2$;methanol 9:1) to yield 25 mg (82%) **55**. ^{1}H NMR (CDCl$_3$) δ

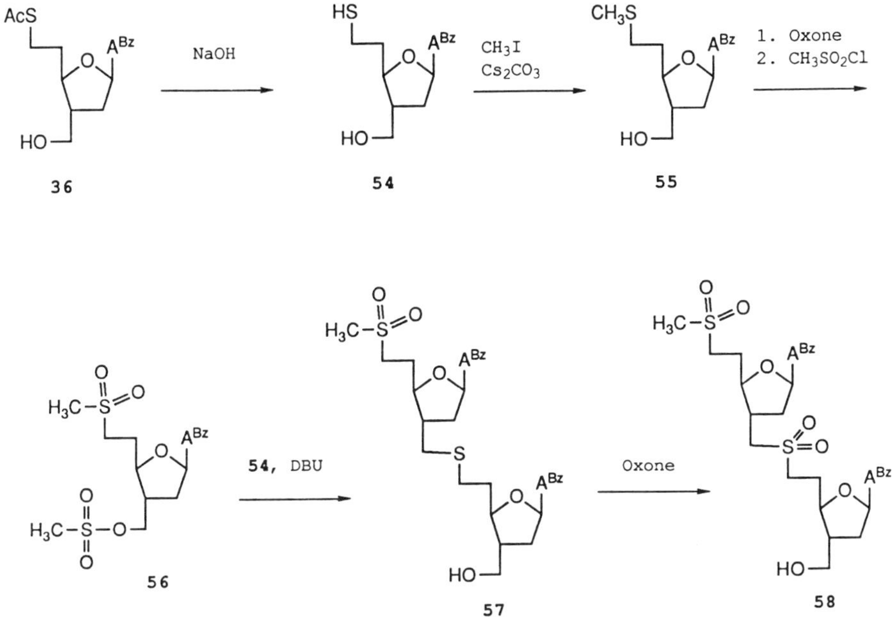

Oligonucleotide Analogs

Scheme 6.

1.90–2.05 and 2.42–2.80 (2 m, 8 H, OH, 3'-H, 4'-H, MeS-C$\underline{H}_2$-C$\underline{H}_2$), 2.10 (s, 3 H, C$\underline{H}_3$), 3.80 (m, 2 H, HO-C$\underline{H}_2$), 4.14 (m, 1 H, 5'-H), 6.32 (dd, J = 3.5, 6.5 Hz, 2'-H), 7.55 (m, 3 H, *m,p*-ar-H), 8.05 (m, 2 H, *o*-ar-H), 8.20 (s, 1 H, 8-H), 8.78 (s, 1 H, 2-H), 9.18 (s br., 1 H, NH); MS, m/e 413, 310, 294, 239, 211, 156, 105.

3.2.11. N^6-Benzoyl-9-([2R,4R,5R]-4-[Methysulfonyloxymethyl]5-[2-Methylsulfonoethyl] Tetrahydrofuran-2-yl)Adenine (**56**)

Oxone (400 μL of a 0.625M solution in 1M sodium acetate buffer, pH 4.5) was added at room temperature to a solution of **55** (20 mg, 48 μmol) in methanol (400 μL). The mixture was stirred at room temperature for 30 min. Methanol (1 mL) was added, and the mixture filtered. The filtrate was applied to a TLC plate (1 mm), which was eluted three times with CH$_2$Cl$_2$:methanol 9:1. Yield 25 mg (81%) of the corresponding sulfone. The compound obtained was dissolved in pyridine, the solution cooled 0°C, and MsCl (14 μL, 3.0 eq) was added. After 1 h at 0°C the solvent was removed *in vacuo*, and the residue applied to a silica gel TLC plate (0.5 mm), which was eluted with CH_2Cl_2:methanol 9:1 to furnish **56** (24.5 mg, 77%). ^{1}H NMR ($CDCl_3$) 2.18–2.60 (2 m, 5 H, 3'-H, 4'-H, SO_2-C$\underline{H}_2$-C$\underline{H}_2$), 2.88 (s, 3 H, O_2SCH_3), 2.90–3.05 (m, 1 H, SO_2-C$\underline{H}_2$), 3.10 (s, 3 H, O_3SCH_3), 3.15–3.24 (m, 1 H, SO_2-C$\underline{H}_2$), 4.26 (m, 1 H, 5'-H), 4.38 (m, 2 H, 4"-H), 6.26 (dd, J = 3.5, 8 Hz, 2'-H), 7.55 (m, 3 H, *m,p*-ar-H), 8.03 (m, 2 H, *o*-ar-H), 8.12 (s, 1 H, 8-H), 8.77 (s, 1 H, 2H), 9.13 (s br., 1 H, NH); ^{13}C NMR δ 27.1 (C-5'), 34.5 (C-2'), 37.8, 41.1 (C$\underline{H}_3$), 42.6 (C-3'), 51.3 (C-3"), 68.5 (C-6'), 81.4 (C-4'), 128.3, 129.3, 133.3, 133.7 (ar-C), 142.4 (C-8), 150.1, 151.5 (C-4, C-6), 153.2 (C-2), 165.1 (CO); MS, m/e 427, 398, 348, 334, 320, 278, 253, 239, 211, 174, 135, 122, 105.

3.2.12. Preparation of 57, the Thioether-Linked Dimer of the Protected Adenosine Analog

To a solution of **56** (4.7 mg, 8.98 μmol), and **54** (4.3 mg, 1.2 eq) in DMF (50 μL) was added diazabicycloundecene (DBU, 1.8 μL, 1.3 Eq), and the solution stirred for 72 h at room temperature. Acetic acid (0.5 μL) was added, and the solvent removed *in vacuo*. The residue was applied to a silica gel TLC plate (0.25 mm), and the plate eluted three times with CH_2Cl_2:methanol 9:1. Isolation of the product yielded **57** (5.8 mg, 78%). FAB$^+$-MS, m/e 825 (M$^+$–1).

3.2.13. Preparation of 58, the Sulfone-Linked Dimer of the Protected Adenosine Analog

To a solution of **56** (3.2 mg, 6.13 μmol), and **54** (2.9 mg, 1.2 eq) in DMF (40 μL) was added DBU (1.2 μL, 1.3 eq), and the solution stirred for 96 h at room temperature. Acetic acid (0.5 μL) was added, followed by Oxone (30 μL of a 0.625*M* solution in 2*M* sodium acetate, pH 4.5). The mixture was stirred for 30 min, and the solvent removed *in vacuo*. The residue was applied to a silica gel TLC plate (0.25 mm), and the plate eluted three times with C$\underline{H}_2Cl_2$:methanol 9:1. Isolation of the product afforded **58** (3.5 mg, 70%). ^{1}H NMR ($CDCl_3$) δ 2.42–2.98 (m, 8 H), 3.09 (s, 3 H, O_2SCH_3), 3.32 (td, 1 H), 3.44–4.03 (m, 10 H), 4.24 (m, 1 H), 4.43 (m, 1 H), 6.56 (dd, J = 3.5, 7.5 Hz, 1H), 6.62 (dd, J = 3, 7 Hz, 1H), 7.24–7.56 (m, 6H, *m,p*-ar-H), 8.36 (m, 4 H, *o*-ar-H), 8.81, 8.92, 8.94, 9.04 (4 s, 4 H).

4. Discussion

In the first route (Schemes 1–4), the carbon skeleton arises from a Diels-Alder reaction, optical activity is introduced by an enzymatic hydrolysis, and the α- and β-anomers of the nucleoside analogs are separated chromatographically. The primary advantage of the route is cheap starting materials available in bulk, a relatively easy resolution based on pig liver esterase (PLE) *(26–31)*, and accessibility of all four nucleoside bases in any state of protection. The primary disadvantages of this route arise from difficulties in the workup of the ozonolysis reaction and the fact that mixtures of anomers that must be separated chromatographically are obtained by the glycosylation reactions. Since a specific 3'-acyl derivative was not needed to control anomeric specificity, a 3'-protecting group could be chosen to allow selective deprotection of the 3'-oxygen of the target molecules in the presence of the amide-protected bases. The benzoate **22** is optimal with respect to stability and was therefore used.

The assignment of structure of these nucleoside analogs focuses on three details: the absolute configuration, the configuration at the anomeric center, and the sites of the attachment of the purine ring of guanine and adenine (e.g., N^7 or N^9) were obtained by a combination of NMR spectroscopy, correlation with substances with known structure, and crystallography. They are reviewed elsewhere *(18,19)*.

The primary advantage of the second route is that the glycosidation reaction yields a single anomer. Its primary disadvantage is that the 2'-oxygen must subsequently be removed to obtain the deoxyribo analog. The photochemical deoxygenation route used here works well for the uracil and unprotected adenine derivatives, but not with protected derivatives of adenosine, guanosine, and cytosine. Barton-type deoxygenation routes *(32)* seem, however, to be satisfactory for all four nucleoside bases regardless of the protecting groups that they bear.

The synthesis of a dimer was based on the S_N2 reaction between the free thiolate anion of one building block and a mesylate of the second. The 5'-thiol of **36** was capped and the 3'-hydroxyl group activated for nucleophilic displacement by the thiol group of a second monomer (Scheme 6). Thus, **36** was deprotected to **54** and converted to the methyl sulfide **55** by reaction with methyl iodide (75% yield for two steps). Oxidation of the resulting sulfide with $KHSO_5$ (81% yield), and subsequent mesylation with mesyl chloride in pyridine furnished starting monomer **56** in 77% yield. Coupling of **56** to **54** occurred smoothly in the presence of diazabicycloundecene in dimethyl formamide to furnish thioether-linked dimer **57**, which could be oxidized *in situ* to the synthetic target, sulfone-linked dimer **58**, which was isolated in 70% yield. Oligonucleotide analogs octamer containing uridine have been prepared by a set of analogous steps (Z. Huang, unpublished).

5. Notes

1. Although this route appears to be circuitous, it proved to be more reliable than routes based on efforts to obtain *cis*-2-acetoxy-carboethoxy-cyclohex-4-ene, and to maintain the *cis* stereochemistry during the Diels-Alder reaction.
2. Addition of organic cosolvent at this point is critical. The esterase reaction, when run in water as the sole solvent, yields a mixture of products. In water at pH 7 (maintained by automatic titration with dilute NaOH), PLE gave (–)**8** with an enantiomeric excess (ee) of 67.5% after 40 mol% of hydroxide had been added. In mixtures of water and *t*-butanol (9:1) *(33,34)*, the acid obtained after addition of 45 mol% sodium hydroxide was enantiomerically pure, as determined by gas chromatographic (GC) analysis of the l-isoborneol ester of **8** and by NMR analysis of the methyl ester of **8** (from esterification of **8** with diazomethane) in the presence of the chiral shift reagent Eu(hfc)$_3$ (hfc = heptafluorohxdroxymethylene-D-camphorato). The remaining mixture (ca. 10:1) of (+)**7** to (–)**7** could then be subjected again to PLE-catalyzed hydrolysis to yield enantiomerically pure (+)**7**, as determined by NMR in the presence of Eu(hfc)$_3$. The absolute configuration of (–)**7** was assigned by direct correlation to glucose *(vide infra)*.
3. This step is the most difficult of the sequence. After a series of experiments using a variety of acidic catalysts, it was found that diacetal **17** and monoacetal **16** (Scheme 3, in an approx 1:3 ratio) could be obtained from **11** simply by stirring the crude ozonolysis product in methanol in the dark for 14 d. Presumably traces of acid that formed during ozonolysis catalyze this selective acetalization. The crude reaction mixture could be reduced with NaBH$_4$ to furnish **18** in 63% overall yield. It is recommended that the material from the previous step be carried forward in small portions, so that one can develop a intuition for the reaction before committing much material. In a model reaction (Scheme 2), **12** was ozonized in methanol, treated with dimethylsulfide, and then treated with acidic cation-exchange resin to give **13** in 90%

yield. The 6-dimethylacetal could be selectively deprotected by treatment with 50% trifluoroacetic acid in chloroform to yield the aldehyde **15** in 85% yield *(35)*. Remarkably, this sequence of reactions applied to **10** failed to yield any of the analogous products. If harsher conditions were applied, several unidentified products were formed. The difference in reactivity of the two isomers presumably arises from participation of the carboxylate group via a six-membered ring (**14**) during hydrolysis of **13**.

4. To obtain an intermediate suitable for introduction of a base, the hydroxyl group of **18** was internally protected by acid-catalyzed cyclization of **18** to **19** in refluxing toluene (85%). The somewhat strained bicyclic system thus formed is thought to make the introduction of the base more facile, since the formation of the C-1-carbocation *(36)* should be fast compared with its formation from a methyl furanoside.

5. A variety of conditions were examined to introduce the nucleoside base effectively. In all cases, the nucleosidations were faster and cleaner if acetonitrile, rather than dichloroethane, was used as solvent. In general, higher yields were obtained if trimethylsilyl trifluoromethanesulfonate (TMSOTf) was used instead of $SnCl_4$ as the catalyst. In the case of uracil and adenine, catalytic amounts of TMSOTf were sufficient to ensure completion of the reaction within hours, whereas cytosine and guanine required a molar excess of TMSOTf, presumably because the base and the Lewis acid form a complex *(37)*.

6. This reaction works extremely well on every analog that has been examined.

7. The introduction of nucleoside bases is today generally performed by the Hilbert-Johnson reaction *(38)*, modified by Wittenberg *(39)*, to use the silyl group and by Niedballa and Vorbrüggen *(40)* to use Friedel-Crafts catalysts. It is well known that the presence of a 2'-acyl group stabilizes the intermediate C-1-carbocation from the α-face, thereby yielding the β-anomer. We initially hoped that the 3'-pivaloyl group might similarly influence the stereochemical outcome of the reaction in favor of the β-anomer, since the additional carbon atom in the 3'-position could conceivably allow a Lewis-basic group attached to the 3'-position to protect the α-face of the intermediate C-1 carbocation (**31**). Unfortunately, treatment of **19** with *bis*-trimethylsilyl uracil with a variety of Lewis-acids in dichloroethane or acetonitrile led consistently to a 1:1 mixture of anomers of **25** (Scheme 3). The stereochemical course of the reaction of derivatives **21–25**, listed in the order of increasing basicity (Scheme 3), with *bis*-trimethylsilyl uracil was investigated using both $SnCl_4$ and TMSOTf as catalysts, and both dichloroethane and acetonitrile as solvents. Although no stereoselectivity for one of the anomers of **25–28** was observed for reactions with the acyl

derivatives **19** and **21–23**, sulfinate **24** reacted with silylated uracil in acetonitrile using TMSOTf as catalyst to afford in moderate yield (45%) a 1.3:1 mixture of anomers of nucleoside **29**, analyzed by GC following hydrolysis with sodium hydroxide to the alcohol **30** and perimethylsilylation with *N*-methyl-*N*-trimethylsilyl trifluoroacetamide (MSTFA). Although the β-anomer was presumably the predominant product, the moderate stereoselectivity proved on a routine basis to be inadequate to justify the additional synthetic effort. A final approach toward a stereoselective nucleosidation reaction relied on an S_N2 displacement of a furanosyl derivative possessing an α-leaving group. It has been reported that the chloride in 3'-deoxychlororibosides can be stereoselectively displaced by an S_N2 reaction by *bis*-trimethylsilyluracil *(41)*. The highest selectivities are obtained when the reaction is carried out in chloroform, since the rate of S_N2 displacement is high relative to the rate of anomerization of the chloride in this solvent. Further, the C-1-acetate groups of ribosides can be displaced by iodine through the action of trimethylsilyliodide *(42)*. This suggested that if it were possible to react **22** stereoselectively with trimethylsilyl iodide to give corresponding α-iodide, and if this compound did not anomerize under the reaction conditions, a second displacement of the iodide should give the β-anomer of uridine analog **27**. Following this scheme, a mixture of **22** and silylated uracil in chloroform was treated at 0°C with 0.5 eq trimethylsilyl iodide. The readily formed uridine analog again was a 1:1 mixture of anomers of **27**.

8. Regarding the chromatographic separation of anomers, although a simple silica gel column does not produce baseline separation with a variety of solvent systems for many of the anomeric mixtures examined, saturation of the organic eluants with water made it possible to achieve baseline separation via High Performance Liquid Chromatography (HPLC) for at least one derivative of each of the analogs, specified in the individual procedures, even for relatively large amounts (up to 100 mg/injection on a 30×250 mm silica gel column), and this approach provides a reliable route to diastereomerically pure building blocks.

6. Summary

Two sets of experimental protocols are given for the synthesis of 3',5'-*bis*-homodeoxyribonucleosides, building blocks for the synthesis of oligodeoxynucleotide analogs where the $-O-PO_2-O-$ groups are replaced by $-CH_2-S-CH_2-$, $-CH_2-SO-CH_2-$, and $-CH_2-SO_2-CH_2-$ units. Conditions are presented for joining these building blocks to create short nucleic acid analogs. Since isosteric, achiral, and nonionic analogs of natural oligonucleotides stable to both enzymatic and chemical hydrolysis, such molecules have potential application as probes in the laboratory, in studies of the biological function of individual genes, and as "antisense" oligonucleotide analogs.

References

1. Hirashima, A. and Inouye, M. (1973) Specific biosynthesis of an envelope protein of *Escherichia coli. Nature* **242,** 405–407.
2. Green, P. J., Pines, O., and Inouye M. (1986) The role of antisense RNA in gene regulation. *Ann. Rev. Biochem.* **55,** 569–597.
3. Inouye, M. (1988) Antisense RNA: Its functions and applications in gene regulation—A review. *Gene* **72,** 25–34.
4. Paterson, B. M., Roberts, B. E., and Kuff, E. L. (1977) Structural gene identification and mapping by DNA–mRNA hybrid-arrested cell-free translation. *Proc. Natl. Acad. Sci. USA* **74,** 4370–4374.
5. Stephenson, M. L. and Zamecnik, P. C. (1978) Inhibition of Rous sarcoma viral RNA translation by a specific oligodeoxyribonucleotide. *Proc. Natl. Acad. Sci. USA* **75,** 285–288.
6. Rich, A. (1962) On the problems of evolution and biochemical information transfer, in *Horizons in Biochemistry* (Kasha, M. and Pullman, B., eds.), Academic, New York, pp. 103–126.
7. Letsinger, R. L., Bach, S. A., and Eadie J. S. (1986) Effects of pendant groups at phosphorus on binding properties of δ-ApA analogues. *Nucl. Acids Res.* **14,** 3487–3499.
8. Ts'o, P. O. P. and Miller, P. S. (1984) Nonionic nucleic acid alkyl and aryl phosphonates and processes for manufacture and use thereof. US Patent number 4,469,863, Sept. 4.
9. Miller, P. S., Agris, C. H., Aurelian, L., Blake, K. R., Murakami, A., Reddy, M. P., Spitz, S. A., and Ts'o, P. O. P. (1985) *Biochimie* **67,** 769–776.
10. Miller, P. S., McParland, K. B., Jayaraman, K., and Ts'o, P. O. P. (1981) Biochemical and biological effects of nonionic nucleic acid methylphosphonates. *Biochemistry* **20,** 1874–1880.
11. Murakami, A., Blake, K. R., and Miller, P. S. (1985) Characterization of sequence-specific oligodeoxyribonucleoside methylphosphonates and their interaction with rabbit globin mRNA. *Biochemistry* **24,** 4041–4046.
12. Johnsson, K., Allemann, R. K., and Benner, S. A. (1990) Designed enzymes: new peptides that fold in aqueous solution and catalyze reactions, in *Molecular Mechanisms in Bioorganic Processes* (Bleasdale, C. and Golding, B. T., eds.), Royal Society of Chemistry, Cambridge, UK, pp. 166–187.
13. Benner, S. A. and Allemann, R. K. (1989) The return of pancreatic ribonucleases. *Trends Biochem. Sci.* **14,** 396,397.

14. Gonnet, G. H. and Benner, S. A. (1991) Computational biochemistry research at ETH, Technical Report 154, Departement Informatik, March 1991.
15. Moody, H. M., Quaedflieg, P. J. L. M., Koole, L. H., van Genderen, M. H. P., Buck, H. M., Smit, L., Jurrianns, S., Geelen, J. L. M. C., and Goudsmit, J. (1990) Inhibition of HIV-1 infectivity by phosphate-methylated DNA: retraction. *Science* **250**, 125,126.
16. Nambiar, K. P., Stackhouse, J., Stauffer, D. M., Kennedy, W. P., Eldredge, J. K., and Benner, S. A. (1984) Total synthesis and cloning of a gene coding for the ribonuclease S protein. *Science* **223**, 1299–1301.
17. Schaller, H., Weimann, G., Lerch, B., and Khorana, H. G. (1963) Studies on polynucleotides XXIV. *J. Am. Chem. Soc.* **85**, 3821–3827.
18. Schneider, K. C. and Benner, S. A. (1990) Building blocks for oligonucleotide analogs with dimethylene-sulfide, -sulfoxide, and -sulfone groups replacing phosphodiester linkages. *Tetrahedron Lett.* **31**, 335–338.
19. Huang, Z., Schneider, K. C., and Benner, S. A. (1991) Building blocks for analogs of ribo-and deoxyribonucleotides with dimethylene-sulfide, -sulfoxide, and -sulfone groups replacing phosphodiester linkages. *J. Org. Chem.* **56**, 3869–3882.
20. Blancou, H. and Casadevall, E. (1976) Reaction d'elimination-1,3 action du n-butyl lithium sur les iodomethyl-2 tosyloxy-1 cyclohexanes (enes) cis et trans. stereochimie et mecanisme. *Tetrahedron* **32**, 2907–2913.
21. Mitsunobu, O. (1981) The use of diethyl azodicarboxylate and triphenylphosphine in synthesis and transformation of natural products. *Synthesis* **13**, 1–28.
22. Volante, R. P. (1981) A new highly efficient method for the conversion of alcohols to thiolesters and thiols. *Tetrahedron Lett.* **22**, 3119–3122.
23. Mazur, A., Tropp, B. E., and Engel R. (1984) Isosteres of natural phosphates. 11. Synthesis of a phosphonic acid analog of an oligonucleotide. *Tetrahedron* **40**, 3949–3956.
24. Gurjar, M. K., Patil, V. J., and Pawar, S. M. (1987) Sythesis of (1R, rR)-2,6-dioxabicyclo [3.3.0] octan-3-one from D-glucose. *Carbohydrate Res.* **165**, 313–317.
25. Saito, I., Ikehira, H., Kasatani, R., Watanabe, M., and Matsuura, T. (1986) Selective deoxygenation of secondary alcohols by photosensitized electron-transfer reaction. A general procedure for deoxygenation of ribonucleosides. *J. Am. Chem. Soc.* **108**, 3115–3117.
26. Boland, W, Niedermeyer, U., and Jaenicke, L. (1985) Enantioselective syntheses and absolute configurations of viridiene and aucantene, two constituents of algae pheromone bouquets. *Helv. Chim. Acta* **68**, 2062–2073.
27. Gais, H. J. and Lukas, K. L. (1984) Enantioselektive und enantiokonvergente Synthese von Bausteinen zur Totalsynthese cyclopentanoida Naturstoffe. *Angew. Chem.* **96**, 140,141.
28. Schneider, M., Engel, N., Honicke, P., Hdnemann, G., and Görisch, H. (1984) Enzymatische Synthesen chiraler Bausteine aus prochiralen) *meso*-Substraten: Herstellung von Methyl(hydrogen)-1,2-cycloalkandicarboxylaten. *Angew. Chem.* **96**, 55,56.

29. Wilson, W. K., Baca, S. B., Barber, Y. J., Scallen, T. J., and Morrow, C. J. (1983) Enantioselective hydrolysis of 3-hydroxy-3-methylalkanoic acid esters with pig liver esterase. *J. Org. Chem.* **48**, 3960–3966.
30. Mohr, P., Waespe-Sarcevic, N., and Tamm, C. (1983) A study of stereoselective hydrolysis of symmetrical diesters with pig liver esterase. *Helv. Chim. Acta* **66**, 2501–2511.
31. Huang, F. C., Lee, L. F. H., Mittal, R. S. D., Ravikumar, P. R., Chan, J. A., Sih, C. J., Caspi, E., and Eck, C. R. (1975) Synthesis and characterization of the fluxional species $H_2OS_3 (CO)_{10}L$. The crystal structure of $H_2OS_3 (CO)_{11}$. *J. Am. Chem. Soc.* **97**, 4144,4145.
32. Barton, D. H. R. and McCombie, S. W. (1975) A new method for the deoxygenation of secondary alcohols. *J. C. S. Perkin I* 1574–1585.
33. Cambou, B. and Klibanov, A. M. (1984) Preparative production of optically active esters and alcohols using esterase-catalyzed stereospecific transesterification in organic media. *J. Am. Chem. Soc.* **106**, 2687–2692.
34. Jones, J. B. and Mehes, M. M. (1979) Effects of organic cosolvents on enzyme stereospecificity. The enantiomeric specificity of α-chymotypsin is reduced by high organic solvent concentrations. *Can. J. Chem.* **57**, 2245–2248.
35. Ellison, R. A., Lukenbach, E. R., and Chiu, C.-W. (1975) Cyclopentenone synthesis via aldol condensation. Synthesis of a key prostaglandin intermediate. *Tetrahedron Lett.* **16**, 499–502.
36. Vorbruggen, H., Bennua, B., and Su, T.S. (1981) Nucleoside von 3,6-dihydro-2H-1,2,6-thiadiazin dioxiden. *Chem. Ber.* **114**, 1269–1286.
37. Watanabe, K. A., Hollenberg, D. H., and Fox, J. J. (1974) Mechanisms of nucleoside synthesis by condensation reactions. *J. Carbohydr. Nucleosides Nucleotides* **1**, 1–37.
38. Johnson, T. B. and Hilbert, G. E. (1929) The synthesis of pyrimidine—Nucleosides. *Science* **69**, 579,580.
39. Wittenburg, E. (1986) Synthese von Thymin-nucleosiden ueber Silyl-pyrimidin-Verbindungen[3]. *Chem. Ber.* **101**, 1095–1114.
40. Niedballa, U. and Vorbruggen, H. (1976) A general synthesis of *N*-Glycosides. 6. On the mechanism of the stannic chloride catalyzed Silyl Hilbert-Johnson reaction. *J. Org. Chem.* **41**, 2084–2086.
41. Hubbard, A. J., Jones, A. S., and Walker R. T. (1984) An investigation by HNM spectroscopy into the factors determining the b:a ratio of the product in 2'-deoxynucleoside synthesis. *Nucl. Acids Res.* **12**, 6827–6837.
42. Tocik, Z., Earl, R. A., and Beranek, J. (1986) The use of iodotrimethylsilane in nucleosidation procedures, in *Nucleic Acid Chemistry* (Townsend, L.B. and Tipson, R. S., eds.), Section III, New York, pp 105–111.

CHAPTER 16

Oligonucleotide Analogs Containing Dephospho-Internucleoside Linkages

Eugen Uhlmann and Anusch Peyman

1. Introduction

Nonionic oligonucleotide analogs have attracted much interest in recent years as potential candidates for oligonucleotide-based therapeutics. Compared to their natural parent molecules, they are expected to have the advantage of:

1. Being stable against degrading nucleases,
2. Showing enhanced cellular uptake, and
3. Having the ability to form more stable complexes with complementary sequences owing to reduced charge repulsion.

Structural criteria for effective use of oligonucleotide analogs include proper spacing and orientation of bases to allow hybridization and possibly a stereoregular backbone for homogeneous binding of the analogs to complementary nucleic acid sequences. Two classes of nonionic oligonucleotide analogs have been synthesized and investigated, those with modified internucleoside phosphate bridges and those having the phosphodiester bridges replaced by a suitable "dephospho" moiety. In the first class, the negatively charged oxygen has formally been substituted by a methyl group or an *O*-alkyl group leading to methylphosphonates or phosphate esters, respectively. This type of compound owns as a common feature a new center of chirality at phosphorus

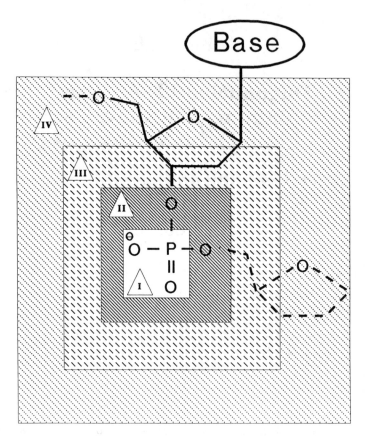

Fig. 1. Displacement of different structural elements in oligonucleotide analogs.

and is described in elsewhere in this volume. Here we will focus on the synthesis of dephospho oligonucleotide analogs whose internucleoside bridges are electroneutral and, in most cases, in which no additional stereoisomerism is introduced.

Depending on the depth of structural variation, the dephospho oligonucleotide analogs may be divided into four groups (Fig. 1, Table 1). In group I oligonucleotide analogs the PO_2^- residue is replaced by a neutral moiety that preferably is isosteric to the tetrahedral phosphate group as, e.g., the siloxane- or methylene-bridged analogs. Replacement of PO_4^- by an appropriate noncharged linkage, such as the carbamate or dimethylensulfide bridge, is represented in group II oligonucleotide analogs. The carbamate-bridged morpholino-type oligomers can

Dephospho-Internucleoside Linkages

Table 1

The Four Groups of Dephospho Oligonucleotide Analogs

	Replaced group	Examples	
(I)	$PO_2^{\ominus}$	—Si(R)$_2$—	siloxane
		—CH$_2$—	formacetal
		—C(O)—	carbonate
		—CH$_2$C(O)—	carboxymethyl
(II)	$PO_4^{\ominus}$	—OC(O)NH—5'	carbamate
		—OCH$_2$C(O)NH—	acetamidate
		—OCH$_2$S—	thioformacetal
		—CH$_2$SCH$_2$— / —CH$_2$CH$_2$S—	thioether
(III)	$PO_4^{\ominus}$ +part of sugar	morpholino structure	morpholinocarbamate
(IV)	$PO_4^{\ominus}$ +complete sugar	—CH$_2$—CH$_2$—(B) / CH$_2$CH$_2$N—	polyethylenimin ("plastic")
		—NHCH$_2$CH$_2$N CH$_2$C(O)— C(O)—CH$_2$—(B)	aminoethylglycine (P N A)

be regarded as oligonucleotide analogs in which the PO_4^- bridge, as well as part of the deoxyribose have been exchanged (group III). The strongest molecular changes were made in group IV analogs in which the complete sugar-phosphate backbone is substituted for a polymeric backbone, like polyethylenimine or polyaminoethylglycine. The last two examples certainly represent extremes of dephospho oligonucleotide analogs. The terms "plastic DNA" for the chain-polymerized compounds and "polyamid nucleic acids" (PNAs) for the peptide-like compounds may be more appropriate for assigning their chemical structure.

For most of the different dephospho oligonucleotide analogs, no extensive investigations were reported so that no detailed practical procedures can be given in this chapter. The main scope of this chapter is therefore to show which types of dephospho oligonucleotides have been synthesized so far, what their chemical properties are, and whether their physical characteristics provide a basis to design new inhibitors of gene expression of potential therapeutic value.

2. Methods for the Synthesis of Dephospho Oligonucleotide Analogs

2.1. Siloxane-Bridged Oligonucleotide Analogs

2.1.1. Synthesis

Preparation of siloxane-bridged oligonucleosides as initially reported for dimers is reminiscent of phosphorus-III-chemistry to synthesize the natural phosphodiester oligonucleotides (Fig. 2). In a first step, the free 3'-hydroxy group of a 5'-*O*-Dimethoxytrityl (DMTr) nucleoside is reacted with dichlorodialkyl/aryl silane (X = Y = Cl) at –78°C in the presence of base to give the monochlorosilane intermediate. The reactive nucleoside-3'-*O*-silylchloride is then trapped at ambient temperature by a 3'-*O*-protected nucleoside to give the desired 3'5'-siloxane-bridged dimer. Ogilvie and Cormier were the first to report on the synthesis of a diphenylsilyl-linked thymidine dimer from 5'-*O*-DMTr-thymidine by successive reaction with dichlorodiphenyl silane and 3'-*O*-levulinoyl-thymidine *(1)*. The major problem in this reaction is formation of a 3'3'-linked side product; a similar problem occurs in phosphitylation with dichloro alkylphosphane. Slow addition of a solution of the 3'-hydroxy nucleoside at –78°C to *bis*-(trifluoromethanesulfonyl)-diisopropyl-silane in the presence of pyridine followed by

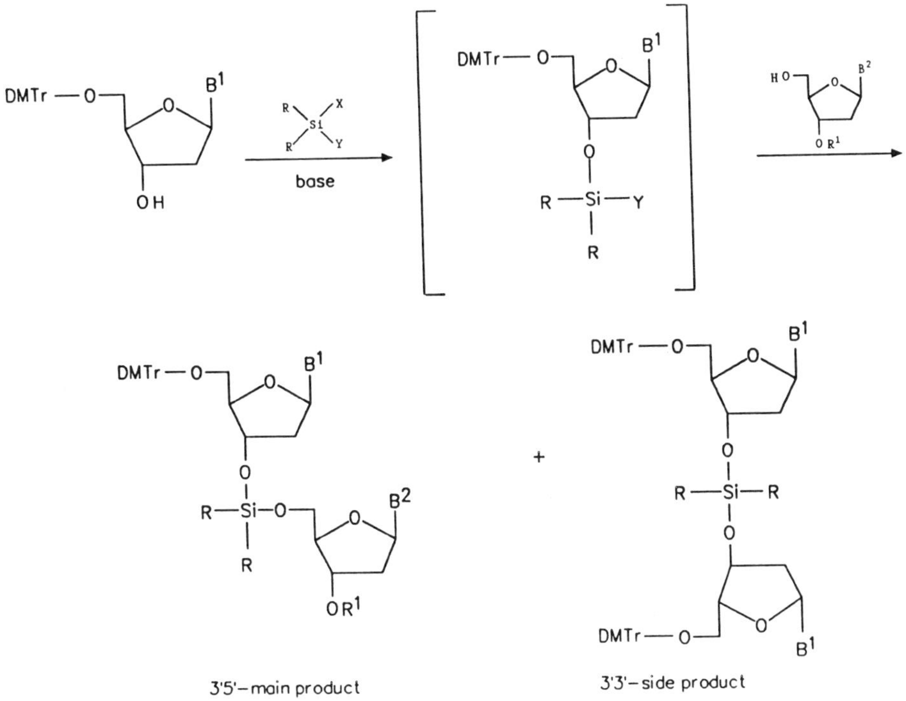

Dephospho-Internucleoside Linkages

Fig. 2. Synthesis of a 3'5'-dinucleoside monosiloxane and the potential formation of a 3'3'-side product (R = C_6H_5 *(1)*, i-C_3H_7 *(2)*, CH_3 *(3)*, R–R = cyclohexyl; R^1 = acetyl or levulinoyl; base = pyridine or imidazole).

condensation with the 5'-hydroxy nucleoside at room temperature afforded the desired 3'5'-bridged dimer after purification by silica gel chromatography in 74% yield *(2)*. Seliger and Feger successfully used the bifunctional reagent chloro-diisopropylamino silane (X = Cl, Y = N[i – C_3H_7]2, R = CH_3 in Fig. 2) to prepare a 3'5'-dimethylsiloxane-linked thymidine dimer *(3)*, whereas dichlorodimethyl silane resulted in a complex reaction mixture. Thereby, the coupling reaction was carried out at room temperature in THF:pyridine (10:4), and resulted in 51% of the dimer. They also prepared thymidine-3'-yl-(1,1silacyclohexyl)-5'-thymidine by means of 1,1-dichlorosilacyclohexane.

Since the siloxane-bridged oligonucleotide analogs exhibit considerable sensitivity to acid, removal of the DMTr group was effected with $ZnBr_2$ or brief treatment with 3% trichloro acetic acid. Short hydrazine treatment (0.5*M* hydrazine hydrate in pyridine:acetic acid [3:2] for 2.5 min at room temperature) allowed for cleavage of the 3'-*O*-levulinoyl group, whereas long reaction times led to decomposition *(1)*. A hexanucleoside-penta-(diisopropyl)siloxane, prepared in solution by block condensation, was freed from levulinoyl and benzoyl groups by treatment with ammonia/methanol/dioxane (3:1:5) in a sealed flask for 6 h. Use of more labile phenoxyacetyl (PAC) *(4)* or amidine *(5)* protecting groups should allow even milder deprotection conditions that could be combined with the labile oxalyl-linker *(6)* in solid-phase synthesis.

2.1.2. Properties

The siloxane-bridged oligonucleotide analogs are susceptible to acid hydrolysis. From the group of phenyl, isopropyl, ethyl, and methyl siloxanes, the diisopropylsiloxane linkage appeared to give the best compromise between stability and reactivity *(2)*. Stability against base of siloxane-bridged oligonucleosides seems to be high enough to allow for their synthesis using base-labile protecting groups as mentioned above. Stability against nucleases would not be a limiting factor for in vivo application of the siloxane analogs. The CD spectra of hexathymidine-penta-(diisopropylsiloxane) showed marked similarity to the natural hexathymidine-pentaphosphate, whereas the siloxane-linked adenine hexamer revealed a reduced base–base interaction as compared to the phosphodiester analog *(2)*. A major problem with the diisopropyl-siloxane-linked hexanucleosides is their poor solubility of 0.01–0.06 mM in water, at which concentration attempts to measure melting curves with complementary sequences failed. It has been suggested that either terminal or internal phosphodiester moieties, or other ionic or highly hydrophilic groups, be introduced to overcome the solubility problem *(2)*. Finally, it still has to be shown if longer sequences having siloxane bridges are able to form duplex or triple-helical structures with complementary RNA or DNA sequences, respectively.

2.2. Methylene-Bridged Oligonucleotide Analogs

2.2.1. Synthesis

In 1988, Zon considered isostructural cognates of phosphodiester linkages, which are represented by the general formula 3'-*O-M(X)(Y)-O*-5', where M = C, Si, or S, as structural classes of oligonucleotide analogs of potential interest *(7)*. It took two more years, however,

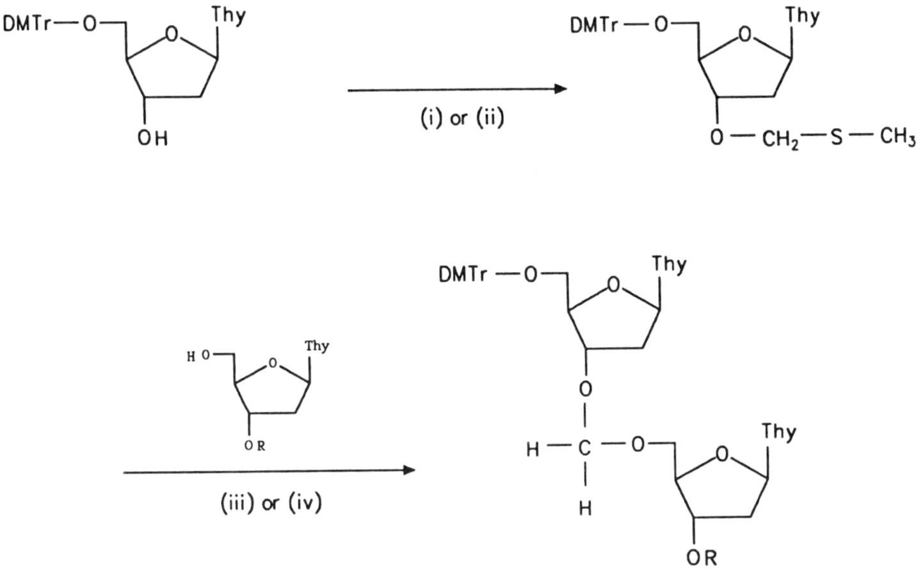

Fig. 3. Synthesis of a 3'5'-methylene-bridged dithymidine. (i) CH_3SCH_2Cl, NaH, THF (80% yield) *(8)*. (ii) CH_3SCH_3, benzoyl peroxide, 2,6-lutidine (75% yield) *(10)* (iii) NBS, 2,6-di-tert-butyl pyridine in CH_2Cl_2, R = dimethylthexylsilyl (45% yield) *(8)*. (iv) IDCT, R = $C(O)CH_2OCH_3$ (yields only traces of dimer) *(10)*.

until the first syntheses of methylene-bridged oligonucleotide analogs were reported by Matteucci *(8)* and Veeneman et al. *(9)*. The formacetal linkage (3'-*O*-CH_2-*O*-5') like the dialkylsiloxane bridge (3'-*O*-Si(R)$_2$-*O*-5'), is an achiral isostere for the tetrahedral phosphate bridge. One way to build up these molecules consists of a modified version of disaccharide synthesis. It involves conversion of a 3'-hydroxy group of a 5'-protected nucleoside to the corresponding 3'-methylthioacetal (Fig. 3) by alkylation with methylthiomethyl chloride in THF in the presence of sodium hydride in 80% yield *(8)*. The methylthioacetal is then activated by *N*-bromo succinimide in the presence of 2,6-di-tert-butyl pyridine in methylene chloride for subsequent coupling with a 3'-protected nucleoside to give the methylene-linked dimer in 45% yield. This reaction sequence proved to be less successful in other hands *(10)*. Van Boom and coworkers succeeded in the preparation of 5'-*O*-DMTr-3'-*O*-methylthiomethylthymidine by reacting 1 Eq of 5'-*O*-DMTr-thymidine at 0°C with 10 Eq of dimethylsulfide and 4 Eq benzoyl peroxide in acetonitrile in the presence of 2,6-lutidine (75% yield). Attempts to couple the 3'-methylthioacetal with a 3'-protected thymidine by activation with iodonium dicollidine triflate (IDCT) resulted only in traces of the desired dimer. Use of a *n*-pentenyloxymethylene (POM) acetal instead of the methylthioacetal gave under the same conditions 15% of the dimer.

From the poor results of this reaction, it has been concluded that N-3 of thymidine has to be protected in order to avoid side reactions. Thus, N-3-protected 3'-thiomethylacetal, activated with 6 Eq of *N*-iodosuccinimide (NIS), gave 33% of the desired dimer on reaction with N-3,3'-protected thymidine, however, with concomitant formation of 50% of a 3'-*O*-CH2-succinimide side product (Fig. 4). IDCT activation of the same reaction afforded only traces of the methylene-bridged dimer. However, NIS-mediated reaction (sixfold excess) of the 3'-*O*-POM-*N*-3-benzoyl-thymidine resulted after 72 h in 1,2-dichloroethane in 80% of the desired dimer (Fig. 5) together with only 10% of side product. In order to shorten reaction times, the stronger activator NIS/TfOH (*N*-iodo succinimide/trifluoromethane sulfonic acid) was used, which required substitution of the acid-labile DMTr group by the levulinoyl group (Fig. 6). The 5'-*O*-levulinoyl-3'-*O*-methylthiomethyl-thymidine on reaction with 0.8 Eq of 3'-*O*-methoxy-acetyl-thymidine promoted by NIS/TfOH (1 Eq/0.15 Eq) at 0°C afforded in an immediate reaction the desired dimer. Thereby, an *N*-3-benzoylated thymidine building unit gave almost identical yield (85%) as compared to the N-3 unprotected starting materials (84%) *(10)*.

Veeneman et al. prepared a trimer having two methylene bridges, a tetramer containing two terminal methylene, and one phosphodiester linkages by solution chemistry, as well as a decamer harboring one methylene-linked dimer by solid-phase chemistry. Matteucci synthesized in solution a thymidine trimer with two methylene bridges that after coupling to a CPG suppport was elongated using H-phosphonate chemistry to give a 13 mer having two 3'-terminal methylene linkages.

Synthesis of a dimer containing a 5'-thioformacetal bridge (Fig. 7) was achieved by activating a 3'-methylthioacetal nucleoside with Br_2 in the presence of 2,6-diethyl pyridine/4Å molecular sieves in benzene prior to condensation with a 5'-mercapto-5'-deoxy-nucleoside *(11)*. Furthermore, 5'-*O*-C(R)$_2$*O*-3'-bridges were suggested, where R bears electron-withdrawing substituents; some examples of cyclic analogs

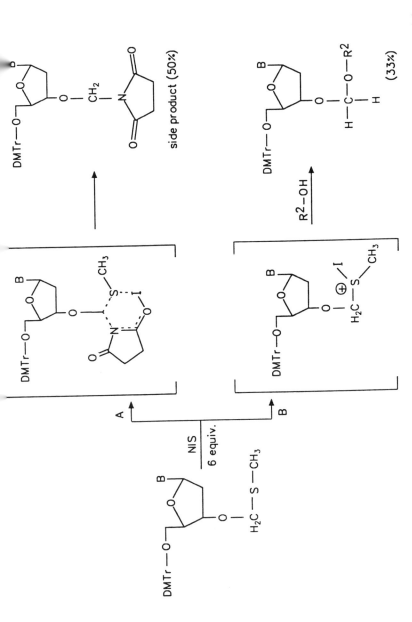

Fig. 4. Activation of methylthioacetal using sixfold excess of NIS and coupling with a 3'-protected nucleoside. (B = N-3-benzoyl-thyminyl, R^2 = C(O)CH$_2$OCH$_3$ (A) Sideproduct formation. (B) Dimer formation (9,10).

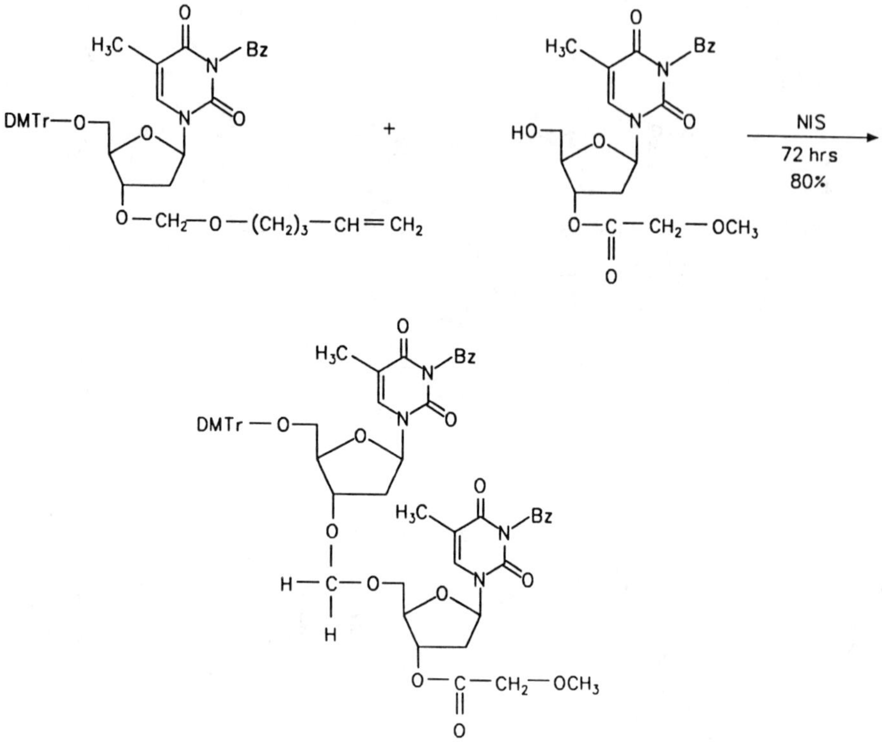

Fig. 5. Synthesis of a 3'5'-methylene-bridged dimer using NIS-activation of a 3'-POM nucleoside (*N*-3-protected thymidine).

are shown in Fig. 7 *(12)*. So far, all methylene bridges were preformed in solution as the corresponding dimers or trimers. They were either hooked to a solid support to gain access to 3'-terminally modified oligonucleotide analogs or were built into oligonucleotides via their 3'-phosphoramidates or H-phosphonates at any desired position. It is likely that synthesis of oligonucleotide analogs having all phosphodiester linkages replaced by formacetal linkages should be possible by applying the fast iodonium ion promoted condensation reported by Veeneman et al. *(10)*.

2.2.2. Properties

The formacetal linkage has been reported to be stable against moderate acid treatment (80% acetic acid for 3 h at 45°C), allowing removal of the DMTr group without additional problems. Treatment of a meth-

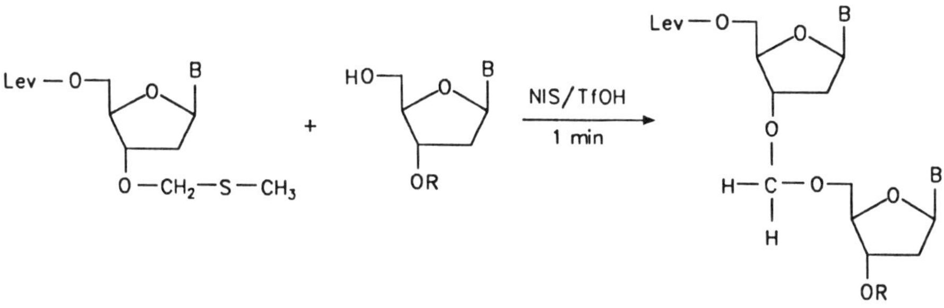

Fig. 6. NIS/TfOH-promoted formation of a 3'5'-methylene-linked dimer. (Yields are 85% for B = *N*-3-benzoylthyminyl, 84% for B = thyminyl; R = C(O)CH2OCH3, lev = C(O) $CH_2CH_2C(O)CH_3$).

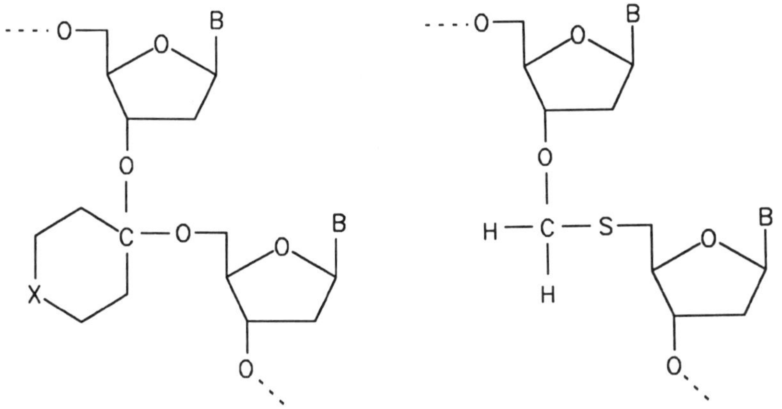

Fig. 7. Potential ketal linkages (left) according to reference *12* (X = O, N-R, SO, SO_2), and 5'-thioformacetal linkage (right).

ylene-bridged trimer with 80% formic acid at 80°C for 1 h resulted in complete decomposition of the formacetal linkage *(8)*. Interestingly, the formacetal linkage is reasonbly stable under conditions used for removal of levulinoyl (1*M* hydrazine in pyridine:acetic acid (3:2) for 1 min), tert-butyldimethylsilyl (tetrabutyl-ammonium fluoride in THF), as well as acyl groups (e.g., 25% ammonia:dioxane [2.5:1] at room temperature). As expected, the formacetal linkage is neither degraded by snake venom nor spleen phosphodiesterase *(8,10)*.

Hybridization properties of methylene-bridged oligonucleotide analogs regarding RNA are quite favorable as compared to other nonionic oligonucleotide analogs, e.g., the methylphosphonates *(13)*. A tetradecamer containing four formacetal linkages showed a T_M value of 59°C when hybridized to a single-stranded RNA as compared to 60°C for the phosphodiester and 51°C for the methylphosphonate analogs, respectively. Surprisingly, binding affinity of the formacetal analogs to single-stranded DNA proved to be considerably lower ($T_M = 39°C$) than for the natural phosphodiesters ($T_M = 50°C$). In addition, triple helix formation is possible with methylene-bridged oligonucleosides as judged by DNA footprint analysis *(11)*. In the relevant experiment, a methylphosphonate analog required a tenfold higher concentration for binding relative to formacetal and phosphodiester analogs. Of course, this finding might be explained by the nature of the diastereomeric mixture (16 diastereoisomers for a oligomer containing four methylphosphonate linkages), from which only a fraction presumably possesses optimal binding properties. In this context, it should be noted that thioformacetal-linked oligonucleosides also showed reduced binding *(11)*.

2.3. Carbonate-Bridged Oligonucleotide Analogs

2.3.1. Synthesis

Oligodeoxynucleosides with carbonate bridges have been described up to trimers by Tittensor and Jones *(14,15)*. For coupling, 3'-carbonate esters, especially the 2,2,2-trichloroethylester of 5'-*O*-trityl-deoxyribonucleosides proved to be superior over the originally used 3'-*O*-chloroformates *(16)* as activated monomers in the reaction with 3'-protected nucleosides (Fig. 8). Coupling is carried out in DMF in the presence of sodium hydride at room temperature for 24 h. The yields are low, at best 66% in the reaction with 2',3'-*O*-isopropylideneuridine, because only unprotected bases were used, causing side reactions with the exocyclic amino functions. The authors therefore predict much higher yields, especially for C and G with the use of acid- or ethanol-labile protecting groups, such as *N,N*-dimethylaminomethylene *(17)*; however, such an experiment has not been carried out yet. The synthesis in the reverse direction using 5'-carbonate esters offered no advantage and was not followed further. Carbonate-linked dimers have been selectively detritylated and reacted with an additional 3'-carbonate ester to form tri(nucleoside) dicarbonates *(14)*.

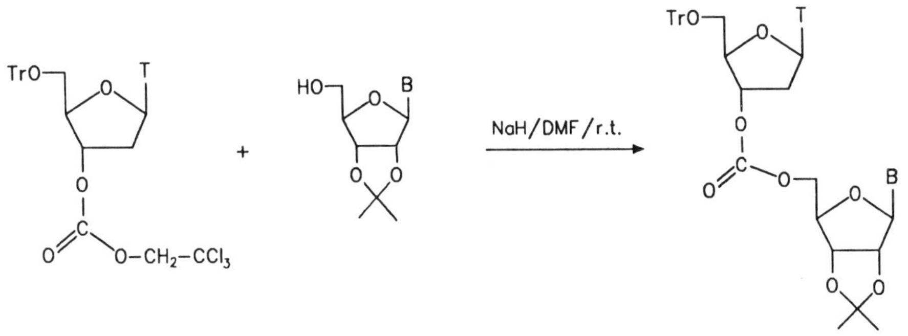

Fig. 8. Preparation of carbonate-linked oligonucleotide analogs using 3'-O-2,2,2-trichloroethylcarbonates (Yield <66%).

2.3.2. Properties

The carbonate linkage is resistant to hydrolysis by acid, and it is rather easily cleaved by base. Detritylation is possible with 80% aqueous acetic acid at 100°C for 4 h *(14)*. Preliminary biological results indicate a fast hydrolysis of the carbonate bridge under physiological conditions *(16)*. Since only trimers were synthesized, there are no reports on the base-pairing capabilities of carbonate-bridged oligonucleotide analogs.

2.4. Carboxymethyl-Bridged Oligonucleotide Analogs

2.4.1. Synthesis

Carboxymethyl-bridged polynucleotide analogs (poly-U and poly-T) were synthesized as early as 1968 by Halford and Jones *(17,18)* under the assumption that although this linkage contains one more atom than the natural phosphodiester linkage, the carboxymethylester should be fairly flexible so that the nucleoside units can become similarly spaced to those of natural polynucleotides. The synthesis was carried out by polymerization of 3'-O-(carboxymethyl) thymidine in pyridine with dicyclohexylcarbodiimid (DCC), and 0.1 Eq of the 5'-O-trityl-protected unit as chain terminator (Fig. 9). In 1975, the poly-dA *(19)* and poly-dC *(20)* analogs were prepared in the same way, with protection being unnecessary for adenine (only 2% of all internucleoside linkages were formed through the N-6 position of adenine), but neces-

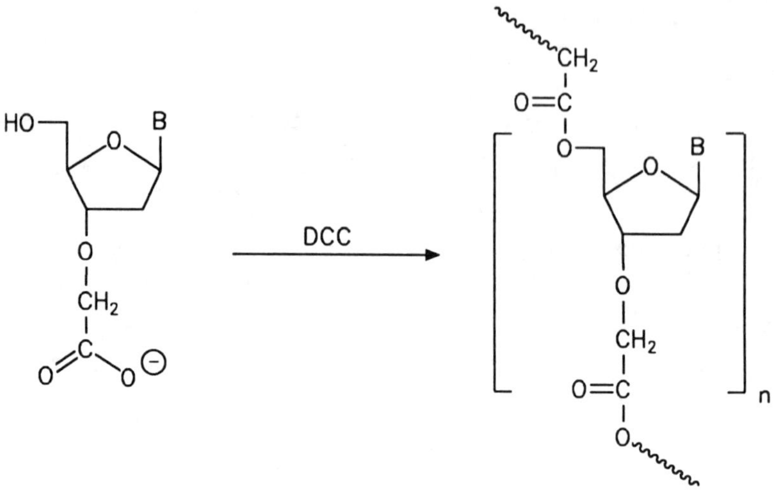

Fig. 9. Preparation of poly(3'-O-carboxymethyl-2'-deoxynucleoside) by polycondensation of 3'-carboxymethylnucleosides in pyridine with DCC, 20°C, 24 h.

sary for cytosine as 4N-phenoxyacetamide. The resulting oligomers had a minimum chain length of 15–25 bases. Using an orthogonal strategy for protecting groups, Edge and Jones were able to replace the simple polymerization step by stepwise synthesis *(21)*. Because of the sensitivity of the carboxymethylester group to basic conditions (the half-life at pH 7.5 is 7 h *[19]*), only protective groups that could be eliminated under acidic or mildly basic conditions could be used. 5'-OH groups were protected with dimethoxytrityl, 2'- and 3'-OH groups were blocked as anisylidene and isopropylidene compounds, and the carboxyl group was in the form of the 2-cyanoethyl or the 2,2,2-trichloroethylester *(22)*. Cytosine required 4-N-(di-methylamino)methylene protection, whereas the other bases were not acetylated under the reaction conditions (DCC in pyridine). According to the reactions outlined in Fig. 10, Jones and collegues were able to perform the specific synthesis of dimers *(22)*, trimers *(21)*, and tetramers *(23)*. Removal of the terminal cyanoethyl group was achieved by treatment with tert-butoxide in DMF for 2 h at 100°C. The 2,2,2-trichloroethylester proved to be superior, since it could be removed without any internucleoside linkage cleavage by heating with the zinc–copper couple in DMF

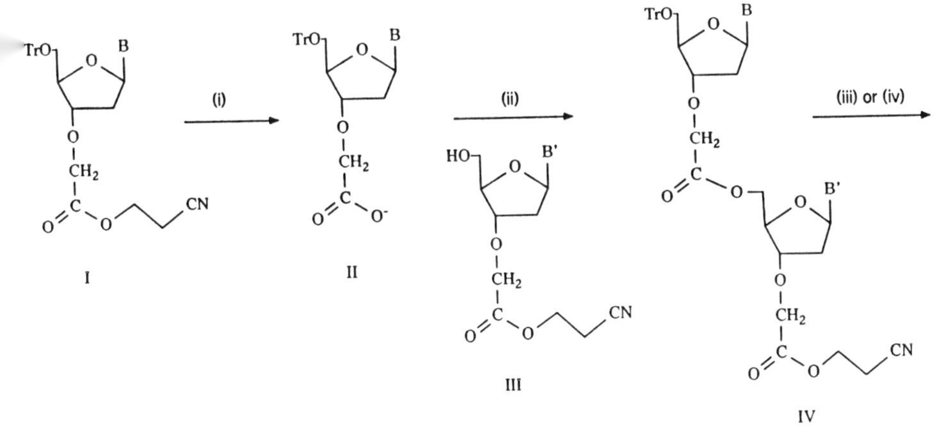

Dephospho-Internucleoside Linkages

Fig. 10. Stepwise synthesis of carboxymethyl-bridged oligonucleosides. (i) Sodium tert-butoxide in DMF, 100°C, 2 h; (ii) DCC in pyridine, 20°C, 24 h; (iii) chain extension via 5'-end: detritylation of IV followed by condensation with II; (iV) repetition of steps (i), and (ii).

(22). Isopropylidene and trityl groups were removed with dilute acid, and the dimethylaminomethylene group was removed with boiling ethanol.

All of the compounds mentioned above contain 3'-5'-directed carboxymethyl ester linkages. Only dimers have been synthesized with a 5'-3' linkage (17), so that the properties of those compounds are still fairly unknown.

2.4.2. Properties

As mentioned above, the stability of oligonucleotide analogs containing the carboxymethylester linkage depends strongly on the pH: the half-life at 37°C at pH 6.0 is 127 h and only 7 h at pH 7.5 (19). Another unfavorable property is their low solubility in water. Stable duplexes have been observed between poly(dA) carboxymethyl analogs (15–25 bases length), and poly(U) (19). The melting temperature of the duplex was 29°C in 0.30M NaCl and 0.03M sodium citrate buffer, which is by far lower than that of the duplex between poly(A), and poly(U), 55°C in 0.1M salt. A duplex between a poly(dA) carboxymethyl analog and a poly(dT) analog melts at 28°C in 1M

salt. It is worth mentioning that Jones and coworkers were able to inhibit the poly(U)-directed synthesis of polyphenylalanine in a cell free system, an equimolar amount causing almost 50% inhibition after 5 min at 37°C, indicating an inhibition mechanism according to the antisense principle.

2.5. Acetamidate-Bridged Oligonucleotide Analogs

2.5.1. Synthesis

A possibility to circumvent the instability of carboxymethyl bridges at physiological pH is to replace the 5'-ester by an amide. The acetamidate bridge ought to provide a suitable internucleotide distance, too. First dimers were synthesized by Gait et al. *(24)* in exact analogy to the preparation of the carboxymethyl esters, by condensation of 3'-*O*-carboxymethyl-5'-*O*-tritylthymidine with 5'-amino-5'-deoxythymidine and DCC. However, polymerization proved to be far more difficult: condensation of 5'-amino-5'-deoxy-3'-*O*-carboxymethylthymidine with DCC yielded no polymer, only the 3'5'-bridged lactam as main product (Fig. 11). Polymerization of 5'-chloroacetamidyl-5'-deoxythymidine with sodium hydride in DMSO was equally unsuccessful, because alkylation took place not only at the 3'-oxygen, but also at the N^3 position of thymidine *(25)* (Fig. 11). On the other hand, when the corresponding 5'-chloroacetamidyl-5'-deoxy-4-*N*-acylcytidine was employed, polymerization was successful, but most of the product was lost on deprotection of the base, so that the overall yield was only about 1%. Because the lactam formation, which has been mentioned above, was certainly owing to the favorable entropy for intramolecular ring closure, Gait et al. *(25)* successfully used an appropriate dimer for the polymerization with the aid of triphenylphosphine and 2,2'-dipyridyl disulfide (Fig. 12). It was possible to isolate a 54% yield of the highly dispersed polymer, whose mean chain length was reported to be 10–13.

Recently, a stepwise synthesis of acetamidate-bridged oligonucleotides has been published by Nyilas et al. *(26)*. They used the 5'-azido-compounds as "5'-aminonucleoside synthons" (Fig. 13). 3'-*O*-carboxymethylation is carried out with sodium chloroacetate in the presence of NaH at 20°C. 5'-Azido-5'-deoxythymidine does not need to be protected, for the other bases, the monomethoxytrityl protection

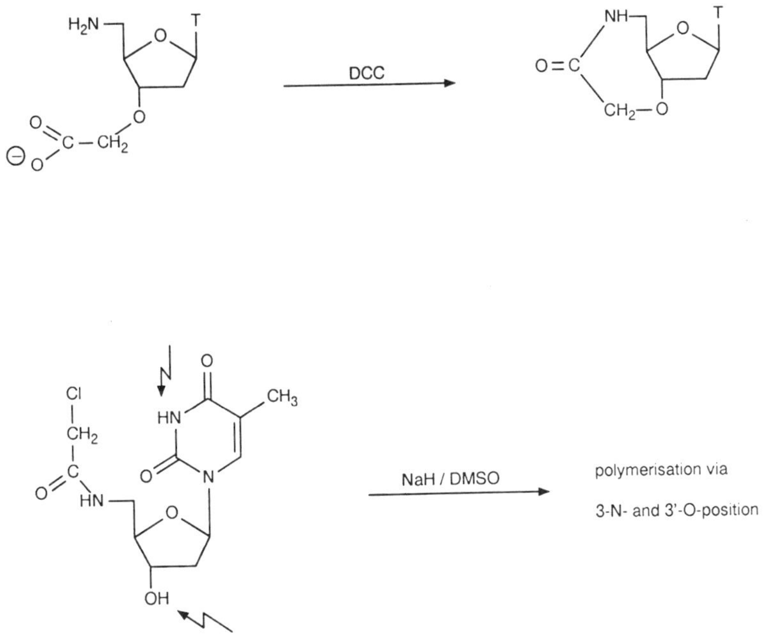

Fig. 11. Unsuccessful attempts to prepare acetamidate bridged polythymidine.

was used for the exocyclic amino groups. It could be conveniently removed under nonbasic conditions at the end of the synthesis with ZnBr$_2$ in nitromethane (68% yield). Coupling was performed by the peptide condensation method of König and Geiger *(27)* with DCC, 1-hydroxybenzotriazole, and *N*-ethylmorpholine (50–70%), followed by the reduction of the 5'-azido group with triphenylphosphin and treatment with methanolic ammonia (85%). The synthesis was carried out up to the trimers.

2.5.2. Properties

The acetamidate bridge is stable over a large pH range; no hydrolysis is found at pH 6–7.5 at 20°C for 7 d. At pH 14 and 37°C, the half-life is 3.4 h *(25)*. However, other properties of these compounds are less advantageous, such as the low solubility in water, the strong tendency to adsorb onto glass and plastic surfaces and, most unfortunately, they do not show any tendency to hybridize with their complementary natural polynucleotides *(25)*.

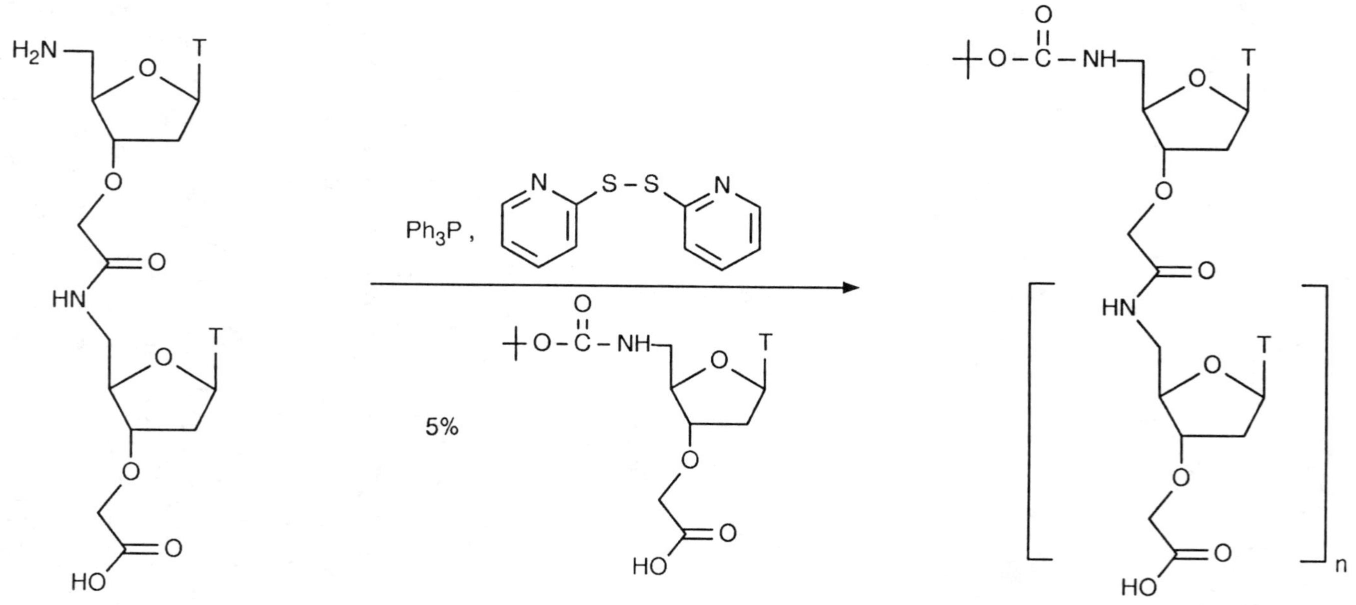

Fig. 12. Preparation of acetamidate-bridged oligonucleosides. Polymerization is effected by triphenylphosphine, 2,2'-dipyridyldisulfide in pyridine, 14 h, 54%.

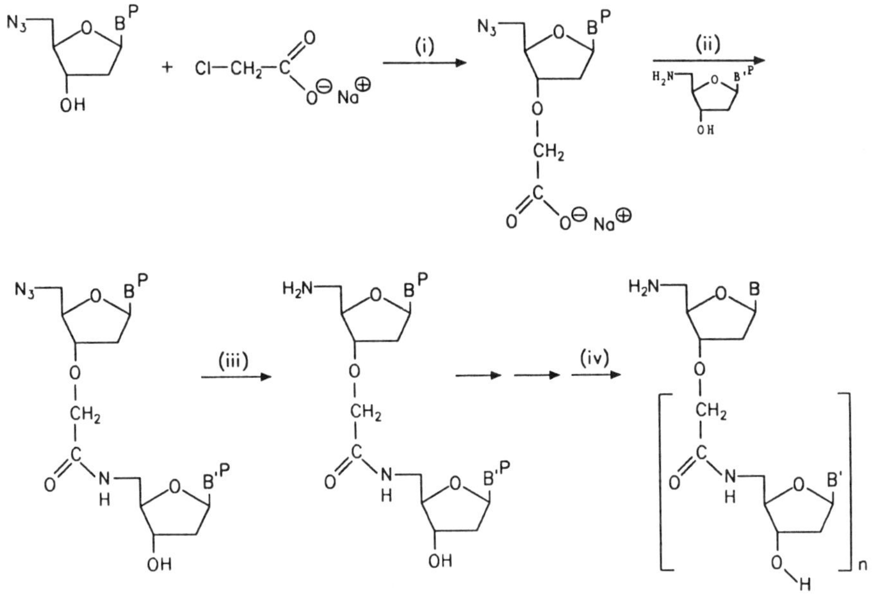

Fig. 13. Stepwise synthesis of acetamidate-bridged oligonucleosides. B^P: Thymine, 4-N-MMTr-cytosine, 2-N-MMTr-guanine, 6-N-MMTr-adenine; B: unprotected bases; cat: HOBT, N-ethylmorpholine; (i) NaH in DMSO at 0°C for 1–3 d (85%); (ii) DCC, HOBT, N-ethylmorpholine (27) in DMF, O-20°C, 24h (50–70%); (iii) a. PPh3 at 0°C, 24 h, b. methanolic NH$_3$ 24 h (85%); (iv) ZnBr$_2$ in nitromethane, 20°C, 24 h, 70%.

2.6. Carbamate-Bridged Oligonucleotide Analogs

2.6.1. Synthesis

Carbamate bridges were supposed to be superior to the isosteric carbonate analogs by having greater pH stability. The first 3'-O-5'-N-carbamate-bridged dimer of thymidine was synthesized in 1974 by Gait et al., who obtained it in 38% yield by condensation of 3'-O-(2,2,2-trichloroethyl)carbonate of 5'-O-tritylthymidine with 5'-amino-5'-desoxythymidine *(24)* (Fig. 14).

In 1977, the synthesis was improved by Mungall and Kaiser *(28)*, who used 3'-O-p-nitrophenylcarbonate esters in place of the trichloroethylesters. The dinucleoside carbamate was obtained in 70%

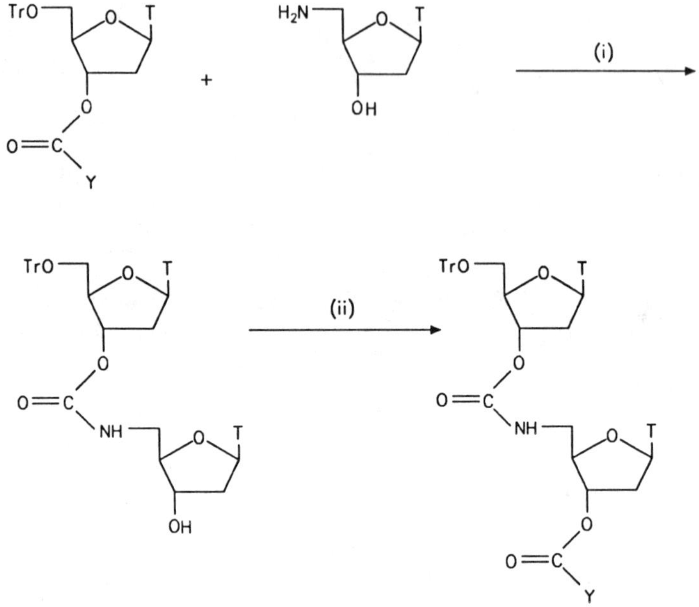

Fig. 14. Preparation of carbamate-bridged oligonucleosides using 3'-activated carbonate esters and 5'-amines. For Y = imidazole: (i) pyridine, 20°C, 20 h, 87%; (ii) CDI, THF, 2h, 20°C, 97%. For Y = 4-nitrophenoxy: (i) pyridine, 0°C, 30 min, 70%; (ii) 4-nitrophenoxycarbonylchloride, 2,6-lutidine in dioxan, 0–20°C, 2 h, 86%. For Y = 2,2,2-trichloroethoxy: (i) DMF, 20°C, 48 h, 38%; (ii) 2,2,2-trichloroethoxycarbonylchloride, pyridine, 4°C, 30 min, 30–60%.

yield. Mungall and Kaiser synthesized a trimer by derivatizing the dimer again with *p*-nitrophenylchloroformate and condensing it with 5'-amino-5'-deoxythymidine. The same activating agent was used ten years later by Stirchak et al. *(29)* who synthesized a hexamer carbamate-linked cytosine oligomer using dimer blocks for homologization. The *N*-benzoyl-protected oligomer was deprotected with aqueous ammonia in DMSO for 24 h at 30°C. The hexamer was obtained in 40% yield. Without giving the details of its synthesis (the same synthetic scheme as described above was used), Griffin et al. *(30)* describe the characterization of a blocked 11 mer, consisting of all four bases by FAB mass spectroscopy. Also in 1987, the synthesis of a hexamer thymidine was described by Coull et al. *(31)*. They used

1,1'-carbonyldiimidazole (CDI) as carbonyl synthon. 5'-O-(DMTr)-thymidine was alternately reacted with CDI in THF and then, after workup, with 5'-amino-5'-deoxythymidine in pyridine.

Both monomer building blocks do have advantages: The 3'-O-carbonylimidazoles are stable in pyridine, a solvent known to decompose the 3'-O-(4-nitrophenyl)carbonates *(31)*. On the other hand, the latter compounds can be purified by chromatography on silica gel, a procedure that is not possible for the 3'-O-carbonylimidazoles *(29)*.

2.6.2. Properties

The carbamate-bridged oligonucleotides exhibit low solubility in water. The carbamate-bridge is very stable to hydrolytic conditions (0.1M NaOH or 0.1M HCl solutions for periods of at least 24 h), and they are resistant to nucleases *(28)*. There are contradicting reports on their ability to enter base-pairing with the natural complementary nucleic acids. Coull et al. *(31)* report no interaction between A_6 or $(dA)_6$ and their carbamate-linked hexamer of thymidine. On the other hand, Stirchak et al. *(29)* report melting temperatures for the carbamate-linked hexamer of cytidine with $p(dG)_6$, and poly(G) of 70 and 79°C. For comparison: $p(dC)_6$ paired with $p(dG)_6$, and poly(G) shows melting temperatures of 32 and 40°C.

2.7. Carbamate-Bridged Morpholino-Type Oligonucleotide Analogs

2.7.1. Synthesis

A further variation of carbamate-bridged oligonucleotide analogs has been reported by Stirchak et al. *(32)* with morpholino-type oligonucleotides. In these compounds, phosphate backbone and sugar moieties are modified. The synthesis of the N-trityl-6'-(p-nitrophenyl)carbonate protected monomer cytosine from cytidine is outlined in Fig. 15. Coupling with the free amino compound occurs regiospecifically in 90% yield. This dimer was used as a building block for the synthesis of a hexamer (Fig. 16): activation with *bis*-(p-nitrophenyl) carbonate with triethylamine as catalyst, coupling in DMF, again with triethylamine as catalyst, and deblocking with 80% acetic acid in methanol. Deprotection was carried out with DMSO/NH_4OH for 24 h at 30°C. An octamer thiocarbamate-bridged morpholino-type

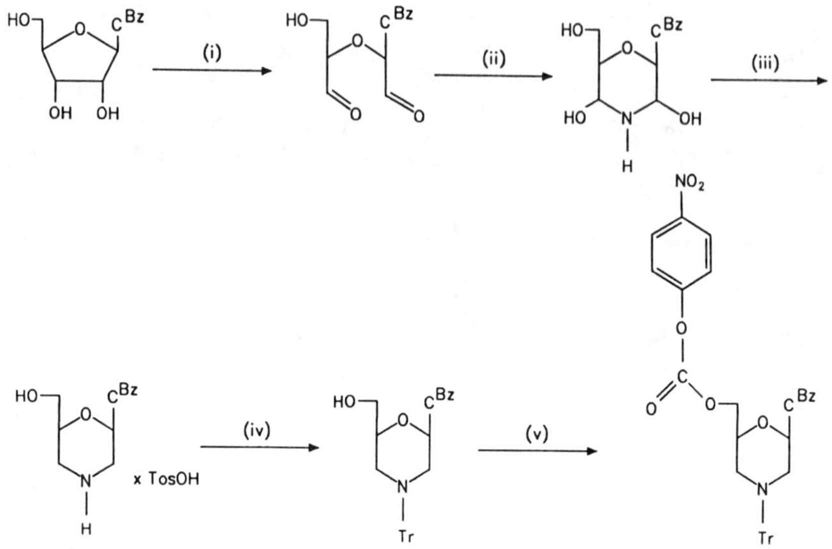

Fig. 15. Preparation of morpholino-type nucleoside monomers. (i) NaIO$_4$; (ii) (NH$_4$)$_2$B$_4$O$_7$; (iii) 1. NaCNBH$_3$, 2. TosOH; (iv) TrCl, Et$_3$N, DMF; (v) *bis*-(*p*-nitrophenyl)carbonate.

analog of adenosine has been prepared by Ohgi et al. *(33)* using comparable coupling techniques. Preliminary results show no advantage compared to their parent compounds.

2.7.2. Properties

Neither under deprotection conditions nor under any acidic conditions used was any cleavage of the carbamate link observed. The solubility of the hexamer analog of cytidine is only 4 µ*M* in pH 7.5 buffer. In order to increase the water solubility, the hexamer was conjugated at the N4'-end with polyethylene glycol (PEG), again using the carbamate coupling procedure. The PEG conjugate is freely soluble at pH 7.5 in concentrations up to 50 µ*M*.

The morpholino-type oligonucleotide analogs show interesting base-pairing abilities: They bind to DNA much stronger than their natural parent molecules, but they show no affinity to RNA. The T_m value of the hexamer-PEG conjugate with p(dG)$_6$ is 62.5°C in comparison to the T_m value of 26.5°C for binding of p(dC)$_6$ to p(dG)$_6$.

Dephospho-Internucleoside Linkages

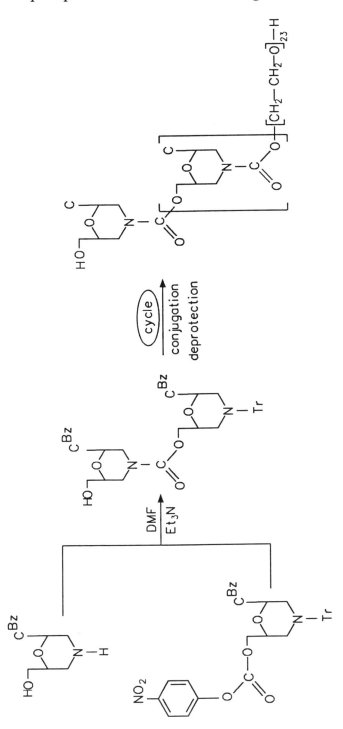

Fig. 16. Stepwise synthesis of carbamate-bridged morpholino-type oligonucleosides. Cycle: activation with *bis*-(*p*-nitrophenyl)carbonate, deblocking with 80% AcOH in methanol; conjugation with *p*-nitrophenyl-*O*-CO-*O*-PEG1000; deprotection with NH_3/DMSO 30°C, 24 h.

However, preliminary results indicate a more complex binding stoichiometry than for a simple Watson-Crick duplex. The fact that there is no specific binding to RNA was explained by the authors by modeling the effective backbone length that matches DNA rather than RNA: In the morpholino-type analogs, the C1'-C1' distance is shortened by 1.3 Å in comparison with RNA in the A-conformation, but only by 0.8 Å compared to the natural phosphodiester linkage of DNA in the B-conformation.

2.8. "Plastic" DNA

2.8.1. Synthesis

Besides the carbamate-bridged oligonucleotides of the morpholino type there are many other ways to design oligonucleotide analogs in which the sugar, as well as phosphate residues have been replaced by new polymeric backbones. The first monomers and polymers containing the pyrimidine moiety were already reported in 1958 by Overberger and Michelatti *(34)*. Since then, a large number of such analogs have been prepared (Fig. 17). These include poly(*N*-vinyl), poly(methacryloxyethyl), poly(methacrylamide), and poly(ethylenimine) derivatives, to mention just a few. Further examples in this series are the polyamide nucleic acids (PNAs), which are treated separately under Section 2.9. Except for PNA*s* the mean chain length of the compounds mentioned above was about 300 *(35)*, so it is appropriate to refer to these molecules as polymers rather than oligomers. Giving the details of the synthesis of these polymeric compounds, mainly free radical polymerization, would go far beyond the scope of this chapter. Their chemistry and use have been reviewed in several excellent articles, some of recent date *(36–38)*. In principle, the chemistry involves free-radical polymerization or copolymerization of the corresponding monomers, such as *N*-vinyl-bases, *N*-(methacryloyloxyethyl)bases, or *N*-(methacryloylaminoethyl)bases *(36,37)*. An interesting area of research is the attempted template-directed polymerization, e.g., of methacrylate monomers in the presence of polymer-containing complementary nucleic acid bases *(36,37)* (Fig. 18). However, besides an acceleration of the reaction, there is no strict control, because the interaction of the monomer and the polymer is not realized specifically. Another synthetic route is provided by grafting-activated derivatives of nucleic acid bases onto preformed polymeric backbones, e.g., grafting of carboxyethyl derivatives of adenine and thymine onto polyethyleneimine *(36,37)*.

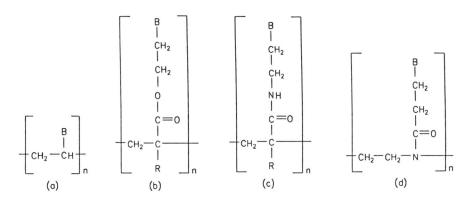

Fig. 17. Examples of "plastic-DNA." (a) Poly-N-vinyl derivatives; (b) poly(acryloyloxyethyl) derivatives (R = H), poly(methacryloyloxyethyl) derivatives (R = CH_3); (c) poly(acryloylaminoethyl) derivatives (R = H), poly (methacryloylaminoethyl) derivatives (R = CH_3); (d) polyethylenimine derivatives.

2.8.2. Properties

Most of the compounds mentioned above are electrically neutral and therefore poorly soluble in water at neutral pH. There are exceptions, such as poly (9-vinyladenine), and poly(1-vinyluracil). All compounds are stable to chemical and enzymatic hydrolysis, and they form complexes with complementary polynucleotides. However, these complexes do not have the simple stoichiometry found in natural nucleic acids. Stability and stoichiometry of complexes do depend on solvent, temperature, and pH of the system *(36)*. Again, because of the complexity of the field, the reader is referred to the review articles mentioned above *(36–38)*. Poly(N-vinyl) derivatives have been carefully tested for their biological and antiviral activity by Pitha *(38)* in enzyme assays, cell cultures, and in organisms.

2.9. Polyamide Nucleic Acids (PNA)

2.9.1. Synthesis

A polyamide nucleic acid chimeric molecule has been computer modeled by Nielsen et al. by replacing the deoxyribose phosphate backbone of DNA through an achiral polyamide backbone *(39)*. The new backbone, supposed to be homomorphous to DNA, consists of aminoethylgylcine units, which can be regarded as reduced diglycine. Since the nucleobase is connected to the backbone via nitrogen, the

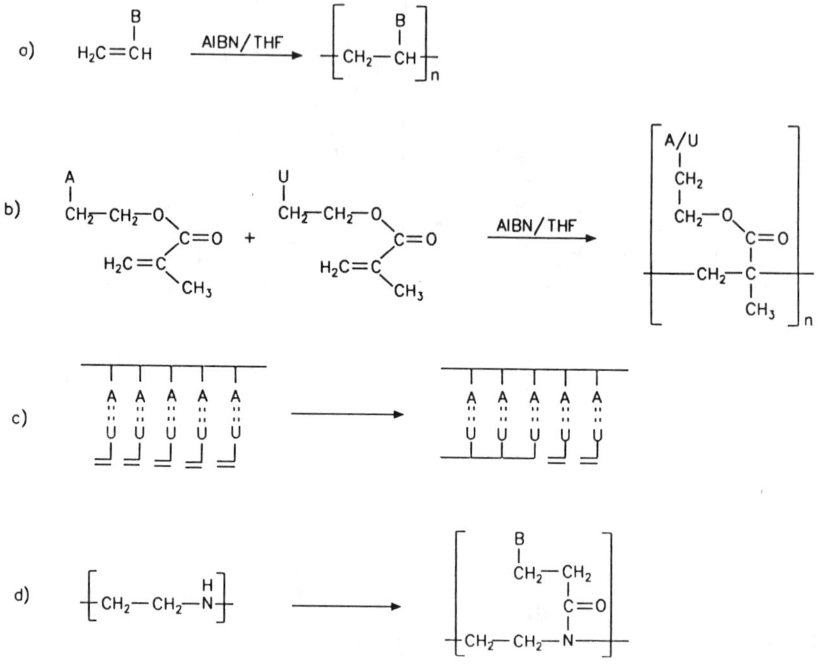

Fig. 18. Principal chemical steps in the preparation of "plastic-DNA." (a) Free-radical polymerization; (b) free-radical polymerization; (c) template derived polymerization; (d) grafting of nucleic acid derivatives onto preformed polymeric backbones.

building blocks have the advantage of being achiral. Up to now, the synthesis of only homopolymers of this type were reported. For the construction of monomeric building blocks, thymine was alkylated by bromoacetic acid ethylester to give after saponification thyminylacetic acid. After activation of the acid with pentafluoro phenol/DCC, it could be coupled with N-(2-Boc-aminoethyl)-glycine to the Boc-(tert-butoxycarbonyl)-protected monomer (Fig. 19). This monomeric building block is suitable to building up oligomeric PNAs according to standard peptide chemistry. Thus, a decamer PNA having one additional lysine residue at the C terminus has been synthesized by standard Merrifield peptide synthesis following Boc-strategy on a 4-methylbenzhydrylamino resin. For coupling, the pentafluoro phenyl-activated esters were prepared, which were condensed in >99% yield

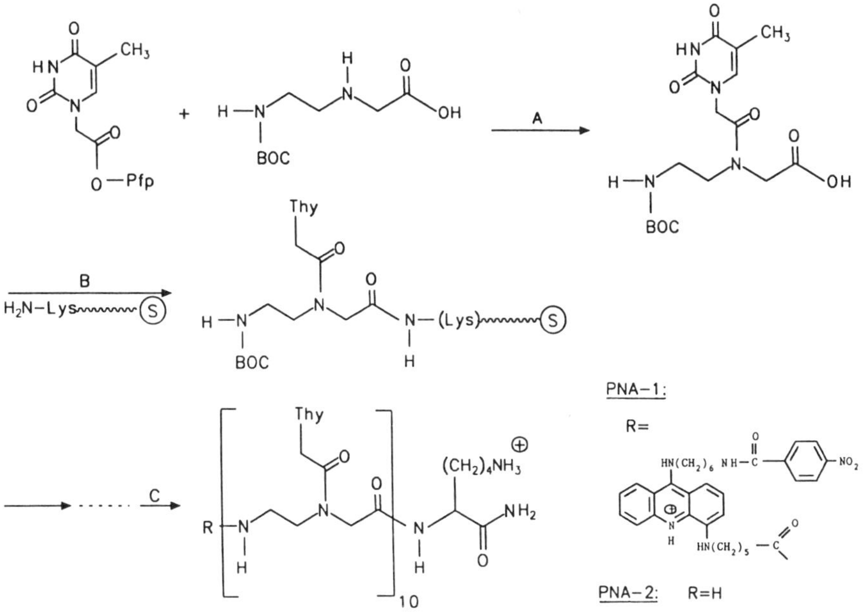

Fig. 19. Synthesis of polyamide nucleic acids (PNA). (A) Generation of a monomeric thyminyl building block. (B) Coupling of the monomer to the solid support. (C) Merrifield synthesis using Boc-strategy. (Boc = tert-butoxy carbonyl, Pfp = pentafluoro phenyl, S = 4-methylbenzhydrylamine resin).

with the amino component after removal of the Boc group from the latter. After HF cleavage of PNA resin, the decathymidylate analog with a lysine carboxamide at the C terminus was obtained in 80% purity. The reverse phase HPLC-purified product (>98% purity) was characterized by plasma-desorption mass spectrometry *(39)*.

Two other types of polyamide backbone were reported by De Koning and Pandit in 1971 *(40)*. The first type of peptide nucleic acid has a nucleobase affixed at every peptide unit (Fig. 20 a; m = 0, x = 4), whereas in the second type two nucleobase-containing units are separated by a glycine (Fig. 21 a; m = 1, x = 4). However, only a tetrapeptide corresponding to two nucleobases has been synthesized. Polymers in which each peptide unit contains a base were made by polymerization of the racemic *p*-nitrophenyl esters of the monomers at room temperature in the presence of triethylamine.

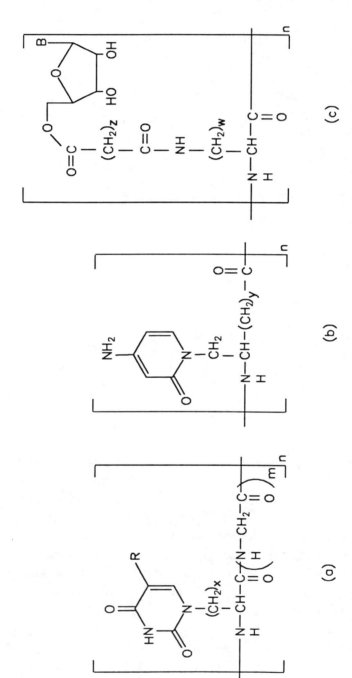

Fig. 20. Examples of oligonucleotide analogs with a polyamide backbone. (a) Thyminyl-alkyl polyamides with (m = 1, x = 4), and without glycine spacers (m = 0, x = 4) (40); poly-thyminyl-alanin (m = 0, x = 1) (41). (b) Cytosylmethyl derivatives of 4-amino butyric acid (y = 1), and 5-amino valeric acid (y = 2) (43). (c) Poly-L-lysine nucleoside conjugates (z = 2, 3, w = 2,4) (44).

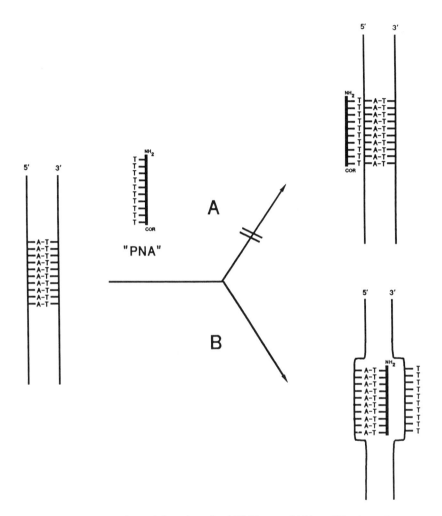

Fig. 21. Hybridization of decathyminyl PNA to a $(dA)_{10}\cdot(T)_{10}$ target sequence on double-stranded DNA. (A) Triplex formation does not take place. (B) Formation of a strand-displacement complex.

Similar analogs, but composed of optically active building blocks, were prepared by Buttrey et al. *(41)*. In this type of analog, the nucleobase was separated from the backbone by one methylene group (Fig. 20 a; m = 0, x = 1). The enantiomeric pure monomers were polymerized by the mixed anhydride procedure using 2-methylproan-1-yl-chloroformate and *N*-methyl-morpholine. A considerable difficulty was the very poor solubility of the β-(thyminyl-1-yl)-alanine monomer. Molecular weights of the polymers were on the order of 2000 to 4000 dalton, and little racemization took place.

Weller and colleagues designed by molecular modeling several other backbone variations of polyamide nucleic acids, which should recognize and bind to complementary nucleic acids *(42)*. The most promising candidates, cytosyl-methyl substituted 4-amino-butyric or 5-amino-valeric acids, were synthesized from the corresponding chiral (*S*)-(5-oxo-2-pyrrolidinyl)-methyl or (*S*)-(6-oxo-2-piperidinyl)-methyl alcohols, respectively, via their tosylates *(43)*. After protection of *N*-H of the cyclic amide by Boc, cytosine could be alkylated by the tosylate in the presence of potassium tert-butoxide. Blocking of *N*-4 of cytosine with the lipophilic tert-butylbenzoyl group was a necessity in view of the high polarity of the monomeric building blocks and their subsequent oligomerization products (Fig. 20b). Hexamers were obtained by block condensation of the 4-amino-butyric acid derivatives (Fig. 20b, y = 2) or, owing to solubility problems, by stepwise elongation in the case of the 5-amino-valeric acid compounds (Fig. 20b, y = 3).

Yet another type of polyamide nucleic acids analogs with a poly-L-lysine backbone (Fig. 20c, w = 4, z = 3) was described by Takemoto et al. *(37)*. The ribose units in this analog mainly served the purpose of enhancing solubility of the polymers in water by the free hydroxy groups. For their preparation, 2'3'-*O*-isopropylidene-5'-*O*-glutaric acid succinimido nucleosides were coupled to prepolymerized L-lysine, a procedure not allowing sequence-specific introduction of nucleobases.

2.9.2. Properties

In contrast to deoxyribonucleotides, PNA oligomers are stable against strong acids as HF. Although PNAs are certainly resistant to nucleases, their stability against peptidases must be checked. Nothing is known about cellular uptake of PNA molecules; however, peptides usually penetrate less effectively than oligonucleotides of the same length. Surprisingly, PNA–DNA affinity is much higher than DNA–DNA affinity. Whereas a $(dA)_{10} \cdot (T)_{10}$ duplex melts at about 23°C, the melting temperature of the hybrid between $(dA)_{10}$ and a thyminyl-PNA decamer (Fig. 19, R = H) is 73°C *(39)*. The same oligomer, but having an acridine nitrobenzamido ligand coupled to the amino terminus, showed a T_m of 86°C. Footprinting experiments conducted with a 248-bp DNA fragment containing the $(dA)10$-$(T)_{10}$ target sequence indicated that binding of the decathyminyl-PNA (R = acridine/nitrobenzamido ligand) to double-stranded DNA results in binding to the complementary strand under displacement of the noncomplementary $(T)_{10}$-containing strand (Fig. 21). Finding of a strand-displacement reaction was surprising in that the PNA oligomer has been originally designed to recognize DNA through Hoogsteen-like base pairing via T•AT triplets, and may be explained by the extraordinarily high stability of the doublestranded PNA•DNA hybrids *(39)*.

The poly-β-(thymin-1-yl)-alanines synthesized by Jones and colleagues (Fig. 20a, x = 1, m = 0) showed no secondary structure and base stacking. On mixing these compounds with polyadenylate, there appeared to be no evidence for base pairing, because no hypochromic effect could be detected *(41)*. In contrast, the oligomers synthesized by De Koning and Pandit (Fig. 20a, x = 4, m = 0) revealed maximal hypochromicity when measured in a 1:1 mixture with polyadenylate *(40)*. Monomers as a control did not give any hypochromic effect. For compounds of type b in Fig. 20, no biophysical data have been reported so far. Poly-L-lysine analogs (Fig. 20c, b = uracil) showed some base–base interaction on hybridization, although the observed decrease in absorption was a slow process *(44)*.

3. Discussion

The properties of nonionic antisense oligonucleotide analogs make them an interesting class of compounds, both for theoretical studies of nucleic acids structure and interactions and, on the other hand, for practical aspects as antisense oligonucleotide inhibitors of gene expression *(45)*. At present, it is not conclusive from the experimental data which analogs will serve best for application as antisense drugs. Criteria, such as stability under physiological conditions, solubility, cellular uptake, strength, and specificity of base pairing, must be considered, not to mention other pharmacological requirements, such as organ distribution and toxicity. With the exception of "plastic" DNA and some types of polyamides, elegant methods of specific, stepwise synthesis have been worked out for most types of compounds treated in this chapter. Chemical accessibility seems to be only a matter of effort. The synthesis of "plastic" DNA is hard to control, so that only highly dispersed mixtures of rather long polymers are obtained. Some classes of compounds will certainly not be considered for use as antisense oligonucleotides because of unfavorable properties: Carbonate- and carboxymethyl-bridged oligonucleotides tend to hydrolyze easily at neutral pH, and acetamidate-bridged analogs lack the ability to form hybrids with natural nucleic acids. Properties of other compounds have not been investigated enough: The different binding properties of carbamate-bridged poly(dC), and poly(dT) advise us to be careful in making general conclusions on characteristics, even within a single class of compounds. Correctly, each sequence should be treated individually. The unclear stoichiometry of binding for the carbamate-bridged morpholino-type oligonucleotides casts some doubt on their mode and specificity of binding. Although PNA oligomers of the aminoethylglycine type show very promising hybridization properties, their stability to peptidases and their uptake by living cells still have to be investigated. Furthermore, the strand-displacement reaction observed with A/T sequences may not be the exclusive mode of binding when mixed or G/C-rich sequences are used.

Already in 1968, Halford and Jones *(17)* concluded, from the fact that carboxymethyl-bridged oligonucleotides specifically hybridize to natural polynucleotides in water at low ionic strength, that "polymers of this type might interfere with the function of messenger RNA in biological systems." So far we do not know of any other biological data obtained with dephospho analogs that could serve as a lead regarding potential nucleic acid therapeutics that act by an antisense mechanism or triple helix formation. Most of the chemistry discussed above must still be optimized to allow for rapid solid-support synthesis of oligonucleotide analogs of appropriate length to carry out biological studies. Preformation of di- to pentameric dephospho building blocks, which are incorporated at different positions of an oligonucleotide to give sequences of the needed length, only partly solves the problem.

Acknowledgments

We thank G. Schluckebier for preparation of the art work, and G. O'Malley for carefully reading the manuscript. We are also indebted to Peter Nielsen and colleagues for making a preprint on PNAs available to us prior to publication.

References

1. Ogilvie, K. K. and Cormier, J. F. (1985) Synthesis of a thymidine dinucleotide analog containing an internucleotide silyl linkage. *Tetrahedron Lett.* **26,** 4159–4162.
2. Cormier, J. F. and Ogilvie, K. X. (1988) Synthesis of hexanucleotide analogs containing diisopropylsilyl linkages. *Nucl. Acids Res.* **16,** 4583–4594.
3. Seliger, H. and Feger, G. (1987) Oligonucleotide Analogs with dialkyl silyl internucleoside linkages. *Nucleosides & Nucleotides* **6,** 483,484.
4. Schulhof, J. C., Molko, D., and Teoule, R. (1987) The final deprotection step in oligonucleotide synthesis is reduced to a mild and rapid ammonia treatment by using base-labile protecting groups. *Nucl. Acids Res.* **15,** 397–416.
5. McBride, L. J. and Caruthers, M. H. (1983) N^6(*N*-methyl-2-pyrrolidine amidine) deoxyadenosine—A new deoxynucleoside protecting group. *Tetrahedron Lett.* **24,** 2953–2956.
6. Alul, R. H., Singman, C. N., Zhang, G., and Letsinger, R. L. (1991) Oxalyl-CPG: a labile support for synthesis of sensitive oligonucleotide derivatives. *Nucl. Acids Res.* **19,** 1527–1532.
7. Zon, G. (1988) Oligonucleotide analogs as potential chemotherapeutic agents. *Pharmaceutical Res.* **5,** 539–549.
8. Matteucci, M. (1990) Deoxyoligonucleotide analogs based on formacetal linkages. *Tetrahedron Lett.* **31,** 2385–2388.
9. Veeneman, G. H., van der Marel, G. A., van den Elst, H., and van Boom, J. H. (1990) Synthesis of oligonucleotides containing thymidines-linked via an internucleosidic-3'5'-methylene bond. *Rec. Trav. Chim.* **109,** 449–451.
10. Veeneman, G. H., van der Marel, G. A., van den Elst, H., and van Boom, J. H. (1991) An efficient approach to the synthesis of thymidine derivatives containing phosphateisosteric methylene acetal linkages. *Tetrahedron* **47,** 1547–1562.
11. Matteucci, M., Lin, K.-Y., Butcher, S., and Moulds, C. (1991) Deoxyoligonucleotides bearing neutral analogs of phosphodiester linkages recognize duplex DNA via Triple-Helix Formation. *J. Am. Chem. Soc.* **113,** 7767,7768.
12. Matteucci, M. (1991) Oligonucleotide analogs with novel linkages. Int. Pat. Application WO 91/06629.
13. Matteucci, M. (1991) Hybridization properties of a deoxyoligonucleotide containing four formacetal linkages. *Nucleosides & Nucleotides* **10,** 231–234.
14. Tittensor, J. R. (1971) The preparation of nucleoside carbonates. *J. Chem. Soc.* (C), 2656–2662.
15. Jones, D. S. and Tittensor, J. R. (1969) The preparation of dinucleoside carbonates. *Chem. Commun.,* 1240.
16. Mertes, M. P. and Coats, E. A. (1969) Synthesis of carbonate analogs of dinucleosides. 3'-thymidinyl-5'-thymidinyl carbonate, 3'-thymidinyl-5'-(5-fluoro-2'-deoxyuridinyl) carbonate, and 3'-(5-flouro-2'-deoxyuridinyl)-5'-thymidinyl carbonate. *J. Med. Chem.* **12,** 154–157.
17. Halford, M. H. and Jones, A. S. (1968) Synthetic analogs of polynucleotides. *Nature (Lond.)* **217,** 638–640.

18. Halford, M. H. and Jones, A. S. (1968) Synthetic analogs of polynucleotides. Part IV. Carboxymethyl derivatives of uridine and thymidine. *J. Chem. Soc.* (**C**), 2667–2670.
19. Jones, A. S., MacCoss, M., and Walker, R. T. (1973) Synthetic analogs of polynucleotides X. The synthesis of poly-(3'-O-carboxylmethyl-2'-deoxyadenosine), and its interaction with polynucleotides. *Biochim. Biophys. Acta* **365**, 365–377.
20. Bleaney, R. C., Jones, A. S., and Walker R. T. (1975) Synthetic analogs of polynucleotides XIV. The synthesis of poly-(3'-O-carboxylmethyl-2'-deoxycytidine), and its interaction with polyinosinic acid. *Nucl. Acids Res.* **2**, 699–706.
21. Edge, M. D. and Jones, A. S. (1971) Synthetic analogs of polynucleotides. Part V. Analogs of trinucleoside diphosphates containing carboxymethylthymidine. *J. Chem. Soc.* (**C**), 1933–1939.
22. Edge, M. D., Hodgson, A., Jones, A. S., MacCoss, M., and Walker, R. T. (1973) Synthetic analogs of polynucleotides. Part IX. Synthesis of 3'-O-carboxymethyl-2'-deoxyribonucleosides and their use in the synthesis of an analog of 2'-deoxyadenylyl-(3'-5')thymidine 3'-phosphate. *J. Chem. Soc. Perkin* **I**, 290–294.
23. Edge, M. D., Hodgson A., Jones, A. S., and Walker, R. T. (1972) Synthetic analogs of polynucleotides. Part VIII. analogs of oligonucleotides containing carboxymethylthymidine. *J. Chem. Soc. Perkin* **I**, 1991–1996.
24. Gait, M. J., Jones, A. S., and Walker, R. T. (1974) Synthetic analogs of polynucleotides. Part XII. Synthesis of thymidine derivatives containing an oxyacetamido- or oxyformamido-linkage of a phosphodiester group. *J. Chem. Soc. Perkin* **I**, 1684–1686.
25. Gait, M. J., Jones, A. S., Jones, M. D., Shephard, M. J., and Walker, R. T. (1979) Synthetic analogs of polynucleotides. Part 15. The synthesis and properties of poly(5'-amino-3'-O-carboxymethyl-2',5'-dideoxyerythropentonucleosides) containing 3'(O)-5'(C) acetamidate linkages. *J. Chem. Soc. Perkin* **I**, 1389–1394 36.
26. Nyilas, A., Glemarec, C., and Chattopadhyaya, J. (1990) Synthesis of [3'(O)-5'(C)]-oxyacetamido linked nucleosides. *Tetrahedron* **46**, 2149–2164.
27. König, W. and Geiger, R. (1970) Eine neue Methode zur Synthese von Peptiden: Aktivierung der Carboxylgruppe mit Dicyclohexylcarbodiimid unter Zusatz von 1-Hydroxybenzotriazolen. *Chem. Ber.* **103**, 788–798.
28. Mungall, W. S. and Kaiser, J. K. (1977) Carbamate analogs of oligonucleotides. *J. Org. Chem.* **42**, 703–706.
29. Stirchak, E. P., Summerton, J. E., and Weller, D. D. (1987) Uncharged stereoregular nucleic acid analogs. 1. Synthesis of a cytosine-containing oligomer with carbamate internucleoside linkages. *J. Org. Chem.* **52**, 4202–4206.
30. Griffin, D., Laramee, J., Deinzer, D., Stirchak, E., and Weller, D. (1988) Negative ion fast atom bombardment mass spectrometry of oligonucleotide carbamate analogs. *Biomed. Environm. Mass. Spectrom.* **17**, 105–111.
31. Coull, J. M., Carlson, D. V., and Weith, H. L. (1987) Synthesis and characterization of a carbamate-linked oligonucleoside. *Tetrahedron Lett.* **28**, 745–748.

32. Stirchak, E. P., Summerton, J. E., and Weller, D. D. (1989) Uncharged stereoregular nucleic acid analogs: 2. Morpholino nucleoside oligomers with carbamate internucleoside linkages. *Nucl. Acids Res.* **17,** 6129–6141.
33. Ohgi, T., Ishiyama, K., Tomi, H., and Yano, J. (1991) Synthesis and its properties of oligonucleotide analog containing thiocarbamate interlinkages. *Nucl. Acids Res. Symp. Ser.* **25,** 17,18.
34. Overberger, C. G. and Michelatti F. W. (1958) *N*-Vinyl derivatives of substituted pyrimidines and purines. *J. Am. Chem. Soc.* **80,** 988–991.
35. Pitha, J., Pitha, P. M., and Ts'O, P. O. P. (1970) Poly(1-vinyluracil): The preparation and interactions with adenosine derivatives. *Biochim. Biophys. Acta* **204,** 39–48.
36. Inaki, Y. and Takemoto, K. (1987) Functionality and applicability of synthetic nucleic acid analogs. *Current Top. Polym. Sci.* **1,** 80–100.
37. Takemoto, K. and Inaki, Y. (1981) Synthetic nucleic acid analogs: preparation and interaction. *Adv. Pol. Sci.* **41,** 1–51.
38. Pitha, J. (1983) Physiological activities of synthetic analogs of polynucleotides. *Adv. Pol. Sci.* **50,** 1–16.
39. Nielsen, P. E., Egholm, M., Berg, R., and Buchardt, O. (1991) Sequence-selective recognition of DNA by strand displacement with a thymine-substituted polyamide. *Science*, **254,** 1497-1500.
40. De Koning, H. and Pandit, U. K. (1971) Unconventional nucleotide analogs. VI. Synthesis of purinyl and pyrimidinyl-peptides. *Rec. Trav. Chim.* **91,** 1069–1080.
41. Buttrey, J. D., Jones, A. S., and Walker, R.T. (1975) Synthetic analogs of polynucleotides-XIII. The resolution of DL-β-(thymin-1-yl)alanine and polymerisation of the B-(thymin-1-yl)alanines. *Tetrahedron* **31,** 73–75.
42. Weller, D. D., Daly, D. T., Olson, W. K., and Summerton, J. E. (1991) Molecular modeling of acyclic polyamide oligonucleotide analogs. *J. Org. Chem.* **56,** 6000–6006.
43. Huang, S.-B., Nelson, J. S., and Weller, D. D. (1991) Acyclic nucleic acid analogs: synthesis and oligomerization of γ,4-diamino-2-oxo-1(2H)-pyrimidinepentanoic acid and δ,4-diamino-2-oxo-1(2H)-pyrimidinehexanoic acid. *J. Org. Chem.* **56,** 6007–6018.
44. Takemoto, K. and Inaki, Y. (1988) Nucleic acids analogs: their specific interaction and applicability. *Polym. Mat. Sci. Eng.* **78,** 250–253.
45. Uhlmann, E. and Peyman, A. (1990) Antisense oligonucleotides: a new therapeutic principle. *Chem. Rev.* **90,** 543–584.

CHAPTER 17

Scale-Up of Oligonucleotide Synthesis

Solution Phase

H. Seliger

1. Introduction

Solution-phase methods have been used for the first synthesis of an internucleotidic bond *(1)* and, some 20 years ago, had a glorious period during the first gene synthesis. At that time, their present-day competitor, the polymer support technique, was also developed and saw some first applications to oligonucleotide preparations *(2)*, but it took nearly 15 further years until the solid-phase methods became a serious rival to solution methods. The efficiency of the phosphoramidite chemistry, combined with the development of mechanization, brought about a dramatic change ca. 10 years ago, putting polymer-support synthesis into first place. Yet, solution methods were never completely replaced, and the following survey will testify to their vitality.

How much promise do they hold for the future, especially for future applications in large-scale synthesis? To make it clear from the beginning, this chapter does not intend to give simple answers to this question. In fact, the topic is in several ways difficult. First of all, what defines a preparation as being "large scale"? Anyone involved in the early gene syntheses in Khorana's laboratory and elsewhere remembers how much labor and expense went into the preparation of very short oligonucleotide segments for enzymatic ligation. Most of these short oligonucleotides were prepared by solution-phase phosphodiester

From: *Methods in Molecular Biology, Vol. 20: Protocols for Oligonucleotides and Analogs*
Edited by: S. Agrawal Copyright ©1993 Humana Press Inc., Totowa, NJ

methods in microgram quantities, and the first two gene syntheses could not have been accomplished without the concomitant development of biological cloning and amplification techniques.

The option of chemical large-scale preparations was first noticed as a feature of phosphotriester synthesis. In this context, the publication that, to my knowledge, was the first to describe a scale-up of oligonucleotide synthesis was the paper by R. L. Letsinger and K. K. Ogilvie *(3)* in 1967, where di- and trithymidylate were shown to be obtained on a scale of several grams. A decade later, such preparations became routine, thus quickly increasing the number of synthetic genes for the bioproduction of peptides. These syntheses were based on the availability of a "library" of short oligonucleotide blocks (e.g., *4*). The bulk production of such short oligomers, mostly not longer than trimers or tetramers, in milligram quantities marks the first period of "large-scale synthesis," and these were mainly syntheses in solution. During the last ca. eight years, the unrivaled efficiency and simplicity of automated solid-phase methods of chemical synthesis, combined with new techniques of enzymatic chain amplification (e.g., PCR), seemed to make large-scale synthesis by chemical methods in most cases dispensable.

Now a new challenge comes up, especially with potential applications of structurally modified oligonucleotides as therapeutics, but also with special requirements of structural and material research. The demand is for much longer oligonucleotides—ca. 20 bases or more—and for larger quantities than ever synthesized previously. Several grams of potential antisense chemotherapeutics are required for pharmacokinetic studies, and preclinical and clinical testing will necessitate quantities up to the kilogram range *(5)*, which constitutes a 1000-fold, later an even 10^6–10^7-fold, increase of the scale of present-day preparations. An industrial production of this kind is without precedent in oligonucleotide chemistry. Therefore—and here is the second difficulty—this chapter deals with a subject that, strictly speaking, does not yet exist. In this situation, the reviewer's choice is to make a projection, i.e., to bring together different bits and pieces of existing methodology, and ask the question of whether they appear promising for future scale-up. Since summaries of older developments are available through several reviews (e.g., *6–8)*, and monographs (e.g., *9–11)*, the focus of this chapter will be on methods elaborated and/or used

Scale-Up Solution Phase

throughout the last decade. Also, the author will define as "solution-phase" all those methods where the condensation of the growing chain with an incoming nucleoside or oligonucleotide block is performed in homogeneous solution, regardless of whether further steps of a chain-elongation cycle also include operations in the heterogeneous phase.

2. Conventional Solution-Phase Methods in the Synthesis of Deoxyoligonucleotides

Considering oligonucleotide synthesis as a multistep process, the term "conventional" will apply to all those approaches where chain-elongation is performed by repeating the three operations:

1. Activation/condensation with a protected monomer or block;
2. Separation of the elongated chain from the starting materials; and
3. Deprotection of the elongated chain prior to next chain lengthening.

Steps 1 and 3 are done in solution, and step 2 usually by chromatographic methods. The protection and activation strategies define different synthetic methods as belonging to the phosphate-diester, phosphate-triester, phosphite-triester, or H-phosphonate type. The change in synthetic methodology in the early 1980s from mainly solution-phase to predominantly solid-phase methods was accompanied by a change in preparative strategies. The phosphodiester method, previously climaxing in the total synthesis of the first genes *(12)*, nearly disappeared (some mechanistic aspects being the last topics under study; *see* e.g., *13)* in favor of the phospate- and phosphite-triester methods using phosphate- and phosphite intermediates (for review, *see* e.g., *6–8).*

Three areas remain where conventional solution synthesis is still of interest:

1. Reactions essential as a test to new protection and activation methods;
2. Mechanistic studies, model reactions, and identification of side products; and
3. The elucidation of stereochemical pathways as well as regio- and stereoregulated syntheses.

Studies on new protecting groups were often initially conducted in solution, but in recent years, generally directed toward immediate application in polymer-support synthesis. An account of recent protecting groups is given by Sonveaux in Chapter 1 of *Proto-*

cols for Oligonucleotide Conjugates. Likewise, stereochemical questions are treated by Stec and Lesnikowski in Chapter 14. This section, therefore, will focus on uses of solution synthesis for the elucidation of reaction pathways and the clarification of mechanistic questions.

2.1. Solution Studies on the Phosphate-Triester Method

A considerable number of papers published during the last decade elaborate on the chemical basis of phosphate- and phosphite-triester synthesis, more recently on the H-phosphonate method. The "modified triester synthesis" *(4)* has been the target of several investigations introducing new phosphate-protecting groups, and condensing agents for one-step and two-step chain elongations *(8)*. The role of phosphoric sulfonic anhydrides and symmetrical dinucleosidylpyrophosphates as intermediates was shown through ^{31}P-NMR and kinetic studies *(14)*. Better knowledge of mechanisms allowed the development of new activation procedures based on the cocatalytic effect of *N*-methylimidazole *(15)*, and similar heterocyclic bases *(16)*, and, more recently, of oxygennucleophilic compounds *(17)*. Alternative routes to phosphotriester formation were described via conversion of the P(O)OH-moiety to P(O)Cl, by tris-(2,4,6-trihalophenoxy-)dichlorophosphanes *(18)*, or through an "activable protection," which has been described for the case of *S,S*-diaryl-nucleoside phosphorodithioates *(19)*. A low, but detectable level of side products arising, from, e.g., irreversible base phosphorylation, was demonstrated in several publications (20), thus emphasizing the necessity for protection of guanine-O^6 and thymine- uracil-N^3 in deoxy- and ribooligonucleotide syntheses following the phosphate-triester route *(21)*.

2.2. Solution Studies on the Phosphite-Triester Method

The phosphite-triester approach, although originally developed through solution synthesis *(22)*, rapidly proved its value in connection with solid-phase techniques. The phosphoramidite chemistry *(23)* was designed by M. H. Caruthers and coworkers especially for use with polymer supports *(24)*. Solution syntheses accompanying this development have tested new activation methods and have clarified mechanistic questions *(25)*. Phosphorothioites were shown to be an alternative to phosphoramidites as activable synthons *(26)*. The problem of formation of byproducts, in particular of 3'3'-isomers, was the topic of several papers discussed by M. H. Caruthers in an earlier review *(24)*.

2.3. Solution Studies on the H-Phosphonate Method

Based again on work originally done in Lord Todd's laboratory *(27)*, the oligonucleotide synthesis via H-phosphonate intermediates *(28,29)* has established itself as the most recent of the major synthetic approaches. Variant protection schemes have been recommended *(30)*, and routes to internucleoside H-phosphonate bond formation alternative to the use of acyl or sulfonyl chloride-condensing agents *(28,29,31)* have been proposed *(32)* through solution studies. Aroyl phosphonates *(33)*, phosphorochloridites *(34)*, activable phosphite-triesters *(32,35,36)*, and last but not least, phosphoramidites *(37)* were shown to be valuable intermediates. In order to ensure quantitative oxidation, alternatives to iodine, such as, e.g., hexachloroacetone or dipyridyldisulfide (with or without presilylation) *(38)*, have been explored and were also conducted as a one-step condensation/oxidation *(39)*. In model studies done by solution methods, P. Garegg and coworkers have shown the mechanism of H-phosphonate internucleoside bond formation to involve acyl hydrogen phosphonates and H-pyrophosphonates *(40)*, and, additionally, trimetaphosphites *(41)*, if chlorophosphates or arene sulfonyl derivatives were used as activating agents. Side reactions arising mainly from an excess of condensing reagents are self-capping *(42)*, and form as well trinucleoside phosphites *(40)*. The latter have been exploited for the preparation and characterization of the unusual trinucleoside monophosphates *(43)* *(see also* Section 4.). O^6-acylation of guanine bases was not found to be a source of side products in H-phosphonate syntheses, since these modifications are reverted on ammoniacal treatment after chain elongations *(44)*. The possibility of combining H-phosphonate with phosphodiester cycles in oligonucleotide synthesis has been shown in a recent paper *(45)*.

2.4. Large-Scale Preparations Combined with Blockwise Phosphotriester Synthesis

Of these "classical" preparative methods, the one that has been most often used for large-scale synthesis is the phosphotriester method. It involves typically a blockwise chain extension. Rationally, the blocks are very short oligomers, mostly dimers or trimers. These are then combined to longer oligonucleotides *(4)*. Mostly, the strategy is being followed to break down a given target sequence into a minimum of multiply used blocks. If necessary, an optimized scheme of block arrangements can be generated with the aid of a computer *(46)*. Most useful for this kind of block condensation are protection schemes that allow a given block to be used for lengthening in either 3' to 5' or 5' to 3' direction. This approach was first described by Catlin and Cramer *(47)*, and is shown in Fig. 1A. Approaches including blocks for essentially unidirectional addition have been described by C. B. Reese and colleague *(48)*, and are shown in Fig. 1B *(8)*.

The scale-up of these "classical" preparative methods is, in the first place, connected with the availability of large-scale separation techniques, such as preparative TLC, microparticulate silica gel chromatography, and HPLC. Phosphodiester synthesis had already benefited from these developments, which reportedly cut to one-half the time required for the preparation of gene segments *(49)*. These new separation techniques, however, were really instrumental for scaling up the preparation of phosphotriester building blocks *(4)*. This, together with improved and high-yielding activation methods *(8)*, allowed preparations of longer oligonucleotides in quantities large enough to satisfy the demands of molecular biology, as well as spectroscopy and crystallography. Yet, the number of papers actually dealing with such large-scale preparations is small *(8)* probably because of the concomitant development of much simpler polymer-support techniques. Oligonucleotides of defined sequence up to 16 bases long, amounting to ca. 200–350 mg, have thus been prepared *(50–53)*. An octaconta-thymidylate (100 mg!) has been reported as the longest oligomer obtained in solution by chemical blockwise synthesis following the strategy outlined in Fig. 1A *(54)*. The high purity of target oligonucleotides that can be obtained via the intermediate purifications inherent to solution synthesis has been emphasized *(55)*.

Scale-Up Solution Phase

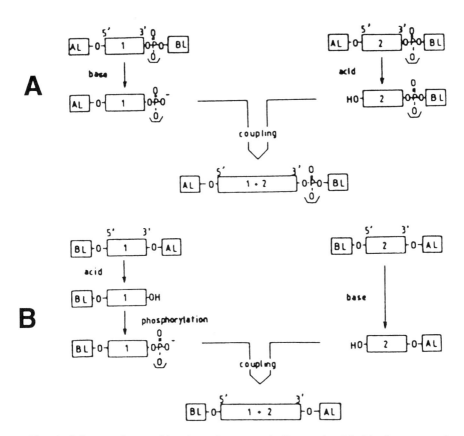

Fig. 1. Schemes for combination of protected oligonucleotide blocks prepared by phosphotriester solution synthesis (from 8). Symbols: BL = base-labile-protecting group; AL = acid-labile-protecting group; ▭ -fully protected oligodeoxyribonucleotide block.

The two most recent preparative approaches, the phosphoramidite and H-phosphonate routes, have, to my knowledge, not been used to any extent to prepare longer oligonucleotides in solution. Several papers describe the use of solution-synthesized blocks (mostly dimers prepared by the phosphotriester route) for solid-phase chain extension *(56a–e)*. The advantage of such blockwise support synthesis is the reduction of cycles and, thus, of overall time required for the preparation of longer chains. However, this may not lead to much decreased labor and expense because of prior block synthesis. Also the purification of long oligonucleotides is not significantly facilitated if truncated sequences arise through random chain breakage after acid or ammoniacal treatments.

Similar considerations apply to oligoribonucleotides. Here, too, preparations in solution in recent time have mainly been done with a focus on protection and activation studies, but in most cases were ultimately directed toward solid-phase synthesis. However, the methodology is far less developed than in deoxyoligonucleotide synthesis. Scale-up problems start already at the level of monomeric synthons, and large-scale production in the sense described in Section 1. cannot yet be envisaged. In view of a more detailed description of the current status *(see* elsewhere in this book), we can confine our discussion to these brief remarks and to the description of some methodic aspects in Sections 4. and 5.

3. "Unconventional" Solution-Phase Synthesis of Deoxyoligonucleotides

During multiple repetitions of protection/deprotection, activation and separation steps, simplifications can, in principle, be introduced by either canceling or combining certain steps. This is the basis of solid-support preparations *(see* Chapters 2, 3, and 4), but it also leads to simplifications of the "conventional" solution-phase synthesis. Such procedures, which can be classified as "unconventional," will be summarized in this section.

3.1. Use of Partly Unprotected Synthons

It is possible, with precautions, to leave the 5' and 3' OH groups of the nucleoside component unprotected *(56e,57),* although this does not completely exclude the formation of 3' 3' isomers *(58).* At least, this can be a feasible approach for the solution-phase preparation of short oligonucleotide blocks, which, after terminal phosphorylation or phosphitylation, serve as reagents for further chain-elongation.

The sensitivity of N-acylated bases to acid 5'-OH deprotection conditions has been a stimulus to investigating N-nonacylated nucleoside phosphate or phosphite synthons *(59,60).* Using the same principle recently in polymer supported synthesis, S. M. Gryaznov and R. L. Letsinger *(61),* however, recommend an additional "debranching" step during each cycle, which makes the use of N-unprotected synthons less advantageous, at least for routine preparations.

3.2. Combination of Activation and Protection Principles

Some of the concepts most attractive for the simplification of oligonucleotide synthesis and, thus, also for reducing cost and effort of potential large-scale preparations, come from a combination of two of the three operations essential for "conventional" chain-elongation. Although the combination of protection and activation principles plays a minor role in phosphate-diester and phosphate-triester chemistry, this is the basis of success of the phosphoramidite synthesis *(23,24)*. The relative stability of nucleoside phosphoramidites at room temperature and their fast and efficient activation by tetrazole *(23)* (*see also 25* for mechanistic studies) are essential to today's most utilized process of internucleotide bond formation.

3.3. Combination of Activation with Fast Separation Methods

3.3.1. Flash Chromatography and Solvent Extraction

Fast and efficient separation methods have been the key to new developments of solution-phase phosphotriester methods (*see* Section 2.), but evidently cannot compete with the simplicity of filtration and washing steps during solid-phase syntheses. Attempting to retain the option of product purification during intermediate stages of solution synthesis while designing chain extensions to equal solid-phase techniques in their simplicity, various "filtration-like" very rapid and efficient separation techniques have been recommended. Thus, C. B. Reese and collaborator have used fast chromatography through silica gel to remove educts from products in phosphate-triester synthesis, a technique for which they have coined the name "filtration method" *(62)*.

G. R. Gough et al. *(63)* have further elaborated on extraction procedures previously introduced for phosphodiester preparations *(64)*. Although in the phosphodiester case highly hydrophobic groups were required to solubilize the anionic chains in organic solvents, the hydrophobicity of protected oligonucleotide phosphotriesters is *per se* sufficient to keep them in chloroform solution *(63)*, whereas detritylated starting materials are extracted with water. The elongation cycle, using dimer blocks for chain extension, is shown in Fig. 2. Using this technique, the preparation of a dodecanucleotide was achieved on a scale of 207 A_{260} U, and further scale-up can be envisaged.

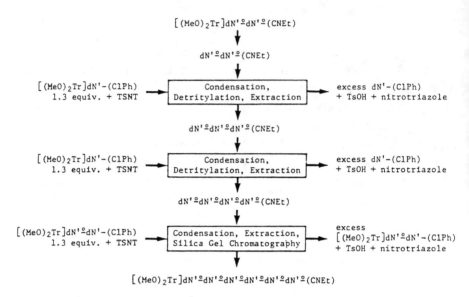

Fig. 2. Scheme for solution-phase triester synthesis of hexanucleotides using solvent extraction of intermediates (from 63).

A further feature of this method was the application of barium salts of protected nucleoside phosphates as synthons. Using a similar approach, H. Schott and H. Ruess *(65)* accomplished the preparation of 1.1 g of icosanucleotide $dC_4A_4C_4A_4C_4$ as shown in Fig. 3 by condensation of tetranucleotide blocks (ca. 1.2-fold excess of phosphate component) in yields ranging from 50–70%.

3.3.2. Precipitation and Crystallization Techniques— The "Liquid-Phase" Synthesis

The precipitation of unreacted excess monomer and dimer synthons as barium salts or through dropwise addition of the reaction mixtures into apolar solvents *(50,52,65,66)* allowed their recycling and reuse. It may be mentioned in this context that a similar recycling strategy has been recently followed for H-phosphonate synthesis *(67)*, but cannot be envisaged for tetrazole-activated phosphoramidites. In this latter case, hydrolysis and reconversion of the resulting H-phosphonates *(37)* or further use of such synthons in H-phosphonate synthesis would seem to be the only possible salvage route.

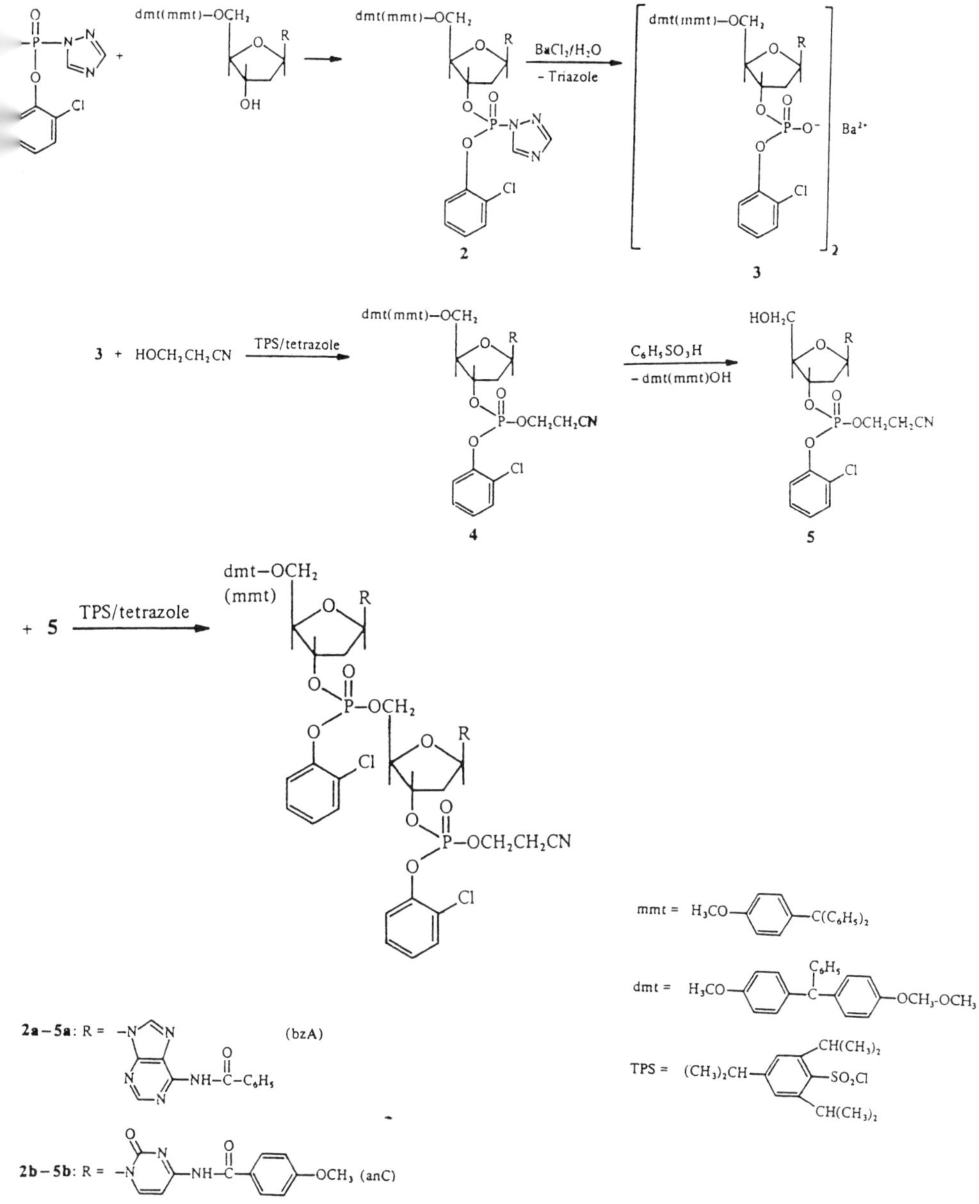

Fig. 3. Reaction scheme for a phosphotriester solution synthesis in gram scale using the osphodiester component as a barium salt.

An apparatus for continuous-flow solution-phase reactions has been described *(68)* combining a flow reactor with a solvent extraction, continuous evaporation, and liquid chromatography system. A scheme for the use of this apparatus in dinucleotide synthesis has been proposed *(46)*, but applications to the preparation of longer oligonucleotides have not been reported.

Crystallization is used as a separation technique in "liquid phase synthesis" using monofunctionalized polyethylene glycol as a solubilizing polymeric-protecting group. Following up earlier publications by E. Bayer and coworkers *(69)*, this approach has recently seen a revival through work done by G. M. Bonora and colleagues *(70,71)* (Table 1). Attractive features are the relatively high loading capacity, as compared to solid-phase carriers, as well as the fact that, although chemical reactions occur in solution, the separations are done in a heterogeneous fashion, i.e., through crystallization of the polyethylene glycol carrier and filtration of the low-molecular reactants. However, this method still has to prove its routine applicability for the preparation of medium-size chains, and automation of this process is an unsolved problem.

Similar advantages and disadvantages can be discussed for other liquid-phase methods applying cellulose acetate carriers *(72,73)* (Fig. 4; *see also 74* for related method) or short bifunctional trityl anchor chains, which allow bidirectional growth of oligonucleotides *(75)*.

3.3.3. Affinity Chromatography

Affinity chromatography is also known as a highly efficient separation technique. It can be used for oligonucleotide separations after applying affinigenic end groups. A simple example is the dimethoxytrityl group, which, when left on the 5' end of otherwise deprotected oligonucleotides, allows the latter to be purified on reverse-phase columns. Since its first publication *(76)*, the "trityl-on" separation, today done by simple cartridge methods *(5)*, has become a routine purification tool for polymer support products and can similarly be applied to sequences synthesized in solution *(74)*. Components capable of forming Hg–S bonds, such as reduced lipoate groups *(77)* or mercurated nucleosides *(76,78)*, can also serve as potential affinigenic substitutes, and their use in combination with (substituted) trityl end groups may lead to an oligonucleotide synthesis in solution with stepwise removal of

Table 1
Crystallizaation in "Liquid Phase Synthesis"

Step	Solvent or reagent[a,b]	Time, min
Detritylation	3% TCA in CH_2Cl_2 (10 mL)	15
Recrystallization	CH_2Cl_2/Et_2O 2× (10/90 mL)	60
Condensation	5'-DMTr-O-nucleoside (3 Eq) MSNT (6 Eq), NMI (10 Eq) in pyridine/2,6-lutidine (5 mL)	60
Recrystallization	EtOH abs. 2× (100 mL)	60
Capping	Ac2O 10% in pyridine (10 mL)	60
Recrystallization	CH_2Cl_2/Et_2O 1× (100 mL)	30

[a]Volumes are for 1.0 g of PEG-oligonucleotide.
[b]Abbreviations: TCA = trichloroacetic acid; MSNT = 2-(mesitylenesulf-onyl-)-3-nitro-1,2,4-triazole; NMI = N-methyl-imidazole; DMTr = 4,4'-dimethoxytrityl.

all educts by affinity techniques *(79)* (Fig. 5). "Classical" components of bioaffinity systems (e.g., biotin), because of their chemical lability, have not been used during chain elongations, but may well serve as "purification handles" *(80)* after terminal labeling.

3.3.4. Use of Polymeric Reagents

To complete this brief overview of "unconventional" solution syntheses, the use of polymeric reagents should be mentioned, either as carriers for the incoming monomers *(81)* or as polymeric activating agents *(82)*. The interest in such polymeric reagents has been mainly connected with mechanistic questions, and their potential for large-scale solution synthesis still remains to be exploited.

4. Solution-Phase Methods in the Synthesis of Structurally Modified Oligonucleotides

The preparative chemistry of oligonucleotides containing other than the four common deoxy- or ribonucleosides, although having started in the early times of nucleic acid synthesis, has, on the whole, developed more slowly than the preparation of sequences of biological structure. This relates partly to increased preparative problems encountered on introducing structural modifications, but it is also as a result of the fact that synthetic genes were, for a long time, in the focus in preparative projects; thus, any modification of the biological structure was rather undesirable. Studies on hybridization properties or on enzyme specificity and related physicochemical or biochemi-

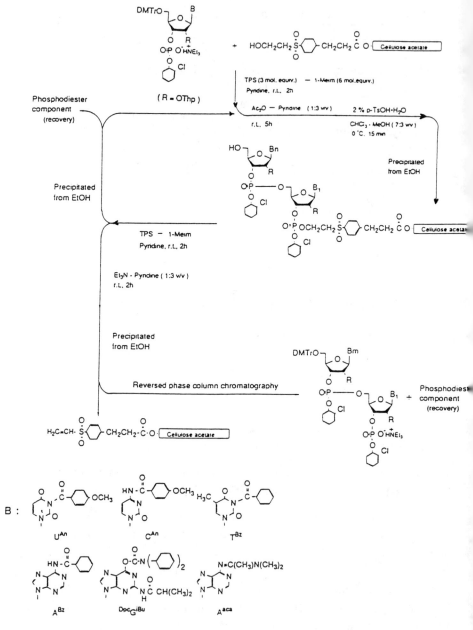

Fig. 4. Reaction cycle for liquid-phase chain elongation using cellulose acetate carrier (from 72).

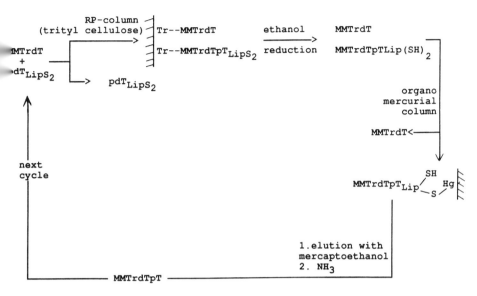

Fig. 5. An example of oligothymidylate chain elongation by phosphodiester solution synthesis on a 30-mg scale using a combination of two affinity end groups (see 78). Abbreviations: T = thymidine; MMTr = 5'-O-monomethoxytrityl-; $LipS_2$ = 3'-O-lipoyl-; Lip $(SH)_2$ = 3'-O-dihydrolipoyl-.

cal questions were, for a long time, the reason for preparing structurally modified oligo- and polynucleotides. Recently, there has been an enormous stimulus to the chemical synthesis of oligonucleotide analogs through new therapeutic concepts that make use of such compounds as inhibitors of transcription or gene expression. These oligonucleotide analogs, while retaining specific hybridization properties, must be so modified as to stabilize them toward intracellular degradation. Additionally, they may contain substituents that facilitate permeation of the cell membrane and promote degradation of the target nucleic acid (83). In view of these new applications as "antisense" drugs, several recent articles (84–86) have reviewed this field. Therefore, a description of possible structural modifications is not necessary in this context. Rather, the question will be raised, where solution methods have been of special value and may be of interest, for scaling up of preparations.

4.1. 2'-5'-Linked Oligoribonucleotides and Their Analogs

Oligoadenylates with 2'-5'-internucleotidic linkage of the general structure ppp(A2'p5')$_n$A (n = 1–4) were isolated from interferon-treated cells and shown to inhibit cell-free protein synthesis *(see* refs given, e.g., in *87).* This discovery of biological activity of oligonucleotides deviating from the canonical structure has triggered an enormous synthetic effort. A large number of compounds relating to the structure shown above have been prepared and their activity studied. The routes to these oligomers, because of their short length and relatively low structural complexity, mostly follow the schemes of solution synthesis. Chemical or enzymatic polymerization *(88)* and the phosphorimidazolide method *(89)* were used early on, but the major part of all these syntheses has been done by the phosphotriester approach, as described in Section 2. *(87,90,91)* using different protection schemes *(see also* 92 for review of different methods).

The use of simultaneous 3'- and 5'-protection by the tetra-isopropyl-disiloxan-diyl groups *(93)* proved to be most profitable in these preparations *(91,94).* In addition to the "core" oligonucleotides, also the respective 5'-(tri-)phosphorylated structures have been studied (89,95). Different modifications of the sugar *(96–100),* and base *(101)* moieties have been introduced in this way. Several recent papers deal with the chemical *(102–105),* but also enzymatic *(106,107)* synthesis of phosphorothioate analogs of 2',5'- oligoadenylate of controlled stereochemistry and their efficiency in binding and activating RNase L.

General interest in the structure of modified oligonucleotides and, in particular, studies modeling the introduction of modifications into tRNA fragments have induced work on the preparation of oligonucleotides with altered bases *(108–114).* Research on chemical mutagenesis and carcinogenesis has been the focus for the preparation of oligonucleotides containing modified bases *(115–118)* or apurinic sites *(119).*

Oligoribonucleotides modified in the 2' position *(120–122)* as well as oligomers of cyclonucleoside phosphates *(123,124)* have been of interest for conformational studies. The first preparations of α-anomeric oligonucleotides were also done by solution phosphotriester synthesis *(125),* although solid-phase phosphoramidite chemistry was used for longer sequences.

4.2. Branched Oligoribonucleotides

Another area that has recently kindled a considerable effort in solution synthesis relates to branched oligonucleotide chains. Although such structures previously were postulated as byproducts of phosphodiester oligonucleotide condensations and have occasionally been prepared as deoxytrinucleoside monophosphates *(43,126,127)*, branched oligonucleotide structures have gained much interest in the ribo series because of the discovery of "lariats" as intermediates in the process of intron splicing *(see* references in *128–131).*

Following these findings, chemical routes to such "lariat" structures have been explored *(128–136)*, recently progressing to branched nona- and decanucleotides as well as cyclic "lariat" structures in work reported from the laboratory of J. B. Chattopadhyaya *(137)* (Fig. 6). Various studies, mainly of a spectroscopic nature, have been done to establish the correct structure and investigate the conformation of these compounds *(138–141)* in order to elucidate effects operational in the splicing mechanism *(131).*

4.3. Oligonucleoside Phosphorothioates and Phosphorodithioates— Ionic Oligonucleotide Analogs

Oligonucleotides containing mono- and dithiophosphate internucleotide linkages are of particular interest as ionic analogs of biological sequences and, thus, as antisense inhibitors of gene expression. Therefore, they are most likely targets for routine large-scale production. Although such modified oligomers are now routinely synthesized on polymer supports, most primary chemical developments have been conducted in solution. The chemistry of oligonucleosidephosphorothioates has been pioneered by F. Eckstein, and an overview of the synthetic approaches can be obtained from his review *(142)* as well as from a more recent article by C. A. Stein in *(83).* Preparative improvements based on the use of phosphotriester *(143)* or H-phosphonate methodology (partly via the intermediacy of acyl- or silyl-phosphonates *[85,144,145]*, in some cases also with the aid of zwitterionic monomers *[146]*) have been the aim of recent solution studies. Highly stereospecific syntheses, pioneered by W. Stec and colleagues *(147)* have been in the focus of several of these papers *(148,149).*

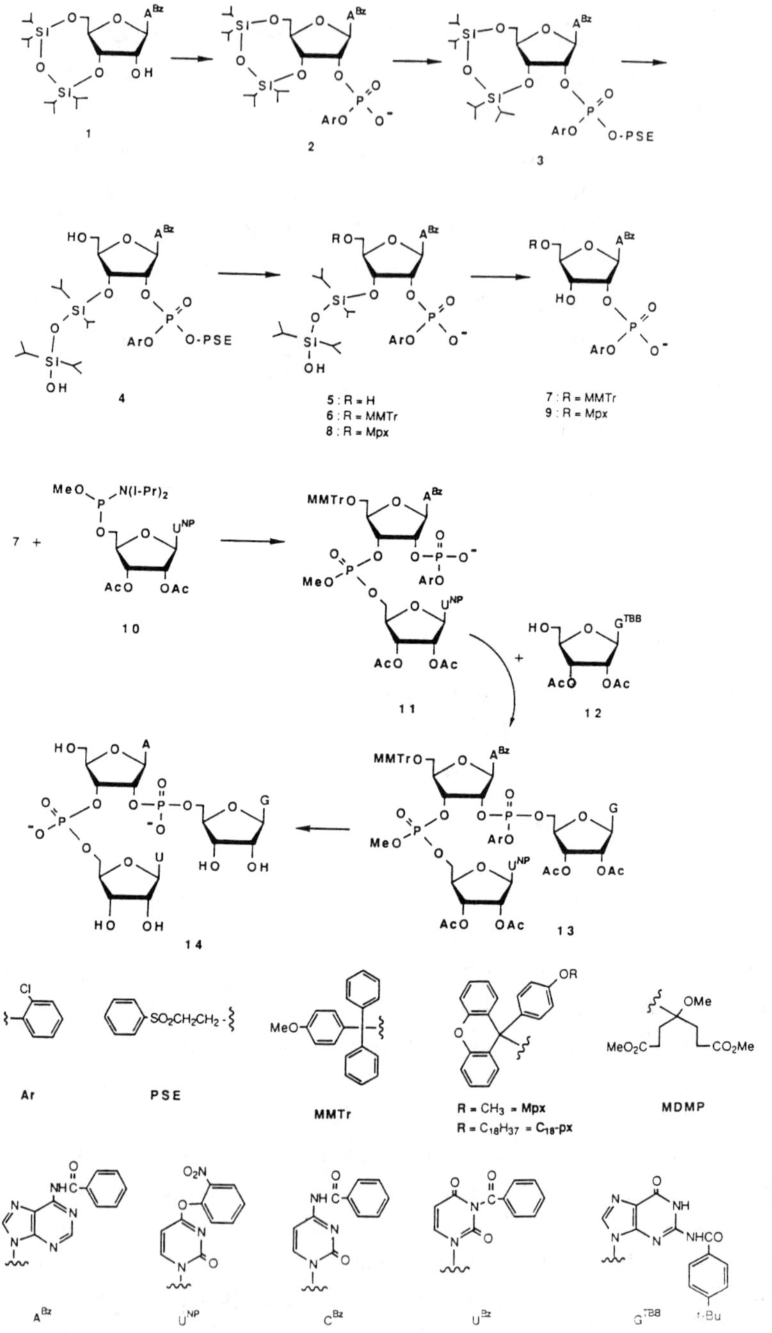

Fig. 6. Construction of a "lariat" triribonucleoside diphosphate by solution-phase phosphotriester synthesis (from *137a*).

Scale-Up Solution Phase

In contrast to phosphorothioates, the chemistry of dithioate analogs has developed only recently. Nucleosidephosphorodiamidites *(150–152)* (the putative intermediate bis-[1,2,4-triazolides] *[153]),* as well as nucleoside-thiophosphoramidites *(152,154–156),* and H-phosphono(di-)thioates *(157,158)* are the preferred intermediates. The use of thiophosphoranilidates *(159)* and protected dithiophosphate N-hydroxybenzotriazole derivatives *(160a,b)* was recently described especially with the aim of up-to-gram-scale syntheses (Fig. 7). The methyl group was found not to be a suitable protection for these compounds, since demethylation by nucleophilic reagents is accompanied by chain scission *(161).*

Examples of alternative ionic oligonucleotide analogs synthesized in solution include compounds with methylene phosphonate *(162,163),* pyrophosphate *(164),* or phosphoramidate *(85,165,166)* linkages. Syntheses of the latter, as well as of phosphorothioates, with the aim of attachment of various conjugate groups *(84,86)* can only be mentioned in this context.

4.4. Oligonucleoside Methylphosphonates—Nonionic Oligonucleotide Analogs

The preparation of oligonucleotide analogs with phosphorus-containing nonionic internucleosidic linkages has been pioneered, and recently reviewed, by P. S. Miller in *(83).* Protected nucleoside-methylphosphonate diesters, methylphosphonic-dichloridites and *bis*-(1-hydroxy-benzotriazolides) have been used for preparations along the phosphotriester route (for reference to earlier papers, *see 85),* the latter also for the synthesis of methylphosphonothioate analogs *(167).* The application of nucleoside-methylphosphonamidite *(85),* and H-methylphosphonate *(168)* synthons shows analogies to the previously described (Section 2.) phosphoramidite and H-phosphonate chemistry. The diastereoselective preparation of methylphosphonate dimer blocks was used for the synthesis of oligonucleotide analogs with stereoregular, alternating methylphosphonate/phosphodiester backbone *(169),* and respective conjugates *(170).* Improvements of diastereoselective syntheses have been reported recently for dimers *(171–173),* and also extended to longer oligomers of the thymidylate series *(174).* Studies that are in some respect analogous to the ones described before have been conducted with oligonucleotides containing alkyl phosphotriester internucleoside linkages *(152,175–179).* However, because of increased

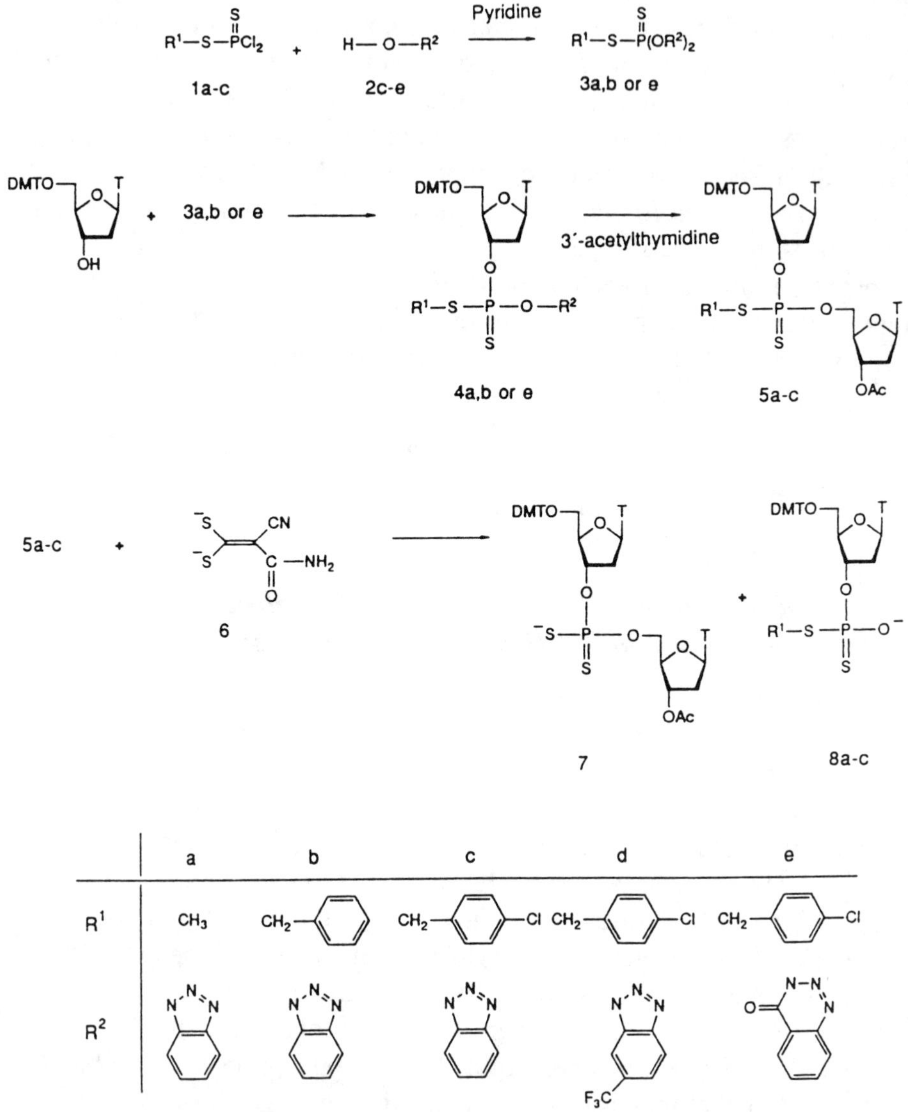

Fig. 7. A phosphotriester method for the large-scale preparation of oligonucleotide phosphorodithioates in solution (from *160a*).

lability of the phosphotriester bond, these compounds have received less attention with respect to application in the antisense field. Dinucleoside phosphorofluoridates have been described *(180)* as the most recent addition to the variety of such nonionic oligonucleotide analogs.

Phosphotriester- and methylphosphonate analogs, like oligonucleoside phosphorothioates, suffer from the fact that the internucleoside bonds are chiral, and only one stereoisomer is likely to hybridize well with biological polynucleotides. In order to overcome this problem, a number of nonionic, achiral internucleoside linkages not containing phosphorus atoms have been designed, and preparative approaches have usually been put to a first test by solution syntheses. Since such compounds are reviewed in another chapter of this book, we will refrain from discussing such compounds in more detail. Although the investigations on such "dephospho" analogs are not only interesting from their potential applications in the antisense field, but also in a more academic sense in probing the question of why nature chose (deoxy)-ribose phosphate linkages to build up nucleic acids *(181)*, the yield of knowledge from the studies conducted so far is relatively meager, and more efforts are worthwhile.

It is also beyond the scope of this chapter to discuss the field of conjugates, i.e., oligonucleotides substituted with reporter groups or containing substituents that promote crosslinking or cleavage on hybridization to a target nucleic acid. The application of such groups often involves additional reaction steps performed in solution, but rarely changes the synthesis schemes for the parent oligonucleotide, since they have been described in this and the preceding sections. Excellent reviews specializing on this topic have appeared recently *(83–86)*, and details will be reported in other chapters of this book.

5. Miscellaneous Methods of Oligonucleotide Solution Synthesis

This section will briefly look at some methods that produce oligonucleotides or their analogs in solution, although not following the schemes of stepwise or blockwise chemical elongation treated previously. The following methods have features that may be of interest for a potential scaleup or, in some instances, have already proven their value in preparations on a scale larger than usual.

5.1. Template-Dependent Chemical Condensations and Ligations

Mimicking the fact that biological DNA and RNA are always produced in template-dependent reactions, nucleotide monomers or oligonucleotide blocks have been chemically condensed in the presence of complementary polynucleotides as templates *(182)*. Targets of syntheses have been:

Homo-oligonucleotides *(183)*;
Oligonucleotides composed of different bases *(184)*; and
Oligonucleotide analogs *(185–187)*.

In a few cases, the synthesis could be directed toward the formation of defined sequences using anionic *(186,188,189)*, but also cationic *(190)* polymer templates.

The condensation of synthons has been carried out using:

Metal ion catalysis *(89,191)*;
Adsorption to mineral surfaces *(192–194)*;
Water-soluble carbodiimide *(195,196)*;
Nucleotide imidazolides *(197)*;
N-Hydroxy-benzotriazole *(198)*; or
Cyanogen bromide *(199,200)*.

Similar activation methods allow the chemical cyclization of (solution or solid-phase synthesized) linear oligonucleotides *(62,201–204)*, which recently have been studied for triple-helix formation to oligonucleotide templates *(204)*.

5.2. Activation of the Nucleoside Hydroxyl Groups

In contrast to the usual activation of phosphate, phosphite, or H-phosphonate components, internucleotide bond formation via selective activation of the alcohol moiety of the nucleoside educt *(205a,b)* is generally less efficient and has received little attention in recent years *(92,206)*, although in some cases, it may prove to be a cheap alternative.

5.3. Isolation of Oligonucleotides from Biological Sources by "Template Chromatography" (207)

Using, e.g., herring sperm DNA *(208)* as a cheap source for the isolation of oligonucleotide blocks of defined sequence, quantities of several hundred grams of material can be handled. However, the yields of individual blocks will quickly decrease with their length, in line with the decrease of occurrence of longer sequences in biological material.

5.4. Template-Independent Chemical and Enzymatic Polymerizations

Statistical considerations are also valid for generating oligonucleotide blocks by cocondensation of nucleotides with one or two "capped" nucleotide components as terminator units *(209)*. In oligoribonucleotide synthesis, coincubation of a binary mixture of ribonucleoside diphosphates with polynucleotide phosphorylase, followed by base-specific ribonuclease cleavage and separation of the resulting series of homologs, will give oligomer blocks of the general sequence B_nC or AB_nC (B, C = any ribonucleotide substrate for polynucleotide phosphorylase and base-specific RNAse; yields can be predetermined from copolymerization parameters *[210]*). Primed synthesis (A = primer) with limited substrate addition gives sequences AB_n *(209,211)*. This has been used as a tool for large-scale oligoribonucleotide synthesis *(211)*. On the basis of immobilized enzyme, a continuous production of homooligonucleotides *(212)* has been described.

5.5. Enzymatic Template-Dependent Polymerization

The enzymatic synthesis of DNA or RNA fragments on DNA templates has recently gained enormous impetus through the development of PCR and related techniques, as well as of modern transcription methods. In both cases, a DNA fragment must primarily be provided, which may require some synthetic effort, but is then present as a long-lived template from which the desired oligonucleotides can be produced. The potential of this technique for the large-scale production of DNA probes has been described in a patent application *(213)*, even before PCR was published (for recent collection of PCR methods and literature, *see* e.g., *214*). Similarly, T7 RNA polymerase has been used to transcribe even small RNAs from DNA templates, and a protocol for the large-scale synthesis and purification of oligoribonucleotides on this basis has just been issued *(215)*.

5.6. Enzymatic Monoaddition Reactions

Enzymatic monoadditions, such as:

The reversal of the equilibrium of ribonuclease-catalyzed reactions *(216);*
The enzymatic monoaddition of blocked substrates to primers catalyzed by polynucleotide phosphorylase *(217)* or deoxynucleotidyl terminal transferase *(218);* and
Single ribonucleotide addition catalyzed by RNA ligase *(219).*

have found little attention recently. A 10-fold upscaling of the latter preparations has been achieved with immobilized enzyme *(220)*.

6. Outlook: Solution-Phase Methods, an Option for Large-Scale Production?

What is the present status of solution-phase synthesis, after the polymer-supported methods of oligonucleotide preparation have experienced a decade of success? The fact that some 400 papers have appeared during this period describing progress in nucleic acids chemistry through solution preparations clearly demonstrates that such approaches are not obsolete, but have retained areas of preferred application. In the preceding sections, such areas have been identified as relating to the test of new protection and activation schemes, the elucidation of reaction mechanisms and the identification of byproducts, as well as the establishment of regio- and stereoselective reaction pathways, especially in the preparation of structurally modified oligonucleotides.

Relatively few of these papers describe the preparation of oligomers longer than 10–12 bases. This indicates that solution methods have (and also previously had) their particular application in the synthesis of short oligonucleotides *(8)*.

A similarly small number of papers declare "large-scale synthesis" as a goal (although a look at the Experimental Section shows that milligram quantities were not obtained so infrequently). In the beginning of the 1980s such papers often related to gene synthesis *(see* discussion in Sections 1–3. and *[4]* as well as previously cited reviews for reference), whereas recently, such strategies of block combination are no longer in use for this purpose. A number of publications relate to such syntheses as a means to provide material for crystallography or conformational analysis (examples have been described handling ca. 10–50 mg of 9–17-mer oligonucleotides of

Scale-Up Solution Phase

defined sequence *[8,221]*, but many papers in these areas do not even give indications about the preparative background), but again, more recently, solid-phase methods compete for such preparations *(222a,b)*. In any case, the previous "large-scale syntheses" in solution are quite diverse in methodology and motivation, and give little indication for the kind of scale-up necessary to meet future demands *(see* Section 1.).

In principle, any of the current methods for internucleotide bond formation and most of the protection schemes described in Sections 2–5. would lend themselves to scale-up. The phosporamidite method appears attractive, since it has the highest coupling efficiency on supports (>99%), but the yields are reported to decrease when this chemistry is used in manual synthesis and on a large scale (e.g., *5,222a*).

Phosphotriester synthesis needs somewhat longer condensation times, although yields have been improved to 90% and beyond. High selectivity is shown even in additions to nucleosides with unprotected hydroxyl functions *(see* discussion in Section 3.), but the avoidance of irreversible base modifications *(20)* may require additional protection. Last, but not least, there is enormous experience in the preparation of protected blocks, which may facilitate the design of routes for scale-up.

The H-phosphonate method has recently been recommended for scaling up automated solid-phase preparations *(5,222b,223)*. It is featured by highly efficient coupling and shortest cycle times (ca. 4 min), although an additional capping step seems indispensable for large-scale preparations *(222b)*. However, its applicability to the stepwise or blockwise synthesis of longer oligomers in solution still has to be demonstrated. An additional consideration is the recycling of excess synthons, which can be more easily done in H-phosphonate or phosphotriester preparations *(see* Section 3.). Finally, desired structural modifications can influence the choice of chemistry for scale-up (for instance, H-phosphonate synthesis is most suitable for all-phosphorothioate oligonucleotides *[5,223]*).

It must be emphasized that large-scale synthesis relies not only on the efficiency of protection and activation procedures, but also of separation techniques. Failures statistically build up, regardless of whether oligonucleotides are synthesized to large length, large scale, or large number. The main problem is the separation of sequence homologs (n mer from $[n-1]$ mer), which show increasingly similar chromatographic or electrophoretic behavior with increasing chain length. A major problem of support synthesis is the accumulation of all these homologs in the product released from the carrier. "Conventional" solution methods usually rely on chromatographic separations after each extension to ensure maximum purity of the growing chains. However, this makes solution synthesis time- and labor-consuming. Several of the "unconventional" methods discussed in Section 3. introduce, with the incoming monomer, "purification handles" that allow a "filtration-like" separation, e.g., by extraction, crystallization, or affinity techniques. If solution syntheses are to be of interest for an industrial production of oligonucleotides, they will most likely make use of such simplified separations.

Solution phosphotriester syntheses have usually been performed by blockwise chain-elongation. The strategy of using recyclable, bidirectionally applicable (47) blocks may help to reduce the cost and facilitate the separations. Eventually, solution and support synthesis may join efforts, the latter providing for a rapid assemblage of solution-synthesized blocks—the problems, which still remain to be solved, have been discussed in Section 2. The reverse option of joining support-synthesized blocks in solution (224) seems realizable in view of recent developments (225,226); however, this alternative will probably be limited to special applications.

A striking feature of polymer-support synthesis is its possibility of mechanization. Although occasionally mentioned in the literature (46,68), there is, to my knowledge, no apparatus in operation to produce oligonucleotides by solution methods. Hence, all efforts for scale-up are now concentrated on designing large-scale solid-phase synthesizers (e.g., 5,223,227,228). However, as has been pointed out by T. Geiser, the extrapolation of current technologies to a 10^6-fold increased oligonucleotide production alone would not allow affordable clinical evaluation of potential antisense drugs. This necessitates "the development of new technology, that permits the cost-effective production, characterization and quality assurance of up to kilogram quantities of oligonucleotide material" (5). Solution methods may play a role in these new developments.

Finally, in Section 5., attention was drawn to methods deviating from the scheme of stepwise chemical preparations. In particular, enzymatic methods of template-independent (211,220) or template-dependent *(213,214,220)* chain elongation have already served for up-scaled preparations and may be subject to further developments. The specificity of enzymatic reactions and the elimination of problems arising from solvent regeneration and waste disposal are factors that may favor biotechnology as an addition to the method of future oligonucleotide production.

Acknowledgment

The author gratefully acknowledges assistance by members of the research group, in particular by G. Gröger and Z. Földes-Papp, as well as by L. Majer and I. Spitzberg, in preparing and typing this article. He is indebted to K. C. Gupta—on leave from the CSIR Centre for Biochemicals, Delhi—for critically reading the manuscript.

References

1. Michelson, A. M. and Todd, A. R. (1955) Nucleotides Part XXXII. Synthesis of a dithymidine dinucleotide containing a 3':5'-internucleotidic linkage. *J. Chem. Soc.* 2632–2638.
2. Letsinger, R. L. and Mahadevan, V. (1965) Oligonucleotide synthesis on a polymer support. *J. Am. Chem. Soc.* **87,** 3526,3527; Cramer, F., Helbig, R., Hettler, H., Scheit, K.-H., and Seliger, H. (1966) Oligonucleotide synthesis on a soluble polymer as carrier. *Angew. Chem.* **78,** 640; *Angew. Chem. Int. Ed. Engl.* **5,** 601; Hayatsu, H. and Khorana, H. G. (1966) Deoxyribooligonucleotide synthesis on a polymer support. *J. Am. Chem. Soc.* **88,** 3182,3183.
3. Letsinger, R. L., and Ogilvie, K. K. (1967) A convenient method for stepwise synthesis of oligothymidylate derivatives in large-scale quantities. *J. Am. Chem. Soc.* **89,** 4801–4803.
4. Crea, R., Kraszewski, A. Hirose, T., and Itakura, K. (1978) Chemical synthesis of genes for human insulin. *Proc. Natl. Acad. Sci. USA* **75,** 5765–5769; Goeddel, D. V., Kleid, D. G., Bolivar, F., Heyneker, H. L., Yansura, D. G., Crea, R., Hirose, T., Kraszewski, A., Itakura, K., and Riggs, A. D. (1979) Expression in *Escherichia coli* of chemically synthesized genes for human insulin. *Proc. Natl. Acad. Sci. USA* **76,** 106–110.
5. Geiser, T. (1990) Large-scale economic synthesis of antisense phosphorothioate analogs of DNA for preclinical investigations. *Ann. NY Acad. Sci.* **616,** 173–183.
6. Rosenthal, A., Cech, D., and Shabarova, Z. A. (1983) Chemische Synthese von DNA-Sequenzen. *Z. Chem.* **23,** 317–327.
7. Itakura, K. Rossi, J. J., and Wallace, R. B. (1984) Synthesis and use of synthetic oligonucleotides. *Ann. Rev. Biochem.* **53,** 323–356 and references therein.
8. Sonveaux, E. (1986) The organic chemistry underlying DNA synthesis. *Bioorg. Chem.* **14,** 274–325.

9. Gassen, H. G. and Lang, A. (eds.) (1982) *Chemical and Enzymatic Synthesis of Gene Fragments. A Laboratory Manual.* Verlag, Weinheim.
10. Gait, M. J. (ed.) (1984) *Oligonucleotide Synthesis—A Practical Approach.* IRL, Oxford.
11. Gait, M. J. (1990) *Nucleic Acids in Chemistry and Biology.* (Blackburn, G. M. and Gait, M. J., eds.), IRL, Oxford.
12. Khorana, H. G. (1979) Total synthesis of a gene. *Science* **203,** 614–625, and literature cited there.
13. Wang, Y., Yang, Z., Wang, Q., Xu, Y., Liu, X., Xu, J. F., and Chen, C. (1986) The role of metaphosphate in the activation of the nucleotide by TPS and DCC in the oligonucleotide synthesis. *Nucl. Acids Res.* **14,** 2699–2706.
14. Zarytova, V. F., and Knorre, D. G. (1984) General scheme of the phosphotriester condensation in the oligodeoxyribonucleotide synthesis with arylsulfonyl chlorides and arylsulfonyl azolides. *Nucl. Acids Res.* **12,** 2091–2110.
15. Efimov, V. A., Buryakova, A. A., Reverdatto, S. V., Chakhmakhcheva, O. G., and Ovchinnikov, Y. A. (1983) Rapid synthesis of long-chain deoxyribooligonucleotides by the *N*-methylimidazolide phosphotriester method. *Nucl. Acids Res.* **11,** 8369–8387.
16. Dobrynin, V. N., Bystrov, N. S., Chernov, B. K., Severtsova, V., and Kolosov, M. N. (1979) Nucleophilic catalysis of phosphorylation by phosphorotriazolidates in the triester synthesis of oligonucleotides. *Bioorg. Khim.* **5,** 1254–1256.
17. Efimov, V. A., Chakhmakhcheva, O. G., and Reverdatto, S. V. (1987) Nucleophilic catalysis in the oligonucleotide synthesis, in *Biophosphates and Their Analogs—Synthesis, Structure, Metabolism and Activity* (Bruzik, K. S. and Stec, W. J., eds.), Elsevier Science, Amsterdam, pp. 23–36.
18. Hotoda, H., Wada, T. Sekine, M., and Hata, T. (1987) Tris-(2,4,6-tribromophenoxy)dichlorophosphorane:a novel condensing agent for rapid internucleotidic bond formation in the phosphotriester approach. *Tetrahedron Lett.* **28,** 1681–1684.
19. Sekine, M., Hamaoki, K., and Hata, T. (1981) Synthesis and properties of S,S-diaryl thymidine phosphorodithioates. *Bull. Chem. Soc. Jpn.* **54,** 3815–3827.
20. Gdaniec, Z., Mielewczyk, S., and Adamiak, R. W. (1987) Side reactions in oligonucleotide synthesis. ^{31}P NMR study of 4-chlorophenyl-phosphorodi/1,2,4-triazolide/preparations and their reactivity toward nucleoside lactam systems, in *Biophosphates and Their Analogs—Synthesis, Structure, Metabolism and Activity* (Bruzik, K. S. and Stec, W. J., eds.), Elsevier Science, Amsterdam, pp. 127–132.
21. Welch, C. J., Zhou, X.-X., Remaud, G., and Chattopadhyaya, J. (1987) Some aspects of reactivity and protection of the imide functions of uridine and guanosine in nucleic acid synthesis, in *Biophosphates and Their Analogs— Synthesis, Structure, Metabolism and Activity* (Bruzik, K. S. and Stec, W. J., eds.), Elsevier Science, Amsterdam, pp. 107–125.

22. Letsinger, R. L., Finnan, J. L., Heavner, G. A., and Lunsford, W. B. (1975) Phosphite coupling procedure for generating internucleotide links. *J. Am. Chem. Soc.* **97,** 3278,3279; Letsinger, R. L. and Lunsford, W. B. (1976) Synthesis of thymidine oligonucleotides by phosphite-triester intermediates. *J. Am. Chem. Soc.* **98,** 3655–3661.
23. Beaucage, S. L. and Caruthers, M. H. (1981) Deoxynucleoside phosphoramidites—a new class of key intermediates for deoxypolynucleotide synthesis. *Tetrahedron Lett.* **22,** 1859–1862.
24. Caruthers, M. H. (1987) DNA synthesis for nonchemists: the phosphoramidite method on silica supports, in *Synthesis and Application of DNA and RNA* (Narang, S. A., ed.) Academic, London, pp. 47–94.
25. Berner, S., Mühlegger, K., and Seliger, H. (1989) Studies on the role of tetrazole in the activation of phosphoramidites. *Nucl. Acids Res.* **17,** 853–863.
26. Nagai, H., Fujiwara, T., Fujii, M., Sekine, M., and Hata, T. (1989) Reinvestigation of deoxyribonucleoside phosphorothioites: synthesis and properties of deoxyribonucleoside 3'-dimethylphosphites. *Nucl. Acids Res.* **17,** 8581–8593.
27. Hall, R. H., Todd, A., and Webb, R. F. (1957) Nucleotides. Part XLI. Mixed anhydrides as intermediates in the synthesis of dinucleoside phosphates. *J. Chem. Soc.* 3291–3296.
28. Garegg, P. J., Regberg, T., Stawinski, J., and Stromberg, R. (1985) Formation of internucleotidic bonds via phosphonate intermediates. *Chemica Scripta,* **25,** 280–282.
29. Froehler, B. C. and Matteucci, M. D. (1986) Nucleoside H-phosphonates: valuable intermediates in the synthesis of deoxyoligonucleotides. *Tetrahedron Lett.* **26,** 469–472.
30. Jäger, A., Charubala, R., and Pfleiderer, W. (1987) Synthesis and characterization of deoxy- and ribo H-phosphonate dimers. *Nucleic Acid Symp. Ser.* **18,** 197–200.
31. Kuyl-Yeheskiely, E., Spierenburg, M., van den Elst, H., Tromp, M., van der Marel, G. A., and van Boom, J. H. (1986) Reaction of pivaloyl chloride with internucleosidic H-phosphonate diesters. *Recl. Trav. Chim. Pays-Bas.* **105,** 505,506.
32. Sakatsume, O., Yamane, H., Takaku, H., and Yamamoto, N. (1990) Use of new phosphonylating and coupling agents in the synthesis of oligodeoxyribonucleotides via the H-phosphonate approach. *Nucl. Acids Res.* **18,** 3327–3331.
33. Kume A., Fujii, M. Sekine, M., and Hata, T. (1984) Acylphosphonates. 4. Synthesis of dithymidine phosphonate: A new method for generation of phosphonate function via aroylphosphonate intermediates. *J. Org. Chem.* **49,** 2139–2143.
34. Wada, T., Hotoda, H., Sekine, M., and Hata, T. (1988) 2–Cyanoethyl nucleoside 3'-phosphonates as novel starting materials for oligonucleotide synthesis. *Tetrahedron Lett.* **29,** 4143–4146.

35. Watanabe T., Sato H., and Takaku, H. (1989) New phosphite method: the synthesis of oligodeoxyribonucleotides by use of deoxyribonucleoside 3'(Bis-(1,1,1,3,3,3-hexafluoro-2-propyl)phosphites) as new key intermediates. *J. Am. Chem. Soc.* **111**, 3437–3439.
36. Eritja, R., Smirnov, V., and Caruthers M. H. (1990) O-Aryl phosphoramidites: synthesis, reactivity and evaluation of their use for solid-phase synthesis of oligonucleotides. *Tetrahedron* **46**, 721–730.
37. Marugg, J. E., Burik, A., Tromp, M., van der Marel, G. A., and van Boom, J. H. (1986) A new and versatile approach to the preparation of valuable deoxynucleoside 3'-phosphite intermediates. *Tetrahedron Lett.* **27**, 2271–2274.
38. Garegg, P. J., Regberg, T., Stawinski, J., and Stromberg, R. (1987) Studies on the oxidation of nucleoside hydrogenphosphonates. *Nucleosides & Nucleotides* **6**, 429–432.
39. Sekine, M., Mori, H., and Hata, T. (1979) New type of chemical oxidative phosphorylation: activation of phosphonate function by use of triisopropyl-benzenesulfonyl chloride. *Tetrahedron Lett.* **13**, 1145–1148.
40. Garegg, P. J., Regberg, T., Stawinski, J., and Stromberg, R. (1987) Nucleoside H-Phosphonates. V. The mechanism of hydrogenphosphonate diester formation using acyl chlorides as coupling agents in oligonucleotide synthesis by the hydrogenphosphonate approach. *Nucleosides & Nucleotides* **6**, 655–662.
41. Garegg, P. J., Stawinski, J., and Stromberg, R. (1987) Nucleosides H-phosphonates. 8. Activation of hydrogen phosphonate monoesters by chlorophosphates and arenesulfonyl derivatives. *J. Org. Chem.* **52**, 284–287.
42. Froehler, B. C., Ng, P. G., and Matteucci, M. D. (1986) Synthesis of DNA via deoxynucleoside H-phosphonate intermediates. *Nucl. Acids Res.* **14**, 5399–5407.
43. Garegg, P. J., Lindh, I., and Stawinski, J. (1987) Synthesis of trinucleoside monophosphates using nucleoside H-phosphonates, in *Biophosphates and Their Analogs—Synthesis, Structure, Metabolism and Activity* (Bruzik, K. S., and Stec, W. J., eds.), Elsevier Science, Amsterdam, pp. 89–92.
44. Regberg, T., Stawinski, J., and Stromberg, R. (1988) Nucleoside H-Phosphonates. IX. Possible side reactions during hydrogen phosphonate diester formation. *Nucleosides & Nucleotides* **7**, 23–35.
45. Dubey, I. Ya., Lyapina, T. V., and Fedoryak, D. M. (1991) H-phosphonate oligonucleotide synthesis in solution. *Nucl. Acids Symp. Ser.* **24**, 270.
46. Hara, S., Okugawa, T., Ohkuma, T., Eguchi, M., and Oka, K. (1982) Programmed flow preparation of DNA-oligomers. *Nucl. Acids Symp. Ser.* **11**, 85–86.
47. Catlin, J. C. and Cramer, F. (1973) Deoxyoligonucleotide synthesis via the triester method. *J. Org. Chem.* **38**, 245–250.
48. Chattopadhyaya, J. B. and Reese, C. B. (1980) Chemical synthesis of a tridecanucleoside dodecaphosphate sequence of SV40 DNA. *Nucl. Acids Res.* **8**, 2039–2053 and literature cited there.

49. Fritz, H. J., Belagaje, R., Brown, E. L., Fritz, R. H., Jones, R. A., Lees, R. G., and Khorana, H. G. (1978) Studies on polynucleotides. 146. High-pressure liquid chromatography in polynucleotide synthesis. *Biochemistry* **17,** 1257–1267.
50. Schott, H., Semmler, R., and Closs, K. (1987) Preparative synthesis of guanylate-rich oligonucleotides using the phosphotriester method in solution. *Nucleosides & Nucleotides* **6,** 407,408.
51. Ohtsuka, E., Shin, M., Tozuka, Z., Ohta, A., Kitano, K., Taniyama, Y., and Ikehara, M. (1982) Synthesis of deoxypolynucleotides interacting with proteins: large scale synthesis of λ OR3 17 mer and CAP site 22 mer duplexes by 3'-phosphoro-p-anisidate method. *Nucl. Acids Symp. Ser.* **11,** 193–196.
52. Denny, W. A., Leupin, W., and Kearns, D. R. (1982) Simplified liquid-phase preparation of four decadeoxyribonucleotides and their preliminary spectroscopic characterization. *Helv. Chim. Acta* **65,** 2372–2393.
53. Mazzei, M., Balbi, A., Sottofattori, E., Abramova, T., Alama, A., and Nicolin, A. (1991) Liquid-phase synthesis and evaluation of antisense oligodeoxynucleotides to DNA polymerase. *Nucl. Acids Symp. Ser.* **24,** 298.
54. Nakahara, Y., and Ogawa, T. (1983) Chemical synthesis of nucleotide oligomers: convergent synthesis of fully protected 80mer of thymidylic acid. *Nucl. Acids Symp. Ser.* **12,** 59–62.
55. Rosenthal, A. and Cech, D. (1983) Chemische Synthese des Pentadecadesoxyribonucleotides d(TTCTTCTA CACACCC) nach der verbesserten Triestermethode. *J. Prakt. Chem.* **325,** 764–773.
56a. Kumar, G. and Poonian, M. S. (1984) Improvements in oligodeoxyribonucleotide synthesis: Methyl-*N,N*-dialkylphosphoramidite dimer units for solid-support phosphite methodology. *J. Org. Chem.* **49,** 4905–4912.
56b. Bannwarth, W. (1985) 200. Synthesis of oligodeoxy-nucleotides by the phosphite-triester method using dimer units and different phosphorus-protecting groups. *Helv. Chim. Acta* **68,** 1907–1913.
56c. Tanaka, T., and Oishi, T. (1985) Chemical synthesis of deoxyribonucleotides containing deoxyadenosine at the 3'-end on a polystyrene polymer support. *Chem. Pharm. Bull.* **33,** 5178–5183.
56d. Wolter, A., Biernat, J., and Köster, H. (1986) Polymer support oligonucleotide synthesis XX: Synthesis of a henhectacosa deoxynucleotide by use of a dimeric phosphoramidite synthon. *Nucleosides & Nucleotides* **5,** 65–77.
56e. Miura, K., Sawadaishi, K, Inoue, H., and Ohtsuka, E. (1987) Blockwise mechanical synthesis of oligonucleotides by the phosphoramidite method. *Chem. Pharm. Bull.* **35,** 833–836.
57. Seliger, H. Bach, T. C., Goertz, H. H., Happ, E., Holupirek, M., Seemann-Preising, B. Schiebel, H. M., and Schulten, H. R. (1982) Synthesis with nucleic acid constituents. Part XI. High-performance liquid chromatography in combination with field desorption mass spectrometry: separation and identification of building blocks for polynucleotide synthesis. *J. Chromatog.* **253,** 65–79.

58. Chattopadhyaya, J. B. and Reese, C. B. (1979) Some observations relating to phosphorylation methods in oligonucleotide synthesis. *Tetrahedron Lett.* **20,** 5059–5062.
59. Hayakawa Y., Uchiyama M., and Noyori, R. (1984) A convenient method for the formation of internucleotide linkage. *Tetrahedron Lett.* **25,** 4003–4006.
60. Fourrey, J.-L. and Varenne, J. (1985) Preparation and phosphorylation reactivity of N-nonacylated nucleoside phosphoramidites. *Tetrahedron Lett.* **26,** 2663–2666.
61. Gryaznov, S. M. and Letsinger, R. L. (1991) Synthesis of oligonucleotides via monomers with unprotected bases. *J. Am. Chem. Soc.* **113,** 5876–5877.
62. Rao, M. V. and Reese, C. B. (1989) Synthesis of cyclic oligodeoxyribonucleotides via the 'filtration' approach. *Nucl. Acids Res.* **17,** 8221–8239.
63. Gough, G. R., Brunden, M. J., Nadeau, J. G., and Gilham, P. T. (1982) Rapid preparation of hexanucleotide triester blocks for use in polydeoxyribonucleotide synthesis. *Tetrahedron Lett.* **23,** 3439–3442.
64. Jones, R. A., Fritz, H. J., and Khorana, H. G. (1978) Studies on polynucleotides. 147. Use of the lipophilic *tert*-butyldiphenylsilyl protecting group in synthesis and rapid separation of polynucleotides. *Biochemistry* **17,** 1268–1278; Mishra, R. K. and Misra, K. (1988) Protecting groups as purification tool in large-scale synthesis of small oligodeoxynucleotides. *Indian J. Chem.* **Sect. B, 27B,** 817–820.
65. Schott, H. and Ruess, H. (1986) Synthesis of fragments of the terminal inverted repeating units of macronuclear DNA from hypotrichous ciliates. *Makromol. Chem.* **187,** 81–104.
66. Gough, G. R., Brunden, M. J., and Gilham, P. T. (1981) Recovery and recycling of synthetic units in the construction of oligodeoxyribonucleotides on solid supports. *Tetrahedron Lett.* **32,** 4177–4180.
67. Seliger, H. and Rösch, R. (1990) Simultaneous synthesis of multiple oligonucleotides using nucleoside-H-phosphonate intermediates. *DNA and Cell Biology* **9,** 691–696.
68. Oka, K., Dobashi, Y., Ohkuma, T., and Hara, S. (1981) Liquid column switching extraction and chromatography for programmed flow preparation. *J. Chromatography* **217,** 387–398.
69. Brandstetter, F., Schott, H., and Bayer, E. (1975) Polymeric phosphate groups as protective groups for the liquid phase synthesis of oligonucleotides. *Makromol. Chem.* **176,** 2163–2175.
70. Bonora, G. M., Scremin, C. L., Colonna, F. P., and Garbesi, A. (1990) HELP (High Efficiency Liquid Phase) new oligonucleotide synthesis on soluble polymeric support. *Nucl. Acids Res.* **18,** 3155–3159.
71. Bonora, G. M., Biancotto, G., and Scremin, C. L. (1991) Use of the amidite chemistry in the PEG-supported, large-scale synthesis of oligonucleotides. The 'HELP Plus' method. *Nucl. Acids Symp. Ser.* **24,** 222.
72. Kamaike, K., Hasegawa, Y., Masuda, I., Ishido, Y., Watanabe, K., Hirao, I., and Miura, K. I. (1990) Oligonucleotide synthesis in terms of a novel type of polymer support: a cellulose acetate functionalized with 4-(2-Hydroxyethylsulfonyl)dihydrocinnamoyl substituent. *Tetrahedron* **46,** 163–184.

73. Kamaike, K., Ogawa, T., Inoue, Y., and Ishido, Y. (1991) Further improvement in protecting method for the oligonucleotide synthesis in terms of a cellulose acetate derivative as a polymer support. *Nucl. Acids Symp. Ser.* **24**, 37–39.
74. Hsiung, H. M. (1983) Isolating oligonucleotide product from a coupling reaction mixture. US-Pat. Appl. US 4417046, 5pp.
75. Biernat, J., Wolter, A., and Köster, H. (1983) Purification oriented synthesis of oligodeoxynucleotides in solution. *Tetrahedron Lett.* **24**, 751–754.
76. Seliger, H., Holupirek, M., and Görtz, H.-H. (1978) Solid-phase oligonucleotide synthesis with affinity chromatographic separation of the product. *Tetrahedron Lett.* **24**, 2115–2118.
77. Seliger, H., Holupirek, M., and Bach, T. C. (1977) The lipoyl affinity group and its use in oligonucleotide synthesis. British Chemical Society Nucleotide Group, 10th anniversary meeting, communications (poster no. 6)
78. Feist, P. L. and Danna, K. J. (1981) Sulfhydrylcellulose: a new medium for chromatography of mercurated polynucleotides. *Biochemistry* **20**, 4243–4246.
79. Görtz, H.-H. and Seliger, H. (1981) New hydrophobic protecting groups for the chemical synthesis of oligonucleotides. *Angew. Chem, Int. Ed. Engl.* **20**, 681–683.
80. Lewis, W., Stout, J., van Heeke, G., Wylie, D. E., Schuster, S. M., Wagner, F. W., and Coolidge, T. R. (1990) Peptide and oligonucleotide purification using immunoaffinity techniques. PCT Int. Appl., WO 9006936 A1, 68 pp.
81. Seliger, H. and Gupta, K. C. (1985) Three-phase synthesis of oligonucleotides. *Angew. Chem. Int. Ed. Engl.* **24**, 685–687.
82. Rubinstein, M. and Patchornik, A. (1975) Poly(3,5-diethylstyrene. sulfonyl chloride: A new reagent for internucleotide bond synthesis. *Tetrahedron* **31**, 1517–1519; Rubinstein, M. and Patchornik, A. (1975) A novel method for phosphodiester and internucleotide bond synthesis. *Tetrahedron* **31**, 2107–2110.
83. Cohen, J. S. (ed.) (1989) *Deoxyoligonucleotides, Antisense Inhibitors of Gene Expression.* Macmillan, London.
84. Goodchild, J. (1990) Conjugates of oligonucleotides and modified oligonucleotides: A review of their synthesis and properties. *Bioconjugate Chem.* **1**, 165–187.
85. Uhlmann, E. and Peyman, A. (1990) Antisense oligonucleotides: a new therapeutic principle. *Chem. Rev.* **90**, 543–584.
86. Englisch, U. and Gauss, D. H. (1991) Chemically modified oligonucleotides as probes and inhibitors. *Angew. Chem. Int. Ed. Engl.* **30**, 613–629.
87. Charubala, R., Uhlmann, E., and Pfleiderer, W. (1981) Synthese und Eigenschaften von Adenylyl-adenylyl-adenosinen. *Liebigs Ann. Chem.* 2392–2406.
88. Martin, E. M., Birdsall, N. J. M., Brown, R. E., and Kerr, I. M. (1979) Enzymic synthesis, characterisation and nuclearmagnetic-resonance spectra of pppA2'p5'A2'p5'A with chemically synthesised material. *Eur. J. Biochem.* **95**, 295–307.

89. Sawai, H., Shibata, T., and Ohno, M. (1979) Synthesis of oligonucleotide inhibitor of protein synthesis: pppA2'p5'A2'p5'A. *Tetrahedron Lett.* **47**, 4573–4576.
90. Ogilvie, K. K. and Theriault, N. Y. (1979) The synthesis of 2',5' linked oligoribonucleotides. *Tetrahedron Lett.* **20**, 2111–2114.
91. Karpeisky, M. Y., Beigelman, L. N., Mikhailov, S. N., Padyukova, N. S., and Smrt, J. (1982) Synthesis of adenylyl-(2'-5')adenylyl-(2'-5')adenosine. *Coll. Czech. Chem. Commun.* **47**, 156–166.
92. Noyori, R., Uchiyama, M., Kato, H. Wakabayashi, S., and Hayakawa, Y. (1990) Organometallic methodologies for nucleic acid synthesis. *Pure Appl. Chem.* **62**, 613–622.
93. Markiewicz, W. T. (1979) Tetraisopropyldisiloxane-1,3-diyl, a group for simultaneous protection of 3'- and 5'-hydroxy functions of nucleosides. *J. Chem. Res.* **(S)**, 24,25.
94. Gioeli, C., Kwiatkowski, M., Oeberg, B., and Chattopadhyaya, J. B. (1981) The tetraisopropyldisiloxane1,3-diyl: a versatile protecting group for the synthesis of adenylyl-(2'-5')-adenylyl-(2'-5')adenosine (2-5A core) *Tetrahedron Lett.* **22**, 1741–1744.
95. Kvasyuk, E. I., Kalinichenko, E. N., Kulak, T. I., Podkopaeva, T.L., Mikhailopulo, I. A., Popov, I. L., Barai, V. N., and Zinchenko, A. I. (1985) Chemical and microbiological 5'-phosphorylation of (2'-5')oligoadenylates. *Sov. J. Bioorg. Chem.* **11**, 670–677.
96. Kvasyuk, E. I., Kulak, T. I., Zaitseva, G. V., Mikhailopulo, I. A., and Pfleiderer, W. (1984) Synthesis of 2',3'-cyclic acetal derivatives of (2'-5')oligoadenylates and affinity sorbents based on them. *Bioorg. Khim.* **10**, 506–514.
97. Charubala, R., Pfleiderer, W., Sobol, R. W., Wu Li, S., and Suhadolnik, R. J. (1989) Chemical synthesis of adenylyl-(2'-5')-adenylyl-(2'-5')-8-azidoadenosine, and activation and photoaffinity labelling of RNase L by I[32P]Ip5'A2'p5'A2'p5'N$_3$[8]A. *Helv. Chim. Acta* **72**, 1354–1361.
98. Mueller, W. E. G., Weiler, B. E., Charubala, R., Pfleiderer, W., Leserman, L., Sobol, R. W., Suhadolnik, R. J., and Schroeder, H. C. (1991) Cordycepin analogues of 2',5'-oligoadenylate inhibit human immunodefiency virus infection via inhibition of reverse transcriptase. *Biochemistry* **30**, 2027–2033.
99. Herdewijn, P., Ruf, K., and Pfleiderer, W. (1991) Nucleotides. Part XXXIV. Synthesis of modified oligomeric 2'-5'A analogues: potential antiviral agents. *Helv. Chim. Acta* **74**, 7–23.
100. Mikhailov, S. N., Charubala, R., and Pfleiderer, W. (1991) Nucleotides. Part XXXV. Synthesis of 3'-deoxyadenylyl-(2'-5')-3'-deoxyadenylyl-(2'-.omega.)-9-(.omega.hydroxyalkyl)adenines. *Helv. Chim. Acta* **74**, 887–891.
101. Herdewijn, P., Charubala, R., De Clercq, E., and Pfleiderer, W. (1989) 191. Nucleotides, Part. XXXII. Synthesis of 2'-5' connected oligonucleotides. Prodrugs for antiviral and antitumoral nucleosides. *Helv. Chim. Acta* **72**, 1739–1748.

102. de Vroom, E., Fidder, A., Saris, C. P., van der Marel, G. A., and van Boom, J. H. (1987) Preparation of the individual diastereomers of adenylyl-(2'-5')-P-thioadenylyl-(2'-5')adenosine and their 5'-phosphorylated derivatives. *Nucl. Acids Res.* **15**, 9933–9943.
103. Charachon, G., Sobol, R. W., Bisbal, C., Salehzada, T., Silhol, M., Charubala, R., Pfleiderer, W., Lebleu, B., and Suhadolnik, R. J. (1990) Phosphorothioate analogs of (2'-5')(A)$_4$: agonist and antagonist activities in intact cells. *Biochemistry* **29**, 2550–2556.
104. Charubala, R., Sobol, R. W., Kon, N., Suhadolnik, R. J., and Pfleiderer, W. (1991) Syntheses and biological characterization of phosphorothioate analogues of (3'-5')adenylate trimer. *Helv. Chim. Acta* **74**, 892–898.
105. Battistini, C., Brasca, M. G., and Fustinoni, S. (1991) High stereoselectivity in the formation of the interribonucleotidic phosphorothioate bond. *Nucleosides & Nucleotides* **10**, 723–725.
106. Kariko, K., Sobol, R. W., Suhadolnik, L., Li, S. W., Reichenbach, N. L., Suhadolnik, R. J., Charubala, R., and Pfleiderer, W. (1987) Phosphorothioate analogues of 2',5'-oligoadenylate. Enzymatically synthesized 2',5'-phosphorothioate dimer and trimer: Unequivocal structural assignment and activation of 2',5'-oligoadenylate-dependent endoribonuclease. *Biochemistry* **26**, 7127–7135.
107. Kariko, K., Li, S. W., Sobol, R. W., Suhadolnik, R. J., Charubala, R., and Pfleiderer, W. (1987) Phosphorothioate analogues of 2',5'-oligoadenylate. Activation of 2',5'-oligoadenylate-dependent endoribonuclease by 2',5'-phosphorothioate cores and 5'-monophosphates. *Biochemistry* **26**, 7136–7142.
108. Charubala, R., Bannwarth, W., and Pfleiderer, W. (1980) Nucleotide, XII. Synthese und Eigenschaften von Trinucleosiddiphosphaten mit Thymidin, 2'-Desoxyadenosin und 1-(2'Desoxy-alpha-sowie 1-(2'-Desoxy-β-D-ribofuranosyl)-lumazinen als Bausteine. *Liebigs Ann. Chem.* 65–79.
109. Flockerzi, D., Silber, G., and Pfleiderer, W. (1983) Nucleotides. XXI. Synthesis and properties of dihydrouridine-containing oligonucleotides. *Helv. Chim. Acta,* **66**, 2641–2651.
110. Ohtsuka, E., Matsugi, J., Takashima, H., Aoki, S., Wakabayashi, T., Miyake, T., and Ikehara, M. (1983) Studies on transfer ribonucleic acids and related compounds. XLI. Synthesis of tRNA fragments containing modified nucleosides. *Chem. Pharm. Bull.* **31**, 513–520.
111. Millican, T. A., Mock, G. A., Chauncey, M. A., Patel, T. P., Eaton, M. A. W., Gunning, J., Cutbush, S. D., Neidle, S., and Mann, J. (1984) Synthesis and biophysical studies of short oligodeoxynucleotides with novel modifications: a possible approach to the problem of mixed base oligodeoxynucleotide synthesis. *Nucl. Acids Res.* **12**, 7435–7453.
112. Altermatt, R. and Tamm, C. (1985) Synthese eines Tridecanucleosiddodecaphosphats, das die unnatürliche Base 2 (1H)-Pyrimidinon enthält. *Helv. Chim. Acta* **68**, 475–483.

113. Uesugi, S., Miyashiro, H., Tomita, K., and Ikehara, M. (1986) Synthesis and properties of d(ATACGCGTAT) and its derivatives containing one and two 5-methylcytosine residues. Effect of the methylation on deoxyribonucleic acid conformation. *Chem. Pharm. Bull.* **34,** 51–60.

114. Kawase, Y., Iwai, S., and Ohtsuka, E. (1989) Synthesis and thermal stability of dodecadeoxyribonucleotides containing deoxyinosine pairing with four major bases. *Chem. Pharm. Bull.* **37,** 599–601.

115. Li, B. F. L., Reese, C. B., and Swann, P. F. (1987) Synthesis and characterization of oligodeoxynucleotides containing 4-O-methylthymine. *Biochemistry* **26,** 1086–1093.

116. Butkus, V., Klimasauskas, S., Petrauskiene, L., Maneliene, Z., Janulaitis, A., Minchenkova, L. E., and Shelkina, A. K. (1987) Synthesis and physical characterization of DNA fragments containing N4-methylcytosine and 5-methylcytosine. *Nucl. Acids Res.* **15,** 8467–8478.

117. Goddard, A. J. and Marquez, V. E. (1988) Synthesis of a phosphoramidite of 2'-deoxy-5,6-dihydro-5-azacytidine. Its potential application in the synthesis of DNA containing dihydro-5-aza- and 5-azacytosine bases. *Tetrahedron Lett.* **15,** 1767–1770.

118. Smith, C. A. (1991) Chemical synthesis of oligonucleotides containing a naphthalene diolepoxide deoxycytidine adduct in solution and using a mixed chemistry semiautomated solid phase approach. *Carcinogenesis* **12,** 631–636.

119. Vasseur, J.-J., Rayner, B., and Imbach, J.-L. (1986) Preparation of a short synthetic apurinic oligonucleotide. *Biochem. Biophys. Res. Commun.* **134,** 1204–1208.

120. Ikehara, M., Takatsuka, Y., and Uesugi, S. (1979) Polynucleotides. LIII. Synthesis and properties of 2'-azido-2'-deoxyadenylyl-(3'-5')-2'-azido-2'deoxyadenosine. *Chem. Pharm. Bull.* **27,** 1830–1835.

121. Uesugi, S., Takatsuka, Y., Ikehara, M., Cheng, D. M., Kan, L. S., and Ts'o, P. O. P. (1981) Synthesis and characterization of the dinucleoside monophosphates containing 2'fluoro-2'-deoxyadenosine. *Biochemistry* **20,** 3056–3062.

122. Damha, M. J., Usman, N., and Ogilvie, K. K. (1987) The rapid chemical synthesis of arabinonucleotides. *Tetrahedron Lett.* **28,** 1633–1636.

123. Ikehara, M., Uesugi, S., and Shida, T. (1980) Polynucleotides. LV. Synthesis and properties of dinucleoside monophosphates derived from adenine 8,2'-S- and uracil 6,2'-O-cyclonucleosides. Further support for the left-handed stacking of oligonucleotides giving high-anti base torsion angles. *Chem. Pharm. Bull.* **28,** 189–197.

124. Uesugi, S., Shida, T., and Ikehara, M. (1980) Polynucleotides. LXI. Synthesis and properties of dinucleoside monophosphates containing 8, 2'-S-cycloadenosine and 8, 2'-S-cycloinosine residues. Sequence dependency of the stability of the stacking conformation. *Chem. Pharm. Bull.* **28,** 3621–3631.

125. Morvan, F., Rayner, B., Imbach, J.-L., Thenet, S., Bertrand, J.-R., Paoletti, J., Malvy, C., and Paoletti, C. (1987) α-DNA II. Synthesis of unnatural α-anomeric oligodeoxyribonucleotides containing the four usual bases and study of their substrate activities for nucleases. *Nucl. Acids Res.* **15,** 3421–3437.

126. Nagyvary, J. (1966) Studies on the specific synthesis of the natural internucleotide linkage by the use of cyclonucleosides. I. The utilization of unprotected nucleotides. *Biochemistry* **5**, 1316–1322.
127. Norman, E. J. and Nagyvary, J. (1974) Synthesis of some trinucleoside monophsophates of biological interest. *J. Med. Chem.* **17**, 473–475.
128. Sekine, M. and Hata, T. (1985) Synthesis of branched ribonucleotides related to the mechanism of splicing of eukaryotic messenger RNA. *J. Am. Chem. Soc.* **107**, 5813–5815.
129. Damha, M. J., Pon, R. T., and Ogilvie,K. K. (1985) Chemical synthesis of branched RNA: novel trinucleoside diphosphates containing vicinal 2'-5' and 3'-5' phosphodiester linkages. *Tetrahedron Lett.* **26**, 4839–4842.
130. Kierzek, R., Kopp, D. W., Edmonds, M., and Caruthers, M. H. (1986) Chemical synthesis of branched RNA. *Nucl. Acids Res.* **14**, 4751–4764.
131. Remaud, G., Balgobin, N., Sandstrom, A., Vial, J.-M., Koole, L. H., Buck, H. M., Drake, A. F., Zhou, X.-X., and Chattopadhyaya, J. (1989) Why do all lariat RNA introns have adenosine as the branch point nucleotide? Conformational study of naturally-occurring branched trinucleotides and its eleven analogues by 1H-, ^{31}P-NMR and CD spectroscopy. *J. Biochem. Biophys. Meth.* **18**, 1–36.
132. Zhou, X.-X., Nyilas, A., Remaud, G., and Chattopadhyaya, J. B. (1987) Regiospecific synthesis of branched tetranucleotides: U3'p5'A$^{2\,P5\,G}$3'p5'U, U3'p5'A$^{2\,P5\,G}$3'p5'C, A3'p5'A$^{2\,P5\,G}$3'p5'U &a3'p5'A$^{2\,P5\,G}$3'p5'C. *Tetrahedron* **43**, 4685–4698.
133. Sekine, M., Heikkila, J., and Hata, T. (1987) A new method for the synthesis of branched oligoribonucleotides using a fully protected branched triribonucleoside diphosphate unit. *Tetrahedron Lett.* **28**, 5691–5694.
134. Hayakawa, Y., Nobori, T., Noyori, R., and Imai, J. (1987) Synthesis of 2'-5',3'-5' linked triadenylates. *Tetrahedron Lett.* **28**, 2623–2626.
135. Fourrey, J. L., Varenne, J., Fontaine, C., Guittet, E., and Yang, Z. W. (1987) A new method for the synthesis of branched ribonucleotides. *Tetrahedron Lett.* **28**, 1769–1772.
136. Huss, S., Gosselin, G., and Imbach, J.-L. (1987) Synthese chimique de nucleotide possedant des liaisons phosphodiesters 2'-5' et 3'-5' vicinales. *Tetrahedron Lett.* **28**, 415–418.
137a. Zhou, X. X., Remaud, G., and Chattopadhyaya, J. (1988) New regiospecific synthesis of the branched tri-, penta-, and heptaribonucleic acids which are formed as the lariat in the pre-mRNA processing reactions (splicing) *Tetrahedron* **44**, 6471–6489.
137b. Sund, C., Agback, P., and Chattophadyaya, J. B. (1991) Synthesis of tetrameric cyclic branched-RNA (lariat) modelling the introns of group II and nuclear pre-mRNA processing reaction (splicing) *Tetrahedron* **47**, 9659–9674.
138. Lee, M., Huss, S. Gosselin, G. Imbach, J.-L., Hartley, J. A., and Lown, J. W. (1987) Strucure and conformation of the branch core triribonucleotide containing 2'-5' and 3'-5' phosphodiester linkages (A$^{2\,P5\,G}$3'p5'C) in solution, essential for yeast mRNA splicing, deduced from ^{1}H-NMR. *J. Biomolec. Struct. & Dynam.* **5**, 651–668.

139. Huss, S., Gosselin, G., and Imbach, J.-L. (1988) Synthesis of various branched triribonucleoside diphosphates by site-specific modification of a diphenylcarbamoyl-protected guanine residue. *J. Org. Chem.* **53**, 499–506.
140. Damha, M. J. and Ogilvie, K. K. (1988) Conformational properties of branched RNA fragments in aqueous solution. *Biochemistry* **27**, 6403–6416.
141. Zhou, X.-X., Nyilas, A., Remaud, G., and Chattopadyaya, J. (1988) 270 MHz 1H-NMR study of four branched tetraribonucleotides:A3'p5'A$^{2\ P5}$ ^{G}3'p5'U, A3'p5'A$^{2\ P5}$ ^{G}3'p5'C, U3'p5'A$^{2\ P5}$ ^{G}3'p5'U and U3'p5'A$^{2\ P5}$ ^{G}3'p5'C which are formed as the lariat branch-point in the pre-mRNA processing reactions (splicing) *Tetrahedron* **44**, 571–589.
142. Eckstein, F., (1983) Phosphorothioate analogs of nucleotides—tools for the investigation of biochemical processes. *Angew. Chem. Int. Ed. Engl.* **22**, 423–439.
143. Marugg, J. E., van den Bergh, C., Tromp, M., van der Marel, G. A., van Zoest, W. J., and van Boom, J. H. (1984) Synthesis of phosphorothioate-containing DNA fragments by a modified hydroxybenzotriazole phosphotriester approach. *Nucl. Acids Res.* **12**, 9095–9110.
144. Fujii, M., Ozaki, K., Kume, A., Sekine, M., and Hata, T. (1986) Acylphosphonates. 5. A new method for stereospecific generation of phosphorothioate via aroylphosphonate intermediate. *Tetrahedron Lett.* **27**, 935–938.
145. Dabkowski, W., Michalski. J., and Wang, Q. (1991) Silyloxyphosphanes. New phosphitylating reagents in nucleotide chemistry. *Nucleosides & Nucleotides* **10**, 601–602.
146. Bogachev, V. S., Kumarev, V. P., and Rybalkov, V. N. (1986) Phosphorothioate analogues of nucleic acids. V. synthesis of 5'-phosphorothioate analogues of oligodeoxyribonucleotides with the aid of zwitterionic monomers. *Sov. J. Bioorg. Chem.* **12**, 64–70.
147. Stec, W. J., Grajkowski, A., Koziolkiewicz, M., and Uznanski, B. (1991) Novel route to oligo(deoxyribonucleoside phosphorothioates) Stereocontrolled synthesis of Pchiral oligo(deoxyribonucleoside phosphorothioates) *Nucl. Acids Res.* **19**, 5883–5888.
148. Cosstick, R. and Williams, D. M. (1987) An approach to the stereoselective synthesis of S$_p$-dinucleoside phosphorothioates using phosphotriester chemistry. *Nucl. Acids Res.* **15**, 9921–9932.
149. Stawinski, J., Thelin, M., and von Stedingk, E. (1991) Studies on sulfurization of nucleoside H-phosphonate and H-phosphonothioate esters using 3H-1,2-benzodithiol-3-one 1,1-dioxide. *Nucleosides & Nucleotides* **10**, 517,518.
150. Nielsen, J., Brill, W. K.-D., and Caruthers, M. H. (1988) Synthesis and characterization of dinucleoside phosphorodithioates. *Tetrahedron Lett.* **29**, 2911–2914.
151. Grandas, A., Marshall, W. S., Nielsen, J., and Caruthers, M. H. (1989) Synthesis of deoxycytidine oligomers containing phosphorodithioate linkages. *Tetrahedron Lett.* **30**, 543–546.

152. Dahl, B. H., Bjergarde, K., Sommer, V. B., and Dahl, O. (1989) Synthetic approaches to oligodeoxyribonucleoside phosphorodithioates using tervalent phosphorus monomers. *Nucleosides & Nucleotides* **8**, 1023–1027.
153. Porritt, G. M. and Reese, C. B. (1990) Use of the 2,4-dinitrobenzyl protecting group in the synthesis of phosphorodithioate analogues of oligodeoxyribonucleotides. *Tetrahedron Lett.* **31**, 1319–1322.
154. Farschtschi, N. and Gorenstein, D. G. (1988) Preparation of a deoxynucleoside thiophosphoramidite intermediate in the synthesis of nucleoside phosphorodithioates. *Tetrahedron Lett.* **29**, 6843–6846.
155. Dahl, B. H., Bjergarde, K., Sommer, V. B., and Dahl, O. (1989) Deoxyribonucleoside phosphorodithioates: Preparation of dinucleoside phosphorodithioates from nucleoside thiophosphoramidites. *Acta Chem. Scand.* **43**, 896–901.
156. Brill, W. K.-D., Tang, J.-Y., and Caruthers, M. H. (1989) Synthesis of oligodeoxynucleoside phosphorodithioates via thioamidites. *J. Am. Chem. Soc.* **111**, 2321,2322.
157. Stawinski, J., Thelin, M., and Zain, R. (1989) Nucleoside H-phosphonates. X. Studies on nucleoside hydrogenphosphonothioate diester synthesis. *Tetrahedron Lett.* **30**, 2157–2160.
158. Porritt, G. M. and Reese, C. B. (1989) Nucleoside phosphonodithioates as intermediates in the preparation of dinucleoside phosphorodithioates and phosphorothioates. *Tetrahedron Lett.* **30**, 4713–4716.
159. Piotto, M.E., Granger, J.N., Cho, Y., Farschtschi, N., and Gorenstein, D. G. (1991) Synthesis, NMR and structure of oligonucleotide phosphorodithioates. *Tetrahedron* **47**, 2449–2461.
160a. Dahl, B. H., Bjergarde, K., Nielsen, J., and Dahl, O. (1990) Deoxynucleoside phosphorodithioates. Preparation by a triester method. *Tetrahedron Lett.* **31**, 3489–3492.
160b. Dahl, B. H., Bjergarde, K., Nielsen, J., and Dahl, O. (1991) Synthesis of oligodeoxynucleoside phosphorodithioates by a phosphotriester method. *Nucleosides & Nucleotides* **10**, 553,554.
161. Dahl, B. H., Bjergarde, K., Henriksen, L., and Dahl, O. (1990) A highly reactive, odourless substitute for thiophenol/triethylamine as a deprotection reagent in the synthesis of oligonucleotides and their analogues. Acta *Chem. Scand.* **44**, 639–641.
162. Heinemann, U., Rudolph, L.-N., Alings, C., Morr, M., Heikens, W., Frank, R., and Blöcker, H. (1991) Effect of a single 3'-methylene phosphonate linkage on the conformation of an A-DNA octamer double helix. *Nucl. Acids Res.* **19**, 427–432.
163. Stawinski, J. and Szabo, T. (1991) Studies directed towards efficient synthesis of oligo-5'-deoxy-5-C (phosphonomethyl)deoxyribonucleosides. *Nucl. Acids Symp. Ser.* **24**, 71,72.
164. van der Woerd, R., Bakker, C. G., and Schwartz, A. W. (1987) Synthesis of Pl,P2–dinucleotide pyrophosphates. *Tetrahedron Lett.* **28**, 2763–2766.

165. Letsinger, R. L. and Mungall, W. S. (1970) Phosphoramidate analogs of oligonucleotides. *J. Org. Chem.* **35,** 3800–3803.
166. Krayevsky, A. A., Azhayev, A. V., Kukhanova, M. K., Scapcova, N. V., and Zayceva, V. E. (1981) Synthesis of oligonucleotides with 5'-3' phosphoamidoester bond. *Nucl. Acids Symp. Ser.* **9,** 203–205.
167. Roelen. H. C. P. F., de Vroom, E., van der Marel, G. A., and van Boom, J. H. (1988) Synthesis of nucleic acid methylphosphonothioates. *Nucl. Acids Res.* **16,** 7633–7645.
168. Stawinski, J., Stromberg, R., and Szabo, T. (1991) Convenient synthesis of dinucleotide methylphosphonates. *Nucl. Acids Symp. Ser.* **24,** 229.
169. Miller, P. S., Annan, N. D., McParland, K. B., and Pulford, S. M. (1982) Oligothymidylate analogues having stereoregular, alternating methylphosphonate/ phosphodiester backbones as primers for DNA polymerase. *Biochemistry* **21,** 2507–2512.
170. Durand, M., Maurizot, J. C., Asseline, U., Barbier, C., Thuong, N. T., and Helene, C. (1989) Oligothymidylates covalently linked to an acridine derivative and with modified phosphodiester backbone: circular dichroism studies of their interactions with complementary sequences. *Nucl. Acids Res.* **17,** 1823–1837.
171. Loeschner, T. and Engels, J. (1989) One pot R_p-diastereoselective synthesis of dinucleoside methylphosphonates using methyldichlorophosphine. *Tetrahedron Lett.* **30,** 5587–5590; Loeschner, T. and Engels, J. W. (1990) Diastereomeric dinucleoside-methylphosphonates: determination of configuration with the 2-D NMR ROESY technique. *Nucl. Acids Res.* **18,** 5083–5088.
172. Engels, J. W., Loeschner, T., and Frauendorf, A. (1991) Diastereoselective synthesis of thymidinemethylphosphonate dimers. *Nucleosides & Nucleotides* **10,** 347–350.
173. Lesnikowski, Z. J., Jaworska-Maslanka, M. M., and Stec, W. J. (1991) Stereospecific synthesis of *p*-chiral di(2'-*O*-deoxyribonucleoside)methanephosphonates. *Nucleosides & Nucleotides* **10,** 733–736.
174. Lesnikowski, Z. J., Jaworska, M., and Stec, W. J. (1990) Octa(thymidine methanephosphonates of partially defined stereochemistry: synthesis and effect of chirality at phosphorus on binding to pentadecadeoxyriboadenylic acid. *Nucl. Acids Res.* **18,** 2109–2115
175. Miller, P. S., Chandrasegaran, S., Dow, D. L., Pulford, S. M., and Kan, L. S. (1982) Synthesis and template properties of an ethyl phosphotriester modified decadeoxyribonucleotide. *Biochemistry* **21,** 5468–5474.
176. Abramova, T. V., Komarova, N. I., Lebedev, A. V., and Tagai, A. (1984) An investigation of the diastereomers of nonionic analogues of oligonucleotides. II. Assignment of the configurations at the phosphorus atoms in the diastereomers of ethyl esters of oligothymidylates. *Sov. J. Bioorg. Chem.* **10,** 742–746.
177. Asseline, U., Barbier, C., and Thuong, N. T. (1986) Oligothymidylates with alternating alkyl phosphotriester and phosphodiester structure covalently bonded to an intercalating agent. *Phosphorus Sulfur* **26,** 63–73.

178. Kuijpers, W. H. A., Huskens, J., Koole, L. H., and van Boeckel, C. A. A. (1990) Synthesis of well-defined phosphate-methylated DNA fragments: the application of potassium carbonate in methanol as deprotecting reagent. *Nucl. Acids Res.* **18**, 5197–5205.
179. Quaedflieg, P. J. L. M., van der Heiden, A. P., Koole, L. H., Coenen, A. J. J. M., van der Wal, S., and Meijer, E. M. (1991) Synthesis and conformational analysis of phosphate-methylated RNA dinucleotides. *J. Org. Chem.* **56**, 5846–5859.
180. Michalski, J., Dabkowski, W., Lopusinski, A., and Cramer, F. (1991) Stereoselective synthesis of nucleoside phosphorofluoridates. *Nucleosides & Nucleotides* **10**, 283–286.
180a. Michalski, J. (1991) New synthetic approach to nucleoside phosphorofluoridates and fluoridites. Novel type of phosphitylating reagents containing 4-nitrophenoxy group. *Nucl. Acids Symp. Ser.* **24**, 79–82.
181. Eschenmoser, A. (1991) Warum Pentose- und nicht HexoseNucleinsäuren *Nachr. Chem. Tech. Lab.* **39**, 795–807.
182. Naylor, R. and Gilham, P. T. (1966) Studies on some interactions and reactions of oligonucleotides in aqueous solution. *Biochemistry* **5**, 2722–2728.
183. Chen, C. B., Inoue, T., and Orgel, L. E. (1985) Template-directed synthesis on oligodeoxycytidylate and polydeoxycytidylate templates. *J. Mol. Biol.* **181**, 271–279; Grzeskowiak, K. and Orgel, L. E. (1986) Template-directed synthesis on short oligoribocytidylates. *J. Mol. Evol.* **23**, 287–289.
184. Joyce, G. F. and Orgel, L. E. (1986) Non-enzymatic template-directed synthesis on RNA random copolymers. *J. Mol. Biol.* **188**, 433–441.
185. Zielinski, W. S. and Orgel. L. E. (1985) Oligomerization of activated derivatives of 3'-amino-3'-deoxyguanosine on poly(C) and poly(dC) templates. *Nucl. Acids Res.* **13**, 2469–2484; Zielinski, W. S. and Orgel, L. E. (1987) Oligoaminonucleoside phosphoramidates. Oligomerization of dimers of 3'-amino-3'-deoxynucleotides (GC and CG) in aqueous solution. *Nucl. Acids Res.* **15**, 1699–1715.
186. Visscher, J. and Schwartz, A. W. (1988) Template-directed synthesis of acyclic oligonucleotide analogues. *J. Mol. Evol.* **28**, 3–6.
187. Visscher, J. and Schwartz, A. W. (1990) Oligomerization of cytosine-containing nucleotide analogues in aqueous solution. *J. Mol. Evol.* **30**, 3–6.
188. Inoue, T., Joyce, G. F., Grzeskowiak, K., Orgel, L. E., Brown, J. M., and Reese, C. B. (1984) Template-directed synthesis on the pentanucleotide CpCpGpCpC. *J. Mol. Biol.* **178**, 669–676.
188a. Zielinski, W. S. and Orgel, L. E. (1987) Autocatalytic synthesis of a tetranucleotide analogue. *Nature* **327**, 346–347.
189. von Kiedrowski, G., Wlotzka, B., Helbing, J., Matzen, M., and Jordan, S. (1991) Parabolisches Wachstum eines selbstreplizierenden Hexadesoxynucleotids mit einer 3'-5'-Phosphoamidat-Bindung.*Angew. Chem.* **103**, 456–459.
190. Shimidzu, T., Murakami, A., and Konishi, Y. (1979) Template-directed synthesis of oligonucleotides. Part 4. Condensation of nucleotides in the presence of nucleic acid base binding (4-vinylpyridine-styrene)copolymer in homogeneous solution. *J. Chem. Research* **(S)**, 232,233, **(M)**, 2751–2766.

191. Sawai, H. (1988) Oligonucleotide formation catalyzed by divalent metal ions. The uniqueness of the ribosyl system. *J. Mol. Evol.* **27**, 181–186.
192. Schwartz, A. W. and Orgel, L. E. (1985) Template-directed polynucleotide synthesis on mineral surfaces. *J. Mol. Evol.* **21**, 299–300.
193. Ferris, J. P., Kamaluddin, G. E., Agarwal, V., and Hua, L. L. (1989) Mineral catalysis of the formation of the phosphodiester bond in aqueous solution: the possible role of montmorillonite clays. *Advances in Space Res.* **9**, 67–75
194. Egofarova, R. Kh., Vasil'eva, N. V., Otroshchenko, V. A., and Pavlovskaya, T. E. (1990) Simultaneous synthesis of peptides and oligonucleotides on kaolinite with aminoacyladenylates. *Izv. Akad. Nauk SSSR, Ser. Biol.* **1**, 136–140.
195. Dolinnaya, N. G., Gryaznova, O. I., Sokolova, N. IK., and Shabarova, Z. A. (1986) Chemical reactions in double-helical nucleic acids. I. Chemical ligation as a method of introducing phosphoramide and pyrophosphate internucleotide bonds into DNA duplexes. *Bioorg. Khim.* **12**, 755–763.
196. Kuznetsova, S. A., Ivanovskaya, M. G., and Shabarova, Z. A. (1990) Chemical reactions in double-stranded nucleic acids. IX. Directed introduction of substituted pyrophosphate bonds into DNA structure. *Bioorg. Khim.* **16**, 219–225.
197. Isagulyants, M. G., Ivanovskaya, M. G., Potapov, V. K., and Shabarova, Z. A. (1985) Condensation of oligodeoxyribonucleotide phosphorimidazolidates within a complementary complex—a general method for the synthesis of natural and modified DNA duplexes. *Bioorg. Khim.* **11**, 239–247.
198. Shabarova, Z. A., Ivanovskaya, M. G., and Gottikh, M. B. (1991) *N*-Hydroxybenzotriazole esters of oligonucleotides. Phosphorylating agents in an aqueous medium. A new class or reagent for template-directed synthesis of internucleotide bonds and for obtaining oligonucleotide derivatives, in *Nucleic Acids Chem.* vol 4 (Townsend, L. and Stuart, R. S., eds.), Wiley, New York, pp. 386–390.
199. Kanaya, E. and Yanagawa, H. (1986) Template-directed polymerization of oligoadenylates using cyanogen bromide. *Biochemistry* **25**, 7423–7430.
200. Dolinnaya, N. G., Sokolova, N. I., Ashirbekova, D. T., and Shabarova, Z. A. (1991) The use of BrCN for assembling modified DNA duplexes and DNA-RNA hybrids; comparison with water-soluble carbodiimide. *Nucl. Acids Res.* **19**, 3067–3072.
200a. Dolinnaya, N. G., Tsytovich, A. V., Sergeev, V. N., Oretskaya, T., and Shabarova, Z. A. (1991) Structural and kinetic aspects of chemical reactions in DNA-duplexes. Information on DNA local structure obtained from chemical ligation data. *Nucl. Acids Res.* **19**, 3073–3080.
200b. Merenkova, I. N., Sokolova, N. I., and Shabarova, Z. A. (1991) New technology non-enzymatic assembly of a gene. *Nucl. Acids Res.* **24**, 261.
201. Hsu, C.-Y. J., Don, D., and Jones, R. A. (1985) Synthesis and physical characterization of bis 3'-5' cyclic dinucleotides (cycloNpNp): RNA polymerase inhibitors. *Nucleosides & Nucleotides* **4**, 377–389.

202. de Vroom, E., Broxterman, H. J. G., Sliedregt, L. A. J. M., van der Marel, G. A., and van Boom, J. H (1988) Synthesis of cyclic oligonucleotides by a modified phosphotriester approach. *Nucl. Acids Res.* **16**, 4607–4620.
203. Capobianco, M. L., Carcuro, A., Tondelli, L., Garbesi, A., and Bonora, G. M. (1990) One pot solution synthesis of cyclic oligodeoxyribonucleotides. *Nucl. Acids Res.* **18**, 2661–2669.
204. Kool, E. T. (1991) Molecular recognition by circular oligonucleotides: increasing the selectivity of DNA binding. *J. Amer. Chem. Soc.* **113**, 6265–6266.
205a. von Tigerstrom, R., Jahnke, P., and Smith, M. (1975) The synthesis of the internucleotide (phosphodiester) bond by a base-catalysed reaction. *Nucl. Acids Res.* **2**, 1727–1736.
205b. Kimura, J., Fujisawa,Y., Yoshizawa, T., Fukuda, K., and Mitsunobu, O. (1979) Studies on nucleosides and nucleotides. VII. Preparation of pyrimidine nucleoside 5'-phosphates and N^3,5'-purine cyclonucleosides by selective activation of the 5'-hydroxyl group. *Bull. Chem. Soc. Japan* **52**, 1191–1196.
206. Kawana,M. and Kuzuhara, H. (1984) The synthesis of partially-protected 2'-deoxyribonucleotide dimers by the selective phosphorylation of stannylated nucleosides. *Bull. Chem. Soc. Japan* **57**, 3317–3320.
207. Schott, H. (1975) Template chromatography on immobilized oligonucleotides. Synthesis and application of oligodeoxyadenosine-5'-phosphate-DEAE-cellulose. *J. Chromatography* **115**, 461–476.
208. Schott, H. (1985) Nucleobases, nucleosides, nucleotides, in *High Performance Liquid Chromatography in Biochemistry* (Henschen, A., ed.), Verlag Chemie, Weinheim, pp. 413–444.
209. Seliger, H., Haas, B., Holupirek, M., Knäble, T., Tödling, G., and Philipp, M. (1980) Non-stepwise methods in the preparation of building blocks for polynucleotide synthesis. *Nucl. Acids Symp. Ser.* **7**, 191–202.
210. Seliger, H. and Knäble, T. (1978) Parameters and sequence-length distribution of enzymatic nucleotide copolycondensations. *Nucl. Acids Res. Spec. Publ.* **4**, s167–s170.
211. Shum, B. W.-K. and Crothers, D. (1978) Simplified methods for large-scale enzymatic synthesis of oligoribonucleotides. *Nucl. Acids Res.* **5**, 2297–2311.
212. Yamauchi, H. and Machida, H. (1986) Continuous production of homopolynucleotides by immobilized polynucleotide phosphorylase. *J. Ferment. Technol.* **64**, 517–522.
213. Dattagupta, N., Rae, P., Crothers, D., and Barnett, T. (1986) Large scale production of DNA probes. Eur. Pat. Appl., 13 pp. EP 184056 A2.
214. Innis, M. A., Gelfand, D. H., Sninsky, J. J., and White, T. J. (eds.) (1990) *PCR Protocols. A Guide to Methods and Applications.* Academic, San Diego.
215. Wyatt, J. R., Chastain, M., and Puglisi, J. D. (1991) Synthesis and purification of large amounts of RNA oligonucleotides. *BioTechniques* **11**, 764–769.
216. Zhenodarova, S. M., Smolyaninova, O. A., Soboleva, I. A., and Khabarova, M. I. (1987) Stepwise synthesis of oligonucleotides. XXXIV. Preparative synthesis of trinucleoside diphosphates and longer oligoribonucleotides with immobilized ribonucleases. *Bioorg. Khim.* **13**, 1023–1030.

217. Gillam, S., Jahnke, P., and Smith, M. (1978) Enzymic synthesis of oligodeoxyribonucleotides of defined sequence. *J. Biol. Chem.* **253**, 2532–2539.
217a. Ohtsuka, E., Tanaka, S., Hayashi, M., and Ikehara, M. (1979) Polynucleotides 58. A method for the synthesis of oligonucleotide by single addition of 2'-*O*-(*o*-nitrobenzyl)nucleoside 5'-diphosphates using polynucleotide phosphorylase. *Biochim. Biophys. Acta* **565**, 192–198.
218. Schott, H. and Schrade, H. (1984) Single-step elongation of oligodeoxynucleotides using terminal deoxynucleotidyl transferase. *Eur. J. Biochem.* **143**, 613–620.
218a. Heidenreich, O. and Eckstein, F. (1991) Inhibition of the reverse transcriptase of HIV-1 by 3'-azidothymidine triphosphate and 3'-azido-oligothymidylate. *Nucleosides & Nucleotides* **10**, 535–536.
219. Uhlenbeck, O. and Gumport, R. I. (1982) T4 RNA Ligase, in *The Enzymes* (Boyer, P. D., ed,), 3rd Ed., vol. **15**, Academic, New York, pp. 31–58.
219a. Middleton, T., Herlihy, W. C., Schimmel, P. R., and Munro, H. N. (1985) Synthesis and purification of oligoribonucleotides using T4 RNA ligase and reverse-phase chromatography. *Anal. Biochem.* **144**, 110–117.
220. Venyaminova, A. G., Vratskikh, L. V., Repkova, M. N., and Yamkovoy, V. I. (1991) Preparative synthesis of short oligoribonucleotides by immobilized RNA ligase of T4 bacteriophage. *Nucl. Acids Symp. Ser.* **24**, 305.
221. Wang, A. H.-J., Quigley, G. J., Kolpak, F. J., Crawford, J. L., van Boom, J. H., van der Marel, G., and Rich, A. (1979) Molecular structure of a left-handed double helical DNA fragment at atomic resolution. *Nature* **282**, 680–686.
221a. Miller, P. S., Cheng, D. M., Dreon, N., Jayaraman, K., Kan, L.-S., Leutzinger, E. E., Pulford, S. M., and Ts'o, P. O. P. (1980) Preparation of a decadeoxyribonucleotide helix for studies by nuclear magnetic resonance. *Biochemistry* **19**, 4688–4698.
221b. Weiss, M. A., Patel, D. J., Sauer, R. T., and Karplus, M. (1984) Two-dimensional ^{1}H-NMR study of the λ operator site O_L1: A sequential assignment strategy and its application. *Proc. Natl. Acad. Sci. USA* **81**, 130–134.
222a. Gaffney, B. L., Marky, L. A., and Jones, R. A. (1984) Synthesis and characterization of a set of four dodecadeoxyribonucleoside undecaphosphates containing O^6-methylguanine opposite adenine, cytosine, guanine and thymine. *Biochemistry* **23**, 5686–5691.
222b. Gaffney, B. L. and Jones, R. A. (1988) Large-scale oligonucleotide synthesis by the H-phosphonate method. *Tetrahedron Lett.* **29**, 2619–2622.
223. Mueller, B. C. (1989) Automatisierte DNA-Synthese im Milligrammasstab. *BioEngineering* **5**, 44–46.
224. Ikehara, M. (1974) Synthesis of ribooligonucleotides having sequences of transfer ribonucleic acids. *Acc. Chem. Res.* **7**, 92–96.
225. Alul, R. H., Singman, C. N., Zhang, G., and Letsinger, R. L. (1991) Oxalyl-CPG: a labile support for synthesis of sensitive oligonucleotide derivatives. *Nucl. Acids Res.* **19**, 1527–1532.

226. Eritja, R., Robles, J., Fernandez-Forner, D., Albericia, F., Giralt, E., and Pedroso, E. (1991) NPE-resin, a new approach to the solid-phase synthesis of protected peptides and oligonucleotides I: synthesis of the supports and their application to oligonucleotides synthesis. *Tetrahedron Lett.* **32,** 1511–1514.
227. Lyttle, M. H., Cook, R. M., and Wright, P. B. (1988) Large-scale automated DNA synthesis. *BioPharm. Manuf.* **1,** 34–38.
228. Andrus, A., Vu, H., Ramstad, P., and Pallas, M. (1991) Large scale automated synthesis of oligonucleotides. *Nucl. Acids Symp. Ser.* **24,** 41,42.

CHAPTER 18

Large-Scale Oligonucleotide Synthesis Using the Solid-Phase Approach

Nanda D. Sinha

1. Introduction

Synthetic DNA of defined sequences are commonly termed "oligonucleotides," which are primarily composed of four different types of nucleosides linked through well-defined deoxyribose phosphate. Over the past ten years, because of refinement in synthesis chemistries *(1a–e)* and automation of synthesis steps *(2a,b)*, oligonucleotide syntheses of defined sequences are common practices in nonchemists' laboratories. Oligonucleotides are routinely used as DNA sequencing primers *(3a–c)*, probes *(4a,b)*, linkers, adaptors, and gene synthesis *(5)*. In addition to these applications, biophysical studies *(6a,b)* (NMR, X-ray crystallography) for structural information of synthetic oligonucleotide using milligram quantities have been carried out. Newer applications of oligonucleotides are also emerging in the field of clinical diagnosis *(7a–c)*, forensic testing, and disease treatment *(8a–e)*. Investigations are being carried out in many laboratories for potential use of oligonucleotides as therapeutic agents broadly referred to as "use as antisense DNA." Inhibition of viral replication in diseases, such as AIDS, herpes, and human pepiloma virus, and regulation of oncogene expression with oligonucleotides or their analogs are two major potential applications of synthetic DNA. Presently, these studies are moving out of research laboratories and into practical medical applications *(9)*. Several oligonucleotides and their analogs are currently undergoing clinical trials for their uses as potential "antisense DNA therapeutics." These devel-

From: *Methods in Molecular Biology, Vol. 20: Protocols for Oligonucleotides and Analogs*
Edited by: S. Agrawal Copyright ©1993 Humana Press Inc., Totowa, NJ

opments require a large amount of a reasonable length oligonucleotide (20–30 bases long) far in excess of what is feasible on a DNA synthesizer designed for low volume (0.2–15.0-µmol scale). Even for preliminary clinical trials, several gram quantities of oligonucleotides are required. This may be a challenge to synthetic chemists for developing effective chemistry and to engineers for automating the process in order to scale-up from milligram to kilogram quantity of oligonucleotides at a reasonable cost.

The present technology, which allows efficient and routine synthesis of oligonucleotides of defined sequences, is exclusively based on the concept of polymeric-support-mediated synthesis strategy in DNA synthesis pioneered independently by Letsinger and Mahadevan *(10)*, similar to the strategy simultaneously developed by R. B. Merrifield *(11)* in peptide synthesis. The original polymeric supports, polystyrene crosslinked with divinylbenzene, have been replaced with glass beads termed controlled-pore glass beads *(12)* (CPG beads) of defined porosity. Similarly, the original chemistry has also been substituted with efficient chemistries phosphoramidite and H-phosphonate of nucleosides as oligonucleotide chain building blocks.

Almost all oligonucleotide syntheses today are carried out on the solid support, which is prederivatized with a desired nucleoside via a stable covalent linkage. On completion of synthesis, the desired oligonucleotide can be released from the support. Generally, oligonucleotide synthesis is performed in 3'-5' direction by the addition of one nucleoside-phosphorus derivative. In chemical terms, the addition of a reactive nucleoside that grows the chain is comprised for several steps. During the chain assembly, these steps are repeated in a cyclic manner. The steps are defined as:

1. *Detritylation,* i.e., removal of dimethoxytrityl group from the 5' end of nucleoside/nucleotide with dichloroacetic acid solution;
2. *Washing,* to remove the released dimethoxytrityl cation and excess acid used;
3. *Coupling,* that is, addition of reactive nucleoside-phosphorus derivative to grow the desired chain with the help of an activator;
4. *Washing,* to remove excess reagent used for coupling;
5. *Oxidation,* to convert the labile phosphorus linkage between two nucleosides into a stable phosphate triester or diester;
6. *Washing,* to remove excess oxidizer from the support; and

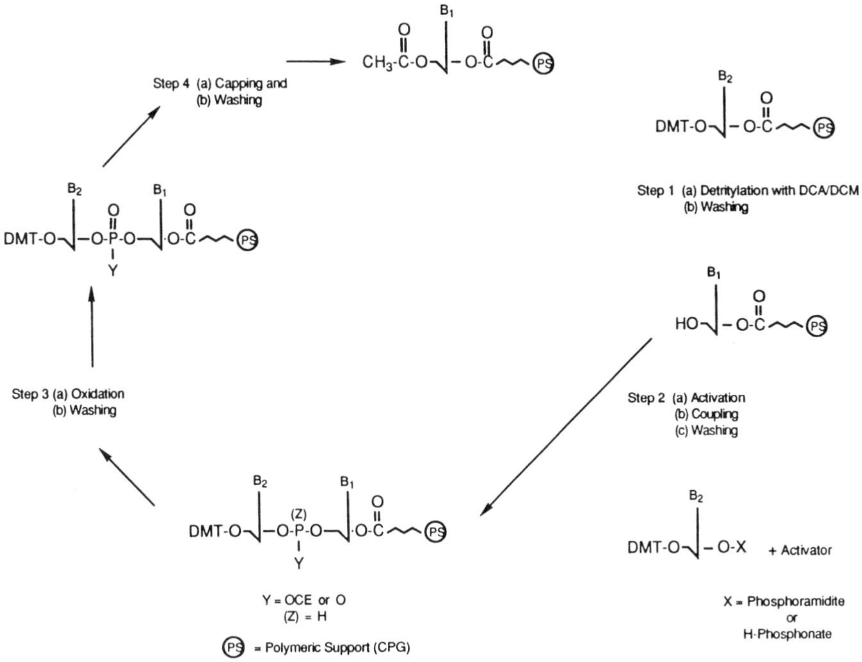

Oligonucleotide Synthesis

Scheme 1. Steps involved in oligonucleotide chain elongation process.

7. *Capping*, to block the unreacted/free hydroxyl group that did not couple to reactive nucleoside-phosphorus derivative during coupling step. Repeat steps 1–7 until the desired oligonucleotide chain is assembled. These steps are summarized in Scheme 1.

The aim of a good synthetic strategy is to obtain almost quantitative reactions at each step. For the synthesis of oligonucleotide in small amounts either based on phosphoramidite or H-phosphonate chemistry, each step is driven to completion at the expense of a large excess of chemicals and reagents, to obtain a good quality and quantity of synthetic oligonucleotides. In the case of oligonucleotide synthesis (0.1–1.0 µmol), use of such excess is not cost prohibitive. However, use of such an excess of chemicals and reagents for large-scale synthesis will definitely be cost prohibitive. Therefore, scale of any synthesis from a small-scale to large-scale is not a simple task. An alternate to solid-phase synthesis, the solution phase synthesis, though extensively used in the seventies because of complexities in isolating the desired product after each chain elongation step and removal of temporary protecting groups, did not achieve the same popularity as solid-phase synthesis among nonchemists. It is highly possible that future large-scale synthesis of oligonucleotides may emerge from solution phase synthesis strategy.

As far as the present need of synthetic oligonucleotide is concerned, because of ease of solid-phase synthesis in manipulating all the steps and efficient coupling during chain elongation, this strategy is still a viable method for scale-up process without being too cost prohibitive, until alternate synthesis strategy emerges. In this chapter, the work carried out in the author's laboratory, to fulfill the present need of synthetic oligonucleotides and their analogs, oligonucleotide phosphorothioates as potential antisense DNA, using an automated DNA synthesizer known as an 8800 DNA synthesizer, is shown in Fig. 1. This instrument can synthesize oligonucleotides from 30- to 600-µmol scale to result in a few hundred milligram to gram quantities. This upper limitation at present may be owing to the amount of supports derivatized with nucleoside that can be used in this synthesis.

2. Materials

Almost all of the chemicals and reagents presently used for DNA synthesis are commercially available. (For details, *see* Chapters 3 and 4 in this volume.) These commercially available materials are of high quality and normally do not require any additional purification. The quality of each reagent and chemical can be easily determined by TLC, HPLC,[31] P-NMR, and Karl Fischer instrument.

3. General Instrument Description

In order to achieve good-quality synthesis on a large scale, proper setting of the instrument is an integral part of the process. Here, large-scale synthesis of oligonucleotides on solid-phase has been discussed using a model 8800 DNA synthesizer. The instrument 8800 DNA synthesizer consists of different components:

1. PC Synthesis Work Station;
2. System Control Module;
3. Reagent and Solvent Delivery System;
4. Reaction Module;
5. Printer; and
6. Dual Stage Waste Management System (Fig. 1). These components of the instrument have to be connected properly.

Oligonucleotide Synthesis

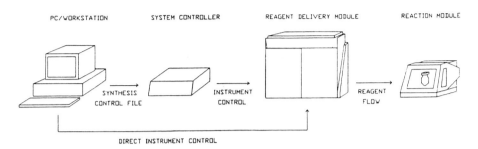

Fig. 1. Large-scale DNA synthesizer Model 8800 instrument configuration.

The PC Work Station is an MS-DOS-based computer. The menu-driven software contains chemical protocols and scaling algorithms that automate the synthesis of a wide range of synthesis scales. When a new quantity of support is specified, the scaling algorithms are automatic. The program calculates the number of excess of monomers, volume of solvents, duration of wash steps, number of times and cycle time to be used in the synthesis and also implements the new parameters. These new or user specified parameters are incorporated into a synthesis control file that contains sequential synthesis steps. The synthesis control file may be executed in either Direct Control (On-Line) or Down-Load mode from the System Control Module.

3.1. System Control Module

The System Control Module is an internal microprocessor that controls and monitors an ongoing synthesis. In Direct or On-Line mode of operation, only one synthesis step at a time is sent to the System Control Module by the Work Station. This mode of operation requires constant communication with the Work Station. In Down-Load mode, the synthesis control file is transmitted to the System Control Module, stored in battery-protected memory, and executed in a sequential manner until all synthesis steps have been performed. This mode does not require constant communication with a PC Work Station.

3.2. Reagent and Solvent Module

This delivery module is an enclosed ventilated cabinet that contains the reagent flow control valves and reagent reservoirs. There are four glass reservoirs of 220-mL capacity for normal active nucleoside monomers, four 135-mL capacity reservoirs for modified or unnatural nucleosides, and six larger glass reservoirs (1- to 6-L capacity) for ancillary reagents, such as acetonitrile, activator, deblock, capping reagents, and oxidizer. Reagents are delivered via positive gas pressure through inert tubing. Each reservoir is pressurized to deliver solvent or reagent at the rate of 0.8–1.5 mL/s.

3.3. Reaction Module

The Reaction Module contains the valves and control system for regulating reagent flow to the reaction vessel and to waste management system. This module consists of:

1. Silianized conical reaction vessel;
2. Fail-safe sensor protection;
3. Zero dead volume control;
4. Support agitation control; and
5. Fraction collector advance circuit and waste management system.

System Control Module is the central commanding station of this 8800 DNA Synthesizer. PC Work Station, Reagent and Solvent Delivery, and Reaction Module are all connected to the System Control Module via proper conductor ribbon cables, according to the instructions provided in the manual. The rest of the details of handling instruments are provided in the machine manual.

4. General Discussion

To scale-up the synthesis from 0.2–1.0-µmol scale to hundreds of micromoles, our efforts *(13)* were to find the optimum condition to obtain almost the same quality of product as obtained from a lower-scale synthesis instrument without using such a large excess of reagents. To achieve this goal, several considerations had to be fulfilled. These conditions are as follows:

4.1. Fluidization

Since the reactor or synthesis vessel used in this instrument is a batch-type reactor unlike a column-type reactor used for small-scale DNA synthesis where the entire amount of solid support is suspended into reaction solution to give efficient coupling, it is very essential to have proper fluidization of the large amount of supports. Fluidization in this approach is effected by bubbling dry gas (argon) through the bottom of a reactor, resulting in complete suspension of the support.

Table 1
Solvent Volumes Required to Fluidize Supports

CPF mass, g	Acetonitrile, mL	Methylene chloride, mL	THF, mL
1	40	20	20
2	50	25	30
4	60	30	50
6	65	40	55
8	70	45	60
10	75	50	65

Failure to complete fluidization of the support will result in poor-quality synthesis. Fluidization depends on solvent density/viscosity, volume/amount of the support, reactor design, and gas pressure. To achieve proper fluidization for a given volume/amount of support, a specific amount of solvent/solution is needed. A general correlation between mass of controlled-pore glass beads (most commonly used support), and solvent/solution is given in Table 1 and Fig. 2.

4.2. Phosphoramidite Approach

The reactive nucleoside derivative and the activator are two main components in DNA synthesis. Proper equivalents of these two reagents are essential parameters for a good synthesis, which depends on the degree of substitution of nucleosides onto CPG beads and the amount of CPG in use. Based on our experiments using a large-scale DNA synthesizer, a summary of amidite requirements for various scales is given in Table 2 and Fig. 3.

For the synthesis of oligonucleotides on a large-scale, two different amidite concentrations are used: 50 mg/mL for syntheses of 300 µmol or less and 100 mg/mL for 300 µmol or more. Concentrations of other reagents do not change with scale of syntheses.

4.2.1. Effect of Moisture on Amidite Solution

The reactive amidite and activator (tetrazole) are generally dissolved in dry acetonitrile. The quality of anhydrous acetonitrile used in dissolving amidite is as important as the equivalent of reagents delivered and amount required for fluidization to achieve good-quality synthesis. Any contamination with water can reduce the reactivity of an amidite, hence lowering the coupling efficiency and resulting in a poor-quality

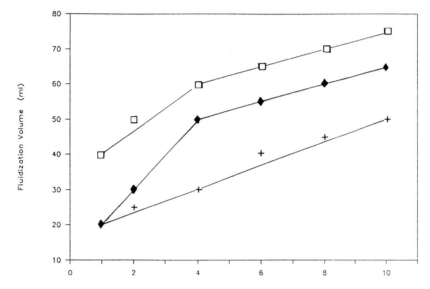

Fig. 2. Fluidization volume vs mass of support: □ acetonitrile, + mass of support (g) methylene chloride, ◆ THF.

Table 2
Equivalence of Amidites Required

Supports	Amidites, no of equivalents			
CPG[a]	A	C	G	T
1.0g	5.5	5.5	8.0	5.0
2.0g	5.39	5.39	7.67	4.89
3.0g	5.28	5.28	7.34	4.78
5.0g	5.05	5.05	6.67	4.55
8.0g	4.72	4.72	6.34	4.22
10.0g	4.50	4.50	5.0	4.0
15.0g	4.50	4.50	5.0	4.0

[a]Based on average substitution 40 µmol of nucleoside/g of controlled-pore glass beads; however, instrument is capable of determining these equivalents depending on the degree of substitution and amount of supports in use.

synthesis. Studies carried out in our laboratory show that effective concentration of an amidite decreases with various amounts of water. The presence of 100 ppm of water reduces the effectiveness of a 50 µmol amidite solution to 35 µmol solution. The reduction of amidite equivalents owing to the presence of water is represented in Fig. 4.

Oligonucleotide Synthesis

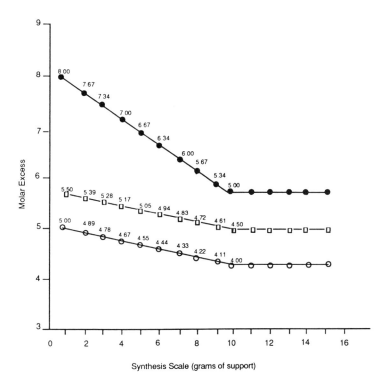

Fig. 3. Relation between equivalence of amidite and amount of CPG. Amidite molar equivalents vs synthesis scale. « A and C; ●—G, ○—T.

Furthermore, the decrease in effectiveness of an amidite in solution owing to the presence of moisture also depends on time. A 50 µmol solution of an amidite with 100 ppm water content over the lapse of time may not even be a 35.0 µmol solution. Therefore, it is recommended that reagents and solutions be freshly prepared in anhydrous solvent. Especially, amidite solutions should be prepared from good-quality and anhydrous acetonitrile, which many vendors supply. The presence of moisture in an activator tetrazole solution also effects the quality of synthesis. The effect of moisture can be eliminated by placing a "trap-bag" in this solution.

Another source of moisture may come from the reagent vessel. It is also suggested that glass vessels be properly dried and cooled in argon atmosphere.

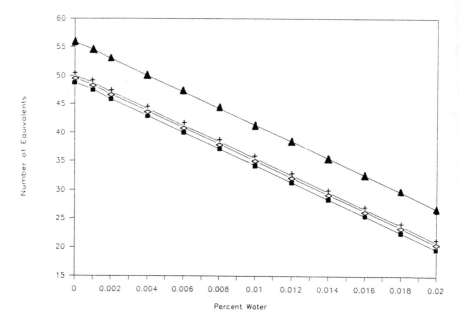

Fig. 4. Reduction of amidite equivalence owing to moisture/w: contamination. ■A, + C., ◇G, ▲T.

Table 3
Amount of Oligonucleotides
Obtained from Large-Scale Synthesis Using Phosphoramidite Chemistry

Length	Synthesis scale, mol	Coupling efficiencies	Crude yield A_{2600} units[a] 260 nm
20 mer	30 (1 g)	97.5	2160
18 mer	110 (5.0 g)	96.3	6900
24 mer	264 (8.0 g)	97.2	24,600
17 mer	330 (10.0 g)	96.6	11,400
28 mer	200 (6.0 g)	95.1	14,238
32 mer	600 (15.0 g)	97	80,000

[a]A_{260} units vary with real sequence.

In the author's laboratory, following the above-described parameters for large-scale synthesis using phosphoramidite chemistry, we achieved average coupling efficiencies comparable to those obtained from a lower-scale DNA synthesizer. The quality of the synthesis product is also comparable. A summary of the coupling efficiency and analysis of the oligonucleotides are given in Table 3 and Fig. 5.

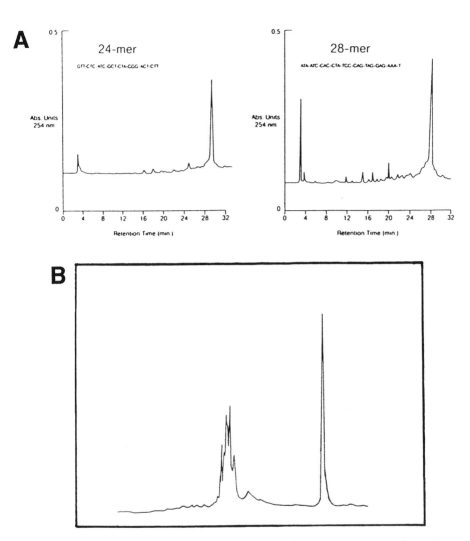

Fig. 5. HPLC analysis of oligonucleotide synthesized on large-scale. **A**. Anion-exchange column analysis of 24 and 28 mers. HPLC condition: Column Ultrasil-AX (4.6 × 250 mm), gradient 0–50% B in 20 min, buffer A = .003M K_2HPO_4 (pH = 6.8)/ 20% CH_3CN buffer B = 0.64M K_2HPO_4 (pH = 6.8)/20% CH_3CN. Flat rate = 1.5 mL/ min. Absorbance at 260 nm AUFS = .64. **B**. By RP=HPLC column analysis of dG-rich 32 mer. HPLC condition: Column water C-18 Delta Pak 300 Å. Gradient 0–40% B in 20 min, buffer A = 0.1M TEAA (pH = 7.0), buffer B = acetonitrile. Flow rate 1.5 mL/ min, absorbance at 260 nm AUFS = 1.0.

4.2.2. Time Dependence Coupling and Detritylation

During large-scale synthesis, we have observed that the rate of phosphoramidite coupling does not only depend on different equivalents of amidite, but also on the nature of nucleoside. Similarly, the rate of removal of dimethoxytrityl group from the nucleoside/nucleotide also depends on the base. Allowing identical duration for either coupling or detritylation at every chain elongation step may not effect a small-scale synthesis, but for a large-scale process this will definitely increase the synthesis time. In addition to this, an unnecessarily longer period for detritylation may cause some depurination. To avoid this side reaction and optimize synthesis protocol, it is therefore necessary to use base-specific programs for coupling and detritylation. This instrument used for large-scale synthesis is capable of performing these base-specific programs successfully and is able to calculate the number of equivalent amounts of reagents required for coupling when variable scales are used. The protocol for preparative scale synthesis is given in Table 4.

4.3. H-Phosphonate Approach

In oligonucleotide synthesis, alternate H-phosphonate chemistry is also being used in many laboratories. This chemistry, owing to limited availability of good-quality building blocks and use of obnoxious smelling pyridine, has not received much attention from nonchemists synthesizing oligonucleotides. However, this approach offers some potential advantages over standard phosphoramidite chemistry. Several advantages are worth mentioning:

1. It is easy to manipulate the phosphate diester backbone modification;
2. It enables the attachment of linkers at desired positions;
3. It creates neutral or cationic backbone *(14);*
4. It allows the use of a small excess of monomers for driving the chain elongation to completion; and
5. It gives faster cycle times.

The use of a lesser excess of monomers and shorter coupling time may also be an important factor to explore this chemistry for synthesizing oligonucleotides in large amounts (if not today, then definitely in the future). Our experience in using this chemistry, although not as extensive as that of the phosphoramidite approach, shows that results obtained are worth reporting.

In order to utilize this chemistry on our large-scale synthesizer, some alterations have to be made with respect to the configuration of reagents and solvent arrangement. The basic chain elongation cycle and reagents used are somewhat different (from standard phosphoramidite chemistry), and are shown in Scheme 2 and Table 5.

Oligonucleotide Synthesis

Table 4
Prep Scale Synthesis Protocol—Phosphoramidite Chemistry

Synthesis step	Reagent	# Repeats/duration scale dependent[a]
Deblock	0.3M DCA in methylene chloride	4–6 rpts/1–2 min
Wash	Acetontrile	5 rpts/1–3 min
Wash	Dry acetonitrile	1 rpt/2–3 min
Couple	50–100 mg/mL amidites 0.46M tetrazole in acetronitrile; molar excess amidites: A 4.5–5.5; C 4.5–5.5; G 5–8; T 4–5	1 rpt/5–7 min
Wash	Acetontrile	2 rpts/1–3 min
Oxidize	0.017M 12, 5.3M H$_2$O 0.05M pyridine in THF	1–2 rpts/3–4 min
or		
Thioate	0.05M Beaucage reagent in acetonitrile	1 rpt/3–4 min
Wash	Acetonitrile	4 rpts/1–3 min
Cap	1.2M methylimidazole, 1.2M pyridine in THF 1.1M acetic anhydride in THF	1 rpt/2–3 min
Wash	Acetonitrile	2 rpts/1–3 min
Repeat cycle sequence is complete		

[a]The amount of each reagent used depends on the amount of CPG (grams) in the synthesis.

In this chemistry, the monomer H-phosphonate is converted into a "reactive mixed anhydride" of phosphonic and carboxylic acids, with a sterically hindered acid chloride (1-adamantane-carboxylic acid chloride). This in turn reacts with a free hydroxyl group generated by detritylation in basic medium. To ensure complete coupling, any trace of acid (left from detritylation step) has to be removed. This is achieved by washing the supports with a neutralizer that is a mixture of pyridine-acetonitrile. The unreacted free hydroxyl group is presumably capped by the presence of excess acid-chloride; however, some reports have shown that inclusion of a capping with H-phosphonate derivative of ß-cyanoethanol improves the purification process. Using 2.5 equivalent of H-phosphonate

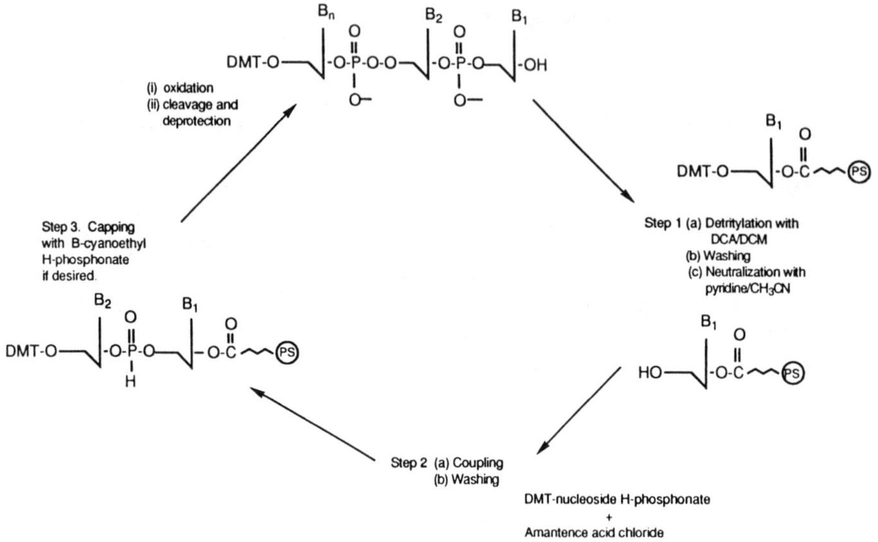

Scheme 2. Steps involved in DNA synthesis following H-phosphonate chemistry.

monomers and 12.5 equivalent of activator, we have carried out several syntheses allowing identical coupling times for each nucleoside valve. However, the removal of 5'-DMT group was performed following a base-specific detritylation program. In one of the syntheses, we introduced capping with ß-cyanoethyl H-phosphonate. Finally at the end of chain elongation, H-phosphonate diester linkages were converted into stable phosphate diester linkages with iodine and N-methylmorpholine water solution.

During the synthesis process, nucleoside monomers are converted into a very reactive intermediate. Moisture or any other impurity, such as a secondary amine or primary amine present in pyridine, may decompose this reactive intermediate and result in a low coupling yield. Therefore, in order to achieve quantitative coupling yield at every cycle, high-quality anhydrous solvents should be used. Another source of moisture may come from the very hygroscopic nature of phosphonate monomers, since they are salt of diazabicycloundecene (DBU). Use of such hydrated monomers in the synthesis will also lower the coupling efficiency. Drying of nucleoside monomers prior to their uses under high vacuum for a few hours helps in improving the coupling efficiencies.

Table 5
Configuration of Reagent Solvent Rearrangement
for H-Phosphonate Chemistry

Reagents	Composition	Location on the instrument in relation to amitdite chemistry
Wash A	Dry acetonitrile	Wash A reservoir
Neutralizer	1:1 Pyridine/acetonitrile	Wash B reservoir
Activator	1-Adamantane carbonyl chloride in 1:1 Pyridine/acetonitrile 70 g/L for scales up to 300 µmol, 140 g/L for scales more than 300 µmol	Activator reservoir
Oxidizer 1	50 g/L I_2 in THF	Oxidizer reservoir
Oxidizer 2	8:1:1 THF/H_2O/N-methyl-morpholine	Cap A reservoir
Cap (if desired)	50 µmol/mL β-cyanoethyl H-phosphonate in 1:1 pyridine/acetonitrile	Cap B reservoir
Monomers	50 mg/mL for scales up to 300 µmol, 100mg/mL for scales more than 300 µmol	

The oxidation, i.e., conversion of H-phosphonate diester linkages into phosphate diester, is relatively slower than phosphite triester. Generally, a higher concentration of iodine and stronger base solutions are used to facilitate this step, either automatically on the instrument or manually off the instrument. Incomplete oxidation results in a poor-quality synthetic product. Very old oxidizing reagents (especially iodine solution) do not convert all H-phosphonate linkages into phosphate diester. Shorter or longer duration of oxidation also affects the final yield. Short duration means incomplete oxidation, which results in degraded product. Similarly, longer iodine solution treatment releases oligonucleotide chains from the support during oxidation. This is because of base-catalyzed hydrolysis of succinic acid ester linkage between the nucleoside and polymer supports. Our studies on the effect of longer incubation of the dimethoxytritylated nucleoside anchored via succinate linkage to the support show that 10–25% loss of material can occur after 1 h. The complete conversion of all H-phosphonate linkages into

Table 6
Large-Scale DNA Synthesis Using H-Phosphonate Protocol

Synthesis	Synthesis scale	Sequence	Est coupling efficiency	Est crude yield
A	45 µmol	20 mer	97.0%	2700 A_{260} units
B	45 µmol	20 mer with capping	96.0%	2500 A_{260} units
C	520 µmol	20 mer	96.5%	33,000 A_{260} units

phosphate diester can be achieved in 45 min of oxidation without significant loss owing to hydrolysis. The result of our large-scale DNA synthesis using H-phosphonate chemistry is given in Table 6 and Fig. 6. If the reagent consumptions and total synthesis time as shown in Table 7 have any influence in deciding the choice of chemistry, then the H-phosphonate approach also deserves attention for large-scale synthesis of oligonucleotide. The protocol used for the synthesis is presented in Table 8.

4.4. Synthesis of Oligodeoxynucleotide Phosphorothioate

Oligomers that are undergoing extensive clinical evaluation as potential DNA therapeutics are unnatural oligonucleotides. One of these unnatural analogs is polyphosphorothioated oligonucleotide, where one of the oxygen atoms of phosphate diester linkages is replaced with sulfur atom. They behave very similarly to actual oligonucleotides with the added advantage that these are not easily degraded by nucleases. Therefore, these analogs, because of their stability in cell media, are more effective in controlling biological reactions compared to natural oligonucleotides. In the past 2 years, large-scale synthesis of these analogs has gained significant attention. As a result of this, several new sulfurizing reagents have emerged for creating phosphorothioate linkages. Synthesis protocol for polyphosphorothioated oligonucleotide follows identical

Fig. 6. *(opposite page)* Reverse-phase HPLC results of 8800 H-phosphonate synthesis: Reverse-phase HPLC conditions—analytical; buffer A—$0.1M$ TEAA, pH 7; buffer—acetonitrile, 5% H_2O; gradient—0–40% B over 20 min, 1.5 mL/min; wavelength—260 nm; temperature—ambient; column—Waters Delta Pak C_{18}—300 Å. **A.** Synthesis A—natural 20-mer 45-µmol scale. **B.** Synthesis B—natural 20-mer 45-µmol scale; capping used. **C.** Synthesis C—natural 20-mer 520-µmol scale.

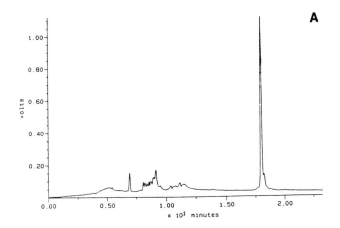

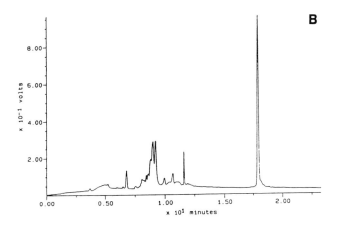

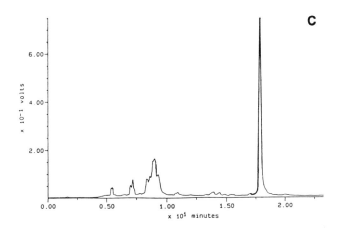

Table 7
Comparison of Active Nucleoside Monmer Consumption
and Cycle Time Required for DNA Synthesis

Scales in μmol	Amidites, mg				Cycle time, min	H-Phosphonates, mg				Cycle time, min
	A	C	G	T		A	C	G	T	
200	874	850	1100	685	50	440	425	430	380	30
400	1545	1500	1680	1170	65	875	850	855	760	40
600	2310	1930	2520	1790	76	1310	1275	1280	1140	50

steps as used in natural oligonucleotide synthesis. In the oxidation step, instead of iodine solution, a sulfur-transferring reagent is used.

Prior to 1990, this step was commonly performed with an elemental sulfur solution in a pyridine-carbondisulfide mixture. The utilization of [3H]-1,2-benzodithiol-3-one-1,1-dioxide by Iyer et al. (15) for converting phosphorus (III) triester into pentavalent phosphorothioate has simplified the process of large-scale polyphosphorothioated oligonucleotide synthesis following phosphoramidite chemistry. This reagent, however, fails to convert H-phosphonate diester linkages into phosphorothioate diester, which is still achieved by treating with 5% elemental sulfur solution in pyridine-carbon disulfide off the instrument.

Using our large-scale DNA synthesizer, we have been able to synthesize large amounts of thioated oligonucleotide using either phosphoramidite or H-phosphonate chemistry. Use of [3H]-1,2-benzodithiol-3-one-1,1-dioxide for converting phosphite triester linkage into phosphorothioate requires some precautions or special treatment of the reagent bottle. The reagent bottle has to be treated with dilute sulfuric acid followed by washing with water and methanol. Prior to its use, the bottle has to be treated with Sigma coat or treated with 2% dichlorodimethyl silane solution in carbon tetrachloride followed by methanol. The metal filter also has to be replaced with a porex filter, and the solution should be covered with aluminum foil to minimize the exposure to light. Using a 50 mM solution of Beaucage reagent in acetonitrile, we have synthesized several phosphorothioated oligonucleotides in good yields. However, following H-phosphonate chemistry, polyphosphorothioate oligonucleotides were obtained by elemental sulfur solution treatment. The yield of these analogs and their analysis by ^{31}P-NMR and RP-HPLC are given in Table 9 and Fig. 7.

Table 8
Prep Scale Synthesis Protocol—H-Phosphonate Chemistry

Synthesis step	Reagent	# Repeats/duration, scale dependent[a]
Deblock	0.3M DCA in methylene chloride	4–6 rpts/1–2 min
Wash A	Acetonitrile	5 rpts/1–3 min
Neutralizer	1:1 Acetonitrile:pyridine	1 rpt/3–4 min
Couple	50–100 mg/mL monomers 0.35–0.70M 1-adamantane carbonyl chloride in 1:1 acetonitrile:pyridine, molar excess monomers: A,C, G, T—2.5 equivalents	1 rpt/1.5–3.5 min
Wash	Acetonitrile	4 rpts/1–3 min
Cap (optional)	0.05M cyanoethyl H-phosphonate in 1:1 acetonitrile:pyridine (5 equivalents), 0.35M acid chloride activatior	1 rpt/1–3 min (optional)
Wash	Acetonitrile	4 rpts/1–3 min
Repeat cylce until sequence is complete		
Oxidize (on or off) instrument in THF	0.2M 12 in THF; 5.6M H$_2$O 0.9M N-methylmorpholine	3–5 rpts/15 min
or		
Thioate	50 g/L sulfur in 1:1 CS$_2$:pyridine	1 rpt/5 h
Wash	Acetonitrile	4 rpts/1–3 min

[a]The amount of each reagent used depends on the amount of CPG (grams) in the synthesis.

5. Postsynthesis Step

At the end of oligonucleotide chain assembly following phosphoramidite or H-phosphonate approach, the desired synthetic DNA is obtained as an either fully protected or partially protected form attached to the solid supports used for carrying out the synthesis. In order to obtain synthetic DNA as a fully deprotected form, the chain has to be released

Table 9
Phosporothioate Synthesis on the 8800

Synthesis	Sequence or base comp	Synthesis scale,^{a} μmol	Crude yield+, A_{260} unit	Coupling efficiency#, %
F-21 mer	a-3, C-7, G-7, T-4	70	4800	98.0
G-21 mer thioated	a-3, C-7, G-7, T-4	470	62,000	97.5
H-20 mer mixed sequence	5'TCCACGTCATCCAGGTCATT underline = thioated linkage	180	16,000	98.0

from the support, and subsequently protecting groups have to be removed from the chain. This process is generally termed as a deprotection step. Proper care should be taken for deprotection and workup process.

5.1. Release and Deprotection

Release from the support and deprotection can be achieved by performing the following steps:

1. Transfer the support into a clean and dry beaker, and cover the beaker with a clean paper towel or Kim Wipe™. Place a rubber band over the paper to keep it tight.
2. Place this covered beaker in a clean vacuum desiccator, and carefully attach to the vacuum pump or vacuum line.
3. Release the vacuum, when the support is dry. (Avoid the contact of any acidic or basic fumes during release of vacuum.) Weigh this material.
4. Before deprotection of the bulk material, take out 20–50 mg of the dried support in a screw cap vial (2- to 5-mL size) for the estimation of yield of crude material (crude material contains short sequences and protecting groups).
5. Add 1.5–2.5 mL of 30% ammonium hydroxide (reagent grade), close the screw cap tightly, and incubate using heat block or water bath for at least 5 h or longer for G-rich sequences at 55°C.

Fig. 7. *(opposite page)* Reverse-phase HPLC purification of a thioated 21-mer: Reverse-phase HPLC conditions—Purification; buffer A—0.1M TEAA, pH 7.5%, acetonitrile; buffer B—acetonitrile, 5% H_2O; gradient—10–50% B over 30 min, 8.0 mL/min; wavelength—296 nm; temperature—30 OC; column—Waters Delta Pak C18, 300 Å, 15 μm, 25 × 100 mm PrepPak. **A.** Synthesis G—thioated 21-mer, 470-μmol scale 3000, A_{260} units injected. B. Purified sample of synthesis G. Sample collected at 19.5–20.5 min. **C.** ^{31}P-NMR spectrum of purified thioated 21-mer.

Oligonucleotide Synthesis

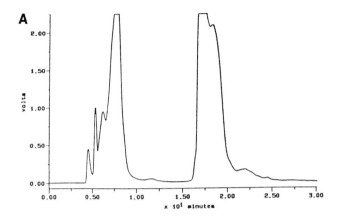

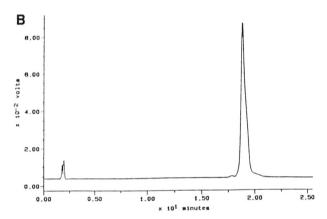

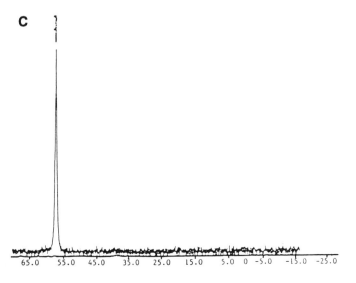

6. Remove the vial from heat block or water bath, and cool it down to room temperature with ice bath for 5–10 min.
7. Open the cap, and transfer the liquid to another clean vial. Rinse the support with HPLC-grade water (2 × 1 mL). Combine all the liquid.
8. Freeze the combined liquids with dry ice–acetone, and lyophilize carefully to remove ammonium hydroxide.
9. Dissolve the dried or lyophilized material in water (1 mL, HPLC-grade). Measure the absorbance at 260 nm, by taking 10.0 µL of this solution and diluting it to 1.0 mL. Record the absorbtion of this diluted solution.
10. Calculate the amount of absorption at 260 nm present in 1.0 mL of the sample and finally in the total synthesis material. For example:
    A. Absorption owing to 10 µL of the above solution after dilution = A unit. Therefore, Absorption owing to 1.0 mL of this solution = A × Dilution Factor = A × 100. Dilution Factor = 100 (since 10 µL of the sample have been diluted to 1.0 mL).
    B. If 20 mg of solid support are used for the above calculation, then amount of material obtained = A × 100 abs. U at 260 nm, and if the weight of supports at the end synthesis = 11.2 g, then total yield = (A × 100 × 11.2 × 1000/20) = A × 100 × 560 A_{260} unit or OD.
11. In order to release the oligonucleotide and remove the protection groups of the bulk material, place the dried solid supports (~10 g) in presilianizied screw-cap Schott bottle (250 mL).
12. Add conc. NH_4OH solution (30%, 100–120 mL), close the cap tightly, and seal the cap with parafilm to prevent any leakage. Place this into a preheated water bath at 55°C, and heat it for 16–20 h.
13. Remove the bottle from the water bath, and cool it down to –20°C with dry ice–acetone.
14. Remove the cap carefully and decant the solution into a clean silianized flash. Wash the support with water HPLC-grade (2–15 mL). Combine total liquid.
15. Freeze the total solution with dry ice–acetone (–78°C), and lyophilize the material under reduced pressure. After lyophilization, the dried solution can be stored or dissolved in HPLC-grade water, and purified by preparative RP-HPLC (discussed elsewhere in this volume).

5.2. Analysis of the Crude Material by HPLC and PAGE

Normal oligonucleotides are analyzed by reverse-phase, anion-exchange, and PAGE. Thioated oligonucleotides can be analyzed by RP-HPLC and PAGE. For anion-exchange analysis, the terminal 5'-DMT group should be removed either with 80% acetic acid after deprotection and cleavage, or on the instrument prior to deprotection and cleavage. General parameters used for analysis by HPLC are given below.

Analytical reverse-phase HPLC conditions:

Buffer A	= 0.1M TEAA pH = 7.0
Buffer B	= Acetonitrile
Flow rate	= 1.5 mL/min
Gradient	= 0–40% B over 20 min, followed by 40% B for 5 min, then to 0% B over 5 min
Column	= Waters Delta Pak C_{18}–300Å
Detection	= 260 nm temperature ambient

Analytical SAX HPLC conditions:

Buffer A	= 0.003M K_2HPO_4 (20% CH_3CN, pH = 6.8)
Buffer B	= 0.64M K_2HPO_4 (20% CH_3CN, pH = 6.8)
Gradient	= 0–50% B over 20 min, flow rate = 1.5 mL/min, followed by 50% B for 5 min, and then 0% B in 1 min
Column	= Beckman Ultrasil AX
Detection	= 260 nm temperature ambient

PAGE conditions: For oligodeoxynucleotides up to 20 bases, 2.0 mm thick 20% polyacrylamide gel should be used, whereas for 20–50 residues, a 15% gel is preferable (8M urea, 1 × TBE, acrylamide/*bis*-acrylamide = 19/1). Gel should be run for 3–6 h at 50–75 mA current.

To determine the amount or percentage of full-length oligonucleotide present in the crude material, an aliquot sample should be analyzed by either RP-HPLC or anion-exchange column HPLC.

1. In order to carry out the analysis by RP-HPLC, clarify the sample solution obtained in Section 5.1., Step 9 by microfuge centrifugation using an Ultrafree-MC0.22-µm filter U (Millipore, UFC30GV00, Bedford, MA), which takes 400 µL solution.
2. Inject about 1.5 OD with material onto HPLC set up with RP-C_{18} column using RP-HPLC parameters described above. The UV-absorbing protecting group and short or truncated sequences elute at a lower percentage of acetonitrile. The full-length desired oligonucleotide with DMT group at the 5' end elutes with a higher percentage of acetonitrile and is usually the last peak in the chromatogram. (A typical RP-HPLC analysis chromatogram is shown in Fig. 5B.) The ratio of peak area owing to oligonucleotide with DMT group to the total peak area gives the percentage of full-length material present in the crude material.

3. In order to perform the analysis by anion-exchange column, lyophilize a 400-μL clarified solution obtained in Step 1.
4. Add acetic acid:water mixture (1 mL, 8:2 v/v), and leave the solution at room temperature for 30 min to remove the 5'-O-DMT group.
5. After this period, concentrate the solution under reduced pressure on a speed vac. Add 500 μL water, and dry it down under reduced pressure to remove acetic acid completely.
6. Extract out DMT-OH released by acetic acid treatment using ether (2 × 500 μL). Remove traces of ether by leaving the microfuge tube open. Finally dissolve in water (400–500 μL, HPLC-grade).
7. Inject about 1.5 OD of this material using anion-exchange column HPLC parameters. The desired full-length product is the last peak (a typical SAX-HPLC analysis chromatogram is given in Fig. 5A). The ratio of the peak area owing to full-length oligonucleotide to the total peak area results in the percentage of desired present in this crude synthetic material. This amount of full-length product can be calculated from the total crude material.
8. In order to analyze the crude material by PAGE, cast a 2.0 nm thick polyacrylamide gel between two glass plates using desired percentage of acrylamide solution containing $8.0 M$ urea and 1 × TBE buffer. Several samples can be loaded in one gel.
9. Dry about 1.0–2.5 OD of the crude material (after detritylation), and dissolve the dried material in a loading buffer (a mixture of formamide, EDTA, and bromophenol blue and xylene cyanol, 20 μL), heat it to 90°C for 5 min, and cool it immediately.
10. Load 10–15 μL of these materials separately in each well of a prerun gel.
11. Run the electrophoresis for 3–6 h at 50–75 mA until the bromophenol blue dye reaches close to the bottom of the gel.
12. Remove gel from both glasses by transferring onto a single sheet of Saran Wrap™. Then shine UV light at 254 nm onto the gel placed on an autoradiography screen, and take a permanent record by Polaroid photography. Although quantitative analysis may not be possible with UV-shadowing gel, this certainly gives an idea of the quality of synthesis.

6. Notes: Pre- and Postsynthesis Precautions

1. Use high-quality and anhydrous solvents for all purposes. Water content should be checked: preferably this should be lower than 50 ppm.
2. If activator solution is not provided by a reputable vendor, purify commercial tetrazole by crystallizing from acetonitrile and drying over phosphorus pentaoxide.
3. Tetrahydrofuran should also be peroxide-free.

4. Dissolve amidite or H-phosphonate under argon atmosphere; prepare these solutions just prior to use.
5. Amidite/H-phosphonate monomer and activator reservoirs should be dried in an oven and cooled in argon-filled desiccator.
6. Do not use old amidite/H-phosphonate monomer and activator solution.
7. DCA solution containing DMT-cation should be collected in a fraction collector placed in a well-ventilated hood.
8. The supports after synthesis should be washed thoroughly with acetonitrile before drying.
9. For deprotection, use good-quality concentrated ammonium hydroxide solution. Use of a previously opened ammonium hydroxide bottle kept at room temperature should be avoided.
10. During removal of ammonia, care should be taken to keep the solution frozen to eliminate spill during evaporation.
11. Use sterile gloves to minimize deprotection due to nucleases.
12. Filter the solution through 0.22-mm membrane filter.
13. Do not incubate oligonucleotide with 80% acetic acid solution more than 30 min at room temperature.

7. Conclusion

It has been demonstrated that a several-gram quantity of synthetic DNA can be obtained successfully based on solid-phase synthesis. Both the existing chemistries (phosphoramidite and H-phosphonate) are capable of producing a large quantity of synthetic oligonucleotide as well as polyphosphorothioated oligonucleotide of good quality to meet the present requirements. However, in the future, large-scale oligonucleotide synthesis may change from the present strategy, based on the nature of supports and chemistry.

Acknowledgment

I would like to thank M. L. Lyttle and P. Wright for their contribution for large-scale DNA synthesis and developing some of the early protocols. I would also like to thank J. Johansen for his encouragement and P. Simpson for taking the trouble to prepare this manuscript.

References

1a. Beaucage, S. L. and Caruthers, M. H. (1981) Deoxynucleoside phosphoramidites—a new class of key intermediates for deoxypolynucleotide synthesis. *Tetrahedron Lett.* **22,** 1859–1862.
1b. McBride, L. J. and Caruthers, M. H. (1983) Nucleotide chemistry, 10, an investigation of several deoxynucleoside phosphoramidites useful for synthesizing deoxyoligonucleotides. *Tetrahedron Lett.* **24,** 245–248.

1c. Sinha, N. D., Biernat, J., and Köster, H. (1983) Beta-CYANOETHYL N,N-dialkylamino/N-Morpholinomonochloro phosphoramidites, new phosphitylating agents facilitating ease of deprotection and work-up of synthesized oligonucleotides. *Tetrahedron Lett.* **24**, 5843–5846.
1d. Sinha, N. D., Biernat, J., McManus, J. P., and Köster, H. (1984) Polymer support oligonucleotide synthesis, 18, use of beta-cyanoethyl-N,N-dialkyamino-/N-morpholino phosphoramidite of deoxynucleosides for the synthesis of DNA fragments simplifying deprotection and isolation of the final product. *Nucl. Acids Res.* **12**, 4539–4557.
1e. Froehler, B. C., Ng, P. G., and Matteuccci, M. D. (1986) Synthesis of DNA via deoxynucleoside H-phosphonate intermediates. *Nucl. Acids Res.* **14**, 5399–5407.
2a. Alvarado-Urbina, G., Sathe, G. M., Liu, W. C., Gillen, M. F., Duck, P. D., Bender, R., and Ogilvie, K. K. (1981) Automated synthesis of gene fragments. *Science* **214**, 270–274.
2b. Hunkapillar, M., Kent, S., Caruthers, M., Dreyers, W., Firca, J., Griffin, C., Horvath, S., Hunkapillar, T., Tempest, P., and Hood, L. (1984) A microchemical facility for the analysis and synthesis of genes and proteins. *Nature (Lond.)* **310**, 105-111.
3a. Sanger, F., Coulson, A. R., Barrell, B. G., Smith, A. J. H., and Roe, B. A. (1980) Cloning in single stranded bacterio phage as an aid to rapid DNA sequencing. *J. Mol. Biol.* **143**, 161–178.
3b. Messing, J., Crea, R., and Seeburg, P. H. (1981) A system for shotgun DNA sequencing. *Nucl. Acids Res.* **9**, 309–321.
3c. Mullins, K. et al. (1986) *Cold Spring Harbor Symp. Quant. Biol.* **51**, 275.
4a. Conner, B., Reyes, A., Morin, C., Itakura, K., Teplitz, R., and Wallace, R. (1983) Detection of sickle cell beta-S globin allele by hybridization with synthetic oligo nucleotides. *Proc. Natl. Acad. Sci. USA* **80**, 278–282.
4b. Di, P., Meldon, P., Skingle, D. C., Lauser, J. A., and Symons, R. H. (1987) Enzyme-linked synthetic oligonucleotide probes—non-radioactive detection of entero-toxigenic *escherichia-coli* in fecal specimens. *Nucl. Acids Res.* **15**, 5275–5287.
5. Khorana, H. G. (1979) Total synthesis of a gene. *Science* **203**, 614–625.
6a. Rich, A. et al. (1979) Molecular-structure of a left-handed double helical DNA fragment at atomic resolution. *Nature* **282**, 680–686.
6b. Drew, A. R. and Dickerson, R. E. (1981) Structure of a B-DNA dodecamer, 3, geometry of hydration. *J. Mol. Biol.* **151**, 535–556.
7a. Chehale, F. F., Doherty, M., Cai, S., Kan, Y. W., Cooper, S., and Rubin, E. M. (1987) Detection of sickle-cell anemia and thalassemias. *Nature* **329**, 293,294.
7b. Greenberg, S. J. Ehrlich, G. D., Abbott, M. A., Hurwitz, B. J., Waldmann, T. A., and Poiesz, B. J. (1989) Detection of sequences homologous to human retroviral DNA in multiple-sclerosis by gene amplification. *Proc. Natl. Acad. Sci. USA* **86**, 2878–2882.
7c. Kogan, S. C., Doherty, M., and Gitschier, J. (1987) An improved method for prenatal-diagnosis of genetic-diseases by analysis of amplified DNA-sequences—application to hemophilia-A. *N. Engl. J. Med.* **317**, 985-990.

8a. Stein, C. A. and Cohen, J. S. (1988) Oligodeoxynucleotides as inhibitors of gene-expression—a review. *Cancer Res.* **48**, 2659-2668.
8b. Heikkila, R., Schwale, G., Wickstrom, E., Loke, S. L., Pluznik, D. H., Watt, R., and Neckers, L. M. (1987) A C-MYC antisense oligodeoxynucleotide inhibits entry into S-phase but not progress from G0 to G1. *Nature* **328**, 445-449.
8c. Zamecnik, P. C., Goodchild, J., Taguchi, Y., and Sarim, P. C. (1986) Inhibition of replication and expression of human T cell lymphotropic virus type III in cultured cells by exogenous synthetic oligonucleotides complementary to viral RNA. *Proc. Natl. Acad. Sci. USA* **83**, 4143-4146.
8d. Cooney, M., Czernuszewicz, G., Postel, E. H., Flint, S. J., and Hoga, M. E. (1988) Site-specific oligonucleotide binding represses transcription of the human C-MYC gene in-vitro. *Science* **241**, 456-459.
8e. Toulene, J. J. and Helene, C. (1988) Antimessenger oligodeoxyribonucleotides—an alternative to antisense RNA for artificial regulation of gene-expression—a review. *Gene* **72**, 51-58.
9. Cohen, J. C. (1989) *Oligodeoxynucleotides, Antisense Inhibitors of Gene Expression.* MacMillan, London.
10. Letsinger, R. L. and Mahadevan, V. (1965) Stepwise synthesis of oligodeoxyribonucleotides on an insoluble polymer support. *J. Am. Chem. Soc.* **87**, 3526.
11. Merrifield, R. B. (1965) Automated synthesis of peptides. *Science* **150**, 178.
12. Köster, H., Biernat, J., McManus, J., Wolter, A., Stumpe, A., Narang, C. K., and Sinha, N. D. (1984) Polymer support oligonucleotide synthesis, 15, synthesis of oligodeoxynucleotides on controlled pore glass (CPG) using phosphate and a new phosphite triester approach. *Tetrahedron* **40**, 103-112.
13. Wright, P., Lyttle, M., Carrol, J., Hudson, D., Warren, W., and Sinha, N. D., Large scale synthesis of oligodeoxynucleotides and polyphosphorothioated oligonucleotides, presented at the International Conference on Nucleic Acid Therapeutics held on January 13-17, 1991 in Clearwater Beach, Florida.
14. Letsinger, R. L. et al. (1988) Cationic oligonucleotides. *J. Am. Chem. Soc.* **110**, 4470,4471.
15. Iyer, R. P., Phillips, L. R., Eagon, W., Regan, J. B., and Beaucage, S. L. (1990) The automated synthesis of sulfur-containing oligodeoxyribonucleotides using 3H-1 2 benzodithiol-3-one 1 1-dioxide as a sulfur-transfer reagent. *J. Org. Chem.* **55**, 4693-4699.

CHAPTER 19

Solid-Phase Supports for Oligonucleotide Synthesis

Richard T. Pon

1. Introduction

The solid-phase strategy for oligonucleotide synthesis has been responsible for much of the widespread utilization of synthetic oligonucleotides. Indeed, without this approach, the chemical synthesis of oligonucleotides would have remained a difficult and tedious task suitable for only the most dedicated chemist. Now of course, as anyone owning an automated synthesizer knows, the solid-phase synthesis of oligonucleotides is as easy as pressing a few buttons. However, despite the obvious advantages of solid-phase synthesis, the development of satisfactory procedures for oligonucleotide synthesis required almost 20 years from the first introduction of this technique. This was primarily because of two obstacles that had to be overcome. The first was the need for rapid and highly efficient coupling reactions, and the second was the need for a suitable solid-phase support. This chapter is devoted to a treatment of the solid-phase supports and the important covalent linkage that binds the oligonucleotide to it.

The insoluble material used to anchor the oligonucleotide cannot be considered as simply an inert carrier, since both its physical and chemical properties can have an important influence on the success of the synthesis. The ideal support should have a uniform surface structure with pores large enough to contain the desired oligonucleotide and should not possess any surface functionality that may produce unwanted side products. The support should also be available in uniformly sized particles

From: *Methods in Molecular Biology, Vol. 20: Protocols for Oligonucleotides and Analogs*
Edited by: S. Agrawal

and be strong enough so that particle fracturing during handling is not significant. Additionally, since chain-extension occurs in the solvated region near the surface of the support, the length and type of spacer used to connect the support to the end of the oligonucleotide are important. A minimum spacer length is required to allow sufficient access for the chain-extension reagents. However, the structural characteristics of the spacers, such as rigidity, hydrogen bonding, and internal dipole–dipole interactions, are also important *(1)*. Finally, the type of covalent linkage joining the first nucleoside to the support is important in determining both the terminal composition of the final product, and the strategy required for the final deprotection and cleavage reactions.

2. Materials

2.1. Nucleosides

Protected nucleosides and nucleoside-3'-succinates are commercially available from a variety of biochemical reagent suppliers (i.e., Sigma Chemical Co., St. Louis, MO, or Penninsula Laboratories, Inc., Belmont, CA), or they may be prepared by following published procedures *(2)*.

2.2. Solid-Phase Supports

Long-chain alkylamine controlled-pore glass (CPG), 125 to 177-μm sized beads with 500-Å pores, is available from the Pierce Chemical Co., Rockport, IL. A similar aminopropyl CPG support with either 1000- or 2000-Å pores is available from Penninsula Laboratories, Inc., Belmont, CA. Fractosil 500 silica gel (300- 400-Å pores, 70–230 mesh) is available from E. Merck, Darmstadt, Germany.

2.3. Reagents

The following materials may be obtained as reagent or other similar quality grade, and used without further purification: chloroform, 2,4,6-collidine, dichloromethane, dicyclohexylcarbodiimide, 1-(3-dimethylaminopropyl)-ethylcarbodiimide, 4-dimethylaminopyridine, ethanol, 1-methylimidazole, ninhydrin, *p*-nitrophenol, pentachlorophenol, piperidine, succinic anhydride, trichloroacetic acid, triethylamine, and toluene. The following reagents may be used as purchased, but must be stored tightly sealed to avoid undue exposure to moisture: acetic anhydride, 3-aminopropyltriethoxysilane, and trimethylsilyl chloride.

2.4. Anhydrous Solvents

Anhydrous pyridine and anhydrous THF may be prepared by refluxing the solvent over calcium hydride. The solvent should then be distilled directly into oven-dried bottles containing type 4A molecular sieves.

3. Methods

3.1. Handling of the Solid Supports

Although controlled-pore glass supports are more rigid than silica gel supports, care must be taken with both materials to minimize any fragmentation. Therefore, it is usually best to perform long derivatization reactions by shaking the flasks on a wrist-action shaker rather than using the more common magnetic stirrers. After reactions are complete, the supports are filtered off using either sintered glass or paper filters (Whatman #1) on top of a Buchner funnel. After washing, the supports can be left in the fume hood to air-dry, or they can be placed in a vacuum desiccator once excess solvent has evaporated off.

3.2. Characterization of the Solid-Phase Supports

3.2.1. Detection of Amino Groups

The presence of primary amino groups can be qualitatively determined by adding ninhydrin solution (0.2% in ethanol, 100 µL) to a small amount of support (2–3 mg) in a glass test tube. The solution is then gently heated until the solvent completely evaporates. The formation of a dark blue color on the support confirms the presence of primary amino groups.

3.2.2. Trityl Analysis

Nucleoside loadings are determined by the colorimetric trityl assay. In this procedure, an aliquot of support (3–4 mg) is accurately weighed directly into a 10-mL volumetric flask. The flask is then filled with 3% trichloroacetic acid in dichloromethane and the resulting orange color, which is formed by the released dimethoxytrityl (DMT) cation, is measured on a spectrophotometer at 503 nm. The nucleoside loading, in µmol/g, can then be calculated as follows:

$$\text{Loading} = (A_{503} \times \text{vol}/76) \times (1000/\text{wt})$$

where A_{503} is the absorbance at 503 nm, wt is the amount of support, in milligrams, and vol is the volume of solution, in millilitres. Monomethoxytrityl-containing supports can be assayed in the same manner if the yellow MMT cation is measured at 470 nm and an extinction coefficient of 56 mL cm^{-1} µmol^{-1} is used in the loading calculation.

3.3. Derivatization of Silica Gel (Fig. 1)

1. Gently stir the silica gel or uncoated CPG (50 g) in a 2*M* hydrochloric acid (500 mL) using a large beaker and a magnetic stirrer. After overnight stirring, allow the silica to settle, and decant off any fine particles in suspension. Resuspend in water, and repeat 2–3 times or until all fine particles are removed. Filter the gel into a large Buchner funnel, and wash with water, and then methanol. Allow the gel to air-dry until free flowing, and then continue drying in an oven (80–100°C).
2. Place the gel into a 500-mL round-bottom flask, and add toluene (200 mL), and then 3-aminopropyltriethoxysilane (50 mL). Fit the flask with a magnetic stirrer and a condenser, and reflux the mixture overnight. Filter off the gel, and wash thoroughly with first toluene, and then chloroform. Allow to air-dry in a fume hood, and then continue drying in an oven. A sample of the silica gel should then be tested to verify the presence of an amino group.
3. The aminopropylated silica gel (50 g) is then stirred in water (600 mL) using a large beaker and the slowest possible speed setting (to avoid creating fines). Succinic anhydride (25 g) is then added, in 2-g portions, over a period of 2 h. During this time, the pH of the solution is continuously monitored, and maintained between pH 4 and pH 5 by the addition of 2*M* NaOH solution. The suspension is then left to stir overnight. Fine particles are decanted off, and the silica gel is washed first with water, and then methanol. A sample of silica gel should be tested for the absence of available amino groups (negative ninhydrin test). If amino groups are still present, the succinylation reaction can be repeated.
4. The silica gel is thoroughly dried by overnight heating in an oven, and then allowed to cool to room temperature. The gel is then transferred into a 500-mL round-bottomed flask and suspended in anhydrous pyridine (100 mL). Trimethylsilyl chloride (10 mL) is added, and the flask is sealed with a drying tube. **Caution:** Avoid eye and skin contact. This liquid reacts violently with water. The silica gel is then stirred overnight, filtered off, and washed first with pyridine and then chloroform.
5. The carboxylated support (1 g), a 5', *N*-protected nucleoside (0.25 mmol), and 4-dimethylaminopyridine, DMAP (6 mg, 0.05 mmol), are placed in a 100-mL round-bottomed flask. Then liquid DCC (2.5 mmol, 515 mg)

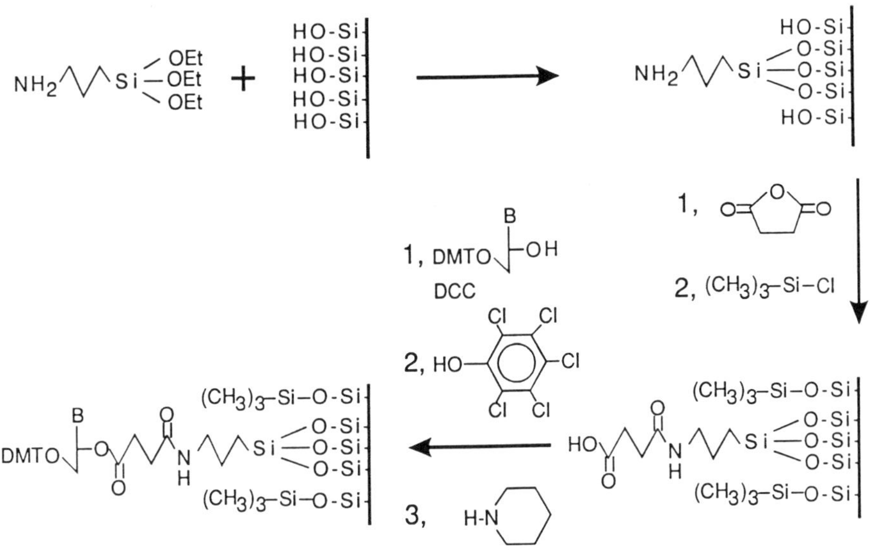

Preparation of Solid-Phase Supports

Fig. 1. The synthesis and derivatization of aminopropylated silica gel or uncoated CPG *(17,18)*.

is weighed directly into the flask. **Caution**: Severe Hazard! Do not use DCC without reading the safety precautions listed under Section 3.4. Anhydrous pyridine (5 mL) is added, and the flask is sealed and shaken, on a wrist-action shaker, for 3 d. *p*-Nitrophenol (2.5 mmol, 347 mg) is added to the flask, and shaking is continued for another 16 h. The silica gel is filtered off, washed with anhydrous pyridine, and then resuspended in piperidine for 5–10 min. The silica gel is then filtered off, and washed first with methanol and then chloroform. The gel is then dried under vacuum and the nucleoside loading determined by a trityl color analysis.

3.4. Derivatization of Long-Chain Alkylamine CPG

Procedure A (Figs. 2 and 3): This method, which uses succinic anhydride first to succinylate the LCAA-CPG, is the easiest route to derivatized CPG supports because it eliminates the synthesis of nucleoside-3'-succinate derivatives.

1. LCAA-CPG (5 g) is slowly stirred or shaken in a solution of 3% trichloroacetic acid in dichloromethane (100 mL; note: This is the same reagent used for detritylation reactions on automated DNA synthesizers)

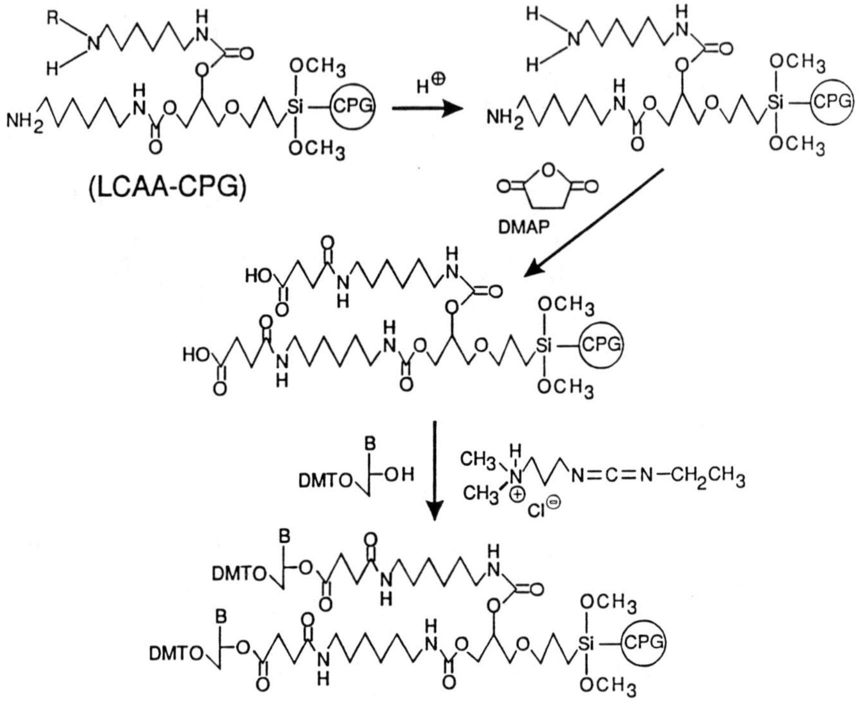

Fig. 2. The synthesis and derivatization of succinylated LCAA-CPG *(31).* This method (procedure A), which uses succinic anhydride to succinylate the LCAA-CPG is the easiest method to prepare derivatized supports, because succinylated nucleosides are not required.

for 2–3 h. The activated LCAA-CPG is then filtered off, washed with dichloromethane, and then left to dry.

2. Activated LCAA-CPG *(5 g)*, succinic anhydride (10 mmol, 1 g), and DMAP (1 mmol, 122 mg) are placed into a 100-mL round-bottomed flask. Anhydrous pyridine (40 mL) is then added, and the flask is shaken on a wrist-action shaker for 24 h. The support is then filtered off, and washed first with pyridine, and then dichloromethane. The succinylated LCAA-CPG may be stored indefinitely at room temperature until required.

3. Succinylated LCAA-CPG *(1 g)*, partially protected nucleoside (0.1 mmol), and DMAP (0.1 mmol, 12 mg) are placed in a 100-mL round-bottom flask. Then 1-(3-dimethylaminopropyl)ethylcarbodiimide (1 mmol, 192 mg) is carefully added. **Caution**: Eye and skin protection is required. Triethylamine (80 µL), and anhydrous pyridine (10 mL) are

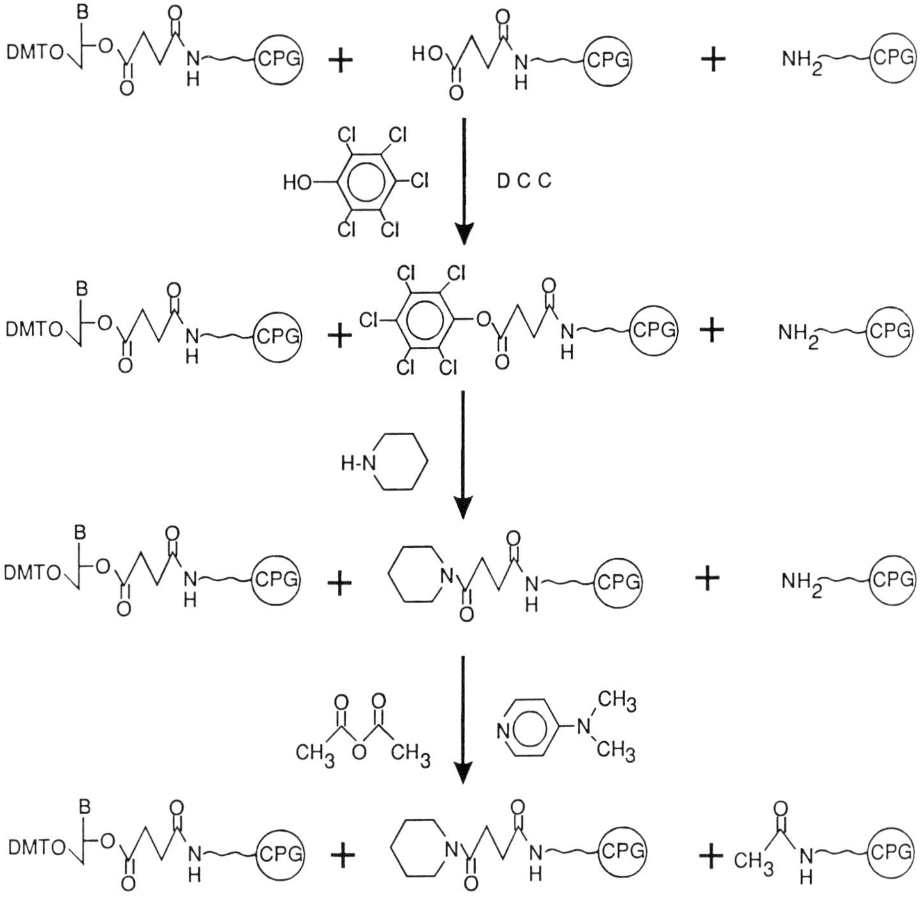

Fig. 3. Capping procedures for blocking unreacted carboxyl and amino groups *(14–16).*

then added. The flask is then tightly sealed and shaken at room temperature for 16 h. Pentachlorophenol (0.5 mmol, 135 mg) is added to the flask, and shaking is continued for another 16 h. Finally, piperidine (5 mL) is added, and after 5 min (important: prolonged treatment with piperidine will also cleave the nucleoside succinate linkages and will decrease the nucleoside loading on the support), the CPG is filtered off and washed successively with dichloromethane and ether. The support is then dried.

4. The derivatized support is then suspended in acetic anhydride capping reagent. The capping reagent used may be made by combining equal volumes of the Cap A and Cap B (acetic anhydride and *N*-methylimi-

dazole or acetic anhydride and DMAP) reagents supplied for use on automated DNA synthesizers. Alternatively, this solution may be prepared by dissolving acetic anhydride (5 mL), 2,4,6-collidine (7 mL), and DMAP (3 g) in anhydrous THF (100 mL). This solution must be used within 24 h of preparation. The support is shaken or stirred for 2 h before being filtered off and washed with dichloromethane. The acetylated amino groups slowly become deblocked during prolonged (>3 mo) storage at room temperature, and excessive trityl colors may occur (i.e., apparent first coupling yields of >100%). These groups may be easily reblocked by using the DNA synthesizer to perform a short (10-min) capping reaction just prior to oligonucleotide synthesis. This capping reaction may be performed using either manual control or a simple automated capping cycle.
5. The nucleoside loading can then be determined by trityl analysis. Loadings of between 20 and 40 µmol/g should be obtained.

Procedure B (Fig. 4): This procedure, which is faster than procedure A, is best used when nucleoside-3'-succinates are available.

1. LCAA-CPG support is acid-activated as in Procedure A, Step 1.
2. Activated LCAA-CPG (1 g), DMAP (0.1 mmol, 12 mg), nucleoside-3'-succinate, and 1-(3-dimethylaminopropyl)-ethylcarbodiimide (2 mmol, 382 mg, **Caution:** Eye and skin protection is required) are added to a 100-mL round-bottomed flask. Triethylamine (80 µL), and anhydrous pyridine (10 mL) are then added. The flask is tightly sealed and shaken at room temperature for 1–24 h, depending on the loading required. Usually a 1 h reaction will provide material in the 20–30 µmol/g range. This is the optimum loading density for oligonucleotide synthesis, and the higher nucleoside loadings (40–60 µmol/g) produced by longer reaction times are neither necessary nor desirable. The support is filtered off and washed first with pyridine, and then with dichloromethane. After drying, the nucleoside loading can be determined by trityl analysis.
3. The support is then capped as described in procedure A, Step 4.

Note: Safety precautions: All of the reagents used should be considered harmful, and good laboratory safety procedures must always be followed. These include, but are not limited to, the use of safety glasses, laboratory coats, and protective gloves. Certain materials, particularly piperidine, pyridine, triethylamine, collidine, and acetic anhydride have noxious odors, and should be used in a fume hood. One particularly hazardous reagent is dicyclohexylcarbodiimide, DCC. This reagent can cause severe eye and skin irritation especially in individuals who

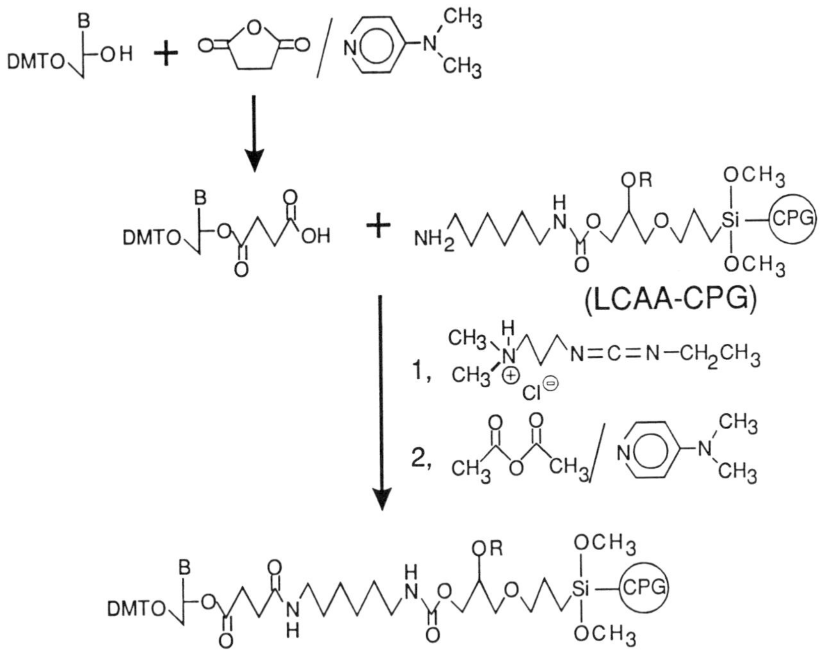

Fig. 4. The derivatization of LCAA-CPG with nucleoside-3'-succinates and DEC *(30)*. This method (procedure B) is faster than procedure A (Fig. 2), and can be used when nucleoside-3'-succinates are available.

have become sensitized to it. This material should not be handled as a solid because of possible airborne release of small particles. The preferred method is to gently warm the reagent, in a water bath, until it melts (mp 34–35°), and then transfer the liquid directly into the reaction flask. All material exposed to DCC should be decontaminated by washing with water immediately after use.

4. Solid-Phase Supports

A large variety of different materials have been tried as insoluble supports during the evolution of oligonucleotide synthesis, and the major materials are described in Table 1. The first solid-phase oligonucleotide syntheses were performed using the same nonpolar "popcorn" polystyrene resin *(3,4)* used for peptide synthesis (Fig. 5). Later, polar polyamide resins were developed that were more suitable for the polar solvents used in early phosphodiester *(5)*/triester *(6)* methods (Fig. 5).

Table 1
The History of Solid-Phase Supports in Oligonucleotide Synthesis

Material	Year of introduction	Properties of the support	Current status
Polystyrene, low cross-linking (3,4)	1966	Nonpolar, swellable support; used in early "diester" and coupling procedures	Not in use; unsuitable for automated synthesis
Polyamide (5,6)	1977	Polar, swellable support; used in early "diester" and "triester" coupling procedures	Not in use; unsuitable for automated synthesis
Polyamide bonded silica gel (8)	1982	Polar rigid support; irregularly shaped and sized particles; very effective in "phosphotriester" coupling procedures; suitable for use in automated synthesizers	No longer used since phosphoramidite coupling procedures have superseded the "phosphotriester" methods
Cellulose (9,10)	1980	Polar, nonswelling support; handling of individual disks can be done manually, but is not easily automated	Not widely used, despite its potential for the rapid synthesis of multiple sequences
Silica gel (14–16)	1972 reintroduced 1981	Rigid, easily derivatizable support available in a wide range of pore and particle sizes; irregularly shaped particles; may be difficult to handle because of ease of particle fracturing	Almost entirely superceded by controlled-pore glass supports
Controlled-pore glass (13,20,23)	1972 reintroduced 1981	Spherical glass beads with increased strength and size uniformity; available in very large pore sizes (500–4000 Å)	The most widely used support of all (>2 million seq/y)
Polystyrene/PEG "Tentacle" copolymer (25,26)	1991	Spherical, very uniformly sized particles with improved swellability in polar solvents; capable of high nucleotide loadings; suitable for use under high pressure and high flow rates	Only recently introduced, suitable for large-scale continuous flow peptide and oligonucleotide synthesis
Polystyrene, high cross-linking (27)	1991	Nonpolar, nonswellable, nonporous beads; lacks surface silanol groups that may contribute to undesirable byproducts during synthesis; used in small-scale (40 nmol) synthesis	Only recently introduced, price and performance advantages may allow these supports to surpass CPG in utility

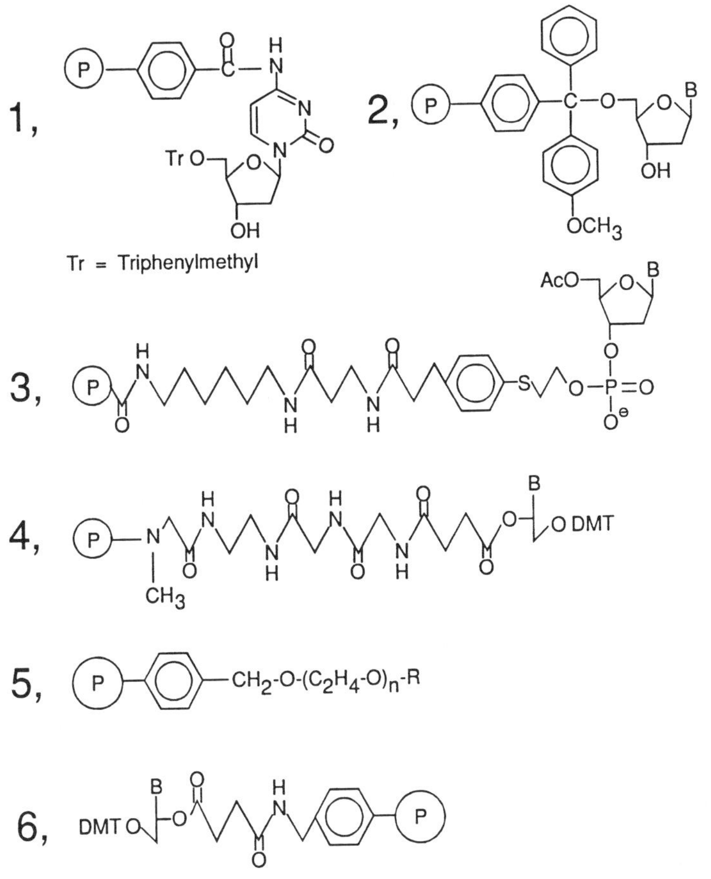

Fig. 5. Solid-phase supports made of synthetic resins. Figs. 1 and 2.: The nucleoside linkages used on the early polystyrene "popcorn" supports (3,4). Fig. 3.: The nucleoside linkage used on a polyamide based resin (5). Fig. 4.: The structure of a polyamide/silica gel composite developed for continuous-flow synthesis (8). Fig. 5.: The structure of a PEG-polystyrene grafted copolymer, "tentacle" support (25). The mean mol wt of the PEG side chain is 3000 daltons. Fig. 6.: The nucleoside linkage used on the recently introduced highly crosslinked polystyrene support (27).

However, the slow diffusion of reagents into and out of these solvent-swollen materials made it difficult to wash reactants off of the supports. Also, the poor rigidity of these resins prevented them from being packed into columns for use under continuous-flow conditions, because compression under solvent pressure led to irregular flow and high back-pressure. These properties, therefore, precluded their use in automated continuous-flow synthesizers, and restricted them to either manual use *(7)* or to slower "batch"-type synthesizers, which could perform extensive shaking and washing cycles.

A rigid polyamide/silica gel composite *(8)* (Fig. 5) was eventually developed for continuous-flow synthesis, which produced quite good results with phosphotriester synthesis chemistry. This support produced quite good results, but it was not widely used despite its commercial availability.

Another interesting material that was shown to produce good results with phosphotriester coupling chemistry was cellulose. This material, in the form of paper filter disks, was readily available and easily derivatized *(9)*. Although the paper disks could be easily manipulated manually to prepare a great number of sequences rapidly using a multiple sorting strategy *(10)*, these manipulations were not easily automated, and these procedures were also not widely adopted.

All of the previous insoluble supports have been superceded by silica-based materials, and phosphite triester reactions on these supports have become the predominant synthetic route. Silica is an ideal material, because it is nonswelling and easily derivatizable using techniques developed by the fiberglass-reinforced plastics *(11)*, and bonded-phase chromatography *(12)* industries. Koster *(13)* was the first to attempt solid-phase oligonucleotide synthesis on glass and silica gel particles in 1972. However, it was not until phosphorochlorodite coupling chemistry was first applied to the solid-phase synthesis of DNA *(14–16)* that silica gel became widely utilized.

The silica gel used was derivatized as shown in Fig. 1. In this procedure, the silica gel was first washed with hydrochloric acid to convert surface siloxane (Si—O—Si) groups into silanol groups. Then the surface was reacted with 3-aminopropyltriethoxysilane to yield an aminopropyl ligand *(17)*. A mixture of single-, double-, and triple-siloxane linkages along with some degree of surface polymerization is formed on the surface of the silica gel. However, this surface heterogeneity has little effect on oligonucleotide synthesis. Any residual silanol groups left unreacted are then blocked by treatment with chlorotrimethylsilane.

The aminopropylated silica gel can be used in a variety of ways. In the most common procedure, the aminopropyl ligand is extended by reaction with succinic anhydride to yield a carboxylated silica gel.

Nucleosides are then coupled to the silica gel by DCC-mediated esterification to yield nucleoside loadings of about 100 µmol/g. Alternatively, nucleoside-3'-pentachlorophenyl esters can be directly attached to the aminopropylated silica *(18)*. Much longer, linker arm extensions can also be added onto the aminopropylated silica gel. Kume et al. *(19)* extended the linker arm by another 24 atoms by adding two consecutive 11-aminoundecanoic acid linkers onto the aminopropyl linker. A total of 23 different linker arms containing a variety of alkylamine, aromatic, amino acid, and polyethylene glycol backbones, of up to 30 atoms in length, were synthesized, and sequences up to 94 bases long were prepared by Katzhendler et al. *(1)*, who showed that the best coupling yields were obtained with a minimum 25-atom-long linker. Linker structures that lacked hydrogen bonding or π-interactions and were less likely to adopt a folded conformation also gave better product homogeneity.

Despite these modifications, silica gel supports still remained relatively fragile and prone to fragmentation. This breakage created fine particles that could plug filters and create high operating pressures in automated instruments. Also, the limited pore size of even the most porous silicas (300–400 Å) soon became a restriction as the synthesis of oligonucleotides >30 bases became possible. Consequently, the use of silica gel supports has been superceded by another silica-based product, controlled-pore glass (CPG) beads.

CPG beads are currently the most suitable and most widely used solid-phase supports for oligonucleotide synthesis. These glass beads retain the same surface silanol groups and hence the same surface chemistry as silica gel particles. The surface of these beads is etched to leave behind a porous surface with mean pore sizes ranging from 75 Å to several thousand angstroms. Unlike silica gel, the beads are uniformly milled into screened spherical particles that provide greatly reduced backpressure when packed into continuous-flow columns. The beads are also much less susceptible to fragmentation than silica gel.

Underivatized CPG beads can be derivatized by reaction with 3-aminopropyltriethoxysilane, in the same way as silica gel, to create an aminopropyl linker arm (cf. Fig. 1). The aminopropyl group can then be coupled directly to a *p*-nitrophenyl ester of a nucleoside-3'-succinate *(20)*. The amount of nucleoside loading achieved is strongly dependent on the pore size of the support, since the surface area decreases with increasing pore size. For example, nucleoside loadings of about 2, 9, and 100 µmol/g were obtained using CPG with pore sizes of 3000, 1400, and 240 Å, respectively *(21)*. The aminopropyl linker arm can also be extended by reaction with 6-aminocaproic acid to create a linker arm intermediate in length between the aminopropyl and long-chain alkylamine-derivatized supports *(22)*.

Although aminopropylated CPG can be easily synthesized, a much better support containing a long-chain alkylamine group (LCAA-CPG) is commercially available. This material contains a primary amino group on the end of a 17-atom-long linker arm (Fig. 2). The long linker arm along with the large (500 Å) pore size of LCAA-CPG greatly improve coupling results. Indeed, in the first report *(23)* of this material's use, a new record for oligonucleotide length (51 bases), which was more than double any previously synthesized length, was achieved. The 500-Å LCAA-CPG supports allow virtually quantitative coupling reactions for oligonucleotides of approx 40 nucleotides or less. After this, the pore size of the support begins to restrict coupling efficiency and yields begin to decline gradually as the sequence length increases to approx 50–60 nucleotides, at which time coupling yields become much less. A maximum sequence length of only 50–60 nucleotides is therefore recommended with the 500-Å LCAA-CPG. However, much longer sequences (50–150 bases) can be routinely prepared on similar 1000- or 2000-Å CPG supports, which are also commercially available. Very long sequences (>200 bases) can also be synthesized on these wide-pore CPG supports, if polymerase chain reaction (PCR) is used to amplify the product before isolation *(24)*. The convenience and effectiveness of CPG supports have been instrumental in improving the ease with which high-quality oligonucleotides can be prepared. These materials are now the support of choice, and they are currently used in the synthesis of more than two million oligonucleotides per year.

However, despite the current popularity of CPG supports, the search for improved materials still continues, and recently two promising new supports based on polystyrene have been introduced. Bayer *(25)* has prepared a "tentacle" support (Fig. 5) that contains polyethylene glycol (PEG) chains grafted onto the surface of monodisperse polystyrene beads (PEG-PS). The hydrophilic PEG tentacles impart greatly improved swellability in the polar solvents required for oligonucleotide synthesis, and the uniform size and spherical shape of the monodisperse polystyrene cores greatly improve the performance of the copolymer in continuous-flow synthesizers. A commercially available PEG-PS support functionalized with either hydroxyl or amino groups has been used for solid-phase oligonucleotide synthesis using H-phosphonate-coupling chemistry *(26)*. This support was advantageous in large (30 µmol) scale synthesis, because a high nucleoside loading (170 µmol/g) that was four- to fivefold greater than typical CPG supports was obtained.

The second new polystyrene based support *(27)* differs from previous polystyrene polymers by containing a very large amount of crosslinker (50%). This creates a rigid, nonswelling and nonporous support. The surface is derivatized with aminomethyl groups, which are then coupled to nucleoside-3'-succinates to yield the linkage shown in Fig. 5. The absence of pores reduces the amount of nucleoside that can be attached onto the resin, and prepacked synthesis columns are only available on a 40-nmol scale. However, the low loading is sufficient for many applications, and the nonporous nature of the support provides rapid reaction kinetics very much like controlled-pore glass (CPG) supports. These resins can therefore be substituted for CPG supports in automated synthesizers without the need for any cycle modifications. The homogeneity of the oligonucleotide prepared on these supports is also improved when compared to CPG supports because of the absence of surface silanol groups, which react to form unwanted side products. This material is therefore very promising, and it may eventually replace CPG as the most widely used solid-phase support.

5. Nucleoside Attachment to LCAA-CPG Supports

In almost all CPG supports, a nucleoside-3'-succinate is attached to the support via an amide bond. The formation of this linkage may be brought about by a number of reagents. However, the huge number of sequences that have been prepared on LCAA-CPG supports has resulted in a very careful examination of the coupling methods used to link the first nucleoside to the primary amino group of the support.

When conditions similar to silica gel derivatizations were used, i.e., using DCC to couple a nucleoside-3'-succinate to the amino CPG, only very low nucleoside loadings (<2 µmol/g) were obtained. This was presumably owing to steric hindrance of the bulky DCC reagent near

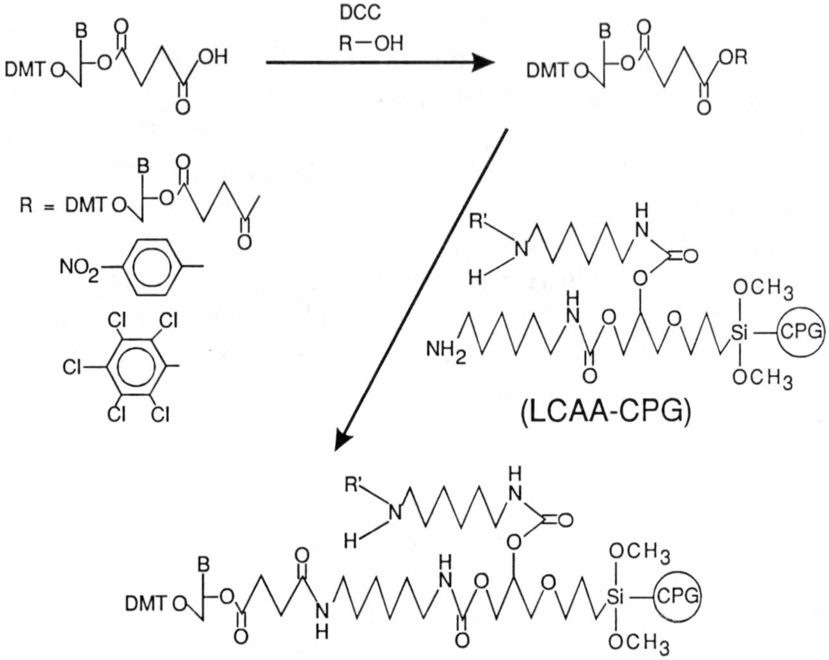

Fig. 6. The derivatization of LCAA-CPG supports using activated nucleoside-3'-succinate compounds *(6,21,28,29)*. This method has been superceded by the improved schemes shown in Figs 2 and 4.

the surface of the support. Consequently, it was necessary to activate the nucleoside-3'-succinate to achieve coupling with the amino group on the LCAA-CPG (Fig. 6). This was done by using DCC to convert the nucleoside-3'-succinate into either the symmetrical anhydride *(6)*, or to convert it into an activated ester using either *p*-nitrophenol *(21)* or pentachlorophenol *(28,29)*. These methods produced satisfactory nucleoside loadings of between 10 and 40 µmol/g, but shared the following disadvantages.

First, the synthesis of the activated pentachlorophenyl succinates required the use of highly toxic materials, such as DCC or pentachlorophenol. Second, the preparation of these compounds was tedious and gave only moderate yields (50–75%). Third, the coupling reactions were quite lengthy with 3–4 d required to make the activated succinates and an additional 4–7 d required for the coupling to the CPG. Fourth, the entire procedure had to be repeated every time an experiment required a nucleoside with a different structure, such as a modified base or a different protecting group. Finally, the CPG loadings obtained were quite variable, and supports with the optimum loading of 30–40 µmol/g were not always obtained.

A major improvement in this procedure, which eliminated the above difficulties, was the observation that direct coupling of a nucleoside-3'-succinate to the amino group on the LCAA-CPG could be performed simply by utilizing a different carbodiimide coupling reagent *(30)* (Fig. 4). This approach was tried, because it was suspected that the bulky DCC molecule was too sterically constrained near the surface of the CPG. The smaller and less rigid carbodiimide, 1-(3-dimethylaminopropyl)-3-ethylcarbodiimide (DEC, sometimes incorrectly called 1-ethyl-3-[3-dimethylaminopropyl]-carbodiimide) was found to give much better results. With this improved approach, nucleoside-3'-succinates could be directly attached to the support to yield nucleoside loadings of up to 50–60 µmol/g in overnight reactions. Lower loadings in the more useful range of 20–30 µmol/g could be easily obtained by restricting the reaction time to only 1 h. Since this method eliminated the synthesis of activated nucleoside-3'-succinate esters, it was therefore much faster and easier to perform.

The derivatization procedure was further improved by Damha et al. *(31)*, who succinylated the LCAA-CPG in a method similar to silica gel derivatization (Fig. 2). The LCAA-CPG was converted into a carboxylated support by reacting it with succinic anhydride. Nucleosides were then directly attached to the support by esterification with DEC. This method was advantageous because the nucleoside-3'-succinate derivatives no longer needed to be prepared, and any 5', *N*-protected nucleoside with a free 3'-hydroxyl group could be used. Using this procedure, nucleoside loadings of 20–40 µmol/g were obtained after a single overnight coupling reaction.

6. Capping Procedures for LCAA-CPG Supports

6.1. General Capping Reactions

The nucleoside coupling reactions used to attach the first nucleoside to the linker arm of the insoluble support never proceed quantitatively, so it is always necessary to block or "cap" the unreacted linker sites. Otherwise, these surface groups may react to produce unwanted contaminants during subsequent chain-extension reactions. This is true for all types of solid-phase supports, although the type of surface derivatization will affect the type of capping required. With controlled-pore glass supports, the following functional groups need to be blocked to prevent unwanted reactions:

1. Surface silanol groups, which are always present on any silica-based supports, may be converted into trimethylsilyl ethers by using chlorotrimethylsilane as the capping reagent. However, new silanol groups can become exposed by acidic hydrolysis during detritylation steps or simply by breakage of the silica support to expose new surfaces.
2. Primary amino groups, which remain after the derivatization of aminoalkyl supports, will also need to be capped. These groups can be easily acetylated by treatment with an acetic anhydride/DMAP reagent capping reagent (Fig. 3).
3. Carboxyl groups, which are present when carboxylated supports, such as succinylated LCAA-CPG, are used, are usually blocked as amides. This blocking is typically a two-step procedure (Fig. 3) in which an activated *p*-nitrophenyl or pentachlorophenyl ester is first prepared. This ester is then treated with piperidine to create a more stable secondary amide. All of these capping reactions are quite simple to perform and very efficient.

6.2. The Elimination of Greater than 100% Coupling Yields

However, when LCAA-CPG supports were first used, an unusual coupling phenomenon was observed, despite the performance of the above capping reactions. This was the apparent formation of coupling yields that exceeded 100% for the first few synthesis cycles performed on the supports. The commonly used colorimetric "trityl" analysis, in which the amount of orange dimethoxytrityl cation released during detritylation steps was compared for successive steps, was used to determine these yields. In these assays, up to twice as much orange color, indicating coupling yields of up to 200%, was observed for the very first coupling reaction. Second and third couplings also showed anomalous results, but to a lesser extent. Phosphoramidite coupling to nonnucleotide groups was the most obvious explanation for the additional nucleotide binding, but pretreatment of the LCAA-CPG with a variety of very powerful capping reagents had no effect on the anomalous results. The failure of these capping reactions combined with the absence of significant impurities in the oligonucleotides caused considerable confusion.

However, it was eventually discovered that an acidic pretreatment of the LCAA-CPG support could double the loading capacity of the LCAA-CPG *(30)*. Ninhydrin testing confirmed that the new functional sites exposed were primary amino groups that had been previously protected with an undisclosed type of acid-labile blocking group *(31)*. It was these blocking groups that prevented effective capping of the support. Apparent yields of over 100% occurred, because the acidic detritylation step of the first synthesis cycle removed these groups and allowed 3'-phosphoramidate linkages to form between the nucleotide and the support. However, once this was known, it was a simple matter to preactivate the LCAA-CPG with acid before derivatization. After attachment of the nucleoside, capping could then be performed as with other supports. This improved procedure greatly reduced the excess trityl color observed. However, a small excess may still occur, because the acetyl groups used to cap the amino groups are slowly lost during storage. Also, surface silanol groups capable of forming 3'-phosphate linkages can become exposed by either surface fragmentation or acidic cleavage of the trimethylsilyl ethers on the surface. The most effective capping procedure is to perform two capping steps. The first is performed on the support just after derivatization. A second shorter (10-min) capping reaction is also performed on the synthesis column just prior to the start of synthesis. Such a step can be performed automatically by many user programmable DNA synthesizers.

6.3. The Importance of Eliminating Unwanted Surface Reactions

The importance of completely eliminating reaction of the phosphoramidites with the surface of the support varies with the type of synthesis performed. Small-scale syntheses, which will be used in biological studies, are probably not affected by this side reaction, since most of the excess material remains attached to the CPG. This is because the phosphoroamidate linkage, just like a phosphodiester linkage, is very resistant to hydrolysis. The remaining impurities are either removed during purification or, because of their 3'-phosphorylated ends, have no biological effect. However, larger-scale syntheses, which use a lower excess of phosphoramidite reagent, can be affected, since the unwanted side products are competing for incoming reagents. Also, purification procedures that select for terminally dimethoxytritylated products may fail, since products with 3' heterogeneity may become enriched. In any case, this side reaction can distort the overall efficiency of the synthesizer, and the >100% first coupling yields should never be used to calculate average or overall coupling yields.

7. CPG Supports Containing Specialized Linkages

7.1. Solid-Phase Supports with Hydrolysis-Resistant Linkages

The succinate linkage used to attach nucleosides onto the CPG is easily cleaved under basic conditions. This is usually desirable, because it allows cleavage from the support to be combined with the phosphate and base deprotections, which also require basic hydrolysis. However, this precludes the synthesis of completely deprotected oligonucleotides on insoluble supports. This shortcoming has been overcome by several strategies that replace the nucleoside succinate attachment with more resistant linkages. These approaches allow easier isolation of the final product, since fewer nonnucleotide impurities are present. They also allow the support-bound material to be used as a solid-phase affinity column.

The first base-resistant linkage on CPG was introduced by Sproat and Brown *(32)*, who used tolylene-2,6-diisocyanate (TDIC) to link 2'-deoxyribonucleosides onto LCAA-CPG via two urethane linkages (Fig. 7). This linkage is sufficiently resistant so that the usual deprotection conditions with ammonium hydoxide (16 h at 50°) can be used to deprotect completely the oligonucleotides while still keeping the majority attached to the CPG *(33)*. An extended treatment with hot aqueous ammonium hydroxide (48 h at 50–60°) will cleave the oligonucleotide from the support. This type of linkage has also been used to synthesize oligonucleotides directly onto magnetic beads (Dynabeads™), so that oligonucleotides can be isolated from cell lysates *(34)*.

Two other approaches, which have not been as widely used, have also been developed. In the first, the primary amino group of LCAA-CPG was attached to the 4-position of a uridine nucleotide (Fig. 7) via reaction of a 4-triazolyl derivative *(35)*. In the other procedure *(36)*,

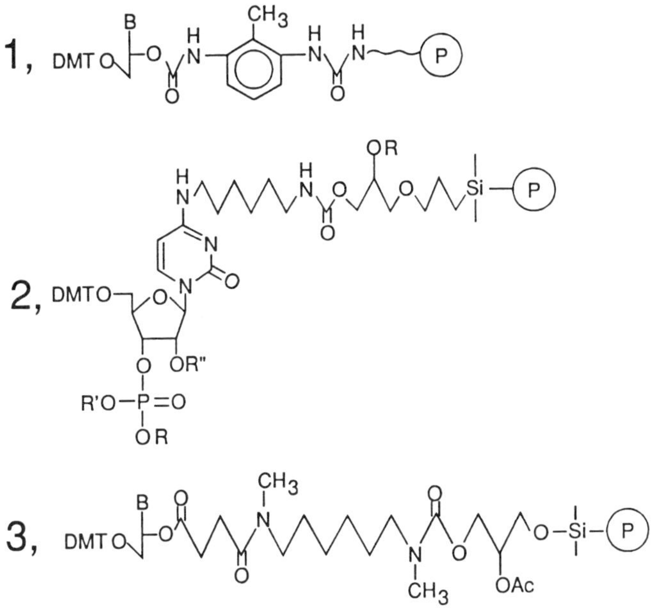

Preparation of Solid-Phase Supports

Fig. 7. Hydrolysis resistant nucleoside linkages. Fig. 1.: The TDIC linkage based on tolylene-2,6-diisocyanate *(32)*. Fig. 2.: A ribonucleotide linked to LCAA-CPG via a secondary amino group *(35)*. Fig. 3.: Nucleoside attachment via a 1,6-*bis*-methylaminohexane linker *(36)*.

dihydroxypropyl-CPG beads were derivatized with *N,N'*-carbonyldiimidazole and 1,6-*bis*-methylaminohexane to yield a different type of alkylamine support (Fig. 7). The modified aminoalkyl linker, which had a tertiary rather than a secondary amide linkage, was now resistant to cleavage by the hindered base, 1,8-diazabicyclo[5.4.0]undec-7-ene (DBU). Therefore, when *p*-nitrophenylethyl (NPE), and *p*-nitrophenylethoxycarbonyl (NPEOC)-protected nucleosides phosphoramidite compounds were used, complete on-column deprotection could be performed with DBU. After deprotection, a treatment with ammonium hydroxide would release the product.

7.2. Solid-Phase Supports with Labile Linkages

In contrast to the above applications, several types of supports have been prepared that use milder conditions to release the oligonucleotide. These more labile supports are also being supplemented by the intro-

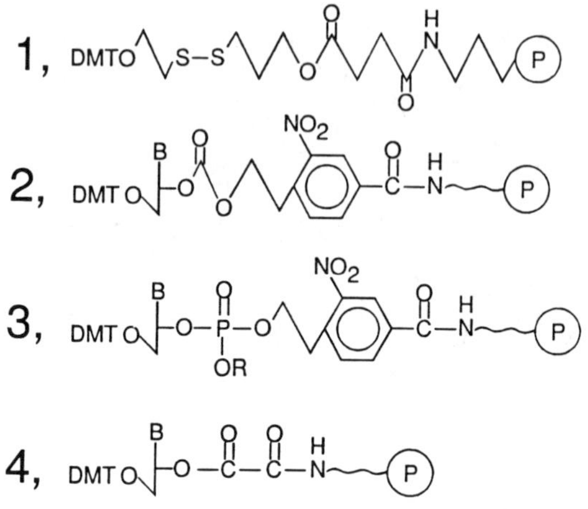

Fig. 8. Easily cleavable nucleoside linkages. Fig. 1.: A disulfide linkage that yields 3'-phosphorylated products *(39)*. Figs. 2. and 3.: Linkages containing a 2-(2-nitrophenyl)ethyl linker, which can be cleaved by DBU, for the synthesis of oligonucleotides with either free or phosphorylated 3'-ends *(40)*. Fig. 4.: The oxalyl-CPG linkage *(41)*, which is the most easily hydrolyzable linkage available.

duction of either more labile-base protecting groups *(37)* or coupling techniques that avoid base protection entirely *(38)*, so that gentler overall conditions for deprotection can be employed.

One possibility for an easily cleavable linkage is to use a disulfide linkage. This has been demonstrated by Asseline and Thuong *(39)*, who performed synthesis on a dimethoxytritylated dithiodiethanol linker (Fig. 8) After synthesis, the final product, which was 3'-phosphorylated, was released from the support by treatment with dithiothreitol.

Another strategy was the use of a 2-(2-nitrophenyl)ethyl linkage *(40)*. In this approach, 4-(2-hydroxyethyl)-3-nitrobenzoic acid was coupled onto LCAA-CPG. Oligonucleotides ending in either 3'-hydroxyl or 3'-phosphate groups could be made depending on the presence or absence of a carbonate linker (Fig. 8). In either case, cleavage from the support was possible by treatment with DBU. Therefore, sequences made from *p*-nitrophenylethyl-protected nucleoside 2-cyanoethyl phosphoramidites could be completely deprotected and cleaved from the support with a single DBU treatment.

An oxalyl linkage has also been used to attach nucleosides to LCAA-CPG *(41)*. In this method, oxalyl chloride replaces succinic anhydride as the linker that joins the nucleoside to the support (Fig. 8). This oxalyl linkage is quite labile and is cleaved in <5 min by 5% ammonium hydroxide in methanol. Other mildly basic reagents, such as wet triethylamine or 40% trimethylamine/methanol, also rapidly cleave the oxalyl linkage. However, it was quite resistant to normal conditions employed in either phosphoramidite or H-phosphonate synthesis and to anhydrous amines, such as pyridine, triethylamine, and diisopropylamine. This support is particularly well suited for use when synthesis is performed without base-protecting groups *(38)*.

7.3. Solid-Phase Supports that Introduce a 3'-Functionality

7.3.1. Supports for 3'-Phosphate End Groups

A wide variety of different linker arms have been designed, so that oligonucleotides bearing 3'-end groups can be synthesized. The most common end group is the 3'-phosphate group. Although 3'-phosphorylated sequences are readily available by periodate oxidation of 3' ribonucleosides *(42)*, several approaches that permit the direct synthesis of 3'-phosphorylated oligonucleotides have been developed. Among these are the disulfide *(39)* and 2-nitrophenylethyl *(40)* linkages already discussed as labile linkages (Fig. 8). A mercaptoethanol linker can also be used. This linker can be attached onto an aminopropyl-CPG support that has been functionalized with either 3-(trimethoxysilyl) propyl methacrylate *(43)* or 3-mercaptopropyltrimethoxysilane *(44)* as shown in Fig. 9. After synthesis, treatment with ammonium hydroxide releases a 3'-phosphorylated oligonucleotide. A 4,4'-diaminobenzidine derivatized CPG (Fig. 9) can also be similarly used if isoamyl nitrite is used as the cleavage reagent *(43)*.

7.3.2. Supports for 3'-Thiol End Groups

A 3'-thiol function can also be incorporated into oligonucleotides to serve as a convenient attachment site for either enzymes or fluorescent dyes. In a multistep method *(45)*, nucleosides or nucleotides may be derivatized with either 1,6-hexanedithiol and mercaptoethanol or 3,3'-dithiodipropanol, respectively, to yield 3'-disulfide linkers (Fig. 9).

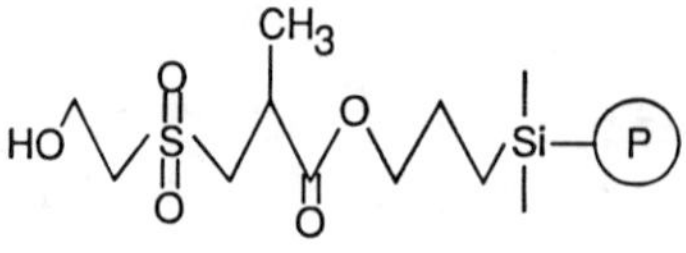

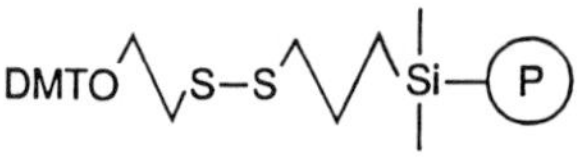

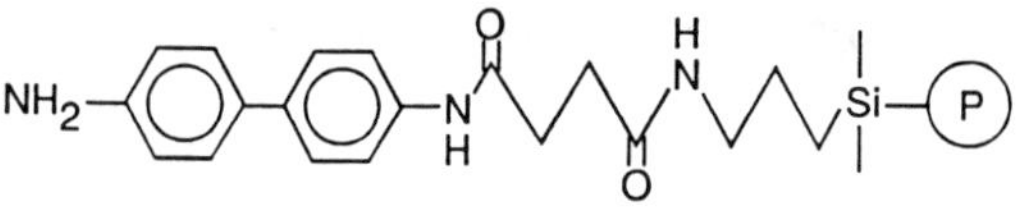

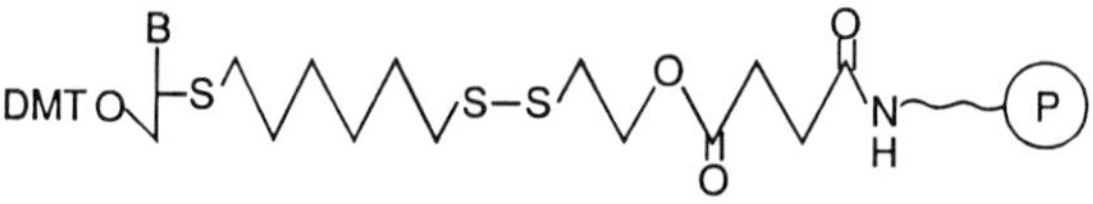

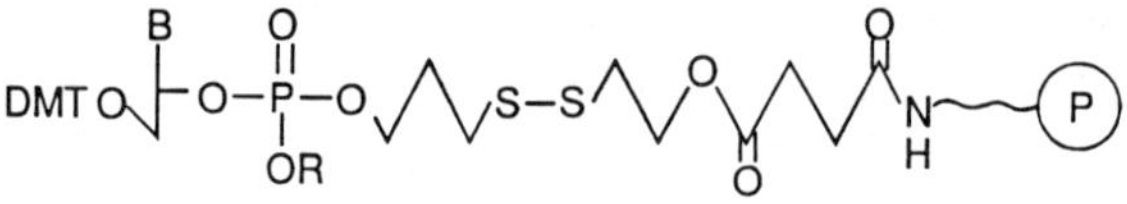

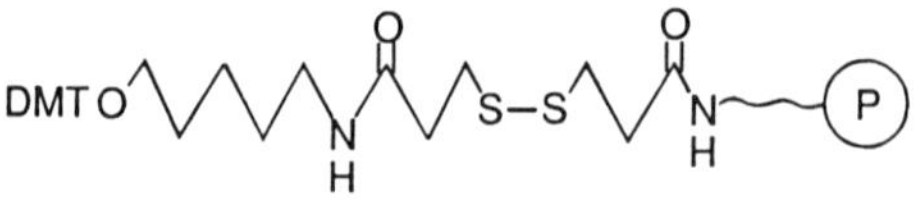

Fig. 9. Solid-phase supports for the preparation of 3'-end-modified oligonucleotides. Fig. 1.: A universal support for 3'-phosphorylated oligonucleotides using a mercaptoethanol linker and an aminopropyl-CPG derivatized with 3-(trimethoxysilyl)propyl methacrylate *(43)*. Fig. 2.: A universal support for 3'-phosphorylated oligonucleotides using a mercaptoethanol linker and an aminopropyl-CPG derivatized with 3-mercaptopropyltrimethoxysilane *(44)*. Fig. 3.: A universal support for the synthesis of 3'-phosphorylated oligonucleotides using a 4,4'-diaminobenzidine linker *(43)*. Figs. 4. and 5.: Linkages for the synthesis of oligonucleotides ending with 3'-thiol groups *(45)*. Fig. 6.: A universal support for 3'-thiol-modified oligonucleotides using a 5-aminopentan-1-ol linker *(46)*. Fig. 7.: A universal support for 3'-thiol modified oligonucleotides using a polyether linker *(47)*. Fig. 8.: A linkage for the synthesis of oligonucleotides ending with a 3'-aminohexyl group *(48)*. Fig. 9.: A commercially available "universal" 3'-amino CPG support *(49)*. Fig. 10.: The 3'-peptide conjugate prepared by a combination of

solid-phase peptide and oligonucleotide synthesis *(49)*. Fig. 11. A support for the synthesis of 3'-vitamin E-modified oligonucleotides *(52)*. Fig. 12.: Four commercially available supports using a serine linker to prepare 3'-modified oligonucleotides *(53)*. Fig. 13.: A trans-4-hydroxy-1-prolinol-containing support used to attach either 3'-cholesterol or acridine groups onto oligonucleotides *(54)*.

These compounds are then succinylated and attached to an aminopropyl-CPG support. Two "universal" approaches that introduce a 3'-disulfide linkage, independent of the 3'-nucleoside, have also been developed. In the first approach *(46)*, commercially available 3,3'-dithio-*bis*-(N-succinimidyl propionate) is used along with an 5-aminopentan-1-ol linker to form the support shown in Fig. 9. In the second approach *(47)*, a longer linker arm is created from 2-(2-[2-chloroethoxy])ethanol using a seven-step synthetic scheme (Fig. 9). After synthesis and deprotection using typical phosphoramidite chemistry, oligonucleotides bearing 3'-disulfide attachments are released. The disulfide groups can then be reduced with dithiothreitol and used to couple other functions, such as fluorescein to the oligonucleotide.

7.3.3. Supports for 3'-Amino End Groups

Primary amino groups are also excellent nucleophiles that can be used to attach a large variety of ligands to oligonucleotides. Oligonucleotides ending with a 3'-hexylamino group can be synthesized using the support described by Asseline and Thuong *(48)* in which 6-aminohexan-1-ol is coupled via a carbamate linkage onto a disulfide-derivatized support (Fig. 9). Nelson et al. *(49)* have also described a support that contains an Fmoc-blocked 3-amino-1,2-propanediol linker attached to LCAA-CPG (Fig. 9), which has the advantage of being commercially available from a number of sources. The Fmoc-blocking group on the amine is released under the same basic deprotection conditions used to cleave the sequence from the support, and no modifications to the synthesis procedure are required.

7.3.4. Supports for 3'-Peptide Sequences

It is also, possible to synthesize oligonucleotides bearing 3'-peptide sequences using a single solid-phase support. This has been demonstrated by Haralambidis et al. *(50)*, who used aminopropyl CPG to synthesize a 30-base-long oligonucleotide with an $(AlaLys)_5Ala$ peptide covalently attached to its 3' terminus (Fig. 9). In this synthesis, the aminopropyl-CPG was first reacted with *p*-nitrophenyl 4-(4,4'-dimethoxytrityloxy)butyrate and then detritylated to yield a linker arm ending in a hydroxyl position. Solid-phase peptide synthesis was then performed to build the peptide chain. The N-terminal was then coupled to a special linker reagent, *p*-nitrophenyl 3-(6[4,4'-dimethoxy-

Preparation of Solid-Phase Supports 491

trityloxy]hexylcarbamoyl) propanoate, and after detritylation, the oligonucleotide was synthesized using the standard phosphoramidite method. A more recent and simpler approach that did not require any solid-phase support derivatization has also been developed by Juby et al. *(51)*, who used a commercially available Teflon™ support to prepare a 3'-oligonucleotide peptide conjugate.

7.3.5. Supports for 3'-Nonisotopic Labels and Carrier Groups

An important trend in biotechnology has been the development of nonisotopic detection methods. Newer methods utilizing fluorescence or chemiluminescence detection are faster and less hazardous than using radioisotopes. They are also, the basis of several important new applications, such as automated DNA sequencing. Another important trend has been the development of antisense oligonucleotides bearing lipophilic carrier groups, such as cholesterol or vitamin E, which aid cellular penetration of the antisense oligonucleotides. A variety of chemical methods have been developed for attaching these modifications onto either the 5' ends or internal positions of the oligonucleotide sequences. However, sequences that do not need to be extended from the 3' end by a polymerase can be conveniently modified on the 3' end by simply attaching the desired label or carrier group onto the insoluble support. Synthesis of the oligonucleotide chain may then proceed normally as long as the support's protecting groups and linkages are compatible with the existing synthesis chemistry.

A trifunctional linker molecule is required to attach the insoluble support, the desired reporter or carrier molecule, and the future oligonucleotide sequence together. A simple glycerol linker has been used to attach vitamin E onto LCAA-CPG as shown in Fig. 9 *(52)*. Another readily available trifunctional linker is the amino acid serine. This linker has been used to prepare commercially available CPG supports containing either biotin, rhodamine, fluorescein, or acridine *(53)* as shown in Fig. 9. A more complex linker with a rigid structure and defined stereochemistry, *trans*-4-hydroxy-L-prolinol, was prepared by the hydrogenolysis of commercially available N-Cbz-hydroxy-L-proline *(54)*. This linker was then used to attach either cholesterol or acridine molecules to LCAA-CPG as shown in Fig. 9.

8. Notes

The solid-phase supports, like the chemistry of the coupling reactions, involved in oligonucleotide synthesis have undergone an intense amount of refinement over the last three decades. The combined efforts of the countless researchers involved in this process now allow several important points to be made that will facilitate the efforts of any novice wishing to perform solid-phase oligonucleotide synthesis. These points are as follows:

1. Handle with care. Although the synthetic steps required to prepare the derivatized supports are much less strenuous than many of the oligonucleotide syntheses performed a decade ago, traditional manipulations and precautions are still required. Most importantly, when anhydrous conditions are required, it is necessary to ensure that solvents and materials are rigorously dried and maintained free of moisture. This will greatly improve reproducibility and yield. Furthermore, all of the solid-phase supports should be handled gently to minimize breakup of the particles. This is especially critical when working with silica gel or any of the large pore CPG supports.
2. Work with low to moderate loadings. Most of the commercially available DNA synthesizers have been designed to provide small quantities of oligonucleotides. Therefore, it is usually not necessary and actually counterproductive to prepare supports with high nucleoside loadings (>40 µmol/g). In fact, the trend is toward the use of supports with low nucleoside loadings (5–20 µmol/g), since the larger excess of reagents produces a better-quality product. The synthesis of very large quantities of oligonucleotide remains an area more suitable to an experienced chemist.
3. Cap all supports carefully. Regardless of the support selected, it is important to ensure that all unreacted functional groups are blocked with an appropriate capping reagent. Otherwise, the formation of unwanted side reactions and byproducts can increase the heterogeneity of the final product, and adversely affect the quality and/or utility of the product.
4. Select the appropriate support for the application. For most applications, the most important factor to consider will be the pore size of the support. However, certain applications, such as those containing modified nucleosides or other covalently attached functionality, may not be compatible with the strongly basic conditions required to hydrolyze the conventional oligonucleotide-support linkage. Expensive or difficult to handle molecules, such as fluorescent dyes, can be easily incorporated onto the 3' end of a sequence by synthesizing special supports, but the possible sequence heterogeneity and the nonextendability of these products by polymerases must be considered.

5. Do not limit your imagination. Presently, supports based on CPG beads are the most widely used with more than 2 million oligonucleotides per year being prepared. However, supports based on other materials may still be useful as new applications for oligonucleotides develop. In particular, as the importance and scope of oligonucleotide conjugates continues to grow, it can be expected that more specialized supports with different end modifications will be developed. Additionally, new techniques promise to increase greatly the ways in which synthetic oligonucleotides are used. One exceptionally promising approach, relevant to this chapter, is the possibility of using solid-phase chemistry in conjunction with photolabile-protecting groups to synthesize huge arrays of defined oligonucleotide sequences immobilized on glass surfaces *(55)*. With such a light-directed approach, it is possible to synthesize all 65,536 possible octanucleotides within a single 1.6-cm^2 matrix. The ready availability of these types of matrices will be extremely valuable in future diagnostic and sequencing strategies. It is also possible that developments in the synthesis and engineering of nanostructures will provide applications for oligonucleotides in ways that are difficult to imagine at the present. In the meantime, it is hoped that the methods described in this chapter will give researchers a better understanding and the opportunity to prepare their own supplies of these important materials.

References

1. Katzhendler, J., Cohen, S., Rahamim, E., Ringel I., and Deutsch, J. (1989) The effect of spacer, linkage and solid support on the synthesis of oligonucleotides. *Tetrahedron* **45**, 2777–2792.
2. Gait, M. J. (1984) *Oligonucleotide Synthesis: A Practical Approach*. IRL, Oxford.
3. Letsinger, R. L. and Mahadevan, V. (1966) Stepwise synthesis of oligonucleotides on an insoluble polymer support. *J. Am. Chem. Soc.* **88**, 5319–5324.
4. Melby, L. R. and Strobach, D. R. (1967) Oligonucleotide syntheses on insoluble polymer supports. I. Stepwise synthesis of trithymidine diphosphate. *J. Am. Chem. Soc.* **89**, 450–453.
5. Gait, M. J. and Sheppard, R. C. (1977) Rapid synthesis of oligodeoxyribonucleotides: A new solid-phase method. *Nucl. Acids Res.* **4**, 1135–1158.
6. Markham, A. F., Edge, M. D., Atkinson, T. C., Greene, A. R., Heathcliffe, G. R., Newton, C. R., and Scanlon, D. (1980) Solid phase phosphotriester synthesis of large oligodeoxyribonucleotides on a polyamide support. *Nucl. Acids Res.* **8**, 5193–5205.
7. Bardella, F., Giralt, E., and Pedroso, E. (1990) Polystyrene-supported synthesis by the phosphite triester approach: An alternative for the large scale synthesis of small oligodeoxyribonucleotides. *Tetrahedron Lett.* **31**, 6231–6234.
8. Gait, M. J., Matthes, H. W. D., Singh, M., Sproat, B. S., and Titmas, R. C. (1982) Rapid synthesis of oligodeoxyribonucleotides VII. Solid phase synthesis of oligodeoxyribonucleotides by a continuous flow phosphotriester method on a kieselguhr-polyamide support. *Nucl. Acids Res.* **10**, 6243–6254.

9. Crea, R. and Horn, T. (1980) Synthesis of oligonucleotides on cellulose by a phosphotriester method. *Nucl. Acids Res.* **8**, 2331–2348.
10. Frank, R., Heikens, W., Heisterberg-Moutsis, G., and Blocker, H. (1983) A new general approach for the simultaneous chemical synthesis of large numbers of oligonucleotides: segmental solid supports. *Nucl. Acids Res.* **11**, 4365–4377.
11. Kaas, R. L. and Kardo, J. L. (1971) The interaction of alkoxy silane coupling agents with silica surfaces. Polymer Engineering and *Science* **11**, 11–18.
12. Cox, G. B. (1977) Practical aspects of bonded phase chromatography. *J. Chromatographic Sci.* **15**, 385–392.
13. Koster, H. (1972) Polymer support oligonucleotide synthesis VI. Use of inorganic carriers. *Tetrahedron Lett.* 1527–1530.
14. Alvarado-Urbina, G., Sathe, G. M., Liu, W.-C., Gillen, M. F., Duck, P. D., Bender, R., and Ogilvie, K. K. (1981) Automated synthesis of gene fragments, *Science* **214**, 270–274.
15. Chow, F., Kempe, T., and Palm, G. (1981) Synthesis of oligodeoxyribonucleotides on silica gel support. *Nucl. Acids Res.* **9**, 2807–2817.
16. Matteucci, M. D. and Caruthers, M. H. (1981) Synthesis of deoxyoligonucleotides on a polymer support. *J. Am. Chem. Soc.* **103**, 3185–3191.
17. Majors, R. E. and Hopper, M. J. (1974) Studies of siloxane phases bonded to silica gel for use in high performance liquid chromatography. *J. Chromatographic Sci.* **12**, 767–778.
18. Kohli, V., Balland, A., Wintzerith, M., Sauerwald, R., Staub, A., and Lecocq, J. P. (1982) Silica gel: an improved support for the solid-phase phosphotriester synthesis of oligonucleotides. *Nucl. Acids Res.* **10**, 7439–7448.
19. Kume, A., Sekine, M., and Hata, T. (1983) Further improvements of oligodeoxyribonucleotides synthesis: Synthesis of tetradeoxyadenylate on a new silica gel support using N^{6}-phthaloyldeoxyadenosine. *Chem. Lett.* 1597–1600.
20. Koster, H., Stumpe, A., and Wolter, A. (1983) Polymer support oligonucleotide synthesis 13: rapid and efficient synthesis of oligodeoxynucleotides on porous glass support using triester approach. *Tetrahedron Lett.* **24**, 747–750.
21. Koster, H., Biernat, J., McManus, J., Wolter, A., Stumpe, A., Narang, Cg. K., and Sinha, N. D. (1984) Polymer support oligonucleotide synthesis—XV. Synthesis of oligodeoxynucleotides on controlled pore glass (CPG) using phosphate and a new phosphite triester approach. *Tetrahedron* **40**, 103–112.
22. Ferretti, L., Karnik, S. S., Khorana, G. H., Nassal, M., and Oprian, D. D. (1986) Total synthesis of a gene for bovine rhodopsin. *Proc. Natl. Acad. Sci. USA* **83**, 599–603.
23. Adams, S. P., Kavka, K. S., Wykes, E. J., Holder, S. B., and Galluppi, G. R. (1983) Hindered dialkylamino nucleoside phosphite reagents in the synthesis of two DNA 51-mers. *J. Am. Chem. Soc.* **105**, 661–663.
24. Barnett, R. W. and Erfle, H. (1990) Rapid generation of DNA fragments by PCR amplification of crude, synthetic oligonucleotides. *Nucl. Acids Res.* **18**, 3094.
25. Bayer, E. (1991) Towards the chemical synthesis of proteins. *Angewante Chemie International Edition in English* **30**, 113–129.

26. Gao, H., Gaffney, B. L., and Jones, R. A. (1991) H-Phosphonate oligonucleotide synthesis on a polyethylene glycol/polystyrene copolymer. *Tetrahedron Lett.* **32,** 5477–5480.
27. McCollum, C. and Andrus, A. (1991) An optimized polystyrene support for rapid, efficient oligonucleotide synthesis. *Tetrahedron Lett.* **32,** 4069–4072.
28. Gough, G. R., Brunden, M. J., and Gilham, P. T. (1981) Recovery and recycling of synthetic units in the construction of oligodeoxyribonucleotides on solid supports. *Tetrahedron Lett.* **22,** 4177–4180.
29. Kierzek R., Caruthers, M. H., Longfellow, C. E., Swinton, D., Turner, D. H., and Freier, S. M. (1986) Polymer supported RNA synthesis and its application to test the nearest-neighbor model for duplex stability. *Biochemistry* **25,** 7840–7846.
30. Pon, R. T., Usman, N., and Ogilvie, K. K. (1988) Derivatization of controlled pore glass beads for solid phase oligonucleotide synthesis. *Biotechniques* **6,** 768–775.
31. Damha, M. J., Giannaris, P. A., and Zabarylo, S. V. (1990) An improved procedure for derivatization of controlled pore glass beads for solid-phase oligonucleotide synthesis, *Nucl. Acids Res.* **18,** 3813–3821.
32. Sproat, B. S. and Brown, D. M. (1985) A new linkage for solid-phase synthesis of oligodeoxyribonucleotides. *Nucl. Acids Res.* **13,** 2979–2987.
33. Atkinson, T., Gillam, S., and Smith, M. (1988) A convenient procedure for the synthesis of oligodeoxyribonucleotide affinity columns for the isolation of mRNA. *Nucl. Acids Res.* **16,** 6232.
34. Albretson, C., Kalland, K.-H., Haukanes, B.-I., Havarstein, L.-S., and Kleppe, K. (1990) Applications of magnetic beads with covalently attached oligonucleotides in hybridization: Isolation and detection of specific measles virus mRNA from a crude cell lysate. *Analytical Biochemistry* **189,** 40–50.
35. Pochet, S., Huynh-Dinh, T., and Igolen J. (1987) Synthesis of DNA fragments linked to a solid support. *Tetrahedron* **43,** 3481–3490.
36. Stengele, K.-P. and Pfleiderer, W. (1990) Improved synthesis of oligodeoxyribonucleotides. *Tetrahedron Lett.* **31,** 2549–2552.
37. Vu, H., McCollum, C., Jacobson, K., Theisen, P., Vinayak, R., Spiess, E., and Andrus, A. (1990) Fast oligonucleotide deprotection phosphoramidite chemistry for DNA synthesis. *Tetrahedron Lett.* **31,** 7269–7272.
38. Gryaznov, S. M. and Letsinger, R. L. (1991) Synthesis of oligonucleotides via monomers with unprotected bases. *J. Amer. Chem. Soc.* **113,** 5876,5877.
39. Asseline, U. and Thuong, N. T. (1989) Solid phase synthesis of modified oligodeoxyribonucleotides with an acridine derivative or a thiophosphate group at their 3'-end. *Tetrahedron Lett.* **30,** 2521–2424.
40. Eritja, E., Robles, J., Fernandez-Forner, D., Albericio, F., Giralt, E., and Pedroso, E. (1991) NPE-Resin, A new approach to the solid-phase synthesis of protected peptides and oligonucleotides I: Synthesis of the supports and their application to oligonucleotide synthesis. *Tetrahedron Lett.* **32,** 1511–1514.
41. Alul, R. H., Singman, N., Zhang, G., and Letsinger, R. L. (1991) Oxalyl-CPG: A labile support for synthesis of sensitive oligonucleotide derivatives. *Nucl. Acids Res.* **19,** 1527–1532.

42. Gough, G. R., Brunden, M. J., and Gilham, P. T. (1981) Recovery and recycling of synthetic units in the construction of oligodeoxyribonucleotides on solid supports. *Tetrahedron Lett.* **22,** 4177–4180.
43. Markiewicz, W. T. and Wyrzykiewicz, T. K. (1989) Universal solid supports for the synthesis of oligonucleotides with terminal 3'-phosphates. *Nucl. Acids Res.* **17,** 7149–7158.
44. Kumar, P., Bose, N. K., and Gupta, K. C. (1991) A versatile solid phase method for the synthesis of oligonucleotide-3'-phosphates. *Tetrahedron Lett.* **32,** 967–970.
45. Zuckerman, R., Corey, D., and Schultz, P. (1987) Efficient methods for attachment of thiol specific probes to the 3'-ends of synthetic oligodeoxyribonucleotides. *Nucl. Acids Res.* **15,** 5305–5321.
46. Gupta, K. C., Sharma, P., Sathyanarayana, S., and Kumar, P. (1990) A universal solid support for the synthesis of 3'-thiol group containing oligonucleotides. *Tetrahedron Lett.* **31,** 2471–2474.
47. Bonfils, B. and Thuong, N. T. (1991) Solid phase synthesis of 5', 3'-bifunctional oligodeoxyribonucleotides bearing a masked thiol group at the 3'-end. *Tetrahedron Lett.* **32,** 3053–3056.
48. Asseline, U. and Thuong, N. T. (1990) New solid-phase for automated synthesis of oligonucleotides containing an amino-alkyl linker at their 3'-end. *Tetrahedron Lett.* **31,** 81–84.
49. Nelson, P. S., Frye, R. A., and Liu, E. (1989) Bifunctional oligonucleotide probes synthesized using a novel CPG support are able to detect single base pair mutations. *Nucl. Acids Res.* **17,** 7187–7194.
50. Haralambidis, J., Duncan, L., and Tregear, G. W. (1987) The solid phase synthesis of oligonucleotides containing a 3'-peptide moiety. *Tetrahedron Lett.* **28,** 5199–5202.
51. Juby, C. D., Richardson, C. D., and Brousseau, R. (1991) Facile preparation of 3'-oligonucleotide-peptide conjugates. *Tetrahedron Lett.* **32,** 879–882.
52. Will, D. W. and Brown, T. (1991) The solid-phase synthesis of 3'-vitamin E containing oligonucleotides. San Diego Conference on Nucleic Acids, San Diego, CA, Nov. 20–22.
53. These products are available from Penninsula Laboratories, Inc., 611 Taylor Way, Belmont CA, 94,002–94,041.
54. Reed, M. W., Adams, A. D., Nelson, J. S., and Meyer Jr., R. B. (1991) Acridine and cholesterol-derivatized solid supports for improved synthesis of 3'-modified oligonucleotides. *Bioconjugate Chemistry* **2,** 217–225.
55. Fodor, S. P., Leighton Read, J., Pirring, M. C., Stryer, L., Lu, A. T., and Solas, D. (1991) Light-directed, spatially addressable parallel chemical synthesis. *Science* **251,** 767–773.

Index

Acetamide bridged oligonucleotide analogs
  properties, 370
  synthesis, 371
Affinity chromatography, 400
Affinity ligands, 52
2'-O-Alkyloligoribonucleotides, 402
  2'-O-allyladenosine synthesis, 127
  2'-O'-allylcytidine synthesis, 124
  2'-O-allylguanosine synthesis, 130
  2'-O-allyluridine synthesis, 118
  deprotection, 134
  purification, 134
  synthesis, 130
Alkaline phosphatase, 88, 106, 168
Amino group detection, 467
5'-aminoalkyl oligonucleotides, 178
3-aminopropyltriethoxysilane, 468
Ammonium persulfate, 39
Antisense oligonucleotides, 167
2,2'-azobis (2-methylpropionitrile) (AIBN), 50

3H-1,2-benzodithiol-3-one 1, 1-dioxide, 55
Biotinylated oligonucleotides, 138
bis-(N,N-diisopropylamino)-2-cyanoethoxyphosphine, 38
bis-trimethylsilyl peroxide, 50
Borane phosphonate, 225
Boranophosphate oligonucleotides
  ^{1}HNMR, 229, 230
  acid base hydrolysis, 236
  boron neutron capture therapy, 240
  conformational studies, 239
  dinucleoside boranophosphate, 227
  nuclease stability, 237
  solid phase synthesis, 235
  solubility and permeability, 237
  toxicity, 238
  trinucleoside boranophosphate, 233
Boron neutron capture therapy (BCNT), 226
Branched oligoribonucleotides, 407

Calf intestinal adenosine deaminase, 117
Capping reagent, 55
Carbamate bridged
  morpholino type oligonucleotide analog
    properties, 375
    synthesis, 376
Carbamate oligonucleotide analog
  properties, 373
  synthesis, 375
Carbonate bridged oligonucleotide analog
  properties, 366
  synthesis, 367
Carboxymethyl bridged oligonucleotide analog
  properties, 369
  synthesis, 367
Catalytic phosphate protecting groups, 27
2-chloro-4H-1,3,2 benzodioxaphosphorin-4-one, 65, 66
Chloro-(N,N-dimethylamino) methoxyphosphine, 33, 38
m-chloroperbenzoic acid, 50
Chlorophosphoramidite derivatives, 41
N-chlorosuccinimide, 2
CPG support
  capping, 481

497

CPG support *(cont'd.)*
  containing labile linkages, 485
  containing nonhydrolytic linkages, 484
  derivatization, 469
  for 3'-end functionalization, 487
  nucleoside loading, 479
Crosslinking agents, 52
Cyanoethyl H-phosphonate, 76
β-cyanoethylphosphate-protecting group, 36
2-cyanoethoxydichlorophosphine, 39

DMT-analysis, 467
a-deoxyadenosine, 263
a-deoxycytidine, 264
a-deoxyguanosine, 267
a-thymidine, 271
Deoxynucleotidyl terminal transferase, 414
Deoxyribonucleoside phosphoramidites, 42, 44
Dephospho-internucleotide linkages, 355
Deprotection of oligoribonucleotides
  cyanoethyl deprotection, 97
  desilylation, 98
  methyl deprotection, 97
  N-deacylation, 97
Derivatization
  of long chain alkylamine CPG, 469
  of silica support, 468
Diagnostic probes, 52
1, 8-Diazabicyclo [5.4.0] undec-7-ene (DBU), 65
1,3-dichloro, 1,1,3,3-tetraisopropyldisiloxane, 118
Dicyclohexylcarbodiimide (DCC), 292
N,N-diisopropylethylamine, 33, 38
Di-*tert*-butylhydroperoxide, 50
Dimethoxytritylchloride, 274
Dimethylsufoxide d_6, 117
Dinitrogen tetraoxide, 50
Dinucleoside
  O-alkyl phosphorothioate, 221
  phosphotriester, 215

DNA sequencing primers, 52
Dowex
  50 W-W8 resin, 86
  Na^+ exchange chromatography, 103
3'-end functionalization of oligonucleotides
  3'-amino group, 490
  3'-nonisotopic labeling, 491
  3'-peptide sequences, 490
  3'-phosphate group, 487
  3'-thiol group, 487

Enzymatic digestion
  of oligonucleotides, 46
  of oligoribonucleotides, 88
Ethanedithiol monobenzoate, 193
1-(3-dimethylaminopropyl) ethylcarbodiimide (DEC), 470
Ethyl H-phosphonate, 67
Ethylenediamine/ethanol, 153

FAB mass spectroscopy, 234
Flash chromatography, 399
Fluorescent attachment, 178, 491

O^6-guanosine protection, 83

N-hydroxysuccinimide ester of carboxyfluoroscein, 168
H-phosphonate approach, 63
  capping, 69, 70
  coupling cycles, 69, 74
  reagents preparation, 68
  synthesis of nucleosides H-phosphonate, 66
H-phosphonate chemistry, 10
History, oligonucleotide synthesis, 1
Hydrogen peroxide, 50
1-hydroxybenzotriazole, 48

Iso-propyl H-phosphonate, 76

^{35}S-labeled oligonucleotide phosphorothioates, 77
Large-scale synthesis of oligonucleotides

Index

solid phase, 437
solution phase, 391
3'-O-levulinylthymidine, 34
2'-5'-Linked oligoribonucleotides, 406
Liquid phase oligonucleotide
    synthesis, 400

Magnetic beads, 484
Methylene bridged oligonucleotide
    analog
  properties, 360
  synthesis, 364
Methylphosphate-protecting group, 36
Methylphosphonate oligonucleoside, 143
MMT assay, 468
MSNT, 20

N-methylanilinium trichloroacetate, 48
N-methylimidazole, 27
  hydrochloride, 48
  trifluoroacetate, 48
  trifluoromethane sulfonate, 48
N-methylmorpholine, 2, 65
^{11}B-NMR, 234
^{31}P-NMR, 73
^{1}HNMR (COSY, HOHAHA
    [Homonuclear Hartman
    Hann]), 234
Nonaqueous oxidizing reagents, 50
Nonionic oligonucleotides, 143, 355
Nonisotopic labeling, 491
Nonphosphate internucleoside
    linkages
  aminothylglycine, 379
  acetamide, 370
  carbamate, 373
  carbonate, 366
  carboxymethyl, 367
  formacetal, 360
  morpholinocarbamate, 375
  polyethyleimine, 378
  siloxane, 358
  thioether, 357
  thioformacetal, 357

Nuclease P1, 88, 106, 168
Nucleoside H-phosphonate, 66
Nucleosides
  as carriers of Boron-10, 240
  phosphoramidites, 42
  phosphotriester, 21
  3'-O-(2-chlorophenylphosphate), 20

α-oligodeoxynucleotides
  5'-DMT-α-deoxynucleosides, 274
  α-deoxynucleoside CPG, 277
  α-deoxynucleoside
    phosphoramidites, 276
  α-deoxynucleosides, 263
  α-oligodeoxynucleotides, 277
  coupling cycle, 277
  deprotection, 278
  purification, 279
Oligonucleoside
  acetamidate analog, 370
  aminoethylglycine analog, 379
  carbamate analog, 373
  carbonate analog, 366
  carboxymethyl analog, 367
  formacetal analog, 360
  morpholinocarbamate analog, 375
  polyethyleimine analog, 378
  siloxane analog, 358
  thioether analog, 357
  thioformacetal analog, 357
Oligonucleotide
  boranophosphate, 225
  dimethylene sulfide, 315
  methylphosphonate, 143
  phosphorodithioate, 191
  phosphorothioate, 165
  sulfones, 315
  sulfoxides, 315
Oligonucleotides
  containing nonphosphate
    internucleotide linkage, 355
  phosphorofluoridate and fluoridite, 244
    diastereoisomeric dinucleoside
      phosphorofluoride, 255

Oligonucleotides; phosphorofluoridate and fluoridite *(cont'd.)*
  nucleoside diesters of monofluorophosphoric acid, 248
  nucleoside monoester of monofluorophosphoric acid, 245
  nucleoside phosphorofluoridate, 253
  phosphotriester, 207
Oligoribonucleotide synthesis, 12, 81
  base composition, 106
  coupling cycle, 95
  deblocking, 96
  deprotection, 86
  desalting, 102
  enzymatic digestion, 106
  HPLC purification, 104
  materials/reagents for synthesis, 84
  PAGE purification, 99
  ribonucleosides phosphoramidite, 89
  silylation of nucleosides, 89
  TLC purification, 103
OPC cartridge, 174
Oxalyl linkage, 487
1-oxido-4-alkoxy-2-picolyl group for phosphate protection, 28

5-(*p*-nitrophenyl)-1H-tetrazole, 48
P-chiral oligonucleotides, 285
p-nitrophenol, 469
PEG, 478
2-(1-methylimidazol-2-yl) phenyl group for phosphate protection, 27
Phosphite and phosphoramidite chemistry, 10
Phosphodiester chemistry, 5
Phosphoramidite approach, 33
  coupling cycle, 54
  purification, 45
  synthesis of nucleosides phosphoramidite, 39
Phosphorodithioate oligodeoxyribonucleotides
  coupling cycle, 198
  deprotection, 201
  HPLC purification, 201
  PAGE purification, 201
  synthesis of nucleoside 3'-phosphorodithioamidites, 195
  synthesis, 196
Phosphorothioate oligonucleotide
  antisense, 167
  attachment of reporter groups, 178
  base composition analysis, 176
  enzyme digestion, 176
  ion exchange HPLC, 179
  manual purification, 174
  reversed-phase purification, 171
  stereochemistry, 183
  synthesis, 168
Phosphotriester approach, 7, 19
  coupling cycle, 23
  deprotection, 26
  nucleoside phosphotriester, 21
  phosphate protecting groups, 27
Phosphotrieste oligodeoxynucleotides
  dinucleoside phosphotriester, 215
  HPLC characterization, 219
  nucleoside protection, 213
  phosphitylation, 214
  synthesis, 217, 220
Pivaloyl chloride, 65
Pixyl group, 21
Plastic DNA
  synthesis, 378
  properties, 379
Polyacrylamide gel electrophoresis, 45, 87
Polyamide nucleic acids (PNA)
  hybridization, 383
  properties, 384
  resins, 20
  synthesis, 379
Polydimethyl-kiselguhr support, 21
Polyethyleneglycol, 478
  (PEG), 402
Polynucleotide phosphorylase, 414
Polystyrene matrix HPLC column, 168

Index

Precipitation and crystallization techniques, 400
Preparative
  polyacrylamide gel electrophoresis, 99, 100
  scale coupling cycle, 449
Purification of oligonucleotides, 45
  by polyacrylamide gel electrophoresis, 45, 99, 201
  ion exchange high-performance liquid chromatography, 47, 104, 179, 201
  reversed phase high-performance liquid chromatography, 46
  thin layer chromatography, 103
Pyridine, 38

Recovery of nucleoside H-phosphonates, 78
Reporter groups, 491
Ribonucleoside
  derivatized CPG, 85
  phosphoramidites, 85, 89
RNA phosphorothioate, 109
RNase H, 191
RNase T2, 88, 106

Scale up, solid phase approach
  analysis by HPLC and PAGE, 458
  coupling steps, 438
  deprotection, 456
  fluidization, 442
  H-phosphonate approach, 448
  instrument, 440
  of oligonucleotide synthesis, 391, 437
  phosphoramidite approach, 443
  phosphorothioate oligonucleotides, 452
Semimanual apparatus for oligonucleotide synthesis, 25
C_{18} sep pak cartridge, 88
Siloxane bridged oligonucleotide analog
  properties, 360
  synthesis, 358
Silyl migration, 83

Silylation, 89
Size exclusion chromatography, 102
Snake venom phosphodiesterase, 88, 106, 168
Sodium
  azide, 117
  methoxide, 273
Solid-phase
  chemistry, 9
  oligonucleotide synthesis
    H-phosphonate approach, 63
    phosphoramidite approach, 33
    phosphotriester approach, 19
Solid supports
  capping of CPG, 481
  containing labile linkages, 485
  containing nonhydrolytic linkages, 484
  derivatization of CPG, 469
  derivatization of silica gel, 468
  for 3'-end functionalization, 487
  nucleoside attachment to CPG, 479
  for oligonucleotide synthesis, 465, 474
  various solid supports, 473
Soluble supports, 402
Solution phase synthesis
  affinity chromatography, 400
  block-phosphotriester synthesis, 396
  branched oligoribonucleotides, 407
  conventional solution phase methods, 393
  enzymatic polymerization, 413
  H-phosphate method, 395
  2'-5'-linked oligoribonucleotides, 406
  liquid-phase synthesis, 400
  methylphosphonates, 409
  modified oligonucleotides, 403
  phosphate triester method, 394
  phosphite triester method, 394
  phosphorodithioates, 407
  phosphorothioates, 407
Solution phase oligonucleotide synthesis, 391
Spleen phosphodiesterase, 88, 106

Stereochemistry of oligonucleotide
    phosphorothioates, 183
Stereocontrolled synthesis of chiral
    analogs of oligonucleotide
  oligonucleotide
    methylphosphonate, 293
    characterization, 298
    deprotection, 299
  oligonucleoside phosphorothioates, 300
    characterization, 302
    nuclease digestion, 306
Stereospecific synthesis, 285
Succinic anhydride, 470
Succinic-sarcosinyl CPG, 303
Succinylated CPG, 470
Sulfoxide, sulfone oligonucleotide analog
  synthesis and characterization of building blocks, 322
Sulfone linked dinucleosides, 346
thio-ether linked dinucleosides, 346

TBDMS protection, 89
TEAB, 66
Template chromatography, 412
Template-dependent chemical condensation and ligation, 411
Template-independent chemical and enzymatic polymerization, 413
Tert-butyl hydroperoxide, 50
1,1,3,3-tetramethylguanidine, 26, 117
Tetra-*n*-butylammonium periodate, 50
Tetraethylthiuram disulfide (TETD), 166
Tetrahydrofuranyl phosphoramide, 52
N,N,N^1,N^1-tetramethylethylene diamine (TEMED), 39
1-H-tetrazole, 38
TLC, 87
TLC on cellulose, 103
Tolylene-2, 6-diisocyanate (TDIC), 484
Triethylbicarbonate, 66
5-trifluoromethyl-1H-tetrazole, 48
Trimethylacetylchloride, 65
Tris-borate-EDTA (TBE), 39
Tris (pyrrolidino) phosphine, 194
Trityl analysis, 96, 467
"Trityl on" separation, 402

Urethane linkage, 484